THE *ARTES* AND THE EMERGENCE OF A SCIENTIFIC CULTURE IN THE EARLY ROMAN EMPIRE

This book offers the first full-scale, synthetic account of the Latin technical treatises called *artes*, arguing that their flourishing in the early Roman Empire represents the emergence and development of a uniquely Roman scientific culture. It introduces the Roman *artes* on architecture, agriculture, land-surveying, medicine, and the art of war to those without specialist knowledge of the disciplines, and advances a new argument for their significance vis-à-vis a common intellectual culture. It unpacks the sociopolitical, literary, and especially philosophical and scientific dimensions of these writings. It characterizes the scientific culture which the *artes* constitute and traces significant themes in their construction of disciplinary expertise, examining the effects of the tension between theory and practice as well as their systematic, explanatory, and interdisciplinary presentation of specialized knowledge. In presenting a novel interpretation of the *artes*, this book aims to add a new chapter to the history of science in Greco-Roman antiquity.

JAMES L. ZAINALDIN is an assistant professor in the Department of Classical and Mediterranean Studies at Vanderbilt University, Tennessee. He has published on the scientific and technical traditions of Greece and Rome; ancient philosophy; Latin literature, especially of the Roman Empire; and comparative Greco-Roman/Chinese studies. His first book (2020) was a study of the agricultural writings of the Roman author Gargilius Martialis for Cambridge Classical Texts and Commentaries.

THE *ARTES* AND THE EMERGENCE OF A SCIENTIFIC CULTURE IN THE EARLY ROMAN EMPIRE

JAMES L. ZAINALDIN

Vanderbilt University

CAMBRIDGE
UNIVERSITY PRESS

CAMBRIDGE
UNIVERSITY PRESS

Shaftesbury Road, Cambridge CB2 8EA, United Kingdom

One Liberty Plaza, 20th Floor, New York, NY 10006, USA

477 Williamstown Road, Port Melbourne, VIC 3207, Australia

314–321, 3rd Floor, Plot 3, Splendor Forum, Jasola District Centre, New Delhi – 110025, India

103 Penang Road, #05–06/07, Visioncrest Commercial, Singapore 238467

Cambridge University Press is part of Cambridge University Press & Assessment, a department of the University of Cambridge.

We share the University's mission to contribute to society through the pursuit of education, learning and research at the highest international levels of excellence.

www.cambridge.org
Information on this title: www.cambridge.org/9781009501637

DOI: 10.1017/9781009501651

When citing this work, please include a reference to the DOI 10.1017/9781009501651

First published 2025

A catalogue record for this publication is available from the British Library

Library of Congress Cataloging-in-Publication Data
NAMES: Zainaldin, James L., author.
TITLE: The *artes* and the emergence of a scientific culture in the early Roman empire / James L. Zainaldin, Vanderbilt University, Tennessee.
DESCRIPTION: Cambridge, United Kingdom ; New York, NY : Cambridge University Press, 2025. | Includes bibliographical references and index.
IDENTIFIERS: LCCN 2024029088 | ISBN 9781009501637 (hardback) | ISBN 9781009501644 (paperback) | ISBN 9781009501651 (ebook)
SUBJECTS: LCSH: Science – Rome. | Technology – Rome. | Medicine, Greek and Roman. | Science, Ancient. | Philosophy, Ancient.
CLASSIFICATION: LCC Q124.95 .Z35 2025 | DDC 509.3–dc23/eng/20240808
LC record available at https://lccn.loc.gov/2024029088

ISBN 978-1-009-50163-7 Hardback

For Katie and Cyrus

Contents

PART III

PART IV

Figures

Tables

Acknowledgments

I did not set out to write a book arguing that the now-seldom-read Latin technical treatises of the early Roman Empire constituted an interesting and unified scientific culture because I did not know that they did. In the earliest days of this project, I merely had a set of puzzles that I had encountered in readings outside of the canon and a conviction that these texts – treasured and transmitted for millennia while so much else of great worth has been lost – had more to say to us today. I am fortunate many times over to have had mentors then who did not extinguish the first sparks and steer me back to more conventional territory, but rather kindled my intellectual curiosity and encouraged me to follow where it led.

The debts I have incurred in writing this book are many. A first harvest of gratitude is owed to my dissertation advisors, Kathleen Coleman and Mark Schiefsky, and committee members Jared Hudson and Rusty Jones. They offered a wealth of advice that significantly improved this project. Mark's penetrating insight and lucid expression meant that even brief meetings yielded much for me to contemplate in the following weeks and months. I am in print in several places already hymning Kathy (see particularly the prefaces to my book on Gargilius Martialis and to my dissertation), who has been to me the veritable Platonic form of an advisor and mentor. As time passes, I only marvel the more at her industry and generosity, and what I owe to her goes far beyond her help on this book.

Friends and colleagues during my years at Harvard offered support for my work, as well as rest and refreshment when needed. Special thanks to Marco Romani Mistretta, a fellow enthusiast of Greco-Roman scientific and technical literature, with whom I co-organized a workshop between 2016 and 2018 on "Technical Traditions in Greece and Rome" that enabled many fruitful exchanges with scholars at Harvard and beyond. The whole Harvard Department of the Classics was indeed a wonderful nurturing ground for the beginning of this project. I am grateful to have benefited from the knowledge and wisdom of a distinguished generation of scholars, including many who did not advise my dissertation formally but were generous with their time and interest: Emma Dench, †Albert Henrichs, Richard Tarrant, and Richard Thomas, among others. Alyson Lynch and Teresa Wu ensured all went smoothly during my time there.

Mentors, now also friends, from my undergraduate days have followed this project with interest and offered much encouragement and advice. Conversations with Jon Master are always stimulating, and I am glad now to be working on another project with him that gives many excuses for exchange. Garth Tissol has served as a kind of shadow advisor, bringing his good sense and enormous erudition – cloaked always in humility and restraint – to whatever I have shared with him, which is almost everything. George Conklin is a good friend, always ready to share advice seasoned with learning from many years of unpretentious study of the classics. New friends made in Cambridge – including Li Lihua, Michael Shear, and Adam Trettel, among others – have also lent their continual support.

This book has necessitated much aid, and I am grateful to Harvard University and the Harvard Department of the Classics for support during my graduate study, during a year as Lecturer afterwards (2020–1), and then in subsequent years (2021–3) as Postdoctoral Fellow. I am profoundly indebted to the trustees of the Loeb Classical Library Foundation, which offered me the incredible gift of two years of time for writing and research (2021–3), during which much of this book was completed. I am also immeasurably grateful to the faculty of the University of Oklahoma for enabling this research leave, and then providing a collegial and intellectually stimulating environment when I arrived on campus: thanks especially to Jeremy Bailey, Daniela Garofalo, Ellen Greene, Kyle Harper, Scott Johnson, Will Selinger, and Ben Watson, plus the wonderful staff members, especially Paula Richmond. Finally, I am grateful to Vanderbilt University, and especially the Department of Classical and Mediterranean Studies, for warmly welcoming me and providing such a comfortable and supportive place to continue and finish this book.

It has been an honor to work with Dr. Michael Sharp and Cambridge University Press again. I am grateful for Michael's expert captaining of the publication process; it is a big ask to bring a book of this scale to print, so I am doubly indebted. The Press's readers offered many incisive comments on both early and late versions of this project that have greatly improved it; to the extent that I have not done justice to all of their valuable suggestions, I take full responsibility, and of course any errors and infelicities that remain are mine alone. I am also grateful to my copyeditor, Helen Cooper, for her expert assistance in preparing the manuscript for publication. Thorsten Fögen, Marco Formisano, and Lars Mielke shared work or advice with me that improved this book. I read Mielke's dissertation (2019) while revising this book and found it most helpful; his subsequent monograph (2023) appeared too late for me to make a fresh accounting of all its substantive changes, but as a partial measure I replaced in most instances existing references to Mielke 2019 with references to Mielke 2023 (with updated page ranges). Thanks to Jens-Olaf Lindermann for a tip on diagrams for the *agrimensores* and for calling my attention to his new edition of Hyginus' work on *limites* (2022); I regret that it did not come into my hands in time to account for it, especially because even cursory consideration shows that it trenchantly extends his earlier work. The caveats and qualifications I offer in studying the *agrimensores* apply all the more forcibly in light of his study. Thanks to Brian Campbell for allowing me to redraw a figure from his article, and to Thomas Noble Howe for permission to use his helpful drawings from a recent book on Vitruvius. I am grateful to Connor Gilmore of Vanderbilt for assistance with reproducing the noted figures. Thanks finally to Grace Shen and Nick Kahn for assistance in proofreading.

Ingrid, Jamil, Carl, Isabelle, Joan, Doug, and Nick have championed me along the way and offered manifold kinds of help. This would have been much harder without their generous grants of love and time. Katie and Cyrus are my inspiration and greatest joy, Katie as ever an intellectual companion and interlocutor of the first rank (with Cyrus fast joining her while this project matured), and it is to them that I dedicate this book.

Note on Texts and Abbreviations

I Ancient Works

In general, editions of Greek texts follow the *Thesaurus Linguae Graecae*. Editions of Latin texts follow those in the corpus provided by the Packard Humanities Institute (http://latin.packhum.org/), or, failing that, those in the Library of Latin Texts provided by BREPOLiS. In general, ancient works are abbreviated according to the conventions of the *Oxford Classical Dictionary* (4th ed.). For the abbreviation of Latin authors and texts not given in *OCD*, I follow the *Thesaurus linguae Latinae*. Note, however, that I always print "u" in place of lowercase "v" in abbreviations of Latin authors and works (e.g. "Ou." not "Ov." for "Ovid"). For Greek authors not in *OCD*, I follow the conventions of LSJ (see §II), except for Galen's works, which are abbreviated according to the appendix printed in the volumes of the Cambridge Galen Translations series.

The principal authors and texts handled in this book have been abbreviated differently than according to the aforementioned rules and are often cited in editions other than those in the standard corpora. Abbreviations are presented at the introduction of each author in Chapter 2 (except for Onasander, introduced in Chapter 6) and are repeated in subsequent chapters. The principal edition used for each text is introduced in a footnote at the first extended quotation of the author (usually, but not always, in Chapter 2); in this same note, other editions or translations consulted are cited. For convenience, I gather the abbreviations and principal editions used for core authors (author–date references are to the general references), with a parenthetical reference to the footnote in this book detailing editions consulted:

[–]	*CAR*	*Corpus agrimensorum Romanorum* (Ch. 2, n.135). All texts in *CAR* are cited by page and line number of Lachmann et al. 1848–52 ("L") and, where applicable, Thulin 1913 ("T"), for example "L 176.7–8 = T 141.2–3."
Cels.	*DM*	Celsus, *De medicina*, ed. Marx 1915, with Ollero Granados 1973 and Capitani 1974 for the material in Book IV transmitted by MS Toledo 97.12 (Ch. 2, n.72).
Col.	*RR*	Columella, *Res rustica*, ed. Rodgers 2010 (Ch. 4, n.4) [note also [Col.] *Arb.* in next list].
Frontin.	*Agr. qual.*	*De agrorum qualitate*, ed. Thulin 1913 (Ch. 2, n.145) [for citation see earlier entry on *CAR*].
	Art. mens.	*De arte mensoria*, ed. Thulin.
	Contr.	*De controuersiis*, ed. Thulin.
	Lim.	*De limitibus*, ed. Thulin.
	Strat.	*Strategemata*, ed. Ireland 1990 (Ch. 2, n.123).
Hyg.	*Const.*	"Hyginus Gromaticus" or "Hyginus minor" or "Hyginus 2," *Constitutio limitum*, ed. Thulin 1913 (Ch. 2, nn. 150, 151, 157) [for citation see earlier entry on *CAR*; see later entries for other works ascribed to a "Hyginus"].
Onasander	*Str.*	Onasander, Στρατηγικός (*Stratêgikos*), ed. Illinois Greek Club 1923 (Ch. 6, n.29).
Vitr.	*DA*	Vitruvius, *De architectura*, ed. Rose and Müller-Strübing 1867 (Ch. 2, n.17).

The following list contains additional references or abbreviations (some standard but given here anyway for convenience) that pertain to ancient authors or texts cited in this book. As noted earlier, bare author–date references are to general references.

[Col.] *Arb.*	[Columella], *incerti auctoris liber de arboribus*, ed. Rodgers 2010.
D–K	H. Diels and W. Kranz, *Die Fragmente der Vorsokratiker*, 3 vols. Berlin, 1952^{6}.
E–K	Fragments of Posidonius, ed. Edelstein and Kidd 1972–99, cited as "F#."
Gal. or [Gal.]	Galen or [Galen]. For abbreviations of titles see earlier remarks. Works are cited by chapter/section numbers plus volume, page, and line numbers in the edition of K. G. Kühn ("K") (Leipzig, 1821–33). *Med. exp.* is cited by section number and page of Walzer and Frede 1985 ("W–F").

	SI is cited by Kühn plus page and line numbers of G. Helmreich ("H") (Leipzig, 1893). *Subf. emp.* is cited by section number and page and line numbers of Deichgräber 1965 ("D").
Garg. Mart. *Pom.*	Gargilius Martialis, *De arboribus pomiferis*, ed. Zainaldin 2020a. Cited by chapter, section, and sentence numbers.
Hippocr.	Hippocrates, plus abbreviated Latin title (for which, see E. Craik, *The "Hippocratic" Corpus: Content and Context.* London, 2015). Works are cited by chapter/section numbers plus volume and page in the edition of É. Littré ("L") (Paris, 1839–61). The work *De hebdomadibus* is cited in the edition of W. H. Roscher (Paderborn, 1913).
Hyg. *Cond. agr.*, *Gen. contr.*, *Lim.*	"Hyginus maior" or "Hyginus 1," fragments *de condicionibus agrorum*, *de generibus controuersiarum*, and *de limitibus castrorum* (Ch. 2, n.140) [for citation see earlier entry on *CAR*; see also earlier entries for other works ascribed to a "Hyginus"].
[Hyg.] *Mun. castr.*	"Pseudo-Hyginus," *De munitionibus castrorum* or *De metatione castrorum* (Ch. 2, n.140) [see earlier entries for other works ascribed to a "Hyginus"].
L–M	A. Laks and G. W. Most, *Early Greek Philosophy*, 9 vols. Cambridge, MA, 2016.
Ph. *Bel.*	Philo, *Belopoeica*. Cited by Thévenot ("T") page and line numbers.
SVF	H. von Arnim, *Stoicorum veterum fragmenta*, 4 vols. Stuttgart, 1964 (rpt). Cited by volume and von Arnim's numbers.
U	H. Usener, *Epicurea*. Leipzig, 1887. Cited by Usener's numbers.

II Reference Works

The following abbreviations are used for modern works of reference. Works of reference not included on this list are cited in author–date style with corresponding entries in the bibliography.

CIL — *Corpus inscriptionum Latinarum*. Berlin, 1863–.

DPA — R. Goulet (ed.), *Dictionnaire des philosophes antiques*. Paris, 1989–.

DGE — *Diccionnario Griego-Español*. Madrid, 1980–.

H–S — J. B. Hofmann, *Lateinische Syntax und Stilistik* (revised by A. Szantyr). Munich, 1965.

IGRRP — R. Cagnat, *Inscriptiones Graecae ad res Romanas pertinentes*, 3 vols. Paris, 1906–27.

L&S — C. T. Lewis and C. Short (eds.), *A Latin Dictionary Founded on Andrews' Edition of Freund's Latin Dictionary*. Oxford, 1879.

LSJ — H. G. Liddell, R. Scott and H. S. Jones, *Greek-English Lexicon* (with revised supplement). Oxford, 1996^{9}.

Maltby — R. Maltby, *A Lexicon of Ancient Latin Etymologies*. Liverpool, 1991.

NP — H. Cancik and H. Schneider (eds.), *Brill's New Pauly: Encyclopaedia of the Ancient World. Antiquity* (English ed., edited by C. F. Salazar). Boston/Leiden, 2002–11.

OCD — *The Oxford Classical Dictionary*, online ed. Oxford, 2015–^{5}.

OED — *The Oxford English Dictionary*, online ed. Oxford, 2000–^{2}.

OLD — *The Oxford Latin Dictionary*. Oxford, 2012^{2}.

OLS — H. Pinkster, *The Oxford Latin Syntax*, 2 vols. Oxford, 2015–21.

PIR^{2} — *Prosopographia imperii Romani saec. I. II. III.* Berlin/Leipzig, $1933\text{–}2015^{2}$.

RE — A. Pauly and G. Wissowa (eds.), *Real-Encyclopädie der classischen Altertumswissenschaft*. Stuttgart, 1893–1980.

Schwyzer — E. Schwyzer, *Griechische Grammatik*, 3 vols. Berlin, 1953–66.

S–H — M. Schanz, *Geschichte der römischen Litteratur bis zum Gesetzgebungswerk des Kaisers Justinian* (revised by C. Hosius), 4 vols. Munich, $1927\text{–}59^{4}$.

Smyth — H. W. Smyth, *Greek Grammar* (revised by G. M. Messing). Cambridge, MA, 1956.

TLL — *Thesaurus linguae Latinae*. Leipzig/Berlin, 1900–.

INTRODUCTION

The Idea of the artes

> tot nos praeceptoribus, tot exemplis instruxit antiquitas, ut possit uideri nulla sorte nascendi aetas felicior quam nostra, cui docendae priores elaborarunt. M. igitur Cato, idem summus imperator, idem sapiens, idem orator, idem historiae conditor, idem iuris, idem rerum rusticarum peritissimus fuit . . . quam multa, paene omnia tradidit Varro! quod instrumentum dicendi M. Tullio defuit? quid plura? cum etiam Cornelius Celsus, mediocri uir ingenio, non solum de his omnibus conscripserit artibus, sed amplius rei militaris et rusticae et medicinae praecepta reliquerit, dignus uel ipso proposito ut eum scisse omnia illa credamus. (Quint. *Inst.* 12.11.22–4)
>
> Antiquity has provided us with so many teachers and so many examples to emulate that that there may seem to be no age more fortunate to be born into than our own, for whose education earlier generations expended so much effort. So M. Cato was – one and the same man! – a general of the highest distinction, a man of wisdom, an orator, an historian, an extremely skilled lawyer and agriculturist . . . How much did Varro hand down to us – indeed, nearly everything there is to know! What means for speaking did M. Tullius Cicero not already possess? What more need I say, when even Cornelius Celsus, a man of moderate ability, not only composed works on all these arts, but also left precepts at even greater length on the art of war, agriculture, and medicine? He was worthy, simply in virtue of his intention, of our credence that he knew all of these things!

In his *Institutio oratoria*, written not long before the end of the first century AD, Quintilian looked back with gratitude on the inheritance that the ancients had bequeathed to his own age. Seeking to inspire his Roman readers to strive for the pinnacle of learning in his own field – rhetoric – he brings forth four exemplary Roman authors – Cato, Varro, Cicero, and Celsus – who, he declares, mastered, and transmitted through writing almost every discipline worth knowing something about.[1] This book is about the intellectual ideal to which Quintilian alludes in this passage; an ideal now largely forgotten, but which once stood very near the center of what the Romans knew, or thought they knew, about the world. This book is about the sprawling, richly varied, and intellectually lively literary phenomenon that gave shape to this ideal. This book is about the *artes*.

I.1 Recovering the *artes*

The *artes*, in the sense of the systematic treatises on various specialized disciplines, were once the limbs of an ancient image of Roman polymathy.[2] But this image now comes to us in a state of dismemberment, like so many of the monuments and statues of antiquity that stood in their own time as glorious unities. Its pieces, scattered widely across fields of modern study, have been carefully examined in their fractured isolation over the decades and centuries: each *ars* thus has its scholarly specialists, who have set their findings into the sweep of disciplinary histories. We have gained a great deal by such work; indeed, understanding each *ars* is indispensable for understanding the *artes*. But it is not enough. To grasp Quintilian's polymathic ideal, the *artes* must be studied together, as the parts of the ancient image to which they once belonged. Only in this way can we cease to view the *artes* through modern eyes as the several domains of specialists and begin to see them as an expansive but unified – and uniquely Roman – way of knowing the world. This book aims to restore the *artes* as a whole to their place in intellectual history, and so also to open them to a wider audience beyond their specialists.

This project of renovating the *artes* has three components. For one, we must understand what it meant for a branch of specialized knowledge to be an "*ars*" as such. What did the notion "*ars*" imply about thinking, knowing, doing, speaking, and writing as regards some subject? Two, we need to understand what it meant for a particular branch of knowledge to be an *ars*, both by itself and in relation to other *artes*. What went into the *ars* of agriculture, for instance, above and beyond the know-how of agricultural practice? And what went into the *ars* of agriculture that did not go into the *ars* of architecture or of medicine? Three, we need to understand what it meant for an *ars* to be only one of many *artes*. How was a given *ars* related to other *artes* in its contents, methods, and aims? These mutually

[1] For these exemplars of Roman polymathy, see Ch. 1.6 and Appendix.
[2] For the double signification of *ars* as both "discipline" and "systematic treatise," see Ch. 1.1.

elucidating questions can be studied in several ways. We can look at explicit statements about what constitutes an "*ars*" as well as at implicit reflections on the concept through the choices and interests of authors of the *artes*. We can inquire into how authors represented or constructed their field of expertise *as* an *ars*. And we can consider how authors located their own *ars* within a broader landscape of *artes*. All these questions and modes of inquiry are pursued throughout this book.

I.2 The *artes* in the Early Empire

This book focuses on the *artes* of the early Roman Empire, a period that we will define roughly as the long first century between Octavian's reception of the title "Augustus" in 27 BC and the death of the Roman senator Sextus Julius Frontinus in AD 103/4. This was undeniably the outstanding epoch for the *artes* at Rome, witnessing the best and most ambitious works, and in some cases also the first, for many disciplines. Vitruvius' *De architectura*, published sometime between 35 and 25 BC, was the first full-scale Latin treatise on its subject. Of Celsus' *Artes*, dating from the first half of the first century AD, there remain only the eight celebrated books on medicine, but his work once also included agriculture, the art of war, rhetoric, and more. In agriculture, Columella's *Res rustica* appeared near the middle of the century, the most ambitious treatise on farming preserved from Greco-Roman antiquity. In the art of war, Frontinus left behind two influential works: one supplying precepts for generalship, and the other, called *Strategemata*, presenting a collection of historical *exempla*; both likely date to the latter years of Domitian's reign. Under the Flavians and Trajan, the writings which would form the nucleus of the Roman land-surveying writings were penned by Frontinus (as it seems), among others. Quintilian's *Institutio oratoria* appeared near the end of the century. Finally, there is Pliny's *Historia naturalis* – not an *ars*, but embracing in the vast sweep of its ambition the spirit and subject of many *artes*, especially agriculture and medicine.

It is no coincidence that so many notable *artes* appeared in the same epoch, nor are the limits of this book marked arbitrarily around this period. Rather, I argue that the *artes* of this age represent a coherent intellectual movement, or, to put it in wider historical terms, the emergence of a distinctive and unified intellectual culture, which it is the aim of this whole book to investigate. Not infrequently it is useful to consider earlier *artes* (those of the Republican period), besides the Greek *technai* (τέχναι), to appreciate the flowering of the *artes*; but the long first century AD remains the age of the *artes par excellence*.[3] Because of the outstanding character of many early Imperial *artes*, which remained unsurpassed in important respects through the end of Roman antiquity and indeed often into the medieval and modern periods, it should hardly surprise to find that they profoundly influenced many later ages. These later writings represent an important episode in the history of *artes*, but they fall outside the scope of this book.

I.3 The Unity of the *artes*

A key aim of this book is to demonstrate *that* and *how* the *artes* of the early Empire (hereafter simply "*artes*") are fundamentally unified. Scholars have long recognized various affinities among the *artes*, although not under this description. The relevant texts have generally been understood, implicitly or explicitly, in relation to notional genres or text-types that embrace writings across many different branches and kinds of specialized knowledge. Thus, scholars have elucidated the style, structure, and contents of the *artes* by appealing to features typical of ancient literature of specialized knowledge, or ancient "technical literature."[4] This is not the only way to understand the relationship among the *artes*, however. Throughout this book, various other kinds of unity will be posited and explored. The *artes* of the early Empire are unified in that they participate in a shared set of intellectual ambitions established by Greek and late-Republican Roman thinkers. They are unified in that they represent a collective effort to articulate the sorts of expertise on which Roman power depended. They are unified in that they construe the facts and goals of their activity through the eyes of Roman culture. The most important source of unity that concerns us, however, is that the *artes* held in common the ambition of constructing a richly theoretical framework for expertise in their fields. To be sure, all the authors of *artes* were concerned, to different degrees, to communicate the practical elements of expertise in a narrower sense. But that was not the only or even the most significant purpose of their writings: the *artes* aimed above all to construct what we may call "images of expertise" that located their

[3] See Ch. 1 on the prehistory of the Imperial *artes*.

[4] For "technical literature," see also Ch. 2, n.88. I have learned much from previous studies, though I do not adopt any frameworks wholesale: cf. Fuhrmann 1960; Kullmann et al. 1998; Meißner 1999; Formisano 2001; Horster and Reitz 2003; Fögen 2005; Paniagua Aguilar 2006; Asper 2007; Fögen 2009; Taub and Doody 2009; Asper 2013; Holmes and Fischer 2015; Roby 2016; Formisano and van der Eijk 2017; König and Woolf 2017; Taub 2017.

respective disciplines between nature and human society and connected them with a greater body of Roman knowledge about the world.

These larger ambitions of the *artes* can be understood as a response to one of the central problems of ancient technical literature: the problem of relating *theory* and *practice*. For present purposes, by "practice" we mean (more concretely) the personal experience that embodies and (more abstractly) the methods, precepts, and rules that articulate know-how, skill, or capability; by "theory" we mean the principles, presuppositions, concepts, and beliefs that are adduced to inform, explain, or justify practice. The attempt to negotiate the theory–practice relationship manifests itself across many kinds of ancient technical writing, not just the *artes*. What knowledge distinguishes someone who has expertise in a discipline from someone who does not? What knowledge distinguishes expertise in one discipline from, or affiliates it with, expertise in another? How exactly does someone acquire the knowledge which is appropriate to a specific discipline? What knowledge proximately guides practice, and what knowledge more remotely grounds or justifies it? Is there a grasp of a certain kind of knowledge that would guarantee a correct approach to practice? If so, what kind of knowledge? What are the limits of the contributions that theoretical knowledge can make to practice? And so on. These questions are coextensive with those we asked more specifically of *ars* and the *artes* – naturally, since the theory–practice problem amounts, in essence, to an invitation to investigate the basis and character of specialized knowledge in its many forms. In a primary sense, the unity of the *artes* derives from the sharing of *questions* centered on the theory–practice problem. In a secondary and more significant sense, the unity of the *artes* derives from the sharing of *interests, strategies, and answers* framed in response to these questions.

The various kinds of unity among the *artes* had multiple efficient causes. They were in part the result of influence both direct and indirect, as is illustrated concisely in the Appendix. But they also stem from what we might call "genetic affinity." That is, authors of the *artes* in the early Roman Empire inherited shared intellectual concepts, models, and goals that suggested paths for their activity more definite than the opportunities and challenges common to all technical writing. In short, the *artes* were unified not only because authors read or otherwise came to know and reacted to one another, but also because they were the products of a specific, shared intellectual culture. This formulation may sound vague or tautological at present but will be made rich and specific throughout the course of the book.

I.4 The *artes* as a Roman Scientific Culture

The shared interests, strategies, and answers stemming from the theory–practice problem opened the *artes* to the natural and human worlds, because the attempt to work out the epistemic underpinnings of the *artes* led naturally into reflection on their place in Roman knowledge more broadly. While the *artes* are far from univocal, significant areas of agreement evolved from such reflection. Authors afforded considerable attention to articulating, implicitly and explicitly, *theories of knowledge* and *theories of nature*. An *ars* can only present a complete "image of expertise" in a given field by including such theories to motivate, frame, and explain its premises, its means, and its ends. The epistemological strand of thought – a *theory of knowledge* – follows from the effort to work out the character of *ars* as such and manifests itself in many ways, as we will see throughout this book. Authors mark out the peculiar kinds of knowledge required in their own disciplines; establish criteria, theoretical and practical, for expertise; explain and justify their methods in various fashions; analyze different modes of acquiring knowledge; scrutinize and critique competing methods and principles; and adumbrate approaches for inquiring into nature and society. The natural-philosophical strand of thought – a *theory of nature* – emerges from the epistemological strand, as a function of the effort to ground the *artes* more fully. That is because knowing an *ars* requires knowing how to influence the natural world to produce the desired end of the *ars*. Minimally, this is knowledge of the qualitative properties of things, of regular patterns of cause and effect, and so forth. In more developed forms, it can be full-blown explanatory knowledge based on natural first principles. The *artes* reveal different kinds of engagement with nature along this spectrum. Authors assign natural properties to things; use analogies and models; offer physical accounts dependent on material and qualitative principles; gather and interpret empirical data; apply reason to the elucidation of ambiguous phenomena; and set out the principles and results of numerous branches of natural inquiry. We will consider all these cases and more, each in its proper place.

The richly theoretical sensibility of the *artes* fully warrants calling their intellectual culture a "scientific culture." And because the early Imperial *artes* were a fundamentally novel development at Rome, we may say that they represent the *emergence of a scientific culture*. By "scientific" I mean a specific and minimal understanding of "science" functional for Greco-Roman antiquity: science as systematic inquiry into the world. In this sense, science is more than merely putting knowledge into order, but less than a prescriptive method governing the character of this inquiry.[5] The *artes* are not sciences, or not merely sciences, because their purpose is not inquiry alone. But they are scientific in that they engage with science, implicate science into their own aims, and exhibit the fruits of science. To say that the *artes* are a scientific culture is another way of saying that they articulate and even argue, through their images of expertise, a systematic conception about how the world is and how we know it is that way. In other words, the image of expertise that constitutes each *ars* always contains theories of both knowledge and nature.

There are good reasons to insist on understanding the intellectual culture of the *artes* as a scientific culture. First, the label is informative if "scientific culture" is not merely a synonym for "culture of knowledge" or "intellectual culture." A scientific culture involves systematic inquiry into the world; a culture of knowledge or intellectual culture need not. To say that the *artes* represent a scientific culture is therefore not a tautological statement of their unity: it is the expression of a surprising and important finding, that the *artes* are also vehicles for the Roman understanding of the world beyond their disciplinary material. Second, the label works against an entrenched but misleading picture of Roman intellectual life. In the eyes of modern scholars and the public alike, the history of Greco-Roman science remains an essentially Greek phenomenon, associated above all with cities such as Athens and Alexandria and only accidentally with Rome, through Greek residents such as Galen. Romans are stereotyped as people of practical genius: inventive and efficient builders, engineers, and problem-solvers, but with little interest in theoretical inquiry. Recognizing the Roman *artes* as a scientific culture provides a welcome corrective to this generalization. The modern cult of scientific genius may pass over Rome in its search for an Aristotle, Archimedes, or Ptolemy; but that does not mean Romans did not care to understand, and articulate their understanding of, the world.[6] The *artes* are one of the principal ways they did so.

It is worth dwelling on this point because it is easy to misunderstand the significance of the *artes* as a scientific culture. The *artes* are neglected in histories of science precisely because they have not seemed to anticipate "science" today: the theorizing that I have suggested lends the *artes* their scientific cast does not essentially mirror practices or habits of empirical inquiry integral to the disciplines themselves that might evoke the

[5] Obviously, this is not modern science, yet it is not so capacious as "the whole industry of ancient knowledge ordering" (König 2017: 4), a reframing which for other purposes helpfully broadens the notion.

[6] See generally also Lehoux 2012.

contemporary practice of science. To be sure, authors frequently offer empirical tests or proofs for their claims, but investigation into the natural world is rarely productive of new techniques, let alone of hard-won "facts" that modern scientists can point to in anticipation of their own work. The theories of nature and knowledge in the *artes* are largely *a posteriori*, applied (as it could seem to the uncharitable modern eye) disputatiously to justify or ornament the author's preferred techniques. This impression is no doubt why, when scientific dimensions of the *artes* are recognized at all, they are almost always described as Roman authors' efforts to elevate the intellectual respectability of their disciplines and to seek social status for themselves and their profession. Social striving is not irrelevant, but this characterization distorts the intellectual agenda of the *artes* by foisting upon them an overly restrictive and modern conception of science. The inquiry that belongs to the *artes* is not that of modern science and its experimental methods, but it nonetheless represents a means of penetrating and compassing the natural world with human understanding. The *artes* do not share the purpose of winning new knowledge for a recognized totality of scientific understanding, but they do embody an effort to intellectually master and explain the empirical elements within their domain. Once again, it is salutary to insist on the label "scientific" for the *artes* not just because this is a recognizably scientific disposition (as it is), but because the *artes* help to stretch the historical category of "science" by revealing the different emphases and contours that scientific thought can take in different cultural settings. The predominantly theoretical interest of the *artes* is also why this scientific culture is above all a literary culture – that is, manifested in the emergence and development of a body of texts.

The *artes* are a specifically and uniquely *Roman* scientific culture. These were Roman authors writing Latin treatises with the shared concepts, models, and goals of the intellectual culture within which they worked. Beyond that, however, the characteristically Roman conjunction of the interests of science and those of disciplinary expertise also shaped the scientific character of the *artes*. It is a significant fact that the *artes* parceled out understanding of the world not according to branches of natural inquiry (astronomy, life sciences, physical theory, and so on) but according to branches of human activity (agriculture, medicine, land-surveying, and so on). These branches of activity were not of course arbitrary: they were a mixture of the fields considered traditionally appropriate by the Romans for study and those which became increasingly significant as supports for Roman *imperium*.[7] In this way, science was channeled along the lines of

[7] Besides Ch. 1 and Ch. 2, see also Ch. 5.1 on Celsus' *Artes*.

Roman interest and yoked to the purposes of Roman mastery over the natural and human worlds. This book is only incidentally concerned with the *artes* as an expression of Roman power or ideology or as props for Roman politics or society, but the influences of Roman Empire cannot be ignored, for these often set the parameters for the scientific character of the *artes*.

In speaking of the "Roman" character of the *artes*, the question that most naturally comes to mind is the distinctness of this scientific culture from that of the Greeks. Is "Roman" merely a chronological and geographic label for the continuation of Greek scientific thought and practice? As Chapter 1 and subsequent chapters show, it is not wrong to say that the theoretical premises of the *artes* are thoroughly Greek – although their synthesis in the *artes* is crucially novel to Rome. But this is far from reducing the *artes* to a chapter in the reception of Greek thought. Scientific knowledge is sparsely distributed over the natural and human worlds: we only find where we look, but we only look where we have reason to be interested. The Roman looked where the Greeks did not and found what the Greeks did not find. And, of course, the Romans found most of all there where their own ingenious means of shaping humanity and nature illuminated the world around them. The lights of Roman Empire marked out the intellectual geography of the *artes*, opening fresh territory into which ancient scientific thought could advance. We will accordingly pay attention to the Roman background as it makes itself felt in the authors' understanding of nature, knowledge, and society.

I.5 The Plan of the Book

The chapters of this book are addressed to the various questions, problems, and themes proposed in the foregoing pages: the significance of *ars* and the unity of the *artes*; the theory–practice relationship; and the theories of nature and knowledge manifested in the *artes*. The ultimate aim is to demonstrate the existence and character of the Roman scientific culture of the *artes*. The project has seven chapters, which divide into four parts.

Part I, comprising Chapters 1 and 2, is propaedeutic. Chapter 1 relates the pre-Imperial history of *ars* and develops a notion of the "higher *artes*" that had taken shape by the end of the Republic. It identifies four features of the higher *artes* which would form the premises for the scientific culture of the early Empire. Chapter 2 sketches the social, historical, and political backdrops of the *artes* and introduces the authors and texts of Chapters 3 to 7, besides addressing related global issues such as Roman attitudes toward the

Greeks and the Roman audience of these works. This chapter also unpacks the specifically *Roman* character of the scientific culture of the *artes* by proposing to understand them in terms of the "Romanization" of specialized knowledge as such.

Chapters 3 to 7 (Parts II–IV) each offer a self-contained treatment of one *ars* through one or two central texts. A special concern of these chapters is to present stand-alone treatments of each *ars* that integrate the authors' more self-conscious prefatory remarks with the precepts and methods presented in their works. I thus aim to go beyond formal or generic similarities to appreciate the intellectual and, more specifically, scientific forces that animated the *artes* individually and collectively. While it is possible to read each chapter on its own, they have been written to build upon one another in gradually establishing a rich conception of this scientific culture. Accordingly, the chapters fall into three loosely chronological parts that exemplify different moments and tendencies in the development of the *artes*.

Chapters 3 and 4 form Part II of the book. Chapter 3 examines Vitruvius' *De architectura* as a paradigmatic early example of the scientific culture of the *artes*. It explores his epistemological analysis of *architectura* as an "architectonic" art and traces its ramifications into the explanations of architectural methods throughout the treatise. Chapter 4 jumps ahead almost a century to consider Columella's *Res rustica* as representing the transposition of an old and well-established *ars* – agriculture – into the scientific culture of the early Empire. It studies his development of certain kinds of explanatory knowledge of nature as a basis for agricultural method. This second part reveals the *artes* at their most scientifically exuberant, as both Vitruvius and Columella elaborate sophisticated and highly interdisciplinary epistemological positions, embed well-developed theories of nature in their expositions of method, and invite their readers to participate in certain kinds of scientific reasoning to become consummate experts.

Part III consists of Chapter 5, which moves chronologically backwards from Columella to study Celsus' *Artes*, principally the medical books (*De medicina*). Against Vitruvius and Columella, Celsus adopts a more reserved attitude toward nature and specialized knowledge in his treatment of medicine – a retrenchment in explanatory ambition, as it were. Hence, this single-chapter part represents a turning point in our study of the scientific culture of the *artes*. Even so, this is very much an *ars* of the early Empire: Celsus carefully stakes out a medical epistemology to justify his moderate and pragmatic approach to the discipline and develops a robust, descriptive conception of nature functional for his project of expounding medical therapies. Celsus' *ars* demonstrates the variety that the scientific culture could sustain against a backdrop of shared interests and concerns.

Part IV consists of Chapters 6 and 7 and examines two later *artes* that sustain themes from Parts II and III but also reveal interesting differences. Chapter 6 reconstructs developments in the art of war through Onasander's *Stratêgikos* and Frontinus' *Strategemata*, showing how epistemological and other pressures led to the generic diversification of the *ars*. Finally, Chapter 7 looks at the strategies used by two early authors in the *corpus agrimensorum Romanorum*, Frontinus and Hyginus, to elevate land-surveying into the ranks of the higher *artes*. These authors adopt certain approaches familiar from earlier chapters but also develop new strategies tailored to the unique challenges of their discipline.

A brief conclusion returns to the larger questions and problems posed earlier, and reconsiders major themes that have emerged across this study of the *artes*.

I.6 A Synthetic View

As the foregoing discussion and plan make clear, the aim of this book is to set the *artes* of the early Roman Empire into synthetic view, by demonstrating the larger scientific culture to which the *artes* belong and at the same time reading them in the terms of that culture. The value of this book lies essentially in this effort to read the *artes* together – as many elements of a single ancient phenomenon, rather than as isolated texts (or disciplines) with scattered episodes of affinity and influence – and to bring the *artes* as a whole to an audience beyond the modern specialists in each field. It is appropriate here to acknowledge my gratitude to the experts in the fields under study who have laid the foundations for the broader study proposed in this book. This effort would have been far more difficult, if not impossible, only two decades ago. Specialists will find much familiar in the respective chapters and will indeed recognize the fruits of their own and others' work. At the same time, I hope that the synthetic perspective adopted in this book provides a new way of framing this work and has raised and perhaps answered many questions that will interest experts in these field. Scholars specialized in these areas will have no trouble recognizing where I am following and where I am departing from conventional lines of interpretation, and I have documented my engagement fully in the footnotes. But for the sake of the larger readership that I hope this book finds, I have refrained from entering into the fray in the main text, instead weighing past discussions in the notes and presenting to the reader "what may seem to be closest to the truth" (pr.45 *quae proxima uero uideri possint*), to borrow the words of Celsus, who faced a similarly challenging task of assimilating a learned tradition and presenting it to a wide audience.

PART I

CHAPTER 1

The Scientific Premises for the Imperial artes
A View from the Late Republic

The early Imperial *artes* are a potent expression of the pursuit of Roman self-knowledge at a time when Romans set themselves to investigating the world that centuries of continual warfare had won them and sought to understand what caused the predominance of their own civilization. They offered the Romans a vehicle for sifting the contents of their collective know-how and accounting for the kinds of expertise which they believed explained their ascendancy. The mirror that the *artes* held up to the Romans was at the same time a window onto the human and natural worlds which they aimed to compass with their understanding. In this perspective, their dramatic proliferation in the early Empire represents an intellectual effort of almost incredible scope and ambition: nothing short of making explicit the knowledge and practices that structured the Roman experience of the world. In the sustained and extensive character of this project, the Imperial *artes* were novel in Roman intellectual history. Yet they grew from deep roots. The foundation for this scientific culture was set in the late Republic, when the core elements of *ars* that would allow it to serve as a vehicle for reflection on the natural world as well as on Roman civilization took shape through the reception and elaboration of Greek ideas of *technê* (τέχνη). In this chapter, we will come to grips with the late-Republican Roman and Greek heritage of the Imperial *artes*. To do so, we will adopt a vantage point near the end of the Republic – say, around 35 BC – and consider how the *artes* were constructed in Roman intellectual culture. By this time, the premises most significant for the development of the *artes* in the early Imperial period were already set.

1.1 The "Higher *artes*"

In the late Republic, as throughout all Roman history, Latin "*ars*" could be used of any branch of specialized knowledge that was discrete, unified, and whole.[1] Cooking, cobbling, and fulling were *artes*, as were law,

[1] These are minimal, nontheoretical criteria.

oratory, and generalship; astronomy and agriculture, medicine, music, and mathematics, and even courtesanship and love-making could be called *artes*.[2] From this it is clear that *ars* could mean not just "art," whether in the narrow or broad sense of the English world, but also "skill," "craft," "science," or "profession," and could generally pick out expertise in any distinct field of thought, activity, or production.[3] It had other meanings too, but this is the widest notion of *ars* in the sense that concerns us.[4]

The Latin word *ars* never lost its capacious usage in ordinary speech, but by the late Republic it had acquired a special significance in the circles of the educated and philhellenic Roman elite such as M. Tullius Cicero and M. Terentius Varro. For these men, *ars* was often subject to numerous qualifications and could be more restricted in its meaning. One important distinction was framed in essentially social terms, between those *artes* which had a place in the education or studies of a free and upright Roman and those which did not. It was the former which were *artes* in the fullest sense. This distinction was neither originally Roman (there were important Greek precedents[5]), nor did it remain static in Republican times. Consider Varro's *Disciplinae*, a work which covered nine *artes* in nine books. The first seven books are usually thought to have handled the disciplines which by the end of late antiquity made up the canonical "liberal arts" (*artes liberales*): grammar, dialectic, rhetoric, geometry, arithmetic, astronomy, and music.[6] And these *artes* indeed appear among those which Cicero speaks about as belonging to the pursuits of educated Romans.[7] Varro's *Disciplinae* continued with medicine and architecture,

2 Ter. *Haut.* 226 for *ars meretricia*; Ou. *Ars am.* 1.1 for *ars amandi*. See *TLL* II 663.40–667.63 (Klotz) for an alphabetized list of *artes*, only a partial conspectus because "*ars*" could be varied by other words (e.g. *disciplina, doctrina, scientia*) or omitted altogether (e.g. *agri cultura, astrologia, geometria*). See also Rawson 1985, part II; cf. Lloyd 2011: 14–16 (on τέχνη).

3 Cf. *OLD* 1, 5–7.

4 Like τέχνη (LSJ A.2), it could mean "deceit" or "trick" (*OLD* 3).

5 According to a standard view, the Greek ἐγκύκλιος παιδεία anticipates the Roman *artes liberales*, both notions by the first century BC representing curricula centered on the seven canonical arts: Marrou 1937: 211–35, 1948: 266–7, 568–9; Kühnert 1961: 3–42; Marrou 1968; Rawson 1985: 117; cf. Bonner 1977: 77–9, and on ἐγκύκλιος παιδεία, also De Rijk 1965. However, Hadot 2005 argues that the liberal arts were not well defined in the late Republic and denies that the ἐγκύκλιος παιδεία was their forerunner (esp. 44–59, 263–93). Even so, the social grading of *artes* has Greek precedents, for example, Xen. *Oec.* 4.2–3 on the τέχναι βαναυσικαί (with Pomeroy 1994: 235–6).

6 Ritschl 1877 is fundamental; see also Rawson 1985: 88; Cardauns 2001: 77–8. Hadot 2005, esp. 156–90, 333–72, argues the skeptical case – that is, we know almost nothing about their content and order (190); Shanzer 2005, esp. 84–103, defends Ritschl's reconstruction.

7 They appear among *artes* that Cicero calls, with various nuances, *bonae* (*De or.* 1.158, 3.136, *Vat.* 8, *Cael.* 45, 77, *Sen.* 29, *Off.* 3.6), *honestae* (*Brut.* 213, *Off.* 1.151), *humanae* (*De or.* 3.21, *Rep.* 1.28), *ingenuae* (*De or.* 1.73, 3.21, *Fin.* 3.4, 5.48, *Tusc.* 5.63), *liberales* (*Inu.* 1.35; cf. *De or.* 1.11, 3.127, *Off.* 1.150), *optimae* (*Diu. Caec.* 39, *Arch.* 1, *Cael.* 24, *Rep.* 3.6, *Fin.* 2.111), and so forth. See generally Kühnert 1961: 28;

however – fields which Cicero regarded as of doubtful worthiness for Romans of the best rank.[8] For his part, Cicero never presented any "liberal arts" as a closed set, and Varro's disciplines did not exhaust the *artes* which might be considered generally "liberal," since it was always decent for upper-class Romans to take an interest in other fields such as agriculture, law, and the art of war (generalship).[9]

Hand in hand with these social distinctions went two further qualifications. The first was a certain theoretical sensibility, regarding the notion of *ars* both in general and in reference to a given discipline. This sensibility was a result of the adoption, and elaboration, of an epistemologically demanding conception of specialized knowledge through the Greek *technai* and the associated philosophical and scientific traditions. Understanding these latter developments will be the main aim of this chapter. But a related second qualification deserves mention: the capacity for textualization. The *artes* cannot be defined without reference to the branch of literature of the same name, for their social and epistemological ambitions could be fully realized only by written treatment.[10] This textualization took the shape of the systematic prose treatise, also called "*ars*."[11] There is no need to stress the fluidity of this textual form, since as much will become obvious in our study. Yet, it should be emphasized that the *ars* proper was a literary form distinct from other kinds of didactic writing, such as the poetic traditions which handled similar subjects (astronomy or agriculture, for example) but whose concerns were only partially shared.[12] Hence, when we speak of the *artes* we are speaking not only about a social and intellectual ideal, but also about a form of literature that was its expression: For the remainder of this book, it should be borne in mind that "*ars*" refers both to the *discipline* and to the *systematic treatise* concerning the discipline. These three qualifications loosely define what we might call the "higher *artes*." The social, epistemological, and textual differentiation of the *artes* entailed that, despite the wide extension of the word that

Marrou 1968: 20, who try to assemble the seven *artes liberales* from various passages but are compelled to admit other disciplines (e.g. philosophy, law, art of war); Hadot (2005, n.5) argues persuasively that Cicero's notion of the *artes* was indeterminate. *Bonae artes* can bear a different sense in Republican political language: Earl 1961: 11–12; Oakley 1997–2005: III 552 on Liu. 9.42.4; cf. Powell 1988: 120–1 on Cic. *Sen.* 9.

8 Cic. *Off.* 151 with Dyck 1996: 336 (also Rawson 1985: 84); Panaetius' Περὶ τοῦ καθήκοντος may be Cicero's source (Brunt 1973: 26–34), but Dyck 1996: 331–3 argues that it is Cicero's own contribution.

9 Cf. n.7. Cato the Elder wrote on all these subjects (Ch. 5, n.18).

10 Cf. Mayer 2005: 228.

11 *OLD* 9b; *TLL* II 671.54–672.50 (Klotz), especially of rhetoric.

12 There was frequent exchange between didactic poetry and the prose *artes*, however; other rubrics might put them into closer conversation. See, for example, Ch. 4, nn.16, 18 on Columella and Virgil.

made cooking and cobbling *artes* in the ordinary sense, in educated circles these were not worthy of notice, let alone engagement. Such disciplines were socially low, beneath what a person of upper-class stock or pretension might deem worthy, and were also intellectually low, offering little to engage an author.

At least by the first century BC, certain basic marks of an *ars* had been inherited from the Greek notion of *technê*, the essential features of which had already begun to take shape by Homer's time.[13] *Technê* or *ars* implies determinate and goal-directed expertise; by means of appropriate methods it allows for reliable achievement of an aim which those without it could not achieve, or could achieve only by chance; its products are recognizable even to those without expertise; it is teachable; and so forth. These features were shared by all *artes*, high and low. Apart from them, certain further premises were crucially significant for what we called the "higher *artes*." (Hereafter, by "*artes*" we mean the "higher *artes*," unless otherwise noted.) In this chapter, we will characterize four such premises that were molded in the late-Republican period by Roman thinkers such as Varro and especially Cicero and which had, in varying degrees, already begun to be instantiated in the *artes* of their age. First, the contents of an *ars* had to be logically organized and articulated. Second, an *ars* was naturally related to other *artes* by the sharing of certain principles and methods. Third, the methods of an *ars* could be explained in terms of the natural principles of its subject matter. Fourth, an *ars* was a composite of experiential know-how and generalized (explanatory) knowledge; complete and successful practice was a function of the balancing of both elements. For convenience, we will say that the *artes* were *systematic*, *interdisciplinary*, *explanatory*, and *balanced*. As we will see, these premises can all be derived from one strand or another of the Greek philosophical and scientific traditions and are therefore not novel in an absolute sense. What is novel in the Roman situation, however, is the synthesis of these premises into a coherent view of specialized knowledge. While this synthesis was in some measure already underway in the writings of Cicero and Varro, it remained in blueprint until the beginning of the Empire, when it was assumed as the basis for the treatment of a range of different fields, especially (but not only) those predicated on manipulation of the physical world (*rerum natura*). These later developments are the subject of this book as a whole; for now, we will examine the four premises for the *artes* more closely.

[13] These pretheoretical marks of τέχνη were fixed by Plato's age: see Heinimann 1961; Kube 1969: 9–114; Schneider 1989: 11–103; Roochnik 1996: 17–88; Schiefsky 2005: 5–18; Cuomo 2007: 7–40.

1.2 Systematicity

The *artes* were *systematic*. The first – and, by some lights, most fundamental – feature of an *ars* was the systematic and logical organization of the contents of a given field of expertise. A quick way into the matter is Cicero's *De oratore*, wherein Cicero's Crassus casts the analytical method that allows for logical ordering as the essential ingredient of an *ars* (*De or.* 1.187–8):

> omnia fere quae sunt conclusa nunc artibus dispersa et dissipata quondam fuerunt; ut in musicis numeri et uoces et modi; in geometria lineamenta, formae, interualla, magnitudines; in astrologia caeli conuersio, ortus, obitus motusque siderum... [188] adhibita est igitur ars quaedam extrinsecus ex alio genere quodam, quod sibi totum philosophi adsumunt, quae rem dissolutam diuulsamque conglutinaret et ratione quadam constringeret.
>
> Nearly all (sc. subjects) that are nowadays covered by the *artes* were once disconnected and scattered, such as rhythms, notes, and tones in music; in geometry, lines, figures, distances, and sizes; in astronomy, the revolution of the heavens, the risings, settings, and movements of the heavenly bodies... [188] And so a certain *ars* was invoked from outside, from some different field that the philosophers claim is entirely theirs, in order to cement together material that had previously been disjointed and kept apart, and to tie it together with the help of a certain method. (transl. May and Wisse, modified)

By *artibus* Cicero means what we have called the "higher *artes*," as is clear from the examples of music, geometry, and astronomy. Of special importance is the *ars quaedam*, which assembles other *artes* by "cementing" and "tying together" their contents. This art is identified elsewhere as dialectic (*dialectica*), the method of logical analysis which allows, inter alia, for the division, definition, appraisal, and description of (the parts of) a subject.[14] In this passage, Crassus represents the organization of a body of knowledge by the application of dialectical methods as a necessary condition for the constitution of an *ars*. In what follows (*De or.* 1.189–90), it also appears to be sufficient. Thus, discussing the prospects of reducing civil law to an *ars*, he says that it could be accomplished by marking out classes (1.190 *genera*), dividing these

[14] Cic. *Brut.* 153 (discussed later in this section); for the philosophical provenance, cf. *De or.* 1.68 (with Leeman et al. 1981–2008: I 158). Putting aside its philosophical heritage (within Stoic λογική, see Ierodiakonou 2019), it is most relevant here for its logical apparatus; for *dialectica* in the late Republic, see also Rawson 1985: 132–42; Moatti 2015: 197–201 and *passim*.

into branches (*quasi quaedam membra*), and then defining the proper meaning of each thing – that is, each term (*propriam cuiusque uim definitione declaret*). It is implied that if a branch of knowledge can be so handled, it can be made into an *ars*. Cicero returns to dialectic, again in the context of civil law, in the later *Brutus*, where he says that the application of *dialectica* (153), "that art which is the greatest of all arts" (*artem omnium artium maximam*), enabled Servius Sulpicius Rufus to make an *ars* of previously disjointed juridical material.[15] Here he fleshes out the role of dialectic in the constitution of other *artes*, describing it as (*Brut.* 152):[16]

> eam... artem, quae doceret rem uniuersam tribuere in partes, latentem explicare definiendo, obscuram explanare interpretando, ambigua primum uidere, deinde distinguere, postremo habere regulam, qua uera et falsa iudicarentur et quae quibus propositis essent quaeque non essent consequentia.

> that art, which taught (sc. Sulpicius) how to divide the whole of a subject into its parts, to draw out what is implicit by defining, to clarify what is obscure by explaining, to discern and then distinguish what is ambiguous, and finally to possess a rule whereby things true and false could be appraised as well as what are and are not the consequences of what premises.

Cicero discusses the importance of dialectic for the *artes* in other writings, too.[17]

Philosophical dialectic aside, from where did the principles of logical organization vital to the *artes* derive? Likely the most significant, though not the only, source were the Greek prose treatises, called *technai* (τέχναι), devoted to the exposition of specialized knowledge.[18] "*Technê*" in this sense refers not merely to any composition on a technical subject (for example, a Hippocratic text), but a specific kind of systematic treatise that exhibits a formal and linguistic apparatus rigorously applied to the organization and definition of a subject matter.[19] The earliest such *technê*, in the rhetorical tradition, was penned not long after 344/3 BC, probably by Anaximenes of Lampsacus, a contemporary of Aristotle and teacher of Alexander.[20]

15 Cf. *De or.* 1.186–90; Cicero himself wrote a now-lost *de iure ciuili in artem redigendo* (S–H I 526–7). See Schulz 1961: 73–84; Rawson 1985: 201–14; Moatti 2015: 265.

16 The translation follows the notes of Douglas 1966: 119–20.

17 For example, *Or.* 16; see also the remarks on *Inu.* later in this section.

18 See Fuhrmann 1960. Beyond the *technai*, see Rawson 1978; Moatti 2015, esp. 227–70.

19 Of course, early technical writings such as Hippocratic treatises did not necessarily lack organizational principles of their own.

20 Kennedy 1963: 114–24; Chiron 2007; for its authorship, Parmeggiani 2015, esp. 9–12.

(We set aside Aristotle's *Rhetoric* itself.[21]) Anaximenes' *Art of Rhetoric* (Τέχνη ῥητορική) is carefully structured by the self-conscious use of division (διαιρεῖν), classification (into genera variously called γένη, εἴδη, ἰδέαι, τρόποι), differentiation (διαφέρειν), and definition.[22] The analytic framework conspicuous in this work was at the heart of the whole *technê*-tradition and is also evident in later treatises on other disciplines.[23] Dionysius Thrax's *Grammatical Art* (Τέχνη γραμματική), at least part of which dates to the second century BC, is a good example, presenting a treatment of grammatical doctrine stripped to the mechanical particulars of division, definition, and enumeration.[24] Similar in its own field is the *Introduction to Harmonics* (Εἰσαγωγὴ ἁρμονική) of Cleonides, a handbook that summarizes Aristoxenus' music theory.[25] Such works take logical organization to an arid extreme but well illustrate the impulse toward systematicity: indeed, they are the distillation of the approach to *ars* vis-à-vis dialectic to which Cicero alludes.

The spirit of systematization that characterizes the *technê*-tradition becomes conspicuous in the *artes* at Rome in the early first century BC. Before this period technical works seem to have lacked the characteristic interest in classification and definition.[26] Thus, while Cato the Elder's *De agri cultura* (first half of the second century BC) reveals the influence of Greek science and medicine,[27] it is wholly innocent of the logical schemata conspicuous in the *technai*.[28] It is Cicero's *De inuentione* (c. 91–88 BC) and the anonymous *Rhetorica ad Herennium* (c. 86–82 BC), both heavily influenced by Greek precedents,[29] that represent the earliest examples of

21 Its philosophical originality makes it less useful here, which is not to downplay its influence on the τέχνη-tradition. For Aristotle's work, see Kennedy 1963: 82–114; 1994: 51–63; Fortenbaugh 2007.

22 See n.20 and Fuhrmann 1960: 11–28.

23 For rhetorical τέχναι before Anaximenes and Aristotle, see Kennedy 2007: 293–306.

24 Fuhrmann 1960: 29–34. Di Benedetto 1958 questioned the treatise's authenticity, and it is uncertain whether it represents substantially the work of Dionysius (sometime c. 180/170–c. 90 BC) or is a late-antique compilation; at least part is Dionysius' work (*NP* "Dionysius" III 17; Lallot 2003: 19–26).

25 Fuhrmann 1960: 36–40; Solomon 1980: 177–84. The date is uncertain besides the *post quem* of Aristoxenus; the traditional view has Cleonides in the second century AD, but without much support (Solomon 1980: 168–9).

26 Besides Cato's writings (n.28), there are other hints of disorder, for example in the agricultural books of the Sasernae written after 146 BC (Ch. 4, nn.7, 8) and in Aelius' writings on dialectic (cf. Gell. *NA* 16.8.1–3); Cicero notes the lack of a systematic work on law before his time (*De or.* 1.186–90 in n.15). See Rawson and Moatti (esp. 231–2) in n.18.

27 Boscherini 1970; von Staden 1996: 385–93; Mudry 2004a; Diederich 2007: 18–20.

28 On the work's structure, see Ch. 4, n.6; attempts to clean up the text with hypotheses concerning its composition, revision, and transmission (Suerbaum 2004: 250) do not explain away all difficulties. For Cato's other technical writings, which do not change the picture, see Ch. 5, n.18.

29 Inter alios, Hermagoras of Temnos (Kennedy 1963: 303–21).

systematic Latin *artes*.[30] Consider only Cicero's self-conscious discussion of rhetoric that promises a treatment "of the character of the art itself, its function, its end, its material, and its parts" (1.5 *de genere ipsius artis, de officio, de fine, de materia, de partibus*), claiming that the reader will thus grasp "the actual principle and method of the art" (*ipsam rationem ac uiam artis*). This is only a foretaste of the concern with division and definition that marks Cicero's treatise as well as *Rhet. Her.*[31] Varro also exhibits intense interest in logical methods for organization throughout his writings, something no less true of *De re rustica* (37 BC). His treatment of agriculture – a discipline advertised as both *ars* and *scientia* (1.3.1) – is marked by various dialectical techniques: division and definition are employed relentlessly and with varying degrees of success, from the departments of the *res rustica* (1.5.3–4) down to particular subjects such as vineyards (1.8.1–4), farm instrumentation (1.17.1), and farm animals (2.1.12).[32] While Roman authors were not immune to the dreary excesses of systematization, almost from the beginning they tended to balance the utility of logical organization with other literary, didactic, and philosophical concerns. In his later writings Cicero did not abandon dialectical methods, but his approach in *De oratore* stands in stark contrast to his earlier rhetorical writings; and we must not forget that Varro's *Rust.* was in dialogue form, with its own literary and philosophical aspirations.[33] In essence, the systematicity of the *technai* was absorbed into the *artes*, and with salubrious effect, but did not distort the expansive view of their task maintained by Roman authors.

1.3 Interdisciplinarity

While systematicity could be regarded as not only a necessary but also a sufficient condition for the constitution of an *ars*, other premises established in the late Republic were also destined to be significant in the early Empire. The *artes* were thought to be *interdisciplinary* – that is, related to one another in their principles and methods. Two passages from Cicero demonstrate the point. In the first, Cicero defends his acquaintance with

[30] See Kennedy 1972: 103–48; Corbeill 2002: 29–46; Gaines 2007.

[31] Fuhrmann 1960: 58–69 (41–58 on *Rhet. Her.*); Rawson 1978: 18–20, and previous n.

[32] Skydsgaard 1968, esp. 10–37; Diederich 2007: 36–44, 410–19. Varro's systematizing approach to agriculture was probably anticipated by Scrofa, who appears as an interlocutor in *Rust.*: see Ch. 4, nn.9–11.

[33] See Ch. 4, n.15.

the poet Archias' literary pursuits by claiming a connection with the oratorical skills he has put to good political ends. He asserts (*Arch.* 2):

> etenim omnes artes quae ad humanitatem pertinent habent quoddam commune uinclum et quasi cognatione quadam inter se continentur.
>
> Truly, all of the *artes* which belong to human culture share a certain common bond and are held together as if by a certain kindred birth.

Again in *De or.*, Cicero has Crassus make a similar claim (3.21) in discussing the relationship between *eloquentia* and other fields of knowledge:

> omnem doctrinam harum ingenuarum et humanarum artium uno quodam societatis uinculo contineri
>
> All the teachings of these noble and cultivated *artes* are held together by a certain single bond of association

In both passages (others might be adduced[34]), Cicero restricts his statements to the higher *artes* and describes their interrelation vis-à-vis a "bond" (*uinculum*). In *De or.*, this notion is labeled a "well-known saying" (3.21 *non inaudita uox*) and attributed to Plato; the reference has long been recognized as pointing to a passage in the (spurious) *Epinomis.*[35] There, it is argued that geometry, stereometry, astronomy, and harmonics are revealed to be "unified" or "one" (991e3 μίαν) for those who approach them correctly. The author claims (991e5–992a1):

> δεσμὸς γὰρ πεφυκὼς πάντων τούτων εἷς ἀναφανήσεται διανοουμένοις.
>
> For a single bond among all these (sc. fields) will become apparent to those cognizing them (sc. in the correct way).

Cicero's *uinculum unum* is a close rendering of δεσμὸς εἷς, with the adjective *commune* in *Arch.* merely working out the implication of a single bond shared by multiple disciplines.

Yet Cicero's view of the *uinculum* differs strikingly from that of the *Epinomis*, where the δεσμός applies only to the mathematical sciences and

[34] For example, *De or.* 3.136.

[35] Although many ancient authors thought the *Epinomis* Plato's, it is not; it was probably not by Philip of Opus either, another traditional candidate (cf. Diog. Laert. 3.37): see Tarán 1975: 3–47, 115–39; Brisson 2005. For similar passages in authentic works, however, cf. Pl. *Resp.* 531c9–d4, 537c1–3, *Leg.* 967e2–3. Cicero's is the earliest attribution of the *Epinomis* to Plato that includes a citation (Tarán 1975: 7), though not the earliest simpliciter (3–4, pace Leeman et al. 1981–2008: IV 140). Cicero may have learned of the bond directly from the *Epinomis* (Tarán 1975: 165–6 n.701) or via an intermediary (Leeman et al. 1981–2008: IV 134–6, further n.37).

refers probably to the theory of proportions (ἀναλογία).[36] For Cicero, there is no question of the *uinculum* being mathematical or even meta-mathematical in character, because it underwrites the connection among all the higher *artes*, not just the exact sciences. Whether Cicero has broadened the *uinculum* himself or knew a version of the "Platonic" doctrine already generalized to other *technai*, the attitude underlying his interpretation was of great significance for the *artes*.[37] In *De or.*, Cicero is interested in the *uinculum* chiefly for explaining how *eloquentia* has the power to traverse the subject matter of far-flung disciplines and retain its unity in speech.[38] But this interdisciplinary view of the *artes* was consequential far beyond rhetoric and oratory, and would indeed become fundamental for the Imperial *artes*, starting with Vitruvius' attempt to work out the implications of the doctrine for architecture.[39] The basic idea of a *uinculum unum* or *commune* was no doubt more widely diffused in late-Republican thought (even if not under that description),[40] but Cicero's emphasis on it probably lent further impetus. At any rate, his claim that it was just the *ingenuae et humanae artes* which were so related to one another could be read with prescriptive force: in other words, an author of an *ars* might have felt the necessity of demonstrating the connection of the principles and methods of his own discipline with those of other *artes* to prove it worthy of the name.[41]

Cicero's generalization of the *uinculum* to all the higher *artes* is not the full extent of his departure from the *Epinomis*. When laying the groundwork for his more expansive view, Crassus asserts that "all things above and below are a unity and are bound together by a single force and

36 The meaning of the bond was much queried by Platonic commentators: Tarán 1975: 165–7, 345–6; Rabouin and Vitrac 2010: 324–35. Theon of Smyrna (early second century AD) rightly interpreted the passage as referring to proportion (Hiller 84.7–84.14), though mistakenly reading ἀναλογίαν for ὁμολογίαν (991e3).

37 Schulte 1935: 87–93; Leeman et al. 1981–2008: IV 135–6 argue that the reinterpretation of δεσμός points to Posidonius or his "vicinity" (*Umgebung*). The suggestion is attractive, given that Cicero makes *causae rerum atque exitus* the basis for the *uinculum* (as discussed later in this section) and that Posidonius investigated the relationship between philosophy and the natural sciences vis-à-vis αἰτιολογία (E–K F18, 90 with Edelstein and Kidd 1972–99: II 130–6, 364–5; Kidd 1978). Bees 2010: 199–202, however, argues that the δεσμός while Stoic is not Posidonian.

38 *De or.* 3.22. Crassus is insisting (3.19–24) on the orator's need for a knowledge of *res*, without which *uerba* will be empty: see Leeman et al. 1981–2008: I 42–3 (on *De or.* 1.16–20); Barwick 1963: 10–13; Douglas 1966: xli (on *Brut.*); May and Wisse 2001: 10–11; Keeline 2018: 219–21.

39 See Ch. 3.1 on *DA* 1.1, the most elaborate discussion in both the Republican and Imperial periods of the interdisciplinary character of the *artes*.

40 Besides its philosophical currency (n.37), something like it is implicit even in vague notions of "liberal" *artes* (Beagon 2018: 659; cf. nn.5–7).

41 A likely motivation not only for Vitruvius (Ch. 3, n.37 on his "encyclopedism"), but also for Columella and his *agricola perfectus* (Ch. 4, n.63).

the agreement of nature" (3.20 *omnia haec, quae supra et subter, unum esse et una ui atque [una] consensione naturae constricta esse*). This claim posits a Stoic view of the world in which *natura* plays an active role in unifying the cosmos as an interconnected whole.[42] The metaphysical sketch allows Crassus to explain what exactly underpins the *uinculum*, as he completes the sentence that we left hanging earlier (*De or.* 3.21):

> ubi enim perspecta uis est rationis eius, qua causae rerum atque exitus cognoscuntur, mirus quidam omnium quasi consensus doctrinarum concentusque reperitur.
>
> for when the force of that principle is grasped in conformity with which the causes and ends of things are understood, then a certain wondrous agreement as it were and harmony among all the branches of learning comes to light.

The passage is complicated by the meaning of *rationis eius*, which scholars have sometimes construed as "theory," "method," or "reason."[43] Grammar and context demand, however, that it mean the "account" or "principle" of the unity of the cosmos that Crassus expounded earlier:[44] hence, *qua* means "in conformity with" or "on the basis of," not "by means of."[45] As the *qua*-clause shows, this *ratio* is the causal nexus that conditions the physical world (*res*) in virtue of the *una uis atque consensio naturae* (cf. 3.20), and which is thus exhibited in the orderly patterns of phenomena (*causae ... atque exitus*).[46] The significance of Crassus' cosmology is not only metaphysical, however: it also has an epistemic payoff in the interdisciplinary relationship among the *artes*, named here as the "wondrous

[42] *Consensio* translates Stoic συμπάθεια (cf. *Diu.* 2.34); for *ui atque consensione naturae*, cf. *Diu.* 2.144. Other Stoic resonances are documented by Leeman et al. 1981–2008: IV 139; there is likely an allusion (IV 137–8) to the early Greek monists in *ueteres illi* (3.20), whose beliefs Crassus recalls with the predication *omnia esse unum* (the same language of Xenophanes at Cic. *Ac.* 2.118), but even so *unum* refers here not to matter but to the causal nexus (as dicussed later in this section). Bees 2010: 197–8 would explain it differently by way of Antiochus.

[43] For example, "theory" (Wilkins; Rackham's Loeb), "theorie" (Bornecque and Courbaud's Budé), "Vernunftprinzips" (Schulte 1935: 80, cf. 93), "scienza" (Norcio), "method" (May and Wisse), "reason" (Watson; Keeline 2018: 220–1). Sometimes *uis* is also taken as "power," a less consequential error (following n.).

[44] Grammar: *eius* picks up 3.21 *haec ... ratio*, which refers to Crassus' cosmology immediately preceding (3.20). Context: An appeal to human rationality or else method or theory *in abstracto* is beside the point. The correct interpretation is put by Leeman et al. 1981–2008: IV 139; Mankin 2011: 117; Li Causi et al. 2015: 537. *Vis* is best taken as "force" in the sense of "meaning" or "significance" (*OLD* 18b).

[45] That is, it is *ablativus normae* (H–S §76.b, *OLS* I 871–3) rather than *instrumenti*.

[46] Earlier (3.20) Crassus said (again in Stoic language, n.42), *nullum est enim genus rerum quod aut auulsum a ceteris per se ipsum constare aut quo cetera si careant, uim suam atque aeternitatem conseruare possint.*

agreement as it were and harmony among all the branches of learning" (3.21). The connections among different disciplines are discerned only when one acquires knowledge (cf. *percepta* and *cognoscuntur*) of the shared causal principles that relate them, presumably specifically of the physical phenomena in their respective domains. Thus, Crassus may account for the unity of the *artes* as discrete but related domains of intellectual inquiry. His further remarks on the *uinculum* support two related conclusions: first, the *artes* have a physical (or metaphysical) and hence also epistemological foundation in nature; second, natural causes and principles are the common currency that relate *ars* to *ars*. It would be difficult to overstate the significance of these conclusions for the Imperial *artes*.

1.4 Explanation

Crassus' Stoic-colored passage in the *De oratore* is in fact just one vantage point, and that not the most direct, from which to consider another issue of the highest importance for the *artes*: the notion that *technê* or *ars* entails an explanatory knowledge of its methods in terms of natural causes – that is, that the *artes* are *explanatory*. This attitude is already found developed to a high degree in late-fifth-century Greece in texts belonging to the Hippocratic corpus.[47] For example, the author of *On the Art* (Περὶ τέχνης, *De arte* = *Art.*) responds to an attack on the medical *technê* by placing *technê* in opposition to chance (τύχη) and spontaneity (τὸ αὐτόματον):[48] what comes about through the *technê* of medicine is the necessary result of prior causal relations, as is the case for everything in nature (φύσις), and spontaneity plays no role. Medicine depends on explanatory knowledge of these causal relations in nature – that is, in diseases, the body, and so on (cf. *Art.* 6 L 6.10, 11 L 6.20, and *passim*). In another Hippocratic work, *On Ancient Medicine* (Περὶ ἀρχαίης ἰατρικῆς, *De uetere medicina* = *VM*), the author attacks other medical writers who have attempted to base the medical *technê* on overly broad theoretical hypotheses (ὑποθέσεις).[49] The issue is not whether the physician should base his *technê* on knowledge of natural causes, however, only what kind of first principles (ἀρχαί) are appropriate and how they may be discovered. Thus, the author works up his own theory of nature (φύσις) on which the method (ὁδός) of the

[47] Lloyd 1979: 49–55, 1991a; Jouanna 1999: 243–58; Nutton 2013: 63–4, 74–6.

[48] The attack may come from Protagoras (Mann 2012: 40–3) or other sophists. For τέχνη vs. τύχη, see *Art.* 4–6 L 6.6–10 with Jori 1996: 317–32; Mann 2012: 116–56; cf. *VM* 1 L 1.570–2, *Loc. hom.* 46 L 6.342–4, and further Edelstein 1931: 105–8; Schiefsky 2005: 5–13.

[49] Cf. *VM* 1–2 L 1.570–4, with Schiefsky 2005: 120–52 on the philosophical background; the hypotheses are the standard ones of hot, cold, wet, and dry.

art can be securely based, distinguishing himself from his opponents chiefly by his assessment of the empirical evidence.[50]

The most significant development in this story would come from Plato, however, who also gave it a polemical edge by establishing a sharp division between *empeiria* (ἐμπειρία, "experience"), which he redefines as a sort of pre- or nontheoretical skill or empirical knack (also called τριβή), and *technê* proper, which (as he maintains) requires explanatory knowledge of causes based on natural first principles.[51] This distinction is developed especially in the *Gorgias* and *Phaedrus*.[52] In the *Gorgias*, cookery (ἡ ὀψοποιική) is opposed to medicine as *empeiria* to *technê*.[53] Cooking is disqualified as a *technê* because it proceeds correlatively, attempting to match its products with the subjective measure of the pleasant (τὸ ἡδύ); Socrates denies that the relation of the cook's activities to the pleasant is a matter of knowledge (γιγνώσκειν), insisting that the cook only "aims" or "guesses at" (στοχάζεσθαι) his mark.[54] He puts the crux of the matter thus (*Gorg.* 465a2–6):

> τέχνην δὲ αὐτὴν οὔ φημι εἶναι ἀλλ' ἐμπειρίαν, ὅτι οὐκ ἔχει λόγον οὐδένα ᾧ προσφέρει ἃ προσφέρει[55] ὁποῖ' ἄττα τὴν φύσιν ἐστίν, ὥστε τὴν αἰτίαν ἑκάστου μὴ ἔχειν εἰπεῖν. ἐγὼ δὲ τέχνην οὐ καλῶ ὃ ἂν ᾖ ἄλογον πρᾶγμα.
>
> I say that it [sc. cooking] is not an art but rather a knack, because it has no account by which it administers the things which it administers, as to what they are by nature, so that it is unable to state a cause for each thing. And I do not call any activity which lacks an account an art.

A true *technê* must possess an account (λόγος) of its methods on the basis of the nature (φύσις) and causes (αἰτίαι) of the things concerned.[56] On this

[50] See Schiefsky 2005: 19–25; the author expounds his theory of "the nature of the human being" (ἡ τοῦ ἀνθρώπου φύσις) at 14 L 1.600–4, where he claims to follow the "first discoverers" (οἱ πρῶτοι εὑρόντες) of the τέχνη.

[51] There is no real pre-Platonic precedent for the sharp τέχνη–ἐμπειρία divide: Allen 2020.

[52] Schiefsky 2005: 349–50 adds relevant passages from the *Laws*.

[53] The comparison serves a larger analogy opposing true τέχναι with false τέχναι that imitate them as species of flattery (κολακευτική); cookery opposes medicine (464d3–5).

[54] At 464c6 κολακευτική imitates the true arts οὐ γνοῦσα ... ἀλλὰ στοχασαμένη, and the verb στοχάζεσθαι is used of ὀψοποιική at 465a2; at 464d4–5 cookery "pretends to know" (προσποιεῖται ... εἰδέναι) what is good for the body. Cf. also 501a3–b1, where cookery proceeds "entirely without art" (κομιδῇ ἀτέχνως) and "completely irrationally" (ἀλόγως ... παντάπασιν). A standard scholarly solution is that pleasure is indeterminable and thus cannot be the object of exact knowledge: Dodds 1959: 229; Irwin 1979: 209–10; Roochnik 1996: 186; Angier 2010: 28, citing *Phlb.* 31a8–9, 65d8–9; for other takes, Woolf 2004; Franklin 2005.

[55] For the reading ᾧ προσφέρει ἃ προσφέρει, see Dodds 1959: 229–30, who argues for adding ἢ between προσφέρει and ἅ.

[56] *Grg.* 501a1–3 is even more direct: ἡ δ' ἰατρική ... ἡ μὲν τούτου οὗ θεραπεύει καὶ τὴν φύσιν ἔσκεπται καὶ τὴν αἰτίαν ὧν πράττει, καὶ λόγον ἔχει τούτων ἑκάστου δοῦναι.

view, an activity which cannot explain the necessary relations between its methods and goals in virtue of the nature of the relevant things – even if the methods are generalized, perhaps quite successfully so, from experience – fails to qualify as a *technê*. In the *Phaedrus*, Plato enforces a similar contrast between *empeiria* and *technê*. A craftsman must "discriminate the nature" (270b4 διελέσθαι φύσιν) of his objects, such as the body in medicine and the soul in rhetoric or dialectic, if he is to possess true *technê*;[57] a systematic knowledge of nature will enable the physician, for example, to base his medical practice on an explanatory grasp of the causal relations that inhere in the body and those things which affect it.[58] Plato further suggests that, without this knowledge, the craftsman will suffer in practice: lacking knowledge of *why* he uses the methods he does, he will founder when confronted with the questions *when*, *for whom*, and *to what extent* they should be employed.[59] Hence, experience, when taken apart from the appropriate explanatory knowledge, is inadequate for *technê*, though in conjunction with *technê* it may have a complementary function in guiding the application of theoretical principles.[60]

After Plato, Aristotle advanced the debate regarding the criteria for *technê*, although the situation is complicated by the plurality of attitudes toward *technê* and *empeiria* found across his writings.[61] One strand of Aristotle's thinking, developed especially in *Metaphysics* A1, stands close to Plato's position. Aristotle thus makes the decisive difference between *technê* and *empeiria* a certain kind of account (λόγος) or knowledge (γνῶσις) – namely, the "why" (981a29 διότι) and the "cause" (981a30 αἰτία). Furthermore, he agrees with Plato that such explanatory knowledge requires a grasp of natural principles.[62] On this view, *technê* is generally superior to *empeiria* because it has more the character of

[57] Without such knowledge the practitioner has τριβὴ μόνον καὶ ἐμπειρία (270b5–6).

[58] Compare the "true account" (ὁ ἀληθὴς λόγος) at 269c9–270d7, attributed to Hippocrates (for the inconclusiveness of this allusion, see Lloyd 1991b: 194–7, 199–208).

[59] Cf. *Phdr.* 268b7–8 (of medicine) οὕστινας δεῖ καὶ ὁπότε ἕκαστα τούτων ποιεῖν, καὶ μέχρι ὁπόσου, and what follows through 272b6.

[60] Cf. *Phdr.* 271d7–272b4, where the practitioner must contemplate theoretical principles ἐν ταῖς πράξεσιν ὄντα τε καὶ πραττόμενα so as to be able to perceive (αἰσθάνεσθαι) the right situations in which to apply them.

[61] A large topic: for example, Frede 1996; Butler 2003; Gregorić and Grgić 2006; LaBarge 2006; Cambiano 2012, esp. 13–36; Hasper and Yurdin 2014; Miller 2014; Johansen 2017. Most prioritize *Metaph.* A1, but Bolton 2018 and 2021 deemphasizes the passage, arguing that Aristotle recognizes multiple notions of τέχνη.

[62] Cf. esp. *Metaph.* 981a10–12, where the doctor with τέχνη recognizes natural kinds (εἴδη) such as phlegmatic (φλεγματῶδες) or bilious (χολῶδες). For ἀρχαί and τέχνη, cf. also *Resp.* 480b22–30, *Sens.* 436a16–b1. See Johansen 2017: 121–8 for limitations on the practitioner's principles (in contrast to the philosopher's).

wisdom (σοφία).[63] Furthermore, *qua* knowledge of universals (981a16 τῶν καθόλου), *technê* has practical value in determining appropriate action for the craftsman in situations where the particulars are novel to him.[64]

However, another strand of Aristotle's thinking represents a relaxation of the requirement for causal knowledge, if not for *technê tout court* then at least so far as concerns successful practice.[65] Thus, even as he emphasizes the distinction between *empeiria* and *technê* in *Metaph.*, Aristotle allows that "with a view to action, experience seems in no way inferior to art" (981a12–13 πρὸς μὲν οὖν τὸ πράττειν ἐμπειρία τέχνης οὐδὲν δοκεῖ διαφέρειν). That is because *empeiria* is defined as (actionable) knowledge of particulars (981a16–17 τῶν καθ' ἕκαστον). But knowledge of particulars does not preclude knowledge of their associations: hence, it is possible to acquire a host of (nonexplanatory) generalizations that are drawn from and corroborated by experience of particulars. Even if the practitioner lacks knowledge of the διότι, his generalized knowledge of particulars will frequently suffice for action; conversely, a person in possession of *technê* (that is, a τεχνίτης) who lacks experience will find his knowledge of the universal of limited value in action.[66] Aristotle's conception of *technê* thus places great weight on *empeiria* as a vital component in practice (τὸ πράττειν), departing from Plato's more depreciatory attitude. At the same time, Aristotle does not relinquish the significance of explanatory knowledge as a prerequisite for a more epistemically complete notion of *technê*.

While the requirement that *technê* be based on explanatory knowledge of natural causes could be downplayed or simply ignored (see Ch. 1.5), it was consequential for some *technai*. The most conspicuous example is medicine, which was fractured in the Hellenistic period by methodological disputes along these lines.[67] Later medical historiography divided physicians into those who believed that knowledge of the nature of the human was required

[63] *Metaph.* 981a25–7, 981b30–982a2. See Gregorić and Grgić 2006: 4–7; Cambiano 2012: 32–9; Johansen 2017: 105–8.

[64] Cf. *EN* 1180b13–23 with Schiefsky 2005: 353–4; Hasper and Yurdin 2014: 134–5; *Metaph.* 1032b6–9 with Miller 2014: 137–9; and on the usefulness specifically of the λόγος-part of τέχνη, Johansen 2017: 129–34. Bolton 2018: 289–91, 294–5, 2021: 148–50, 160–5 argues that knowledge of universals or principles adds nothing to practice, but this conclusion is difficult to credit.

[65] Bolton 2018, esp. 284–9, and 2021, esp. 153–60, relying on, for example, *Rh.* 1354a1–11, *Soph. el.* 172a11–36, holds that Aristotle recognizes a kind of genuine, "non-scientific τέχνη" not involving demonstrative knowledge of causes (also previous n.). It is reasonable to admit a degree of polysemy in Aristotle's usage of "τέχνη." Cf. Miller 2014: 141, arguing that ἐμπειρία and τέχνη are mutually related on a continuum.

[66] Cf. *Metaph.* 981a20–23, and n.60 on Plato.

[67] For the influence of the Platonic–Aristotelian view of τέχνη on Hellenistic medicine, see Hankinson 2004; Schiefsky 2005: 345–59; cf. Allen 2020: 40.

for medical expertise and those who thought such knowledge impossible or superfluous – or both.[68] The latter held that experience (ἐμπειρία) by itself is "realizable, sufficient, and technical" (συστατικὴν καὶ αὐτάρκη καὶ τεχνικήν) for the constitution of medicine and accordingly called themselves the "Empiricists" (ἐμπειρικοί).[69] *Empeiria* is a term of art among the Empiricists for knowledge acquired by empirical means that is sufficient to ground medical theorems (θεωρήματα).[70] Other physicians maintained, however, that *empeiria* alone was not sufficient for *technê*. These doctors did not form a unified group and embraced diverse and sometimes contradictory attitudes,[71] but they would have agreed that knowledge of the nature of the human body was necessary for medical practice, and that accordingly causal reasoning from and to natural principles – some of these at least not immediately evident to perception – was indispensable. Physicians of this persuasion were misleadingly lumped together as a sect (αἵρεσις, *secta*), presumably by the Empiricists themselves, and called "Dogmatists" (δογματικοί) or "Rationalists" (λογικοί).[72] In later accounts, the Rationalists are said to privilege in medical practice the role of explanatory knowledge of the nature (φύσις) of the things involved, not only the body (σῶμα) and its constituents but also externals (climate, food, etc.) germane to health and disease.[73] The stamp of Plato and Aristotle is evident in this epistemological debate, which, besides influencing medical thought, stimulated reflection on the notion of expertise itself.[74] As we will see in Chapter 5, the quarrel would also deeply shape Celsus' approach to medicine. While the philosophical analysis of *technê* did not produce internecine disputes of such intensity in other Roman *artes*, the epistemologically demanding view of *technê* on which it was predicated did have great influence.[75] As we will see throughout this book, Crassus' position in Cicero's *De oratore* is only one

[68] On the Hellenistic medical sects, see von Staden 1982; Walzer and Frede 1985: ix–xxxiv.

[69] Gal. *SI* 5 K 1.76.5 = H 10.1 (transl. after Walzer and Frede). On the Empiricists, Deichgräber 1965 is fundamental; see also von Staden 1975; Frede 1987; Hankinson 1987a; Frede 1988, 1990.

[70] "Theorems" are statements capturing experience about what has happened, graded by degree of frequency: for Empiricist ἐμπειρία and θεώρημα, see Gal. *SI* 2 K 1.66.1–67.18 = H 2.11–3.20, *Subf. emp.* 2–3 D 44.4–46.23, and further Ch. 5.3.4.

[71] Among the "Rationalists" were counted birds of such different feathers as, for example, Hippocrates, Diocles of Carystus, Praxagoras of Cos, Herophilus of Chalcedon, Erasistratus of Ceos, Asclepiades of Bithynia, et al.: cf. Cels. *DM* 1.pr.15–16, 20, Gal. *Med. exp.* 12 W–F 69–70.

[72] The framing "Rationalists" vs. "Empiricists" was well established by Celsus' time (*DM* 1.pr.10, 12, 13, 27).

[73] Cf. Gal. *SI* 3 K 1.69.7–1.72.4 = H 4.18–6.26; see also Ch. 5.3.1 on *natura* in Cels. *DM*.

[74] The "Methodists" later offered a third way in opposition to the Empiricists and Rationalists: Frede 1982; Lloyd 1983: 168–200; van der Eijk 1999; Nutton 2013: 191–206; and Ch. 5.3.3.

[75] The situation with Hellenistic medicine is arguably paralleled by the diversification of the Roman *ars* of war in response to methodological pressures, however: see Ch. 6.5.

Roman reflex of the conception of *ars* as including explanatory knowledge of nature – a notion which is of much greater importance for the Imperial *artes* than has been generally appreciated.

1.5 Balance

We saw in passing that the demand for explanatory knowledge in the *technai* or *artes* was not sovereign; quite apart from the Empiricists, even those like Plato who privileged theoretical understanding allowed that experience was necessary for practice,[76] and Aristotle pointed out that *empeiria* can trump *logos* alone in practice. Such admissions call attention to another fundamental premise for the *artes*: they were *balanced*, in that the demand for theoretical knowledge was tempered by the imperatives of practical experience. Most ancient analyses of *technê* or *ars* awarded a central place to the practical part of expertise that could not be textualized, or at least not easily. Indeed, the contribution of hands-on experience, which no amount of book-learning or theoretical inquiry could replace, was a central problem in written traditions for the very reason that it could *not* be captured in words.[77] In general, the role of experience could be stressed in two ways: proximately, as a significant component of the individual practitioner's personal expertise; or remotely, as an epistemological foundation for the discovery and constitution of the discipline itself. In either case, experience is complementary to theoretical knowledge, not opposed to it. Both sorts of appeal to *empeiria* are attested early, found, for example, in *VM* among other Hippocratic texts.[78] Plato attributes to Polus, a pupil of the sophist Gorgias, what many could have thought an unobjectionable claim about *empeiria* and the invention of *technê*: "many arts have been discovered among men experientially from experience" (448c4–5 πολλαὶ τέχναι ἐν ἀνθρώποις εἰσὶν ἐκ τῶν ἐμπειριῶν ἐμπείρως ηὑρημέναι).[79] Aristotle indeed takes Polus'

[76] See n.60.

[77] Ancient authors frequently emphasize this point to avoid suggesting that what they teach is sufficient to impart expertise. Compare the notion of "tacit knowledge" developed by Polanyi 1966; for an attempt to apply this concept to the *artes* (Vitruvius), Cuomo 2016a: 457–62, 2016b.

[78] (1) The practitioner is distinguished by hands-on experience: *VM* 1 L 1.570–2 (cf. *Flat.* 1 L 6.90–2, *Lex* 4 L 4.640–2, *Mul.* I 62 L 8.126); (2) experience is the origin of medical art: *VM* 3 L 1.574–8, with Schiefsky 2005: 164–6. But *VM* is neither an "Empiricist" work (cf. n.50), nor again "Rationalist": Schiefsky 2005: 345, 357–8; Allen 2020: 44–52; also nn. 51, 67.

[79] It is uncertain whether this is a genuine quotation of Polus, as the scholiast (*ad Grg.* 462b) and Syrianus (*in Hermog.* II 8.22–9.2 Rabe) believe (Radermacher 1951: 113), or merely parody (cf. Dodds 1959: 192). At *Grg.* 462b10–c3, Polus is said to have argued ἐν τῷ συγγράμματι that ἐμπειρίαν ποιῆσαι τέχνην; Aristotle (whether relying on Plato or reading Polus independently: cf. Renehan 1995) says ἡ μὲν γὰρ ἐμπειρία τέχνην ἐποίησεν, ὡς φησὶ Πῶλος, κ.τ.λ. (*Metaph.* 981a3–5).

position as a trustworthy starting point for a cognitive account of the origin of *technê* in *empeiria*, though without collapsing the Platonic distinction between the notions.[80]

Technê remained important in Hellenistic philosophy and was developed variously depending on the interests of the "schools" involved.[81] The definitions that emerged in Stoic and Epicurean circles shed light on attitudes prior to the Roman period that tended to strengthen the view of *technê* as necessarily balanced in its constitution. We are told that the Epicureans defined *technê* as "a method producing something beneficial for life" (μέθοδος ἐνεργοῦσα τῷ βίῳ τὸ συμφέρον).[82] This is an abstract formulation that must be qualified by the Epicureans' circumscribed view of ethical benefit or utility; its conclusion in Philodemus' writings on *technai* reveals an epistemically stripped-down concept of expertise.[83] In the Stoa, Zeno defined *technê*, in what became a very popular formulation, as a "a systematic collection of cognitions acquired together in practice for some goal advantageous in life" (σύστημα ἐκ καταλήψεων συγγεγυμνασμένων πρός τι τέλος εὔχρηστον τῶν ἐν τῷ βίῳ);[84] not insignificantly, the addition ἐμπειρίᾳ is sometimes found after καταλήψεων ("cognitions acquired together in experience").[85] Some sources also credit Zeno with another definition,

80 Previous n. While an arbitrary number of particulars acquired experientially does not amount to τέχνη, experience of particulars is necessary for abstraction to a universalized notion about similars, which becomes a proposition in τέχνη (cf. *Metaph.* 981a5–7). Aristotle can preserve the Platonic distinction because he uses *empeiria* in two ways: (1) as personal experience of the practitioner, and (2) as systematic, cumulative reflection leading to the discovery of art: cf. Alex. *in Arist. Metaph.* 5.9–12 with Striker 2006: 138–9; Cambiano 2012: 21–2; Johansen 2017: 113–16.

81 Many developments (Stoic τέχνη-analogies in cosmology, the idea of a τέχνη βίου, Epicureans' ethical gradation of τέχναι) are irrelevant for present purposes, but for the Stoics on τέχνη, see Sparshott 1978; Mansfeld 1983; Gourinat 2011; Tsouna 2021a; for the Epicureans, Blank 2009; Tsouna 2021b.

82 *Schol. in Dion. Thrac.* 108.27–9 Hilgard = U 227b; cf. 2.23–4 Hilgard. That τέχνη aims at τὸ συμφέρον is uncontroversial (n.13).

83 For Epicureans, τέχνη must be shorn of theoretical content that does not conduce to happiness as they understand it (Blank 1995; 2009; Tsouna 2021b: 218–23): the sage would engage with, for example, γραμματική in the basic sense of knowledge of letters (reading and writing), but would reject any "more profound" (βαθυτέρας) version of the art (Sext. Emp. *Math.* 1.49).

84 Olympiodorus' version (*in Plat. Grg.* 12.1 Westerink = *SVF* I 73); *SVF* I 62, II 93–7 give more than a dozen versions. The translation follows Long and Sedley, save that where they translate συγγεγυμνασμένων (*codd. perperam* -ον) as "unified by practice," I translate "acquired together in practice." The notion of unity comes from συν-, following σύστημα; the verb γυμνάζειν ("train") points to the practitioner's personal, experiential acquisition of the καταλήψεις (cf. Sparshott 1978: 286; Gourinat 2011: 255; Tsouna 2021a: 173–4).

85 David, *Prol. philos.* 44.5–6 Busse; *Schol. in Dion. Thrac.* 108.31–3 Hilgard = *SVF* II 94; Anon. *in Hermog.* 14.262.1–3, 14.320.14–16 Rabe. Anon. *Prol. in Hermog.* 14.262.3–5 knows the definition may be found without ἐμπειρία; Mansfeld 1983: 63–4 concludes that David is the source for the scholiast and other commentators. It remains uncertain whether Zeno's definition originally included ἐμπειρία; Mansfeld suggests it might be Chrysippus' addition.

"a tenor methodically productive" (ἕξις ὁδῷ ποιητική),[86] and Cleanthes is attributed the very similar "tenor accomplishing all things methodically" (ἕξις ὁδῷ πάντα ἀνύουσα),[87] to which Chrysippus added the qualification "with impressions" (μετὰ φαντασιῶν).[88] What can we gather from the two Stoic accounts? *Technê* is systematic or methodical, productive, and inseparable from its practical or experiential content. It is significant that the Stoics do not require that *technê* possess a *logos* (account), and they indeed appear uninterested in opposing explanatory knowledge to mere experience as Plato and Aristotle do. Thus, Zeno's more expansive definition of *technê*, with or without the addition of ἐμπειρίᾳ, does not lay much weight on theoretical knowledge as a criterion for *technê*, but does underscore the necessity of habituation through practice.[89] (At all events, it is clear that *technê* cannot be knowledge, *epistêmê*, *stricto sensu* for the Stoics.[90]) Finally, it is worth quoting in full an account of *technê* reported by the Epicurean Philodemus:[91]

> νοεῖται τοίνυν καὶ λέγεται τέχνη παρὰ τοῖς Ἕλλησιν ἕξις ἢ διάθεσις ἀπὸ παρατηρήσεως τινῶν κοινῶν καὶ στοιχειωδῶν, ἃ διὰ πλειόνων διήκει τῶν ἐπὶ μέρους, καταλαμβάνουσά τε καὶ συντελοῦσα τοιοῦτον, οἷον ὁμοίως τῶν μὴ μαθόντων <οὐδείς>, εἴθ' ἑστηκότως καὶ βεβαίως εἴτε στοχαστικῶς.

> Now τέχνη is conceived and spoken of among the Greeks as a state or disposition arising from the observation of certain common and elementary things extending through many particular cases, which grasps something and which accomplishes something of such a sort that none of those who have not learned it can do in a similar way, whether fixedly and firmly or conjecturally.

86 *Schol. in Dion. Thrac.* 118.14–16 Hilgard = *SVF* I 72 ὁ Ζήνων λέγων «τέχνη ἐστὶν ἕξις ὁδοποιητική», κ.τ.λ. Von Arnim argues that the scholiast's attribution is mistaken, but Mansfeld 1983: 57–62 argues its plausibility and explains Cleanthes' version as a modification of Zeno's. The adjective ὁδοποιητική can stand in the scholiast's text, but for Zeno's definition ὁδῷ ποιητική should be understood (cf. Cic. *Nat. D.* 2.57 *progredientem uia*, reporting Zeno on φύσις, with Quint. *Inst.* 2.17.41 *uia, id est ordine, efficiens*, reporting Cleanthes on τέχνη); see Mansfeld 1983: 61–2 for details; also Festa 1932–5: II 110; Kassel 1971: 118; Gourinat 2011: 252.

87 Olympiodorus, *in Plat. Grg.* 12.1 Westerink = *SVF* I 490.

88 Olympiodorus, *in Plat. Grg.* 12.1 Westerink = *SVF* I 490. Olympiodorus says Chrysippus qualified the definition to distinguish τέχνη from φύσις, which is also methodical but does not rely on impressions.

89 Whether habituation is also *sufficient* for τέχνη depends on how much theoretical content the σύστημα requires (cf. Tsouna 2021a: 174).

90 Τέχνη is a ἕξις ("tenor"), whereas ἐπιστήμη is a διάθεσις ("character"): διάθεσις is stable and unchanging, whereas ἕξις is resistant to alteration but admits of variation and change: cf. *SVF* II 393 with Sparshott 1978: 282–4; Long and Sedley 1987: I 263–4; Tsouna 2021a: 177–9.

91 *Rhetoric* 2a *P. Herc.* 1674.38.3–18 = Longo Auricchio 123.2–15. I print the text and translation of Blank 2003: 70–1.

This is not the Epicurean definition; rather, in accordance with Epicurean method, Philodemus is attempting to articulate the ordinary conception of *technê*, which he will use as a starting point for his own inquiry into rhetoric.[92] Expertise is presented as a matter of empirical generalization, a grasp of particulars that are connected and instrumentalized for practice. While the word *empeiria* itself is not used, this ordinary conception of *technê* is obviously responsive, like the Stoic accounts which it resembles, to the practical dimension of expertise. This brief excursus shows in a word that the Hellenistic schools, while not ignorant of the Platonic–Aristotelian account on *technê*, sidestepped the crux of explanatory knowledge and favored balanced views of expertise rooted in the practitioner's experience.

To put the philosophers aside, the value of experience or practice was frequently maintained in the Hellenistic *technai* themselves, either with or against theoretical knowledge. The Empiricist physicians are an obvious example, but other disciplines provide evidence, too. In his *Technê*, Dionysius Thrax asserts that "grammar is the experience of the ordinary usage of poets and prose authors" (1.1.5 γραμματική ἐστιν ἐμπειρία τῶν παρὰ ποιηταῖς τε καὶ συγγραφεῦσιν ὡς ἐπὶ τὸ πολὺ λεγομένων), before casually referring to the discipline as a *technê* just after (1.1.6);[93] no opposition between *technê* and *empeiria* here.[94] Others, such as the mechanical author Philo of Byzantium, address the issue more self-consciously. While Philo does not appeal to any such dichotomy, he addresses the balance between theory and practice in his artillery treatise (Βελοποιικά) when he stresses the importance of "trial and error" or "experiment" (πεῖρα) in engineering war machines. Philo claims that "it is not possible to grasp all things by theoretical methods based on machines, but many things have been discovered through experiment, too" (T 50.41–3 οὐ πάντα δυνατὸν τῷ λόγῳ καὶ ταῖς ἐκ τῶν μηχανημάτων μεθόδοις λαμβάνεσθαι, πολλὰ δὲ καὶ διὰ τῆς πείρας εὑρίσκεται).[95] Only after trial and error have roughed out the parameters can theoretical methods refine the results. This process

[92] The definition is what Epicureans would call a "preconception" (πρόληψις) or "primary concept" (τὸ πρῶτον ἐννόημα), for which see Blank 1995: 179; Chandler 2006: 88–93.

[93] Sext. Emp. *Math.* 1.57 attributes the same definition to Dionysius, reading ἐπὶ τὸ πλεῖστον for ἐπὶ τὸ πολύ; for the significance of this difference, see Uhlig's *app. crit.*; Di Benedetto 1958: 196–8; Lallot 2003: 69.

[94] Of course, Dionysius was not necessarily ignorant of the epistemological issues – it is just that the equation of γραμματική and ἐμπειρία cuts the knot. For Dionysius' definition in the context of disputes about the technicity of γραμματική, see also Lallot 2003: 69–7; Callipo 2011: 39–46.

[95] "Theoretical methods" is Marsden's translation for the hendiadys; perhaps rightly (cf. T 59.13–15), he also adopts Schöne's μηχανικῶν for μηχανημάτων and translates "pure mechanics."

is illustrated by the discovery of the all-important ratio between the weight of a missile and the diameter of the hole for the spring: the proper relation was discovered "experimentally" (50.27 ἐκ πείρας) and only later furnished with a mathematical treatment.[96] Philo's emphatic claims for *peira* continue earlier views about the contribution of experience (ἐμπειρία) to *technê*. For example, his argument that the "principle" (50.21 ἀρχήν), "measure" (μέτρον), and "a certain established method" (50.24 μεθόδῳ . . . τινι εστηκυίᾳ) of artillery-construction emerged from experimentation recalls the claim in *VM* that medicine possesses a "principle and discovered method" (2 ἀρχὴ καὶ ὁδὸς εὑρημένη) acquired through trial and error.[97]

When we come down to the Republican period at Rome, we find that the *artes* demonstrate a similar concern for the role of experience or practice (*usus*, *exercitatio*, vel sim.) in the makeup of expertise. This is true already in the earliest systematic treatises, such as *Rhet. Her.*, where the author makes *exercitatio*, defined as "constant practice and habituation in speaking" (1.1.3 *adsiduus usus consuetudoque dicendi*), one of the three components of the *ars*.[98] After *Inu.* Cicero, too, would devote considerable attention to this part of oratory as he reevaluated the elements required for mastery in the *ars*.[99] To consider other disciplines, we might look to *Rust.*, where Varro handles experience both as personal dimension of expertise and as a basis for the discovery of the *ars*. As to the former, he announces experience in cultivation as one of his own sources (cf. *Rust.* 1.1.11 *quae ipse in meis fundis colendo animaduerti*).[100] As to the latter, he has Scrofa digress to say that "nature has given us two paths to agriculture, experiment and imitation" (1.18.7 *biuium nobis . . . ad culturam dedit natura, experientiam et imitationem*). Scrofa continues by asserting that the earliest farmers "established most things by trial and error" (*temptando pleraque constituerunt*), and recommends systematic experimentation as a principle for agronomy, "following not chance but some method" (1.18.8 *sequentes non aleam sed*

[96] Cf. T 50.20–37, as well as the comparison with architecture at 50.45–51.11, where too "experiment led the way" (50.47–8 πείρας προσαχθείσης). For calculation of the diameter of the spring hole by doubling the cube, see 51.46–52.29 (with qualifications of Marsden 1971: 158–9).

[97] See n.78.

[98] The other two are *ars* (in the sense of *praeceptio*, precepts or rules) and *imitatio*; Cicero's triad has *natura* in place of *imitatio*. See Leeman et al. 1981–2008: I 209–11, noting the antiquity of the triad φύσις-τέχνη-μελέτη (probably Protagorean).

[99] *Exercitatio* is treated, with *imitatio*, as part of the traditional triad at *De or.* 1.147–59, 2.89–98: see Leeman et al. 1981–2008: I 245–54, II 282–8.

[100] The claim is credible: see Skydsgaard 1968: 117–19; White 1970: 24, 1973: 437–4, 476; Diederich 2007: 22–4.

rationem aliquam).[101] No treatises on the *res militaris* survive from the Republican period, but much circumstantial evidence suggests the supremacy accorded to experience in what Plautus had called the *ars duellica* (*Epid.* 450). At any rate, when authors such as Cicero and Sallust refer to military writings, the balance of theory and practice, or book-learning and hands-on experience, was paramount.[102] The composite nature of *ars* as depending on both theoretical knowledge and practical experience, where the former is regarded as insufficient without the latter, was central for all the Imperial *artes*.

1.6 Toward a Scientific Culture of the *artes*

By the end of the Republic, several premises critical to the higher *artes*, related but each with its own history, had begun to coalesce into a unified conception of specialized knowledge: an *ars* in the fullest sense was systematic, interdisciplinary, explanatory, and balanced, in the senses discussed earlier. As we have seen, these premises emerged from Greek philosophical and scientific traditions and were established in Roman intellectual culture by their reception. Novel at Rome, however, was how these streams of thought, springing from different sources and only occasionally mingling their waters, had begun to flow together. There is indeed no single extant Greek source which unites all the premises we have considered. But, by the time of Vitruvius' *De architectura*, at the hinge of Republic and Empire,[103] the nascent synthesis in thinkers such as Cicero and Varro had been transformed into a coherent vision of *ars*, as we will see in Chapter 3. Here is a hint already of the sense in which the intellectual culture of the *artes* emerging at the end of the first century of BC was distinctively Roman.

The foregoing premises were not the only ones significant for the *artes* in the late Republic, but they deserve special attention as the fundaments for the *scientific* dimension of the Imperial *artes*. These were the premises that allowed an *ars*, both as notional discipline and literary form, to transcend the function of conveying practical expertise and to offer wider reflection

[101] Scrofa's remarks follow criticism of Cato's calculation of the number of slaves required for agricultural operations (1.18.1–7; cf. Ch. 4, n.8). To exemplify the application of *experientia*, he says that if we were to trench more or less deeply, we should examine what effect this would have (1.18.8 *si altius repastinauerimus aut minus quam alii, quod momentum ea res habeat*). Cf. Lucr. 5.1361–78 on the origin of agriculture, esp. 5.1367–8 *inde aliam atque aliam culturam dulcis agelli / temptabant*.

[102] See Ch. 6.5.1. [103] For Vitruvius as a figure of transition, see esp. Ch. 2, nn.8–9.

on nature, knowledge, and society. Because the *artes* were systematic, they allowed the Romans to organize and structure their knowledge and experience of the world in divisions appropriate to the phenomena of their own society and of nature. Because they were interdisciplinary, they held out the promise of an intellectual program in which the full range of Roman expertise and knowledge could be connected in a unified whole. Because they were explanatory, they offered an occasion to begin from first principles and build theories of the world, natural and human. And because they were balanced, they enabled Romans to integrate their own culturally and historically specific practices into a universalizing framework of specialized knowledge. Understanding the late-Republican *artes* is thus essential to the study of the Imperial *artes*, which, in extending the lofty criteria for expertise inherited from earlier traditions to the forms of knowledge and practice on which Roman Empire closely depended, forged a new expression of the Roman understanding of the world.

In conclusion, it is natural to address how the notion of *ars* sketched in this chapter was transmitted from Republic to Empire. This issue also allows for a point about the scope of this book. As will become clear, the intellectual heritage of the early Imperial *artes* is extraordinarily rich: authors engaged directly or indirectly with a wide range of earlier Greek and Roman texts, not only scientific and technical but also philosophical, literary, and historiographic. Yet a case can be made that most roads to the Imperial *artes* run through one or more of three Republican authors: Cato, Varro, and Cicero.[104] There is no *ars* handled in this book for which one or another of these figures does not stand somewhere in the background; the extent of their influence is represented diagrammatically in the Appendix. At the same time, we should resist handling the Imperial *artes* as a series of footnotes to these or any of the other likely candidates (among the Greeks Plato, Aristotle, or Posidonius, for example), as scholars have indeed often done. It is a constant temptation to take earlier and more widely celebrated thinkers as a point of reference, or to formulate theses of the type (put crudely) "X did for Y what Cicero did for oratory." This perspective may have a grain of truth, and many of our authors might be delighted that they would be regarded as a latter-day Cicero for some *ars*.[105] And, to be sure, it has proven, and will continue to prove, a useful task of scholarship to locate later thinkers in the history of the *ars* and *technê* traditions. Nonetheless,

[104] For poetic predecessors (Lucretius and Virgil), see the "Additional Remarks" in the Appendix.

[105] Vitruvius and Columella invoke Cicero's oratorical writings in their treatises, and it is perhaps relevant that moderns dubbed Celsus *Cicero medicorum* or *Cicero medicus* (Jocelyn 1985: 309n90).

once the ideas suggested in earlier periods had passed into the hands of Imperial authors, they were developed in original and creative ways that went far beyond mere reception. We will therefore take a different tack and regard the earlier figures as the footnotes to the Imperial *artes*, not the other way around, and it is therefore mostly in the footnotes where they will be found.

CHAPTER 2

Society, Politics, and Specialized Knowledge in the Early Roman Empire
The artes *and their Authors*

In the remainder of this book, we will study the growth of the Imperial *artes* from the perspective of *ars* itself, by examining the scientific, philosophical, and literary dimensions of the *artes* as a vehicle for Roman understanding. The development of this scientific culture was based on premises established in the late Republic – but that is only part of the story. Another important question is what exactly induced Romans of the early Empire to extend the intellectually demanding model of the higher *artes* to so many different fields, with the result that the long first century AD (roughly 25 BC to AD 100) became the preeminent period of the flourishing of the *artes* at Rome. This question can be handled in different ways. On the one hand, factors intrinsic to the *artes* are vital. It is no embarrassment of explanation to observe that "all humans by nature desire to know," as Aristotle put it (*Metaph.* 980a21 πάντες ἄνθρωποι τοῦ εἰδέναι ὀρέγονται φύσει): as the *artes* became powerful tools for exploring the world and gave rise to varied questions and problems, they would naturally have compelled wide interest. On the other hand, the *artes* as a distinctively Roman phenomenon were also shaped by the society, politics, and ideology of Roman Empire. These extrinsic factors cannot alone explain the character of the *artes*, but they establish the historical context and shed light on the avowed intentions of the authors, who often formulated their ambitions in the terms of Roman power. The intrinsic factors that recommended the *artes* will be studied throughout the whole book; in this chapter we will sketch the most important of the extrinsic factors, which have received comparatively more attention from the scholarly community. At the same time, this effort offers a good opportunity to introduce the cast of authors featured in this book (plus Pliny the Elder, another valuable witness) and offer prolegomena on important general issues, including Roman attitudes toward the Greeks and the Roman audience of the *artes*.

2.1 The Romanization of Specialized Knowledge

The influence of Roman Empire on the *artes* can be understood in terms of the "Romanization" of specialized knowledge. On the one hand, *ars* (like *technê*) could be conceived as a highly generalized, ahistorical form of knowledge. In other words, it was natural to think that possession of expertise should allow for successful practice independently of the practitioner's cultural and historical circumstances:[1] if an *ars* is an *ars*, it should not matter who practices it, or where, or when, provided they have the requisite principles, methods, and experience. Roman authors engaged in the *artes* accordingly based their teachings on natural first principles that would secure their general applicability. Likewise, views on the natural world and, to a certain extent, on human behavior that were argued in the *artes* had pretentions to universality because they were based on principles taken to be true not just for the Romans but true simpliciter, insofar as they were facts about *rerum natura*. On the other hand, the object of an *ars* was a moving target, since the goals were defined by the interests of the practitioners and their society. Indeed, even the choice of which disciplines to develop was culturally dependent in this way. In this vein, it is perhaps unsurprising that authors of the *artes* represented their disciplines as peculiarly bound up with Roman civilization. The interrelations were both temporal and spatial: the *artes* were connected with the historical unfolding of Roman power in the Mediterranean, and they were elaborated and perfected along the geographical axes of Roman power. What is more, the *artes* were imagined as belonging to a contemporary Roman world, and as destined for reception by a Roman future. This deeply historical, subjective perspective might appear to sit oddly with the supposition that the *artes* were also an ahistorical, objective way of cognizing the world. Yet this tension points to a crucial fact about them: precisely, that the *artes* presented universal knowledge of the world *as* a kind of Roman knowledge. The Romans allowed the prerogatives of Empire to set the ends for the *artes* in which they were engaged. The result was a Romanized form of knowledge that understood the world at once objectively, in the terms of its natural principles, and subjectively, in terms of specifically Roman concerns and practices. In this chapter we will consider how the Romanization of specialized knowledge worked itself out as an explicit concern of authors of the *artes*.

[1] Greeks and Romans acknowledged the historicity of disciplines and recognized that the relevant fields were handled differently by different peoples, but they were not thereby committed to any deeper relativism.

2.2 Vitruvius' *De architectura* and Imperial Self-Knowledge

To appreciate the formation of the *artes* in the crucible of Roman *imperium* we can do no better than to begin with Vitruvius and his treatise in ten books, *De architectura* (*On Architecture*: *DA*). M. Vitruvius Pollio is a shadowy figure – even his name is uncertain.[2] We can be sure, at least, that his career spanned the later years of the Republic and that he lived into the time of Augustus' sovereignty, since he tells us that he received remuneration from him (continued by Augustus' sister Octavia) after having earlier won recognition from Julius Caesar, whom he probably served under as staff engineer or architect.[3] Vitruvius' writings afford several clues to his place in Roman society, but the lack of independent evidence to confirm these hints (which themselves cannot necessarily be taken at face value[4]) has made it challenging to characterize him with certainty.[5] A plausible hypothesis would make Vitruvius a member of the *ordo apparitorum* – that is, a professionalized public servant.[6] This suggestion helps us to reconstruct the background of Vitruvius' life and career, though its significance for the interpretation of *DA* itself is less certain.[7]

Vitruvius is a crucial figure for understanding the Imperial *artes* in part because of his unique position, standing as he does at the beginning of their flowering in the early Empire. *DA* takes its impetus from the preoccupations of the late Republic but is addressed to the advent of

[2] Only the *nomen* is secure. The MSS have different *praenomina* (Ruffel and Soubiran 1962: 171–2), M. being only the most common; the best evidence for the *cognomen* comes from Faventinus (Plommer 1973: 40; but see Choisy 1909: III 259n2; Ruffel and Soubiran 1962: 141–2).

[3] Cf. 1.pr.2 *primum parenti tuo . . . fueram notus*. Vitruvius says he received *commoda* from Augustus for *apparatio* and *refectio* of war machines (1.pr.2). *Commoda* probably means a material, possibly monetary, reward: see Morgan 1909: 166–70; also Fleury 2003: 61–2; Oksanish 2019: 34n13; cf. n.6 on the ascription of apparitorial status to Vitruvius. Octavia Minor (d. 11 BC) is known as a patroness to others (Fleury 2003: 63), including possibly the Greek mechanical author Athenaeus (Cichorius 1922: 274–9; Marsden 1971: 4–5; Whitehead and Blyth 2004: 20).

[4] They could be shaped by authorial self-fashioning: see esp. Nichols 2017; Oksanish 2019.

[5] No attempt to match Vitruvius *auctor* with any Vitruvii known from other literary or epigraphic sources (for the latter, Ruffel and Soubiran 1962: 174–6), such as the Vitruvius Mamurra (*CIL* VIII 18913) identified by Thielscher with Caesar's *praefectus fabrum* Mamurra (*RE* (2) IX A.1.419.55–489.18), has been successful (Baldwin 1990, esp. 430–3; Gros 1997: ix–x; Fleury 2003: xi–xii). Another difficulty is that Vitruvius' work was seldom cited in antiquity (Pellati 1921: 318–31; Schuler 1999: 12–46).

[6] Purcell 1983: 155–6, suggesting that Vitruvius was a *decurialis scriba armamentarius*; see further Rawson 1985: 87; Gros 1994, 1997: x–xxvii; Masterson 2004, esp. 390–3; Nichols 2017, esp. 73–9.

[7] Some proponents of the hypothesis (previous n.) read *DA* on the supposition that Vitruvius is anxious about his own (putatively marginal) status as well as "the liminal status of his discipline" (Nichols 2017: 81; see also Ch. 1, n.8), but Oksanish 2019 (esp. 31–58, 119–23) argues that Vitruvius' self-presentation is better explained by the supposition of relatively elite status than by insecurity about his social position.

a new social and political order: in this sense, Vitruvius can be understood as a figure of transition, his intellectual maturation and literary production straddling the regime change that occurred in the last third of the first century BC.[8] *DA* indeed bears the outward stamp of this liminal quality. Vitruvius' leading intellectual lights are, by his own admission, Cicero, Varro, and Lucretius (*DA* 9.pr.17), and he remains firmly oriented toward the firmament of mid- and late-Republican intellectuals, frequently also in dialogue with classical and Hellenistic Greece.[9] Yet *DA*, likely the fruit of an extended composition, did not appear until his Republican heroes were past, probably around 27 BC or within a few years after.[10] The work takes Augustus' sovereign power as a brute fact, and it is clear that Vitruvius, who dedicates his work to the *imperator* and addresses him in all but one of the ten prefaces,[11] intended *DA* to be understood as a response to the seismic developments of the time.

For a Roman architect looking to make his mark, the accession of Augustus indeed proved as good a time as any that might be hoped for. Suetonius reports the famous boast of Augustus that, after finding Rome a city of brick, he left it a city of marble (*Aug.* 28.3 *iure sit gloriatus marmoream se relinquere, quam latericiam accepisset*).[12] As a dramatic expression of Rome's architectural transformation, the vaunt was not wholly overblown: in the *Res gestae* commemorating his benefactions, Augustus extensively catalogues the building projects that he oversaw (*Res gest.* 19–21).[13] Strabo, with an eye to Rome's aesthetic transformation, remarked that "one could easily forget the outside world: such is Rome"

[8] See esp. Romano 2016; also Romano 1994, 2003a: 173–81; McEwen 2003; Wallace-Hadrill 2008: 144–210; König 2009, esp. 34–5.

[9] Vitruvius' Republican intellectual milieu (with attendant Hellenic influence) is well studied in recent scholarship: see esp. Nichols 2017; Oksanish 2019, both exploring elements of transition in Vitruvius' thought but frequently highlighting a late-Republican formation. Besides previous n., see also, for example, Brown 1963; Gros 1982: 669–75; Rawson 1985: 185–93; André 1987a; Romano 1987, 2003b; Gros 1997: xxxii–xl; Courrént 2011, esp. 42–50. Republican influence is also found in Vitruvius' "encyclopedic" education for the architect: Ch. 3, n.37.

[10] Extended composition: Gros 1997: xxvii–xxviii; Fögen 2009: 108; Romano 2016: 338; cf. Boëthius 1939: 116–17. A precise date is hard to establish: Gros 1997: xxvii–xxxii, Fleury 2003: xvi–xxiv, reasonably suggest that it was largely accomplished 35–25 BC; for refinements, Romano 1987: 18–26; Baldwin 1990: 425–30; Anderson 1997: 40–3; Rowland and Howe 1999: 3–5; Novara 2005: 26–34, 162–4. Less probable are dates substantially earlier (40s and early 30s BC: e.g. Rawson 1985: vii, 86n14) or later (around or after 14 BC: e.g. Teuffel and Schwabe 1891: I 548).

[11] See n.24.

[12] For other reports, see Wardle 2014: 222. Cassius Dio (56.30.3–4) interprets Augustus' boast figuratively, but a literal reading is plausible.

[13] Augustan urbanism: Zanker 1987; von Hesberg 1988 (and Hofter 1988 generally); Favro 1996; Galinsky 1996: 141–224; Haselberger 2007.

(*Geogr.* 5.3.8 ῥᾳδίως ἐκλάθοιτ' ἂν τῶν ἔξωθεν· τοιαύτη μὲν ἡ Ῥώμη).[14] The hands of many laborers and, of course, the minds of many architects directed the long-overdue refurbishment of the urban center of *imperium* that would translate the intangible fact of Roman might into physical reality. In the preface to Book I, Vitruvius locates himself at precisely this moment.[15] He says that, while Augustus had been concentrating on "acquiring mastery over the whole world" (1.pr.1 *imperio potiretur orbis terrarum*),[16] he hesitated to issue his writings. When Augustus turned his attention inwards, however, Vitruvius seized the opportunity (1.pr.2):[17]

> cum uero attenderem te non solum de uita communi omnium curam publicaeque rei constitutione habere sed etiam de opportunitate publicorum aedificiorum, ut ciuitas per te non solum prouinciis esset aucta, uerum etiam ut maiestas imperii publicorum aedificiorum egregias haberet auctoritates, non putaui praetermittendum quin primo quoque tempore de his rebus ea tibi ederem.

> But when I marked that you were devoting your attention not only to the public good and the organization of state affairs, but also to the usefulness of public buildings,[18] so that not only should the state have been expanded through your own efforts by (sc. new) provinces, but also that the greatness of its rule acquire outstanding authority through its public buildings, I thought that I ought not neglect the first chance to publish these (sc. volumes) on the subject for you.

DA thus answers to Augustus' efforts to augment the power of the *maiestas imperii* of Rome through public architectural works. Augustus' personal responsibility for this expansion is underscored directly by the phrase *per te* and perhaps further highlighted by etymological play on

[14] See Dueck 2000: 86–7 and esp. Purcell 2017 on Strabo's report; further Roller 2018: 253–9 (noting emphasis on the Augustan program).

[15] On *DA* 1.pr and the Augustan moment, besides McEwen 2003, see Gabba 1980; Romano 1987: 15–45, 2016: 338–42; Elsner 1995: 55–8; Novara 2005, esp. 122–6; Fögen 2009: 139–41; König 2009: 36–40; Gozzoli 2010; Nichols 2017: 23–41; Fögen 2018: 29–33; Courrént 2019: 43–9; Oksanish 2019: 59–93.

[16] Echoed at *DA* 6.1.11 (with Fögen 2009: 141–3; Oksanish 2016: 268–9; Fögen 2018: 33–5), a passage discussed further in Ch. 3.3.3. Baldwin 1990: 426n13; Gros et al. 1997: 58–9 compare *Res gest.* 34.1 *per consensus uniuersorum potitus rerum omnium*; a recently discovered fragment shows that *potens* must be read for *potitus* (Cooley 2009: 257–8), but the parallel remains apt.

[17] I use Rose and Müller-Strübing 1867 as the text of *DA*, discussing other editions as relevant: Rose 1899; Krohn 1912; Granger 1931–4; Fensterbusch 1976; the Budé volumes; Gros et al. 1997.

[18] For *opportunitas* = "usefulness," cf. 1.3.1 (with Morgan 1909: 157–8), where it is defined as *communium locorum ad usum publicum dispositio*.

augere ("increase": cf. *aucta* and *auctoritates*), which would silently name the *imperator* in the dedication.[19]

Of particular interest is the role that Vitruvius imagines his own treatise will play, for he casts *DA* essentially as a contribution to Imperial self-knowledge – that is, Roman knowledge of the works and deeds through which Roman *imperium* is exercised.[20] Thus Augustus, the subject par excellence of such knowledge, will learn through Vitruvius' writings to appreciate the *ars* of design and construction that undergirds the changing Roman cityscape. Vitruvius later says to Augustus (1.pr.3):

> haec tibi scribere coepi quod animaduerti multa te aedificauisse et nunc aedificare, reliquo quoque tempore et publicorum et priuatorum aedificiorum pro amplitudine rerum gestarum ut posteris memoriae traderentur curam habiturum. conscripsi praescriptiones terminatas ut eas adtendens et ante facta et futura qualia sint opera per te posses nota habere. namque his uoluminibus aperui omnes disciplinae rationes.[21]

> I began to write these things for you because I observed that you have built and are now building on a large scale, and that in the future too you would see to it that public and private buildings be handed down to later generations to match the grandeur of your deeds. I have composed these definite rules so that by paying careful attention to them you could understand on your own the character of past and future works. For in the following volumes I have truly disclosed all of the principles of the discipline.

DA is not a blueprint for building, but for understanding.[22] By "disclos[ing] all of the principles of the discipline," Vitruvius will bring it about that Roman architectural works become "known" or "understood" (*nota*) to Augustus: this is fitting because, just as Augustus dignified Rome and its public building through his own means (1.pr.2 *per te* . . . *aucta*), so he ought to comprehend them on his own, too (1.pr.3 *per te* . . . *nota habere*).[23] The focus on Imperial self-knowledge is retained throughout the prefaces of *DA*,

[19] So Gros 1989, esp. 126; Thomas 2007: 22; Wallace-Hadrill 2008: 148. Vitruvius never calls Augustus "Augustus," but the title appears at *DA* 5.1.7 of a shrine at Fanum Fortunae.

[20] See Morgan 1909: 153–5 on the meaning of *imperium* and *imperator* in Vitruvius' time, not yet freighted with their later significance; his observations indirectly signal Vitruvius' own role in shaping the ideology of Roman *imperium* (cf. nn.8, 15).

[21] Fleury's Budé punctuates differently, with a semicolon after *coepi* and comma after *habiturum*. I follow the MSS in reading *traderentur*, which Rose and Müller-Strübing wrongly alter to *tradantur* after Schneider (*animaduerti* is historic).

[22] Cf. Milnor 2005: 117–19; Oksanish 2011: 167, 2019: 10.

[23] "Disclosure" is a key function of *DA*, *aperire* being only one of several relevant words (cf. *dicere*, *exponere*, and esp. *explicare*): Frézouls 1989: 40; Callebat 2005: 69; Novara 2005: 112–26; Oksanish 2011: 91–129, 2019: 12–21. The verb does not imply, however, that this knowledge is secret or guarded (Gros 1997: xx).

often expressed in memorable anecdotes and *exempla*.[24] For example, Vitruvius recounts in the preface to Book VII how the grammarian Aristophanes, thanks to his vast knowledge of the works in the library of Alexandria, aided Ptolemy by exposing plagiarists at a poetry competition (7.pr.4–7).[25] In this episode, the contents of the royal library are only made effective – that is, brought to the useful awareness of the library's own founder and patron – by Aristophanes' expertise. Vitruvius will similarly aid Augustus by unveiling the principles of Roman architecture implicit in what the *imperator* oversees, so that he too will be prepared to throw out sham architects. It is not only Augustus who is to be so educated, however, but "all men of intelligence" (1.1.18 *omnibus sapientibus*), as Vitruvius announces later in Book I. This second, broader dedication identifies educated, upper-class Romans as stakeholders in Roman *imperium*, and hence also as requiring an understanding of the architectural transformation unfolding in their age.[26] Vitruvius thus variously directs his work to Roman heads of household (10.pr.2 *patres familiae*) and officials (10.pr.3 *magistratus*) who, by learning the principles of right architecture, will be prepared to recognize the good architect and reject the charlatan. Socrates, Vitruvius says, wished that human thought be patent to the eye so that men could make informed judgments about whom they ought to trust (3.pr.1–3).[27] *DA* aims to do the next best by offering a public image of architecture that will equip Romans with the knowledge to understand and decide for themselves.

It is not just in the address to Augustus and other Romans of standing that Vitruvius relates his project to Imperial self-knowledge, however. He also does so in figuring architecture as a *corpus* ("body"), a metaphor bound up with the textualization of the discipline through *DA* itself. Vitruvius is not the first to use this figure of a written work, but he elaborates it to a much higher degree than do earlier authors.[28] The notion of *corpus* is valuable to Vitruvius for many reasons, not least

[24] The *imperator* is mentioned in every preface (2.pr.4, 3.pr.4, 4.pr.1, 5.pr.1, 5.pr.5, 6.pr.5, 7.pr.10, 9.pr.18, 10.pr.4, plus 1.1.18) except for Book VIII (for the omission, see Callebat 1973: vii–viii; Novara 2005: 139–45; König 2016, esp. 164–5, 173–4).

[25] See on the episode Fögen 2009: 130–3; McGill 2012: 34–46; Harris-McCoy 2017: 119–21; Nichols 2017: 32–6; Fögen 2018: 26–9.

[26] Cf. Oksanish 2019: 9 (and ibid. n.31) on this "double dedication"; for *omnes sapientes*, see Ch. 2.5 on the audience(s) of the *artes*.

[27] Cf. Romano 1987: 161–6; Long 1999: 83; Novara 2005: 128–30; Laterza 2018; Oksanish 2019: 146–8.

[28] *TLL* IV 1020.62–1021.39 (Lommatzsch), citing earlier only Cic. *QFr.* 2.11.4, *Fam.* 5.12.4; Oksanish 2019: 109n52 would add *Rhet. Her.* 4.58 and *Att.* 2.1.3 (Gk. σῶμα). McEwen 2003: 9 dismisses these earlier uses, but Oksanish 2019: 99–117 shows that they are germane.

because it implies order, unity, and completeness.[29] It furthermore allows for an understanding of the parts and divisions of architecture as "limbs" (*membra*) naturally related to the whole. Vitruvius thus frequently couches his plan for *DA* as the formation or description of a *corpus* of architectural expertise: he wants to "bring the body of so great a discipline to complete order" (4.pr.1 *tantae*[30] *disciplinae corpus ad perfectam ordinationem perducere*), expound its *genera* "as belonging to one body" (5.pr.5 *e corpore uno*), and have "the whole body of architecture be explained in all its limbs" (10.16.12 *totum corpus omnia architecturae membra ... haberet explicata*). Of great importance is the implication that architecture as *corpus* is also architecture as discrete discipline knowable through ordering (*ordinatio*) and explanation (*explicatio*). We will examine these concepts in Chapter 3, but the political significance for Roman Empire is clear. Vitruvius tells Augustus that he thought it a "worthy and very useful thing" (4.pr.1 *dignam et utilissimam rem*) to make a unity of the "disordered beginnings" (*non ordinata ... incepta*) left by previous authors, which were like so many "stray particles" (*particulas errabundas*).[31] By joining these scattered components into a *corpus*, Vitruvius also delimits the discipline (cf. 1.pr.3 *praescriptiones terminatas*[32]) and, in thus rendering it finite, makes it into a knowable whole.[33] This ordering of the *corpus* of the *ars* is analogous to the ordering of the *corpus* of Empire, which has the similar effect of making Roman *imperium* into a knowable whole. While no author after Vitruvius was to exploit the metaphor with such insistent political overtones, the loaded sense of *corpus* as an ordered and knowable whole did pass into other Imperial *artes* and had an afterlife in the first century AD: we can follow its journey through Columella, Quintilian, and Frontinus.[34]

[29] McEwen 2003 investigates the *corpus*-figure in several extended senses; Oksanish 2011, esp. 53–90; 2019, esp. 94–184, is the most detailed examination of the figure itself. The metaphor is most often connected with the organizational ambitions of *DA* (Ch. 3, n.11).

[30] *tantae* is Pontedera's conjecture for the paradosis *antea*; the adverb is defended by Gros et al. 1997: 398–9; Novara 2005: 131–2, but seems at best otiose, and at worst, contradictory with Vitruvius' statements elsewhere (cf. Gros 1992: 42–3). *Tanta disciplina* appears also at *DA* 1.1.11, 6.pr.6.

[31] This is a negative judgment on earlier architectural writings, for which see Ch. 3, nn.1–2.

[32] *Terminatas* is not "complete with technical terms" (Rowland and Howe) but "definite" or "defined" (Morgan 1909: 173–4); Gros et al. 1997: 64 observes that the phrase refers to the "rigore degli enunciati." Oksanish 2019: 19–20 stresses the significance of the notion of "limitation" (following n.).

[33] See Oksanish 2016, 2019 on the "synoptic unity" of *DA*.

[34] Celsus, too, if there is an allusion to his words at Col. *RR* 1.1.14 *Cornelius totum corpus disciplinae quinque libris complexus est*. Otherwise, cf. Col. *RR* 1.pr.21 (*corpus* vs. *partes/membra*); Quint. *Inst.* 4.pr.7 (*corpus* vs. *partes*); Frontin. *Aq.* 2.2, quoted in Ch. 2.6.

2.3 Pliny's *Historia naturalis* and the Measure of Roman *imperium*

Vitruvius' *DA* constructs the *ars* of architecture as a vehicle for an inward-looking Imperial self-knowledge. But in the first century AD, the *artes* also directed the Roman gaze outwards, beyond the *urbs* to the *orbis* over which it had acquired mastery,[35] as they grew to accommodate the totalizing ambitions of a Roman Empire whose dominion was understood, or productively misunderstood, to be coextensive with the habitable world itself (*orbis terrarum* as οἰκουμένη).[36] The best expression of the expanding Roman interest of the world that accompanied the extended *fines* of the Empire is Pliny the Elder's *Naturalis historia* (*Natural History*: *HN*), published in AD 77, about a century after Vitruvius' *DA*.[37] While Pliny's work is not an *ars* proper, it sheds light on the intellectual spirit that animated the extraordinary growth of the *artes* in the early Empire.[38]

In comparison with Vitruvius we know much about C. Plinius Secundus, who was born into an equestrian family in Nouum Comum in AD 23.[39] His personal industry fueled a successful political career,[40] which proved no obstacle, however, to energetic authorial activity producing works on history, rhetorical education, and technical subjects such as javelin-throwing.[41] Famously, he died on 24 August, AD 79, in the eruption of Vesuvius that buried Pompeii.[42] Pliny's dual status as official and author is not uncharacteristic of authors on the *artes* in the first century.[43] Like Vitruvius, Pliny pitches his voice in *HN* directly to the Emperor, addressing Titus, with whom he had served in the military, in the epistolary preface.[44] As has often been emphasized,

35 Vitruvius reveals this outward-looking view even though it remains peripheral in *DA* (1.pr.1; cf. n.16). See also *DA* 2.9.15–17 on the larch, discussed in Ch. 3.4.2.

36 Cf. Ou. *Fast.* 684 *Romanae spatium est Vrbis et orbis idem*; wordplay between *urbs*/*orbis* began in the late Republic (Bömer 1957–8: II 131–2): Bréguet 1969. For the coalescence of *orbis Romanus* and *orbis terrarum* in Pliny (to anticipate the remarks later in this section), cf. Nicolet 1991: 110–11, 114; Naas 2002: 418–21; Murphy 2004: 129–31, 164.

37 The date is inferred from Titus' titles at *HN* pr.3.

38 *HN* is often compared with Varro's *Disciplinae* or Celsus' *Artes* (e.g. Beagon 1992: 13; Naas 2002: 26–7; Murphy 2004: 194–7), although there are important differences.

39 Reynolds 1986; Beagon 1992: 1–22, 2005: 1–20; Healy 1999: 1–30.

40 Syme 1969, 1987, 1991. Beagon and Healy (previous n.) discuss traces of Pliny's provincial appointments in *HN*; see also Serbat 1986a: 2073–5.

41 See Plin. *Ep.* 3.5.1–6; further S–H II 780–3; *RE* (1) XXI 1.284.38–300.62 (Kroll); Serbat 1986a: 2077–81.

42 Serbat 1986a: 2075–7; Beagon 1992: 1n1; Healy 1999: 23n74.

43 For Columella and Frontinus, see Ch. 2.5, 2.6; see Beagon 1992: 5–7; 2005: 16–17 for Pliny's literary activity in connection with his equestrian status. *Otium* vs. *officium* is thematized by Pliny at pr.18–19 (cf. n.45 on Flavian ideology).

44 Military service: Syme 1969: 206–8. For Pliny's preface, see Köves-Zulauf 1973; Pascucci 1982; Howe 1985; Schröder 1999: 50–6; Sinclair 2003; Baldwin 2005; Citroni Marchetti 2011, esp. 31–99; Morello 2011.

HN reflects Flavian ideology in numerous ways – not only in its philanthropic civic ideal of restoring the *mos maiorum* by proving "useful" (cf. pr.16 *utilitatem*) to the Roman people and serving the "public good" (pr.33 *publico bono*), but also as support for Flavian efforts to reestablish Julio-Claudian personages such as Augustus and Claudius as exemplars of Imperial rule, to tout the corresponding renewal of the *pax Romana* after Nero and civil war, and to refurbish the city and reconstitute an archive of Imperial knowledge such as the records lost by the Capitoline fire in AD 69.[45]

Pliny's *HN* is now conventionally seen in its great and dazzling scope as a textual analogue, index, or monument of Roman Empire.[46] This claim runs the risk of reducing the work to a mere political statement, albeit an extraordinary one, and must be tempered by appreciation for the genuine philosophical and scientific aspirations of the work.[47] Yet the thesis is in the main attractive, and the image of *HN* as a textualized and epistemically totalizing likeness of Roman Empire finds support in Pliny's words. The massive work covers the whole of *rerum natura* in thirty-six books, plus an initial book for the preface and indices, proceeding through great and small, inanimate and animate, with many detours along the way: the cosmos (2) is followed by geography (3–6); next animals, beginning from human beings (7–11); plants follow (12–19); then medical properties of plant and afterwards animal products (20–32); finally, metals, earths, and stones (33–7).[48] Thirty-six books seem a great number, but are compendious for a treatment of the whole of nature; Pliny outdoes even this when he compresses their contents into the *summarium* of the prefatory book, which also details his sources.[49]

Throughout *HN*, Pliny invites connection between the immense scope of his work, a novelty worth boasting of (pr.14), and the *Romanum imperium* that has opened the world to Roman knowledge. In a famous passage, he describes the effects of Rome's wide rule (*HN* 27.2–3):

> Scythicam herbam a Maeotis paludibus et Euphorbeam e monte Atlante ultraque Herculis columnas ex ipso rerum naturae defectu, parte alia Britannicam ex oceani insulis extra terras positis, [3] itemque Aethiopidem

[45] See Citroni Marchetti 1982, 1991, esp. 15–21, 68–9; Oliveira 1992; Naas 2002: 70–105; Beagon 2005: 6–14, 2013; Bispham 2007; Ash 2011; Fear 2011; Naas 2011a; Espinosa Espinosa 2013 (and previous n.). For other Flavian stimuli to the *artes*, see Ch. 2.7 on land-surveying.

[46] French 1994: 169–78; Naas 2002; Carey 2003; Murphy 2004; and nn.55–57.

[47] Cf. Doody 2010: 40–91, who connects political readings of *HN* with the history of encyclopedism.

[48] Convenient dispositions at Della Corte 1982: 37; Isager 1991: 48.

[49] Schröder 1999: 107–16; Doody 2001, 2010: 92–131; Naas 2019: 30–3, 2002: 171–95; Beagon 2005: 30–2; Riggsby 2007, esp. 93–8; its political significance is stressed by Carey 2003: 30–1; Murphy 2004: 214; Naas 2019: 32.

> ab exusto sideribus axe, alias praeterea aliunde ultro citroque humanae saluti in toto orbe portari, inmensa Romanae pacis maiestate non homines modo diuersis inter se terris gentibusque, uerum etiam montes et excedentia in nubes iuga partusque eorum et herbas quoque inuicem ostentante. aeternum, quaeso, deorum sit munus istud! adeo Romanos uelut alteram lucem dedisse rebus humanis uidentur.
>
> The Scythian herb from Maeotian marshes, Euphorbia from Mount Atlas and from beyond the Pillars of Hercules, where nature itself begins to fail, and in another quarter Britannica from islands on the ocean that lie beyond the mainland, [3] and likewise Aethiopis from a clime burnt by the heavenly bodies, and other plants besides from other places are conveyed near and far in the whole world for human livelihood, as the immense majesty of *pax Romana* reveals in their turn not only human beings with their different countries and nations but also the mountains and peaks reaching up to the clouds, and their offspring and plants. I pray that this gift of the gods last forever! So true it is that they seem to have given the Romans to the human world like a second sun.

By calling the Romans a "second sun," Pliny suggests that its *imperium* has brought to light (cf. *ostentante*) the contours and contents of the *orbis terrarum*.[50] Indeed, the circulation of medicinal herbs, which allude by their names to their places of origin, metonymically reflects the global exchange of knowledge and goods under the *pax Romana*.[51] While there is a general humanitarianism in the idea that this exchange benefits all peoples (cf. *humanae saluti*, *rebus humanis*), the perspective remains firmly Romanocentric.[52] Another well-known passage handles similar themes (*HN* 14.2):

> quis enim non communicato orbe terrarum maiestate Romani imperii profecisse uitam putet commercio rerum ac societate festae pacis omniaque, etiam quae ante occulta fuerant, in promiscuo usu facta?
>
> Now that the world has been united by the majesty of Roman rule, who would not think that life has improved, both in the exchange of things and in the partnership of joyous peace, and that all things, even those which had previously lain concealed, have now been made available for general use?

50 On the "second sun" and its Stoic resonances, see esp. Beagon 2005: 25–9; also Naas 2002: 424–5; Beagon 2007. Italy is later called *rectrix parensque mundi altera* (*HN* 37.201).

51 Licorice (*Scythica*), spiny euphorbia (*Euphorbea*), water-dock (*Britannica*), and sage (*Aethiopis*) (Morton 1986: 86). It is no accident that three of the four plants show their origin transparently in their names: cf. Doody 2010: 23–30, 63–75, 2011 (esp. 122–6 on plants and places).

52 Cf. Fear 2011, esp. 25–6, for Pliny on the benefits of imperialism; Beagon 1992 on Pliny's "humanitarianism"; Murphy 2004: 131–3 on the Romanocentrism of 27.2–3.

Again, Roman rule has connected the places of the world and allowed for free circulation of things, people, and knowledge. All that can be discovered has been discovered and is now made available for the common use (*promiscuo usu*) of humanity.[53] The Romeward diffusion of this global knowledge explains Pliny's famous claim that "it is storehouses, not books, that we need" (pr.17 *thesauros oportet esse, non libros*).[54] Scholars have traced the influence of Pliny's Imperial and Roman outlook through *HN*, including its botany, ethnography, geography, medicine, mythography, and zoology.[55] His *laudes Italiae* (3.39–42, 37.201–3) powerfully reiterate the centrality of Italy and Rome in the known, and knowable, world,[56] and his insatiable curiosity for wonders (*mirabilia*) of all kinds marks an outward-looking Roman interest that probes, and masters, the possibilities (and paradoxes) at the limits of nature.[57] All this amounts to an inquiry (*historia*) at once totalizing and ineluctably Roman. Though smaller in scale, the Imperial *artes* share Pliny's conception for a world widened by Empire.

The distinctly Roman view of nature on display in *HN* is intimately connected with Pliny's own conceptualization of his theme. He announces that "nature, that is life, is the subject" of his work (pr.13 *rerum natura, hoc est uita, narratur*). "Life" is crucially ambiguous: it names the vital activity of *rerum natura*, but also means "human life" or even "human experience." The claim of identity (*hoc est*) depends on the fact that humanity, too, is a part of nature, and indeed in Pliny's Stoicizing worldview can perhaps be understood as nature's consciousness of itself.[58] His is an anthropocentric account of nature that scrutinizes things for their (providential) usefulness

[53] This passage comes, however, amidst Pliny's complaints that inquiry has faltered because of Roman addiction to *artes auaritiae* (*HN* 14.1–7); *luxuria* names this corrupting force and identifies the ambivalence in *Romanum imperium*. See Citroni Marchetti 1982, 1991; Wallace-Hadrill 1990, esp. 85–96; Isager 1991, esp. 52–5, 70–4; Oliveira 1992: 130–44; Beagon 1992: 75–9, 190–4; Naas 2002: 390–3, 436–8; Carey 2003: 75–101; Murphy 2004: 96–122; Lao 2011.

[54] The quotation is attributed to Domitius Piso and follows Pliny's claim to have collected 20,000 facts (pr.17).

[55] Botany: Pollard 2009; Bertoni 2014: 97–9; Manolaraki 2015; ethnography: Murphy 2004: 77–128; geography: Murphy 2004: 129–64; Evans 2005; mythology: Smith 2017; zoology: Vegetti 1981; Gauly 2018; for medicine, see Ch. 2.4.

[56] Sallmann 1971: 112–19; Naas 2002: 427–32; Carey 2003: 179–83; Fögen 2009: 260–2. For *laudes Italiae* in the *artes*, see Ch. 3.3.3 on Vitr. *DA* 6.1, Ch. 4.4.1 (esp. n.127) on Col. *RR*.

[57] *Mirabilia* (and Empire): Healy 1999: 61–70; Naas 2001, 2002: 243–92, 327–93, 2004, 2011b; Carey 2003: 84–91; Mudry 2004b; Beagon 2005, esp. 17–20, 2007, 2011.

[58] For the (loosely) Stoic background, see esp. Beagon 1992, 2005, 2011, 2013; also Citroni Marchetti 1982; Dumont 1986; Grimal 1986; Sallmann 1986; French 1994: 160–8; Paparazzo 2008, 2011; Naas 2015; Manolaraki 2018.

and finds *rerum natura* where human society meets the ingenious *uarietas* of the world. Pliny continues (pr.16):

> res ardua uetustis nouitatem dare, nouis auctoritatem, obsoletis nitorem, obscuris lucem, fastiditis gratiam, dubiis fidem, omnibus uero naturam et naturae sua omnia.
>
> It is a difficult thing to give novelty to what is antiquated and authority to what is novel; to give a shine to what has faded and to throw light onto the obscure; to make charming what repels and trustworthy what is doubtful; above all, to give everything its nature and to nature everything that is her own.

This passage crowns the synthetic perspective on nature and humanity: present and past, emotion and intellect, judgment and belief stand cheek by jowl in a parade that culminates in the sweeping predication of *natura* as both subject and object. In *HN*, the Roman mind is the agent through which this reflexivity holds. Pliny thus speaks directly to the peculiar fact we identified at the heart of the *artes*: knowledge of nature is constructed as Roman knowledge of the world. The *artes* and *HN* are alike in that they are not merely collections of principles, facts, and methods, but rather represent a way of appropriating the world, natural and human. Crucially, for Pliny this project is informed by an optimistic and humanitarian ideal that sees the dissemination of natural knowledge – which is always knowledge *for human life* – as a vehicle for benefitting others: "god is one human being helping another" (2.18 *deus est mortali iuuare mortalem*).[59] The double meaning of Pliny's *res ardua* speaks to the ambitions of the *artes* as much as his own undertaking: "a thing difficult to accomplish," indeed, but also "a thing lofty" or "sublime."

2.4 Celsus' *De medicina* and Greek Specialized Knowledge in the Roman *artes*

Pliny was not the only first-century author related to the *artes* with wide-ranging ambitions. In fact, he had an earlier Imperial exemplar in the effort to handle multiple fields of knowledge in Aulus Cornelius Celsus. Returning to Celsus will help us to appreciate other aspects of the national character of the *artes*. Celsus the man is shrouded in obscurity, and he is now known

[59] What follows is relevant to Pliny's Flavian backdrop (n.45): 2.18 *haec ad aeternam gloriam uia. hac proceres iere Romani, hac nunc caelesti passu cum liberis suis uadit maximus omnis aeui rector Vespasianus Augustus fessis rebus subueniens.*

almost exclusively for his eight books on medicine.[60] However, this work *De medicina* (*On Medicine*: *DM*) formed only one part (Books VI–XIII) of a larger work called *Artes*,[61] which included books also on agriculture, the art of war, rhetoric, philosophy, and possibly jurisprudence.[62] The *Artes* was likely composed between the second and fourth decades of the first century AD.[63] This was an "encyclopedia" of a different stamp than Pliny's *HN*, if it is appropriate to use that conventional but problematic appellation at all.[64] Although physicians through the ages have claimed Celsus as one of their own, whether he was a doctor, by ancient lights or modern, remains uncertain (see Ch. 2.5). Whatever his profession, his literary output is an indisputable testament to the broad and vibrant interest in the *artes* in the early Empire. The ambition of Celsus' work is obvious, and, from what testimonia exist, his elegant Latin treatment seems to have been successful not only for medicine, as we can judge for ourselves, but also in other disciplines.[65] The scope of the *Artes* makes it probable that, if it had survived in its entirety, we should find that we could apply to it a version of the argument relating the *artes* and Empire put for Pliny earlier, though Celsus appears a more sober assessor of *rerum natura*.[66]

60 A few inscriptions with *Cornelius Celsus* and certain hints from Columella might suggest Gallic origin; that he spent time at Rome is a priori probable and is confirmed if he is the same "A. Cornelius Celsus" as dedicator of *CIL* VI 36285 (Cichorius 1922: 412). For the little evidence, see *PIR*² C 1335; Marx 1915: v–vi; Krenkel 1973: 19–21; Serbat 1995: vii–xii; Schulze 2001: 7–12. His *tria nomina* are preserved correctly in Vat. lat. 5951. For dating, see n.63.

61 The title is found in several MSS and has some indirect support (Krenkel 1973: 21); Κέστοι from a Plautine scholiast (Marx 1915: 1) is a poor candidate (Schanz 1881: 373–4; Marx 1915: vi–vii).

62 The coverage, number, and order of the books of the *Artes* are debated. Certainly it began with five books on agriculture: MSS give Book I of *DM* as Book VI of the *Artes*; Celsus begins *DM* by alluding to the earlier agricultural books (1.pr.1 *ut alimenta sanis corporibus agricultura, sic sanitatem aegris medicina promittit*) and once refers back to them (5.28.16c); and Columella knows his five *libri* on the subject (*RR* 1.1.14; see n.34) and frequently uses them. Afterwards, the order is uncertain. There was rhetoric, possibly in seven books (Quint. *Inst.* 3.1.21, 12.11.24, Iuu. 6.244 with *schol. ad. loc.*); philosophy, possibly in six (Quint. *Inst.* 10.1.123–4, Aug. *Soliloq.* 1.21.2, *Haer.* pr.5); and the art of war, in an unknown number (Quint. *Inst.* 12.11.24, Veg. *Mil.* 1.8, Lyd. *Mag.* 1.47). Jurisprudence is suggested only by Quint. *Inst.* 12.11.24, a passage Capitani 1966 uses to speculate that Celsus also included systematic discussion of historical *exempla*. The *status quaestionis* is shaped by Barwick 1948, 1960 and Krenkel 1959; see Baader 1960: 215 and esp. Krenkel 1973; also Capitani 1966; Jocelyn 1985: 303n.35; Contino 1988: 20–41; Serbat 1995: xi–xiv; Schulze 2001: 13–17. For Celsus' possible connection with a history of the Parthian war distinct from the military books, see Schulze 2011; and for a spurious medical letter to Pullius Natalis, Schulze 2005.

63 Schanz 1881: 374–5; Marx 1915: xiv–xviii; Cichorius 1922: 411–17; Krenkel 1973: 20–1; Contino 1988: 15–19; Serbat 1995: viii–xi; Langslow 2000: 43–4.

64 For this problematic appellation (no such genre existed in Greco-Roman antiquity), see Doody 2009, 2010: 42–58; König 2013b: 1–5, and Ch. 5, n.17.

65 For his warm reception among agriculturalists, see Ch. 5, nn.4–7.

66 Celsus' writings are learned and deeply Roman (see also Ch. 5.2) but lack the moralizing flights, overt appeals to Roman *imperium*, and personification of *rerum natura* that typify Pliny's.

An important aspect of the Imperial *artes* which Celsus' *DM* allows us to approach directly is the Roman reception of Greek specialized knowledge. Here we mean not the adoption of epistemological features of the Greek *technê*-tradition studied in Chapter 1, but the Romans' attitudes toward Greek learning more generally. This is a large subject, much of which falls outside present purposes, but some observations will add texture to the Romanization of specialized knowledge in the *artes.* Before looking at Celsus, however, let us return briefly to Pliny, who expresses a notorious brand of Roman vitriol toward the Greeks (and their *technai*) and indeed cloaks himself in the mantle of hoary tradition while doing so.[67] Book XXIX of *HN* commences with a history of medicine that includes a diatribe against Greek doctors (29.1–28).[68] Pliny alternatively accuses Greek physicians of profiteering, culpable incompetence, and willful malice, and holds them (along with other Greeks) responsible for the moral corruption of Rome (cf. 29.27 *lues morum, nec aliunde maior quam e medicina*). Casting himself as the continuator of an ancient and salubrious Roman antipathy to this *technê*, Pliny quotes from a work of Cato *ad filium* that includes the famous asseveration that Greek doctors have "sworn a pact among themselves to slay all foreigners with the art of medicine" (29.14 *iurarunt inter se barbaros necare omnes medicina*).[69] What is the proper response to this Greek *ars*? Again in Cato's words, "it suffices to gain an acquaintance with Greek ingenuity, not to learn it thoroughly" (29.27 *satis esse ingenia Graecorum inspicere, non perdiscere*). There is no doubt that medicine *qua technê* was overwhelmingly Greek, from its theoretical premises and problems to its conceptualization of the body, health, and disease, and the divisions of its departments and remedies. Pliny's reaction is to foreswear the *ars*, even or especially transposed to Rome, as tainted by its Greek origins, and to propose a return to a pretheoretical, autochthonous Roman medicine. Hence, his history of medicine is the preface to an account of agrarian Roman therapies beginning from such items as eggs and wool (29.29). Yet Pliny's was not the only response for a Roman who wished to retain a measure of cultural independence in the face of a massively Greek inheritance.

[67] Pliny's relationship with Greek culture is complex; he makes extensive use of Greek scholarship, but his attitude is often jaundiced by suspicion or contempt: see Beagon 1992: 18–20; Serbat 1986; Wallace-Hadrill 1990, esp. 92–6; Naas 2011a; Smith 2017.

[68] On Pliny and Greek medicine, see Nutton 1986; Hahn 1991; Beagon 1992: 202–40; Fagan 2006: 196–8.

[69] *Medicina = ars medicinae*, as at Cels. *DM* 1.pr.1 (cf. Ch. 1, n.2). For Cato and Greek medicine, see Ch. 1, n.27 plus von Staden 1996: 375–85; for Cato and the Greeks, Astin 1978: 157–81; Gruen 1992: 52–83.

The long first preface in Celsus' *DM*, which includes a potted history of medicine (1.pr.1–11) followed by an introduction to the Greek medical sects (1.pr.12–75) propaedeutic to the books on regimen, evinces a different strategy that asserts the value of a Roman *ars* without rejecting its Greek foundations.[70] Both parts of the preface are revealing in their own ways. In the first, Celsus de-essentializes the Greekness of medicine by offering a complex history of the discipline. Declaring initially that "medicine promises health to the sick" (1.pr.1 *sanitatem aegris medicina promittit*),[71] he says that "there is absolutely nowhere it does not exist" (*haec nusquam quidem non est*), because all peoples, even the "most unskilled nations" (*imperitissimae gentes*), know how to use herbs for healing wounds and sickness.[72] Medicine in this universal, loosely pharmacological sense is connected with no particular place or people. Immediately afterward, however, Celsus adds that medicine "was cultivated among the Greeks considerably more than among other peoples" (1.pr.2 *apud Graecos aliquanto magis quam in ceteris nationibus exculta est*). Their interest led to multiple further (re-)foundings of medicine. Celsus distinguishes several epochs: Asclepius and his children (1.pr.2–4); Hippocrates and the birth of medicine from philosophy (5–8); the elaboration of the art from Diocles of Carystus to Erasistratus (8–9); and the emergence of the medical sects (10–11).[73] This is a thoroughly Greek history of *medicina*,[74] but it is important for that reason to recall Celsus' opening words, which emphasize that medicine was not *ab ovo* Greek: it was "cultivated" (1.pr.2 *exculta*) by the Greeks with special enthusiasm, but "not even among them from the very beginning, but only a few centuries before we did" (1.pr.2 *ne apud hos quidem a prima origine, sed paucis ante nos saeculis*).[75] This move places some distance between the Greeks and the art of medicine as such: they have a few *saecula* on the Romans, but they are not the sole proprietors of

[70] For the division of the preface, see Fuhrmann 1960: 86–8; Mudry 1982: 45–6; generally on the preface, Ch. 5, n.33.

[71] For *promittere*, see Ch. 5, n.12.

[72] I use Marx 1915, still the standard text for Celsus despite certain inadequacies (inter alia MS Toledo 97.12, which fills a lacuna in Book IV, was unknown to Marx; see Ollero Granados 1973; Capitani 1974). Texts and/or commentaries consulted for individual books: Mudry 1982 on I pr.; Serbat 1995 on I–II; Mazzini 1999 on VII–VIII; Contino 1988 on VIII.

[73] See on the stages Mudry 1982, esp. 47–50; Zurli 1990: 305–14; Stok 1993: 427–39; and esp. von Staden 1999: 253–78.

[74] "Dogmatic [i.e. of the Rationalist sect] in tenor" (Smith 1979: 227; so Mudry 1993b: 815, and cf. von Staden 1999: 269–70), given medicine's origin in natural philosophy; contrast the Empiricist bent of the prefatory remarks on regimen (Ch. 5, n.37).

[75] Celsus is vague on the beginning of the medical art at Rome, but it appears connected with Greek knowledge (cf. Mudry 1982: 48, 52–3).

the discipline. Celsus' concept of progress in *DM* is not straightforward – Hippocrates and the *antiqui* remain crucial sources – but the temporalization of medicine raises the possibility that future (Roman?) epochs may turn the discipline in their own direction.[76]

Celsus thus shows some light between *medicina* and the Greeks without depreciating their contribution. When he digresses to say that the need for a complex art of medicine (such as the Greeks elaborated) arose from the malign effects of "laziness and luxury" (1.pr.4 *desidia. . . luxuria*), however, can we find overt anti-Greek sentiment? Celsus says (1.pr.5):

> siquidem haec duo (sc. desidia et luxuria) corpora prius in Graecia, deinde apud nos adflixerunt ideoque multiplex ista medicina, neque olim neque apud alias gentes necessaria, uix aliquos ex nobis ad senectutis principia perducit.
>
> For the reason that these two things (sc. *desidia* and *luxuria*) have afflicted human bodies, first in Greece and then among us, medicine had to acquire more parts (something which was previously unnecessary nor is necessary now for other nations), yet it still only helps a few of us, and those with difficulty, attain to the threshold of old age.

There is undoubtedly a moralizing element in Celsus' story not wholly dissimilar to what might be found in Pliny.[77] Yet anti-Hellenism is less clearly proposed by the text. There may be a causal relationship between the onset of decadence in Rome and the importation of Greek culture (or medicine),[78] but nothing Celsus writes makes that explicit: the chronology established by *prius* and *deinde* can be read simply as pointing to the congruent (harmful) effects of civilization.[79] For all that Celsus' moralizing language evokes Cato and Pliny, Celsus never maintains that *Greek medicine itself* was a vehicle for Roman moral degeneration; rather, Greeks and Romans are peers in their suffering and (so we gather) also in their practice,

[76] For Celsus' ambivalence on medical progress, see Mudry 1982: 47; von Staden 1999: 253, 255, 272; Gautherie 2017: 81–90. For Celsus and Hippocrates, see Ch. 5, n.48.

[77] For *luxuria*, see n.53.

[78] See Mudry 1982: 57; von Staden 1999: 259–61 (cf. 1994a: 99n41; 1996: 404–6; 2000: 367; 2010: 20), relating these passages to the Roman moralizing tradition of Cato and Pliny; both are careful to stress, however, that Celsus does not reject the value of Greek medicine generally. On the other hand, Stok 1985a, 1993: 433–5; Deuse 1993: 824–8 emphasize instead Greek sources for Celsus' attitude.

[79] As culture-bringers, Greeks were often represented by Roman moralizers as the cause of Roman moral decadence tout court. But the Greeks are only incidental in this story: cf. Sen. *Ep.* 95.13–25, an account of medicine's development into a multifarious discipline (95.15 *multiplicem uarietatem*) accompanying moral decay in which the Greeks play no part. (Deuse 1993: 824n22 argues that Seneca is not dependent on Celsus, contra Stok.)

even if the Greeks anticipated the Romans "by a few centuries" (1.pr.2). The history of *medicina* cannot be told without the Greeks, but it is not for that reason only the Greeks'.

In the second part of the preface, Celsus helps us to understand what exactly the difference between "Greek" and "Roman medicine" might look like in the later history of the discipline. We will return to this in Chapter 5, but some general remarks are useful here. Again in contrast to Pliny, Celsus' solution is not to exile the quarrelsome tribe of Greek doctors, but to adopt and moderate their views. Earlier, Celsus had already remarked that, after Hippocrates, a series of luminary Greek doctors practiced the art in such a way that they "advanced down diverse paths of healing" (1.pr.8 *in diuersas curandi uias processerint*). This refers most immediately to the tripartition of medicine into regimen, drugs, and surgery (1.pr.9), but also foreshadows the methodological disputes that cleft the profession into its sects (1.pr.9–11): Empiricists, Rationalists, and Methodists.[80] This latter subject is the focus of the second part of the preface (1.pr.12–75), which handles the sects in introduction to the books on dietetics (I–IV). Celsus now says that he will sketch the sources of the "disagreement" (1.pr.12 *dissensio*) between Rationalist and Empiricist physicians "so that we can more easily intervene to propose our own opinion, too" (*quo facilius nostra quoque opinio interponi possit*). The history of the Greek medicine that follows is thus already marked as instrumental to the formation of Celsus' (a Roman's) independent judgment.

After careful exposition of the two sects (1.pr.13–44), Celsus turns to critique, adding that his goal is to determine "what may seem to be closest to the truth" (1.pr.45 *quae proxima uero uideri possint*), which is presumptively "somehow midway between (sc. these, the sects') divergent views" (*media quodammodo inter diuersas sententias*).[81] Celsus remains true to his word, and after remarks also on Erasistratus and the Methodists (1.pr.54–73), concludes with a balanced and cautious *opinio* that reflects his own judgment: "I do think that the art of medicine ought to be rational, but it should be informed by evident causes; all obscure (sc. causes) are excluded from the art itself, though not from the practitioner's thought" (1.pr.74 *rationalem quidem puto medicinam esse debere, instrui uero ab euidentibus causis, obscuris omnibus non ab cogitatione artificis sed ab ipsa arte reiectis*). Celsus figures himself as a fair and eminently reasonable umpire – one of the "dispassionate

[80] Cf. Mudry 1982: 66; von Staden 1999: 268. *Via* often means μέθοδος (*OLD* 10) in Celsus (Sconocchia 1997: 225). For the tripartition, see Ch. 5, n.28.

[81] See further Ch. 5.3 (and n.38).

investigators of the truth" (1.pr.48 *sine ambitione uerum scrutantibus*) who hold themselves aloof from sectarian controversies.[82] Greek medicine is handled as something serious and worthy of careful scrutiny, but also something full of "very contentious disputes" (cf. 1.pr.45 *magnae contentionis disputationes*).[83] Celsus' contribution – the Roman contribution – is an act of arbitration from which a moderate and trustworthy consensus may emerge. This Roman synthesis of Greek medicine belies Pliny's declaration that "Roman dignity has not yet exercised this alone of the Greek arts" (*HN* 29.17 *solam hanc artium Graecarum nondum exercet Romana grauitas*):[84] against Pliny's suggestion that the proper Roman response is extirpation of Greek influence, Celsus proposes reception and adjudication. A few times in the *artes* Pliny's vituperative tone surfaces,[85] but more often it is Celsus' confident assimilationism which typifies the Roman attitude toward Greek specialized knowledge.

2.5 Columella's *Res rustica* and the Roman Readership of the *artes*

To return to a problem hinted at earlier, it is uncertain whether Celsus was a physician or a layman. The *communis opinio* leans toward layman, but the evidence remains ambiguous, and the question may indeed be improperly framed.[86] The reality is not important here – Celsus' status as an author of the *artes* is unimpeachable – but recent arguments have connected the problem with the intended audience of *DM*.[87] Because the readership of ancient

[82] Cf. von Staden 1994b on Celsus' construction of a "scientific self."

[83] Marx deletes *disputationes* and reads *magnas contentionis*; Lyngby 1931: 4–5 defends *disputationes* but reads (with **FV** against **J**) *cognitionis* for *contentionis*. I adopt the reading defended by Capitani 1967: 116–7 (followed by Mudry, Serbat).

[84] Pliny's *summarium* has Celsus as a source (though not as *medicus*); it is uncertain whether Pliny knew another Greek-influenced Latin medical work predating *HN*, the *Compositiones* of Scribonius Largus (Claudian): see Sconocchia 1985: 196–205; Baldwin 1992: 74; Sconocchia 1993: 854–5.

[85] Though it is not an *ars* proper, cf. Frontin. *Aq.* 16.1 (Plin. *HN* 36.75 is a source), with Baldwin 1994: 490–1; DeLaine 1996: 124–5; Rodgers 2004: 188–9.

[86] Layman (i.e. not *medicus*): for example, Jocelyn 1985, esp. 303; Contino 1988: 43; Nutton 1993: 50, 2013: 5, 169–70; Langslow 2000: 46; Mayer 2005: 232; Stok 2016: 10; cf. Krenkel 1973: 19, 26–7; Gautherie 2017: 42–5; Beagon 2018: 667–8. Professional (i.e. *medicus*): for example, Ilberg 1907: 378; Spencer 1926; 1935–8: I xi–xii; Scarborough 1969: 59–60 (so 1993: 30–1), 1976: 25; cf. Mazzarino 1992a: 25. Schulze 1999: 11–20 summarizes the dispute before renewing the case that Celsus was a *medicus* (cf. 2001: 78–84, 2002: 230, and following n.).

[87] Schulze 1999 argues that Celsus' identity can be inferred from his audience, and his audience from the style and content of *DM*. He thus aims to show (40–103) that the text indicates an audience of specialists (*Fachleute*) and is hence a specialized treatise (*Fachbuch*), even an "esoteric" work, and that Celsus was accordingly a doctor (*Artz*). But it beggars belief that *DM* could be sandwiched between agriculture, the art of war, rhetoric, and philosophy and yet be intended as an "esoteric"

technical writing in general has proven to be of perennial concern, we should say something about it for the early Imperial *artes*. The question "For whom was Roman technical writing intended?" admits only conditional answer, because "technical writing" is not an ancient category but a modern one that has been used indiscriminately of texts widely divergent in nature, from school exercises, aides-mémoire, and subliterary manuals (on one end of the spectrum) to ornate didactic poetry and popularizing tracts (on the other).[88] If we eschew this imprecise formulation, however, and ask instead "For whom were the *artes* intended?", where the "*artes*" are the "higher *artes*" (Chapter 1), we can say something more definite. On the one hand, a specialist or professional readership (at all events a literate one[89]) is not to be denied: most are richly detailed and expertly informed, and their prescriptions obviously concern extratextual activity. On the other, the *artes* were in no way issued exclusively for an audience of experts. Indeed, most authors imply or explicitly name alternative readerships.[90] Even setting that aside, if we assume the writings themselves as a guide, we must conclude that their avowed social and political intentions; elegant (or at least would-be elegant) style; and broad engagement with major philosophical, historical, and literary traditions all point beyond a narrow audience of professionals to a class of educated Romans who were broadly curious about the (Roman) world that they inhabited. So, even as the *artes* sustain an expert readership, they require that we also understand a larger, elite, and basically nonexpert Roman audience with varying degrees of pragmatic interest in the subject matter.[91]

text for Roman physicians; nor do I find Schulze's internal evidence for the claim that the text was intended exclusively for specialists persuasive (cf. Ch. 5 and n.29).

88 See *NP* s.v. "Technical literature" (Sallmann); cf. Deinlein 1975: 2–11; Fögen 2009: 19–24; Roby 2015, 2016: 26–42. No category was recognized as such in antiquity, although ancient authors appreciated that there were certain desirable features of relevant writings qua didactic (Fögen 2009: 26–66).

89 Most ancient professional education was oral (Meißner 1999: 139–41; Stoll 2001: 295–8; Meißner 2003); literature played a part only in advanced curricula.

90 Vitruvius' multiple addressees are discussed in Ch. 2.2 (cf. *DA* 1.1.18). Pliny (*HN* pr.6) invites judgment from Titus on the same writings which, he self-deprecatingly claims, *humili uulgo scripta sunt, agricolarum, opificum turbae, denique studiorum otiosis* (cf. n.115). Celsus is reticent, but the self-citation of the agricultural books in *DM* (n.62) surely indicates that he did not count only doctors among his readers (even if some doctors read the agricultural part); and if the typical patient in *DM* resembles his readership, he will be writing for rich, lettered Romans (cf. Mazzarino 1992a: 24; von Staden 2000: 359–62). In his *Strategemata*, Frontinus has *duces* in mind (1.pr.1), but he also refers to the *occupati* (1.pr.2) – that is, Roman officials – as his readers, and supposes that his audience contains the sort of men who might peruse histories. For Columella, see later in this section. The land-surveying writings are harder to assess but certainly sustained some popularizing intentions (see Ch. 2.7).

91 Scholarship on technical literature has thus moved away from strict polarities, including that between the *Fachbuch*, a spare composition intended for experts or practitioners, and *Sachbuch*, a more stylistically elaborate form intended for nonexperts and the educated public (cf. Deinlein 1975: 9–11; Sallmann 1984: 12–13). The situation with Vitruvius is representative: cf. Knell 1985: 3–4,

The generalizations just offered will be strengthened in various ways throughout this book, but we can shade in some details already with the authors' help. In particular, Columella gives us an excellent glimpse in his agricultural treatise of the kind of men he imagines would read his work. Lucius Iunius Moderatus Columella was born in Gades while Augustus was still *princeps* and settled in Italy after a stint of military service,[92] only one of several Spanish-born luminaries who made their way to Rome in this century.[93] His *magnum opus*, *Res rustica* (*On Agriculture*: *RR*), in twelve books, was published sometime around AD 60/1–65.[94] Throughout his treatise, which offers a comprehensive account of rural matters, Columella keeps up a steady exchange with a group of Roman men with interests in farming. Chief among these is Publius Siluinus, Columella's friend, neighbor at Caere (3.9.6), and principal addressee.[95] The persona has an important role in *RR*: Siluinus reads Columella's treatise book by book and poses criticisms as well as suggesting topics for further discussion, and Columella, while writing, anticipates and forestalls his questions and objections.[96] Apart from this rhetorical function, the back-and-forth with Siluinus also lends the work a dramatic, dialogical cast, embedding

161, 170; Romano 1987: 174–6; Gros 1994: 75–6; Fögen 2009: 113n20; König 2009: 39n24; Nichols 2017: 10–11; Oksanish 2019: 7–9.

92 An inscription from Tarentum (*CIL* IX 234) attests to Columella's time as military tribune in *Legio VI Ferrata*, stationed in Syria. He was born in Gades in the province of Baetica (8.16.9, 10.185–6), possibly in AD 4 (Richter 1981–3: III 593–4); he recalls his home not infrequently (11×) as well as his uncle Marcus Columella there (n.97). After Columella's military service, he spent the remainder of his life in Italy, where he had several estates, including at Caere, Carseoli, Alba, and Ardea (3.3.3, 3.9.2), though he disposed of the last of these. See further *RE* (1) X 1.1055.10–52 (Kappelmacher); Cichorius 1922: 417–22; Ahrens 1972: 11–14; Richter 1981–3: III 588–602; Martin 1985: 1960–2, García Armendéariz 2004: 8–17; Fögen 2009: 153–8. For a possible allusion to the author's person at *RR* 10.251–4, see Mielke 2023: 74–7.

93 Others include Senecas *maior* and *minor*, Lucan, Martial, Quintilian, and Pomponius Mela (Fontán 1997). For the history of Gades to Columella's time, Rodríguez Neila 1997; for Columella's relationship with Spain, compare the attitudes of the Elder and Younger Senecas: Griffin 1972 (Columella on 17), 1976: 29–34, 253–4; cf. Noè 2002: 63–4; Mielke 2023: 54–5.

94 *Termini post* and *ante* are inferred inter alia from the mention of a vineyard owned by Seneca (Plin. *HN* 14.47–51, and see n.99); a reference to Gallio (9.16.2), Seneca's brother, as though living (n.100); and, less certainly, a *subscriptio* attesting to another work by Columella dedicated to Eprius Marcellus (n.101; Richter 1981–3: III 596 guesses Marcellus' first consulship, AD 62, was the occasion). See further *RE* (1) X 1.1059.13–57 (Kappelmacher); Richter 1981–3: III 595–8; Martin 1985: 1960; Boldrer 1996: 9–11.

95 Siluinus is unknown outside *RR*; some suggest he is a fiction (cf. Carroll 1976: 785, 788; Henderson 2002: 118; Formisano 2018: 512; Mielke 2023: 38n119). For his speaking name ("Woody," Henderson 2002: 130, 2004: 19), see Carroll 1976: 785; Fögen 2009: 157; Formisano 2018: 512; Mielke 2018: 527n42, 2023: 78.

96 Two examples (2.1.1, 3.7.1) are discussed in Ch. 4.2.1, 4.4.1. See further Carroll 1976: 784–7; Martin 1985: 1963; André 1994: 181–4; Santon 1998: 8–10; Noè 2002: 46–8; Diederich 2007: 218–22; Fögen 2009: 156–7, 185–9; Bertoni 2017: 536, 550; Mielke 2019: 51–2, 55–7, 90–4, 2023: 55–6.

Columella's treatise in a vibrant intellectual community that includes several other figures. Among these are M. Columella, Columella's *patruus* residing in Baetica;[97] M. Trebellius Maximus, a general, consular (suff. 56), and governor of Britain;[98] L. Annaeus Seneca, the famous philosopher, tragedian, and statesman;[99] L. Iunius Gallio (Annaeus Nouatus), brother of Seneca, consular (suff. 55 or 56), and proconsul of Achaia;[100] T. Clodius Eprius Marcellus, consular (suff. 62, 74) and proconsul of Asia;[101] and Claudius called Augustalis, otherwise unknown.[102] This is not to count Celsus, Iulius Atticus, and Iulius Graecinus, "men of our time" (1.1.14 *nostrorum temporum uiri*) whose agricultural writings Columella uses but whom he but does not speak of with personal familiarity.[103]

Of course, Columella does also present himself as addressing *negotiosi agricolae* – the men who are so occupied with villa business that they have little leisure for other activities.[104] There can be no doubt that full-time, professional farmers are among his audience – men like Uncle Columella and Siluinus.[105] It is therefore striking that, with the exception of these two, Columella's inner circle is made up not of those identified as *agricolae*, but elite Romans occupied by military and political careers. Yet these men are represented as a natural audience for his treatise, and some, notably Trebellius, are brought into the text as discussants alongside Siluinus.[106] What unites this group most immediately is not a shared professional

[97] Called *uir inlustribus disciplinis eruditus ac diligentissimus agricola Baeticae prouinciae* (5.5.15), he is named 8×.

[98] 5.1.2 *M. Trebellius noster*. Cichorius 1922: 420–1 (cf. Syme 1969: 221–2) identified him with the commander of an army in Syria (Tac. *Ann.* 6.41.1) and with the consular M. Trebellius Maximus, also governor of Britain sometime between 63 and 69 (Birley 2005: 53–6; Woodman 2017: 257–8); see also *PIR*² T 314. As Cichorius observes, Columella would have made Trebellius' acquaintance while serving in Syria (n.92).

[99] Seneca is not addressed directly, but he is called *uir excellentis ingenii atque doctrinae* (3.3.3) à propos of his vineyards. Columella probably moved in the Spaniard's circle (cf. Griffin 1976: 89, 253, 290–1): Trebellius shared the consulship with Seneca, and Columella dedicated a book to Seneca's brother (following n.). Seneca indeed reveals an interest in agriculture in his writings, calling himself *uinearum diligens fossor* (*QNat.* 3.7.1; cf. *Ep.* 12.1–2, 86.14–21, 112.1–2, and Zainaldin 2019).

[100] At 9.16.2, Columella claims he composed Book X *ita ut et tibi* (sc. *Siluino*) *et Gallioni nostro complacuerat*. For Gallio, see Becher 1897: 12–13; *RE* (1) I 2.2236.66–2237.28 (Rossbach); *PIR*² I 757; Griffin 1976: 48, 83–4; Fögen 2009: 154–5.

[101] A *subscriptio* after Book XI attests to a *liber singularis* to Eprius Marcellus *de cultura uinearum et arborum* (probably not *De arboribus*): see S–H II 787; Richter 1981–3: III 595–6. On him, see *PIR*² E 84 with Griffin 1976: 91n2; Birley 2005: 235–6; Mielke 2023: 342–3; he was not in Seneca's circle.

[102] Cf. 11.1.1–2. See Becher 1897: 13–14; Richter 1981–3: III 612n18; Diederich 2007: 370; Mielke 2023: 180; the name indicates a priesthood in the Imperial cult.

[103] See Ch. 4, n.24.

[104] Cf. 9.2.5 with Mielke 2023: 124–5. The *agricola* (in singular or plural) is ubiquitous as the assumed subject of Columella's *praecepta*.

[105] The former is called *agricola* (n.97).

[106] For Trebellius' role, see Ch. 4.2.3.

identity as such but a common interest in learned discourse on the subject. This interest would, naturally, have had a pragmatic or even acquisitive dimension for those of the social and economic standing of Columella's addressees, whose wealth usually depended on their agricultural holdings. His instructions are thus pitched to people of sufficient means to operate the mid- to large-sized farms with mixed cultivation that he prescribes.[107]

Columella in fact has a name for his audience that includes both the *negotiosi agricolae* and those at some remove from the villa: *studiosi agricolationis* or *agrestium operum* (3.3.2, 4.1.1, 5.1.1; cf. 1.1.1, 1.1.17, 2.2.14, 3.2.31, 11.1.1). This appellation picks out the relevant feature of his readership–namely, a serious intellectual interest in the discipline. Columella uses the same adjective, *studiosus*, to refer to those studying other subjects: devotees of astronomy, for example, are *studiosi rerum caelestium* (3.6.4), and those of the natural-philosophical problems connected with bee biology are *studiosi litterarum* (9.2.5).[108] The word can well be rendered by "student" in the older sense of the English word: a person engaged in the earnest pursuit of knowledge in some field.[109] It does not follow that all *studiosi agricolationis* are *agricolae* in the strict sense, even if all good *agricolae* are *studiosi*.[110] So if Columella is speaking in *RR* to, and with, *studiosi* – as he explicitly states that he is[111] – he addresses not only farmers under a narrower professional description, but all whose interest in *res rusticae* is sufficient to motivate discussion, learning, and inquiry.[112] Of course, in the agrarian society of Rome all *studiosi agricolationis* were at least potentially *agricolae*, and the wellspring for interest was the fundamental place of agriculture in ancient life. Against this backdrop, *studiosi* might participate in this conversation for different reasons: most obviously to initiate or improve agricultural operations (or even just to contemplate as much), but

107 As Mielke 2023: 39–41 notes, social rank is less important than *facultates inpendendi* (*RR* 1.1.18). For the socioeconomic status of Columella's audience, see also Gummerus 1906: 77–9; White 1970: 26, 402–4; Flach 1990: 203–4; Diederich 2007: 369–72.

108 Columella carefully distinguishes the interests of *agricolae* from those of other groups: see Ch. 4.2.3. Cf. *studiosi* also at 1.pr.4 (dance and music), 3.9.5 (horse-racing); at 1.pr.5, Columella speaks of notional *doctores* and *discipuli agricolationis*.

109 *OED* 1.

110 A crucial passage is 11.1.1, where the two groups are distinguished (*complurium studiosorum et praecipue agricolarum sermonibus*), though not in such a way as to exclude mutual belonging (pace Mielke 2023: 180). Note thus that when *et praecipue* joins two nouns or adjectives in Columella, the second always stands to the first as species to genus (cf. 2.10.34, 2.14.3, 3.3.3, 4.5.1, 7.6.5, 8.2.4, 10.pr.2 11.2.74, 12.55.1).

111 Cf. 1.1.17 *ideoque haec uelut adminicula studiosis promittimus*, where *haec* refers to the *praecepta* of *RR*.

112 Carroll 1976: 789 calls *studiosi* "intellectual agriculturalists," which is half right: their interest is "intellectual," but *studiosi* need not be full-fledged agriculturalists.

also so as not to be unlearned on a subject of such importance – and perhaps just for the pleasure of it.

The common principles of Columella's *studiosi* are curiosity and diligence, which stimulate a desire for intellectual reflection and exchange. This is only a more concrete notion of the Roman audience for the *artes* imagined earlier, and, should we extend his language, it is not difficult to imagine other groups of Romans whom we might call *studiosi architecturae*, *studiosi medicinae*, *studiosi rei militaris*, and so on.[113] Importantly, for upper-class Romans, these other disciplines would have sustained pragmatic motivations as much as did agriculture: they were all domains in which independent judgment and understanding could be plausibly expected.[114] No doubt some Romans were *studiosi* only of certain *artes*, but many could have been *studiosi* of several. The gathering of disciplines in Celsus' *Artes* and the varied contents of Pliny's *HN* suggest prima facie that this was the case, and indeed Pliny speaks of the *studiosi* simpliciter as those he expects would be interested in the varied material he has compiled.[115] To be sure, the word *studiosus* was not univocal in the *artes*, and authors named their audiences in many ways; the importance of Columella's example is simply to show that Roman authors conceived of an expansive readership in their own terms.

2.6 *Ars* in Service of Empire: Frontinus' Military Writings

Now that we have limned the Roman audience for the Imperial *artes*, we will turn to the politics of the *artes* in a more concrete sense. We have observed that many authors of *artes* held civil and/or military positions (Vitruvius, Pliny, Columella; we do not know enough about Celsus to say). While these authors advertise the political significance of their works differently, this pattern is not coincidental. We can better understand

113 Frontinus does speak of *studiosi* for the *scientia rei militaris* at *Strat* 1.pr.1 (quoted in Ch. 2.6). Quintilian mentions *studiosi eloquentiae* (*Inst.* 5.10.122), and he and Celsus use *studiosi sapentiae* of philosophers (Cels. *DM* 1.pr.29, Quint. *Inst.* 1.pr.14). Vitruvius does not mention *studiosi*, but cf. *DA* 5.11.2 *philosophi, rhetores reliquique, qui studiis delectantur*; the student of Vitruvius' architecture, who must know letters and philosophy (Ch. 3.1.2), could be among these. Compare also Vitruvius' *sapientes* (1.1.18). Consider finally that Pliny's work on rhetorical education (n.41) was called *Studiosi tres* (Plin. *Ep.* 3.5.5).

114 See also Ch. 5.1 on the disciplines of Celsus' *Artes*.

115 Cf. *HN* pr.17, where he says that he has gathered his facts from 200 volumes *quorum pauca admodum studiosi attingunt*. At pr.6 (n.90), Pliny self-deprecatingly addresses himself to *studiorum otiosis* (a phrase whose meaning is uncertain: Naas 2002: 46–53; Darab 2012); Rackham has *studiosorum* as a *varia lectio* (doubtless his own conjecture), commended by Baldwin 2005: 94. The meaning would be "those at leisure among the *studiosi*," though it is not clear that this sense is desired.

how the *artes* were embedded in Roman *imperium* if we look to Sextus Iulius Frontinus, whose writings represent a sort of consummation of the political interests of specialized knowledge. Frontinus was one of the most important Roman figures in the time of the Flavians and Trajan, remaining close to the center of power until his death in AD 103/4.[116] *Praetor urbanus* in AD 70, he was called away to assist in the suppression of the revolt led by Iulius Ciuilis, receiving the surrender of 70,000 Lingones; there quickly followed a suffect consulship (probably 73) and the governorship of Britain (73/4–77). In the 80s he served as legate of the lower German army (81–83/4) and then proconsul of Asia (84/5). Under Nerva he took charge of the Roman water supply (97), and the accession of Trajan saw the distinctions of a second consulship (suff. 98) and a third (ord. 100).[117] He was considered great in his own day: Pliny the Younger called him *princeps uir* (*Ep.* 4.8.3) and ranked him as one of the two most illustrious men of the state (cf. 5.1.5 *duos quos tunc ciuitas nostra spectatissimos habuit*); Tacitus allowed that he was a man as great as the times permitted (*Agr.* 17.3 *uir magnus quantum licebat*).[118] Frontinus was also the author of several technical works: a treatise on warfare (lost),[119] a complementary collection of historical military *exempla* (the *Strategemata*; extant), a work on the aqueducts of the city of Rome (*De aquaeductu urbis Romae*; extant),[120] and probably one or more works on land-surveying (fragmentary).[121] We will handle the military work and treatise on aqueducts first and return to the land-surveying writings later.

In his literary output, Frontinus embodies the ethic of writing as civic duty that we have seen hinted at in other *artes*. Throughout his career, he continually represented his authorship as an extension of the political activity in which he was engaged. To see how he did so, we may begin from his earliest extant work, the *Strategemata* (*Stratagems*: *Strat.*), usually dated to the later years of Domitian's reign.[122] It is worth quoting the

116 For the thumbnail sketch that follows, see Malloch 2015: 1–2; further *PIR*² E 322; Del Chicca 2002 (a literary biography); Saastamoinen 2003: 16–18; Rodgers 2004: 1–5; Birley 2005: 68–71; Fögen 2009: 267–71; Castillo Pascual 2013, esp. 37–42; most fully, Eck 2013.

117 See Eck 2002: 219 on this uniquely rapid succession of consulates.

118 See Sherwin-White 1966: 272–3; and with Tacitus, König 2013: 366–70; Woodman and Kraus 2014: 179.

119 The work on the *scientia rei militaris* is mentioned at *Strat.* 1.pr.1 (quoted later in this section); see Ch. 6, nn.6, 22.

120 Much recent attention: for example, Del Chicca 2004; Peachin 2004; Rodgers 2004; König 2007; Fögen 2009: 265–89; Letzner 2013.

121 See Ch. 2.7, esp. nn.146–148.

122 Bendz 1938: 265–6; Eck and Pangerl 2003: 210; Peachin 2004: 94n19; Rodgers 2004: 3; Birley 2005: 70. Malloch 2015: 95 also entertains a date closer to the Chattan war (AD 83).

opening paragraph in full – a succinct but richly informative statement encapsulating the work's production, intent, and subject matter (1.pr.1):[123]

> cum ad instruendam rei militaris scientiam unus ex numero studiosorum eius accesserim, eique destinato quantum cura nostra ualuit satisfecisse uisus sim, deberi adhuc institutae arbitror operae ut sollertia ducum facta, quae a Graecis una στρατηγμάτων appellatione conprehensa sunt, expeditis amplectar commentariis. ita enim consilii quoque et prouidentiae exemplis succincti duces erunt, unde illis excogitandi generandique similia facultas nutriatur; praeterea continget ne de euentu trepidet inuentionis suae, qui probatis eam experimentis comparabit.
>
> Since I, alone of so many students of this subject (*studiosorum*), have furnished a theoretical treatment (*scientiam*) of warfare and, as it seems to me, completed what I set out to do to the best of my ability, I believe that it is still owed to the task that I have undertaken that I survey in a convenient aide-mémoire (*commentarii*) the clever deeds of generals which the Greeks call "stratagems." For generals will thus be equipped with examples of judgment and foresight that will nourish their faculty for devising and producing similar (sc. deeds). Moreover, (sc. the general) will not worry about the outcome of his plan if he can test it against proven experience.

Frontinus offers these *commentarii* as a pendant to his theoretical treatment (*scientia*) on the *res militaris.*[124] This lost treatise, whose existence is attested independently, would have been an *ars* proper; its relation to the *Strategemata* will be discussed in Chapter 6, where we will revisit other aspects of the project that Frontinus announces here.[125] If we restrict ourselves at present to the politico-historical interests of Frontinus' preface, the high level of abstraction besides the terseness of his explanation of his intentions is striking. This impression could of course be accidental, at least if Frontinus used the preface of the earlier military work (to which *Strat.* is a pendant[126]) to set forth a broader justification for writing on war. But even putting this consideration aside, there are hints of a programmatic political attitude in *Strat.* that can be filled out with the aid of fuller

[123] Text of *Strat.* is Ireland 1990.

[124] The relationship between *commentarius/commentarii* as a genre and technical writings is much studied (Frontinus calls *Aq.* the same); for Frontinus' use of the word, with wider context, see DeLaine 1996, esp. 117–21, 132–6; Formisano 2001: 141–53, esp. 147–8 (cf. 2006: 133–4); Saastamoinen 2003: 19–25; Peachin 2004: 12–34, esp. 14–25; Rodgers 2004: 8–14; Castillo Pascual 2013: 44–6; Roby 2016: 28–31.

[125] See also n.119.

[126] Campbell 1987: 14n9 denies *Strat.* is an "appendix," but Wheeler 2010: 21n55 avers it is. Frontinus evidently did mean *Strat.* to be complementary to the earlier work, but the scale and structure give pause; I call it a "pendant."

remarks in the later *De aquaeductu urbis Romae* (*On the Water Supply of the City of Rome*: *Aq.*).

Frontinus says in *Aq.* that the occasion for its composition was his appointment by Nerva to the post of *curator aquarum* (AD 97). "Every duty assigned by the *imperator* requires a more earnest application of attention" (1 *omnis res ab imperatore delegata intentiorem exigat curam*), he declares, before telling us what the chief part of this *cura* consists in (*Aq.* 1):[127]

> primum ac potissimum existimo, sicut in ceteris negotiis institueram, nosse quod suscepi.
>
> I consider the first and most important thing, as I also undertook in other business, to acquire an understanding of the task that I have assumed.

Frontinus relates his public duties – delegated "by Nerva Augustus, an *imperator* equally diligent and devoted to the public good" (1 *Nerua Augusto, nescio diligentiore an amantiore rei publicae imperatore*) – to the production of the *Aq.*, which fulfills the *primum ac potissimum* part of his *cura*,[128] that of acquiring a personal knowledge of the subject (cf. *nosse quod suscepi*).[129] Frontinus argues that as much is required of any "respectable man" (2.2 *tolerabili uiro*) working in an official capacity (*delegatum officium*... *agere*), for it is "unseemly" (*indecorum*) to be compelled to depend on subordinates for guidance.[130] This reason by no means exhausts the possible motivations for Frontinus' writing,[131] but it underscores his desire to present his authorship as an extension of his public service (*delegatum officium*). Especially important is Frontinus' admission that he is merely following a practice that he "undertook in other business": the phrase *in ceteris negotiis institueram* echoes *institutae* . . . *operae* of *Strat.* (1.pr.1). Hence, by "other business" (*in ceteris negotiis*) here Frontinus will mean his earlier military commands, to which would correspond the treatise on the *scientia rei militaris* and *Strat.*, and perhaps also an administrative assignment (under Vespasian?) for which he would have produced his land-surveying writings (see Ch. 2.7).

127 Text is Rodgers 2004.

128 "Attention" or "care" (*cura*), related to zeal for usefulness, is important in the Imperial *artes*: see Rodgers 2004: 123; also Oksanish 2019: 7–9 (Vitruvius); Citroni Marchetti 2017 (Pliny). See Béranger 1953: 186–217 for *cura* in Imperial ideology.

129 Cf. Col. *RR* 1.1.2 *caput est in omni negotio nosse quid agendum sit* (but lacking Frontinus' bureaucratic overtone).

130 For *tolerabili uiro*, see Rodgers 2004: 127; also Santini 1992: 922–3; Baldwin 1994: 503.

131 Bruun 2013, with bibliography.

Frontinus has more on the political character of his *commentarii.* He continues (*Aq.* 2.2):

> quapropter ea quae ad uniuersam rem pertinentia contrahere potui, more iam per multa mihi officia seruato, in ordinem et uelut [in hunc] corpus[132] deducta in hunc commentarium contuli, quem pro formula administrationis respicere possem. in aliis autem libris, quos post experimenta et usum composui, succedentium res acta est; huius commentarii pertinebit fortassis et ad successorem utilitas, sed cum inter initia administrationis meae scriptus sit, in primis ad meam institutionem regulamque proficiet.

> For that reason, according to the habit I have maintained to now through many official appointments, I have brought into this commentary everything that I was able to collect relevant to the general matter, organizing and putting it into a systematic body, as it were, so that I could consult it as a guide for management. In the other books, which I composed after practical experience, I did so for the benefit of those succeeding me; these notes will perhaps also be useful for my successor in this post, but since I have written it at the beginning of my tenure it will serve in the first place for my own education and guidance.

This autodidactic *commentarius* (*Aq.* 2.2; cf. *Strat.* 1.pr.1 *commentariis*) on water management stems, Frontinus wants us to understand, from a habit maintained throughout his career. The resultant authorship is itself a form of public service, unselfishly performed for his successors in office. Indeed, this pattern of diligence is already announced in *Strat.*, where he says that he "undertook this work, like the rest, rather for the use of others than for the sake of bringing praise to myself" (1.pr.3 *hoc opus, sicut cetera, usus potius aliorum quam meae commendationis causa adgressus sim*).[133] The only difference with *Aq.* is that he prepared it at the beginning of his tenure, whereas the others were produced afterward; yet Frontinus hopes that it, like his other writings, "will also be useful for (sc. his) successor."

Frontinus' retrospective statements in *Aq.* were obviously intended to shape the reception of his work and thus can shed light on the military *commentarii*, including *Strat.*[134] In particular, the claim that he wrote for the benefit of his successors (cf. *succedentium, successorem*) indicates that while his audience may be temporally undefined, it is a certainly Roman

[132] Rodgers follows Bücheler in deleting *in hunc* before *corpus* (n. *ad loc.*, 129).

[133] Since *Strat.* predates *Aq.*, and *sicut cetera* implies a plurality, *cetera* could include the land-surveying writings (Lachmann et al. 1848–52: II 102); Lindermann 2016: 24, who argues against the authenticity of the gromatical fragments (n.146), thus maintains that *opus* refers generally to "assignment" or "task" (*Aufgabe*).

[134] For self-promotion in *Aq.*, see Bruun 1991: 178–9, 186–7, 370; Evans 1994: 57–8, 61–2; DeLaine 1996: 129–30, 136; Fögen 2009: 274–5; König 2013: 373 (cf. 2007, esp. 196–7, 201–2).

future of officialdom to whom he imparts his expertise. Thus, the ethic of civic duty in Frontinus' writings, including the *artes*, proposes a kind of technocratic Roman *imperium* in which specialized knowledge supports governmental administration. Despite his reserve in *Strat.*, Frontinus' posture in the *Aq.* shows that he wanted his military treatises to be understood closely with the commands for and after which they were composed. The military *commentarii* are in this sense coopted by *Aq.*, their implicitly political content drawn out and made thematic: whether or not Frontinus took pains in the treatise on the *scientia rei militaris* to stress that he was writing to bolster Roman government, he now shows us that the military works were political by birth, so to speak, because they were penned by a man obliged by the duties of continual public office and intended as a source of instruction (*institutio*) for the Romans who would follow him. This audience is named in *Strat.* by the *duces* (1.pr.1) who Frontinus hopes will be "equipped" (*succincti*) with his *exempla* as an aid to the conduct of war. The extended technical authorship of Frontinus thus demonstrates the fruition of the rapprochement between *artes* and *imperium* announced a century before by Vitruvius.

2.7 The Surveyors on the History of the Roman Empire

Frontinus' writings show that the *artes* did not gain a political charge in Rome by incidental collision with the structures of Roman power; rather, the *artes* were constitutive of that structure and shaped how the Romans understood the significance of the expertise on which their *imperium* depended. Moving beyond politics proper, the last and perhaps most extraordinary example of the interdependence of specialized knowledge and Roman Empire that we will consider comes from the Roman land-surveying writings. The Roman surveying texts, which date from the late first century AD on, are an invaluable resource, but the peculiar difficulties in their transmission necessitate some preliminary remarks. Most readers now consult the Roman surveyors as a part of the "*corpus agrimensorum Romanorum*" (*Corpus of Roman Land-surveyors*: *CAR*); this "*corpus*" is a modern invention of scholars, though not an arbitrary one.[135] This nomenclature has become a standard way of speaking about the totality of Roman-period Latin surveying writings that are represented in the

[135] This title owes to Thulin's 1913 Teubner, which was not advertised as the "corpus agrimensorum Romanorum" *in toto* but only an initial offering in the series; his work is indeed highly selective and contains only "opuscula agrimensorum ueterum" (i.e. the tidier early treatises).

principal manuscript traditions, plus some odds and ends.[136] I will use it in this broad sense, with the following qualifications.[137] A "*corpus*" may be notionally justified from the transmission by the fact that the bulk of the extant surveying works descends from one or another compilation already prepared in late antiquity (by the late fifth century). But the appellation "*corpus*" also has drawbacks, because it exaggerates the thematic unity of the writings, implies more intention and design in their composition and arrangement than was originally present, and suggests that the tradition was closed when it was very open.[138] An attempt to draw back the curtain on the early history of the *CAR* enables us to consider what ancient unity might actually have obtained among the land-surveying writings and to appreciate their place in the *artes* of the early Empire.

The *CAR* contains dozens of texts attributed to different authors (or often unattributed), which exhibit great variety in their form and contents: Apart from systematic works on different aspects of surveying, there are, for example, legal excerpts, mathematical writings, notes with definitions and descriptions, lists of abbreviations and marks, and accounts of surveyed land.[139] It is very difficult, if not impossible, to ascertain the extent to which many of these writings have been accidentally or intentionally altered in later processes of excerpting, compilation, redaction, and transmission. Nonetheless, it seems possible to identify an ancient core of treatises which date roughly to the period AD 75–110. These include the works of Frontinus, Balbus, and three different authors called "Hyginus" (!) in the manuscripts, with further texts possibly to be added.[140] In recent decades, scholars have argued that the works dating to this early period should be understood as

[136] Introductions to the manuscript traditions: Carder 1978: 1–35; Reeve 1983, forthcoming; see also Toneatto 1994–5, esp. I 1–75. The most complete edition remains Lachmann et al. 1848–52, though this too omits some writings that later editors have (re)attached to the *CAR*, for example, those attributed to Epaphroditus and Vitruvius Rufus (Mortet 1896; Bubnov 1899: 516–51; Guillaumin 1996: 138–97), a work on military surveying (n.140), and certain short mathematical texts (Bubnov 1899: 494–515, 552–3).

[137] Some scholars, reverting to Lachmann, prefer *gromatici ueteres* to *CAR*.

[138] Cf. Folkerts 2014: 131; Lindermann 2014: 199–200 ("*corpus*" is a "Hilfskonstrukt").

[139] Summary: Lachmann et al. 1848–52: I v–vi; Dilke 1971: 227–30; von Cranach 1996: 23–7; Campbell 2000: 450–1.

[140] The dates of many texts in the *CAR* are extremely uncertain. I introduce Frontinus and the Hyginus usually called "Hyginus Gromaticus" later in this section. Balbus is dated by a reference to a Dacian campaign (L 93.6–7), usually interpreted to refer to Trajan's conquests and thus yielding a range AD 102–6: see Toneatto 1994–5: I 5; Guillaumin 1996: 3; Chouquer and Favory 2001: 27. Dilke 1971: 42 allows the possibility of a date under Domitian (cf. Campbell 2000: xxxix), but see Sherk 1974: 541–2n37 *contra*. The treatise attributed to the Hyginus usually called "pseudo-Hyginus" and titled *De munitionibus castrorum* or *De metatione castrorum* has been dated to the time of Trajan (Lenoir 1977: 111–33), in my view plausibly, or alternatively to the third century AD (Grillone 2012: 14–19). Three fragmentary works – (1) *De limitibus* (*Lim.*), (2) *De condicionibus*

originating in a coherent, practical response to the cadastral reforms and land policies of the Flavians.[141] The demands placed upon the *mensores* by Vespasian and his successors provided the impetus for the composition of a formal surveying literature, which would have aided surveyors at the same time as it advertised their activities to a wider public; the rapid appearance of treatises precipitated by this situation in turn established a center of gravity for the written tradition, around which elements of legal and technical interest could accrete. In the growth of surveying literature after the first and early second century, we may thus see the continuation of an intellectualization of the discipline that began in the Flavian period.[142] The details of this story are sketchy and difficult to reconstruct, but it is an intriguing effort to make sense of the origin and growth of the *CAR*. Given the frenetic activity around the *artes* as well as their many connections with Roman *imperium*, it is plausible that this century should see the emergence of the earliest systematic treatises on the *ars mensoria*:[143] "our studies were at a boiling point" (L 92.3–4 *feruentibus studiis nostris*) said the surveyor Balbus, as he looked back from the end of the first century.[144] If the Flavian narrative is even partly true, it would put an exclamation point on the relationship between the *artes* and Roman *imperium* in the first century.

Historical hypotheses aside, we will concentrate in this book on the earliest extant Imperial writings of the *CAR*: those attributed to Sex. Iulius Frontinus and the Hyginus conventionally dubbed "Hyginus Gromaticus." We begin with Frontinus, to whom are attributed four short fragments: (1) *De agrorum qualitate* (*On Types of Fields*: *Agr. qual.*), (2) *De controuersiis* (*On*

agrorum (*Cond. agr.*), and (3) *De generibus controuersiarum* (*Gen. contr.*) – are ascribed to a third Hyginus usually called just "Hyginus" (though not in this book), active around AD 100: Toneatto 1994–5: I 4–5, Behrends et al. 2000: xiii–xxii; Chouquer and Favory 2001: 26. Chouquer and Favory 2001: 26–9 attach others to the early group, including Siculus Flaccus, M. Iunius Nips(i)us or Nyps(i)us, and an anonymous source for Agennius Urbicus, but with less plausibility.

141 Hinrichs 1974: 128–57; Castillo Pascual 1996a; von Cranach 1996, esp. 33–7; Chouquer and Favory 2001, esp. 29–39, 203–16; Chouquer 2004; 2008, esp. 851–7. Among the precipitating historical factors adduced are Vespasian's need to fill the Imperial coffers when he came to power in AD 69; the chaotic state of affairs in land ownership resulting from two years of successional tumult (AD 68–9); and the conflagration of the *tabularium* (in AD 69: cf. n.45), which housed agrarian maps. The Flavian land program comprised several major activities, such as the reconstitution of the archives and reassessment of land for tax purposes. One controversial measure was Vespasian's reclamation of certain marginal public lands (*subseciua*) for the state, forcing occupants to purchase unallocated territory on which they had encroached; the measure was abrogated by Domitian (Griffin 2000: 29–31, 71).

142 See Chouquer and Favory 2001: 20 for the chronological sedimentation of surveying texts, which continued into the medieval period (cf. 39–43); also Campbell 1996: 80, 2005: 180; Lindermann 2018: 22.

143 See Ch. 7, n.1 for surveying writings in the Republic.

144 See also Chouquer and Favory 2001: 33–4; Guillaumin 2010: 162.

[sc. *Land*] *Disputes*: *Contr.*), (3) *De limitibus* (*On* limites: *Lim.*), and (4) *De arte mensoria* (*On the Art of Surveying*: *Art. mens.*).[145] The extracts, none of whose titles is original, deal respectively with land classification, disputes, centuriation (establishment of *limites*), and measurement of irregular territory. It is uncertain in what relation they stand to any original writings Frontinus composed on land-surveying. Indeed, the authenticity of some or all the fragments has been challenged on various grounds; the reasons for doubt are legitimate but, in my view at least, not sufficient to overturn the traditional attribution.[146] It seems probable – especially in view of Frontinus' statements about his writings vis-à-vis official duties – that the occasion for composition was Vespasian's land reforms, for which he would have had an administrative function.[147] Such an appointment, along with his military commands, would have given him ample opportunity to become familiar with the activities of the *mensores*, even if he likely did not have personal surveying experience.[148] Yet there are certainly great textual perils in dealing with his land-surveying fragments: even granting an original work, or works, by Frontinus, some writings presently attributed to him could have been

[145] This is how Thulin distinguishes the texts (L 1–34.13 = T 1–19). Lachmann tries to recover more for Frontinus from later commentators (cf. L 34.15–58.22); although his optimism has been dampened, hopes have persisted that some material in Agennius Urbicus and other later writings could be genuine. Salvage attempts are beset by severe difficulties, however: cf. Lindermann 2016: 25–38. I use Thulin 1913 as the base text for Frontinus; other editions include Behrends et al. 1998; Campbell 2000; Guillaumin 2005.

[146] Expressing (plausible) doubts about authorship are Keppie 1983: 12; Campbell 1996: 76, 2000: xxvii–xxviii; Del Chicca 2002: 216, and esp. now Lindermann 2016. Recent arguments buttressing authenticity include von Cranach 1996: 129–32; Cuomo 2000: 195–7; Chouquer and Favory 2001: 21–3; Guillaumin 2005: 128–9; Paniagua Aguilar 2010–11, esp. 45–9; Castillo Pascual 2013, 2016. Lindermann shows that some external evidence used by the pro-Frontinus camp cannot guarantee his authorship and has furthermore stressed how tenuous the connection between Frontinus' name and any specific surveying work in the *CAR* is. Yet one must still explain attributions to "Iulius Frontinus" in the MSS and thus address the possibility of a Frontinian substrate. (The absence of "Sextus" is not necessarily significant: the *praenomen* is missing in MSS for Frontinus' other works and is known only from epigraphic evidence: Rodgers 2004: 121.) The debate thus turns also on an assessment of how well the gromatical fragments fit with Frontinus' career and interests. Those arguing against his authorship stress the low social status of surveying and the first-person technical detail of the fragments as incongruous with Frontinus' position; those in favor argue that an interest in land-surveying is apt and, despite surface difficulties, that there is nothing in the fragments he could not have written. I lean toward the latter arguments (as discussed also later in this section), but the evidence does not enable certainty.

[147] Cf. Dilke 1983; von Cranach 1996: 131; Cuomo 2000: 196–8; Del Chicca 2004: x–xi; Guillaumin 2005: 128; Eck 2013: 127; Castillo Pascual 2013, esp. 49–52, 2016: 52–3. Various dates have been proposed for its composition: AD 70–90 (Toneatto 1994–5: I 4), 71–73/4 (Castillo Pascual 2013: 51; cf. 2016: 53), after 75 (Behrends et al. 1998: xii n21), 78–82 (Ratti 1996: 230; von Cranach 1996: 131–2), 77/78–82 or early 90s (Chouquer and Favory 2001: 24).

[148] He was no aqueduct engineer, either. Among Frontinus' sources, Castillo Pascual 2013: 47–8, 2016: 53 draws attention besides eye-witness experience to the possibility of his consulting archival materials, commentaries, legal documents, and so forth.

wrongly attached to his oeuvre, or genuine passages might have been altered or interpolated. I will speak of "Frontinus" as the author of these four surveying texts, but it is best not to cling too tightly to his name at every moment; if the words flowed from another author's pen in the same period, it would not vitiate my arguments.[149]

A treatise *Constitutio limitum* (*Establishment of* limites: *Const.*) is attributed to Hyginus, usually called "Hyginus Gromaticus" to distinguish him from other surveying authors of the same name.[150] Throughout this book I will use simply "Hyginus," distinguishing other Hygini as necessary.[151] Unlike Frontinus, Hyginus was a practicing surveyor.[152] His work deals, in an informed and logical fashion, chiefly with the process of centuriation, from orienting the orthogonal grid of *limites* to the sortition and mapping of the territory.[153] Along the way, he attempts to account for the origin and meaning of the elements of centuriation, reports irregularities and variation in older and regional practices, corrects other surveyors' mistakes, and propounds the methods he considers best practice. The date of Hyginus' treatise was long a question mark. The previous *communis opinio* put him in the third century, but renewed scrutiny has suggested that a probable range for composition is AD 75 to 120, with an earlier date perhaps most likely.[154] This dating locates Hyginus among the earlier authors of the *CAR* and connects his work with the conjectured swell of surveying writings during and shortly after the Flavian period. Numerous parallels between Hyginus' *Constitutio* and Frontinus' *De limitibus* indicate one author depends on the other; if the earliest plausible date for Hyginus is accepted, his writings might

[149] Those arguing against Frontinus' authorship also place the fragments in the first or perhaps second century AD (Campbell 2000: xxviii; Lindermann 2016: 39–40).

[150] The title *Constitutio <limitum>* is Thulin's; Lachmann used *De limitibus constituendis* (as some MSS transmit). Although *gromaticus* appears in some MSS, the adjective goes not with the author but the book, (sc. *liber*) *gromaticus*. The name "Hyginus Gromaticus" is maintained now only for convenience; Campbell prefers "Hyginus 2." This Hyginus is distinguished from another Hyginus to whom are attributed several fragmentary works in the *CAR* (n.140) chiefly on stylistic grounds (Gemoll 1876, esp. 168–70), but the matter is not so easily settled: cf. von Cranach 1996: 142–4; Behrends et al. 2000: ix–xi; Campbell 2000: xxxvi n102; Lindermann 2018: 10–11.

[151] When cited in abbreviated form ("Hyg."), these other Hygini are distinguished by the abbreviated title of the work appended (n.140); our Hyginus wrote only "*Const.*"

[152] Pace von Cranach 1996: 123 (see Ch. 7).

[153] See Ch. 7.3 for an overview of centuriation drawing on *Const.*

[154] Ratti 1996; 1998 argues vigorously but inconclusively for composition between AD 75–7. A *post quem* is secured by mention of *colonia Ammaedara* (Haïdra in Tunisia; L 180.1–2 = T 144.9–10) and a citation of Lucan (3.247 = L 188.10 = T 151.18), but the *ante quem* is not as secure: see von Cranach 1996: 137–8; Chouquer and Favory 2001: 25–6; Lindermann 2018: 11. Nonetheless, a dating among the first generation of surveying authors remains most plausible from internal evidence; those who are less confident do not place him later than the second or early third century (Campbell 2000: xxxvi–xxxvii; Toneatto 1994–5: I 6).

be the source, but it is more likely that Frontinus predates.[155] Whatever one believes, the parallels are significant: They not only indicate shared interest in systematic reflection on the *ars*, but also encourage us to view Frontinus' and Hyginus' writings as collaborative in their attempt to produce a textualized, theoretical nexus for the discipline.[156] We will study this point of contact between their writings further in Chapter 7.

Here we will consider one aspect of the early surveying writings relevant to the Romanization of the *artes*: the presentation of the discipline via the history of Empire. We glimpse this subject through Hyginus' explanation of "how *limites*," the boundary lines that were the foundation of Roman land-division practices, "were established by the ancients" (L 175.15–16 = T 140.12 *quem ad modum ab antiquis acti sint limites*). He says (L 176.1–5 = T 140.16–20):[157]

> finitis ergo ampliorum bellorum operibus, augendae rei publicae causa inlustres Romanorum uiri urbes constituerunt, quas aut uictoribus populi Romani ciuibus aut emeritis militibus adsignauerunt et ab agrorum noua dedicatione culturae colonias appellauerunt.

> Once the labors of their great wars were ended, illustrious men of Rome founded cities with a view to expanding the *res publica*. These they granted either to the victorious citizens of the Roman people or to soldiers who had earned their discharge, calling them "colonies" (*coloniae*) because the fields had been newly dedicated to cultivation (*culturae*).

According to Hyginus' history of land-surveying,[158] the discipline grew directly out of great wars (*ampliorum bellorum*) and the associated redistribution of land to Roman citizens.[159] At that time, "the Roman people did not

155 For the sharings, see von Cranach 1996: 145–7; several are discussed *passim* in Ch. 7. Ratti 1996: 230 (followed by Guillaumin 2005: 67–8) has Frontinus using Hyginus, but his argument depends on dating Hyginus' treatise AD 75–7 and Frontinus' 78–82; such narrow ranges are uncertain (n.147, plus previous n.). The examples at Ratti 1996: 229–30 do not prove that Frontinus follows Hyginus (cf. Toneatto 2002: 21), and in some places Hyginus seems rather to summarize Frontinus (cf. Lindermann 2013: 133); his more elaborate treatments of material found in Frontinus may represent independent development (Lindermann 2018: 19–22).

156 Cf. Lindermann 2018: 22.

157 I use Thulin 1913 as the base text for Hyginus; other editions include Clavel-Lévêque et al. 1996, Campbell 2000, Guillaumin 2005, Lindermann et al. 2018 (the most radical). Regrettably, Lindermann 2022 came into my hands too late for to use.

158 The historical "excursus" runs L 175.15–178.9 = T 140.12–142.19, followed by discussion of *limites* adopted by different colonies (L 178.10–180.9 = T 142.20–144.17). A similar passage is found in ps.-Boëthius' *Demonstratio artis geometricae* (L 395.21–396.12), which Lindermann 2018: 44–8 uses (inter alia) to argue that the excursus is interpolated; Lindermann et al. 2018. accordingly eject L 175.15–178.2 = T 140.12–142.12 from their text.

159 Clavel-Lévêque et al. 1996: 43n36 assert that *ampliora bella* "renvoie sans doute aux guerres civiles [doubtless refers to civil wars]" (cf. Guillaumin 2010: 163). Ratti 1996: 222–4 defends this claim and argues that *inlustres Romanorum uiri* are Julius Caesar and Augustus (cf. Guillaumin 2009c: 135),

have enough soldiers for the increase of the *res publica*" (L 176.7–8 = T 141.2–3 *non … tantum militum incremento rei publicae populus Romanus habuit*); the solution was to make land into a reward for military service (cf. L 176.8–9 = T 141.3–4 *erat tunc praemium terra et pro emerito habebatur*). Hyginus singles out two men for their contributions: Julius Caesar and Augustus. Caesar is noted for his military prowess and for the settlement of his soldiers "once peace had been brought about after numerous wars" (L 177.7–8 = T 142.2 *post aliquot bella parta iam pace*).[160] Augustus is praised for finding homes for the veterans of civil war: "with peace allotted to the world" (L 177.8–9 = T 142.3 *adsignata orbi terrarum pace*), he settled his enemies' legions as well as his own in Italy and abroad.[161] Under Augustus, new colonies were established on the site of enemy cities after their people had been extirpated, Roman *oppida* received discharged veterans under the name of colonists, and cities depopulated by civil war were refounded and settled as colonies (L 177.8–178.2 = T 142.6–12). In all this, the common thread was war – and close to the center of the empire-making enterprise were the *gromatici*, measuring, mapping, and allocating the soil redistributed by those in power. In Hyginus' history, the discipline thus emerges as a hinge between war and peace: when legionaries had "successfully concluded their wars" (L 176.9–10 = T 141.5 *bella feliciter transigere*), they turned to the "laborious respite of agriculture" (*laboriosam agri culturae requiem*);[162] the art of surveying turned plundered lands into farms, as the *mensores* traced out fields in the territory of Italy and the ashes of enemy cities alike and handed them over "newly dedicated to cultivation" (L 176.4–5 = T 140.19–20 *agrorum noua dedicatione culturae*). Other surveyors offer similar accounts of the development of the *ars*.[163]

The symbiosis of surveying and warfare helps to explain one of the major foci of the *CAR*: the division of land. But the establishment of

whom Hyginus mentions shortly after. More probably, Hyginus is distinguishing between earlier wars resulting in colonization and later consolidation following civil conflicts. Cf. Peyras 2003: 111–14.

160 The expression reflects the Augustan ideology of "pacification after military victory" (Cooley 2009: 158): cf. *Res gest.* 13 *cum per totum imperium populi Romani terra marique esset parta uictoriis pax*, with Scheid 2007: 48.

161 See Ratti 1996, esp. 221–5, for Augustus in Hyginus' account (for L 176.1–2 = T 140.16–17 *augendae rei publicae* as a paranomastic reference to Augustus, cf. n.19). Ratti (ibid. 233–4) sees Flavian ideology in Hyginus' treatment of Augustus (cf. n.45).

162 The oxymoron *laborosa requies* and juxtaposition of warfare and agriculture reflect idealized agrarian values (cf. Guillaumin 2009c: 137): Ch. 4, n.23.

163 Cf. Siculus Flaccus (*Cond. agr.*), L 136.14–15 = T 100.7–8, L 153.27–8 = T 117.26–118.2; at L 155.3 = T 119.7 he says *causam autem diuidendorum agrorum bella fecerunt*. Cf. also Hyg. *Cond. agr.* L 116.8 = T 79.8–9.

colonies was only one way that the *ars mensoria* served the state. Most extant surveying texts do not in fact deal with the procedures of colonization per se but rather the adjudication of land disputes, the determination of the types of lands and specification of their borders and area, and the creation of maps and historical records.[164] Here it was another Roman institution that was most relevant: law. Lawfulness is indeed a guiding concern of surveying, in that the *mensores* demonstrate an obsession with legal control and exactitude in all aspects of their profession.[165] Mensuration was an instrument of Roman governance aiding in the efficient but centralized management of the vast area under Roman sway. The juridical orientation of the *CAR* may be explained in part by a Vespasianic origin, but more generally reflects the utility of surveying in mapping, quantifying, and thus knowing the land that Rome controlled.[166] In sum, the *ars mensoria* emerges as a sort of intermediate art: here were violence, the military camp, the art of war; there were peace, cultivated fields, and jurisprudence. Hedged in by warfare on one side and the law on the other, land-surveying played an indispensable role in the gap, and it is not surprising that the discipline was represented by its authors as a third support between these twin pillars of Roman Empire. Here is another example of the technocratic Roman *imperium* we found in the works of Frontinus.

2.8 The Sociopolitical Background of the *artes*

This brings us to the conclusion of our effort to set the *artes* of the early Empire into their social and political context and to introduce the major authors and disciplines of this book.[167] At the outset we proposed that the most significant effect of Roman Empire on the *artes*, so far as the *artes* themselves go, could be understood as the Romanization of specialized knowledge. This Romanization had the result that the objective ways of

164 Cf. Chouquer 2008: 851; see n.139 for the contents of the *CAR*.

165 Cf. esp. Hyg. *Gen. contr.* L 134.2–6 = T 97.18–22. Law and surveying is a large subject; see briefly Campbell 2000: 475–7; further, Hinrichs 1974; De Nardis 1994; Gargola 1995; Castillo Pascual 1996b; Maganzani 1997; Chouquer and Favory 2001; Gargola 2017, esp. 154–86; Möller 2018.

166 Nicolet 1991 (with Purcell 1990); Moatti 1993.

167 The extrinsic factors noted earlier are not the only ones that influenced the *artes*, but others are less specifically related to the Roman situation. Lloyd, for example, influentially studied the agonistic character of Greek science, esp. medicine (e.g. 1979, esp. 86–98, 1983, esp. 208–9, 1991a; Lloyd and Sivin 2002 in comparative context). See also Rihll 1999: 8–13. Another extrinsic factor, an aspiration to raise the social profile of a discipline, has been claimed by scholars for virtually every *ars*; it is too often exaggerated, however, and will not be much emphasized here (see also the Introduction).

knowing and acting in the world developed in the *artes* were realized as specifically *Roman* ways of knowing and acting. We have seen abundant evidence in this chapter to support both abstract and concrete interpretations of this thesis. A few generalizations are offered here. First, on the authors. We have observed that authors of the *artes* were largely members of the Roman elite with successful political and/or military careers, although their specific status varied;[168] even those whose social position is in doubt appear comfortable addressing their works to men at the apex of Roman power.[169] The *artes* did not therefore emerge from the milieu of low-status practitioners, but from the literary circles of educated, upper-class Romans. Next, the audience: While the *artes* were probably perused by expert practitioners, the intended audience was certainly much broader, and included all those Romans who had the means and the will to pursue an intellectual interest in the disciplines pertaining to Roman private and public life. These were the ancient "students" (*studiosi*) of the *artes* (in Columella's words). At times, the authors' words reveal more specific political interest vis-à-vis the addressee, as when an *ars* is dedicated to the *imperator* or presented as a contribution to future Roman officialdom. Now, on the ideology and politics of Empire. Authors of the *artes* were aware of political developments under the Empire and frequently cast their works as responsive to the same. This response could range from the symbolic and suggestive (promotion of Imperial values) to the highly concrete and specific (service or implementation of Imperial policies). Few *artes* are deaf to Roman *imperium*, and in these cases the impression may simply be accidental. This is no banal realization when we recognize how many ancient authors, especially Greek, penned technical works almost devoid of traces of their historical context.[170]

168 Senator: Frontinus; *equites*: Columella, Pliny; unknown: Celsus (probably *eques*), Hyginus, Vitruvius (cf. following n.). Hyginus was probably subequestrian, though the social standing of surveyors in the Imperial period improves alongside their own pride in what they call *nostra professio*: Dilke 1971: 34–46; Hinrichs 1974: 158–70; Chouquer and Favory 1992: 15–17, 1993: 256–9; Classen 1994, esp. 166–70; Campbell 1996: 80, 2000: xlv–liii; Guillaumin 2005: 32–35, 2010: 161–2; Fögen 2014: 231–4.

169 For example, Vitruvius, who is frequently thought as an *apparitor* to belong to a kind of Roman "middle class" (cf. n.6); but cf. Oksanish 2019: 35–6 (many Republican *apparitores* reached equestrian rank).

170 Consider Geminus and Hero, who have been dated widely throughout the Hellenistic and Roman periods because of a dearth of historical references that would fix the context of their writings. Modern efforts to date them – Geminus to the first century BC (Evans and Berggren 2006: 15–22) and Hero to the first AD or later (Neugebauer 1938–9: 21–4 with Sidoli 2011) – rely chiefly on astronomical data.

Finally, some remarks on the specific significances that the authors attached to their writings. The *artes* are presented as a form of Roman self-knowledge, to the extent that they explain the products and practices of Roman *imperium* to the Romans themselves. This self-knowledge was centered on Rome and Italy but came to include all the territory under Roman rule: because this space was conceptualized as identical with the habitable world, or at least the habitable world up to the point that it could be known, they offered a totalizing cognition of the world qua Roman world. Furthermore, the *artes* were represented as playing an important role in the growth and maintenance of Roman Empire: certain kinds of specialized knowledge issued from the needs of the state and in turn supported Roman *imperium*. And, last, in response to the rich Greek legacy bequeathed by the *technai*, possibly a challenge to the Romanization of the *artes*, the Romans tended to adopt an attitude of assimilation and advancement. Greek knowledge was to be absorbed and synthesized in the areas where it really mattered, just as other physical and cultural riches of Greece worth preserving flowed to and took their place in Rome. The subtext for all these observations is the deceptively obvious point that the Roman authors were not writing merely (or sometimes at all) to teach people *how* to perform an *ars*, but at least as much *what* an *ars* meant for the Romans. The *artes* were therefore as much an epistemic as practical paradigm of Roman Empire.

2.9 Conclusion to Part I

In this and Chapter 1, we explored two different sets of preconditions for the emergence of the early Imperial *artes* and the formation of a Roman scientific culture. One set, studied in Chapter 1, concerns the premises for specialized knowledge established by Greek and Roman-Republican traditions of *technê* and *ars*. Absent such profound intellectualization of *ars*, it is impossible to conceive of the enlargement and flourishing of the *artes* in the early Empire. Equally, the distinctive scientific character of the Imperial *artes* depended on a synthesis of Greco-Roman attitudes toward expertise which had taken shape by the end of Cicero's and Varro's lives. These attitudes would influence the development of all the *artes* through the end of the period under study in this book and beyond. The other set of preconditions studied in this chapter relates to the transformation of Rome into an Empire and the social, political, and ideological changes attendant on this process. The intellectual culture of the *artes* is uniquely Roman in large part because the *artes* are deeply informed by the facts and prerogatives of Roman *imperium*. An absorbing interest in Empire encouraged

Roman authors to present the universalized ways of knowing and doing that belonged to their *artes* at the same time as specifically Roman ways of knowing and doing. In Chapter 3 we will study Vitruvius, whose *DA* offers the ideal lens through which to examine the emergence of the Roman scientific culture of the *artes* in the early Empire against the backdrop examined in this first part of the book.

PART II

CHAPTER 3

The Architectonic ars *of Architecture: Explanation and Method in Vitruvius'* De architectura

In *De architectura* (*DA*), Vitruvius sets out to order and explain the discipline of architecture. The attempt to impose synoptic unity on the field is advertised as something new. A parade of Greeks left behind books on different aspects of architecture,[1] and a few Romans had written on the subject in Latin, including a book in Varro's *Disciplinae*.[2] But these "precepts and volumes of commentaries" (4.pr.1 *praecepta uoluminaque commentariorum*) had proven insufficient to bring the field "to complete order" (*ad perfectam ordinationem*). Vitruvius has thus gathered what is "useful" (7.pr.14 *utilia*) from earlier compositions and reduced it "to a unified body" (*in unum ... corpus*). What are the parts of this *corpus*? Vitruvius adopts an expansive view of architecture, comprising three departments (1.3.1): *aedificatio*, *gnomonice*, and *machinatio*. The first, "building" or "construction," has several subdivisions. It occupies the bulk of *DA*, eight books which range from city-planning (Book I) and building materials (II), temples (III–IV), and theaters and other public buildings (V), to private houses (VI), finishing (*expolitio*) and painting (VII), and water supply (VIII). "Gnomonics," the construction of sundials, and "engineering," the contrivance of machines, are dispatched with comparative alacrity, in one book each (IX and X).[3] The "architecture" of *DA* is not our modern "architecture"; but, then again, "architecture" was a diffuse and evolving range of expertise in antiquity and did not mean the same thing to all Greeks and Romans before and after Vitruvius' time, either.[4] Vitruvius

1 *DA* 7.pr.11–14 with Romano 2002: 69–72; Liou et al. 2003: xviii–xxiv; Wilson Jones 2014: 43–5.

2 Vitruvius (7.pr.14) mentions (1) one book of Fuficius, unknown (perhaps G. Fuficius Fango, Cic. *Att.* 14.10.2, Dio. Cass. 48.22.1; or perhaps to be corrected to Fufidius, Schneider 1807–8: I xli with Wiseman 1971: 232, or Fufius, Rowland 2005: 24n58); (2) two of P. Septimius, possibly the dedicatee of Books II–IV of Varro, *Ling.* (5.1, 7.109) and/or the author of *libri Obseruationum* (Quint. *Inst.* 4.1.19); (3) one in Varro's *Disciplinae* (Ch. 1, n.6). The only other extant Roman work, the third- or fourth-century *De diuersis fabricis architectonicae* of M. Cetius Faventinus (Plommer 1973), is based on Vitruvius.

3 For a lucid outline-cum-summary of *DA*, see Oksanish 2019: 191–205.

4 Cf. Müller 1989: 60; Romano 2002: 66–7.

was not unique in conceiving of the field in capacious terms, but this delimitation of its contents and the structure was his own.[5] *DA* is thus an assertion of Vitruvius' own understanding of *architectura*, guided by a principle of usefulness (*utilitas*).[6]

Precisely because Vitruvius recognized the need to establish an intellectual unity for his subject matter, one of the central preoccupations of *DA* is ordering (*ordinatio*). In Chapter 2, we saw that the metaphor of the body (*corpus*) plays an important role in figuring Vitruvius' ambition to order (*ordinare*) his subject.[7] *DA* reflects this organizational interest in its literary design. Book I is emphatically a book of beginnings, of overlapping and accumulating definitions, divisions, and grand plans: a complex description of *architectura* is followed by a discourse on the disciplines germane to architectural expertise (1.1); the key propositions or concepts of architecture are enumerated and exemplified (1.2);[8] and its *partes tres* (mentioned earlier) are adumbrated and connected with general principles. In later books, Vitruvius often recalls these schemata to remind us of where we are in his promised disquisition or how this or that discussion conforms to his plan.[9] Linguistic features of *DA*, including the self-conscious manipulation of Greek and Latin technical idiom, likewise reflect Vitruvius' attention to the disciplinary coherence of architecture.[10] Textual and conceptual *ordinatio* is indeed usually regarded by scholars as Vitruvius' chief contribution to *architectura*.[11]

For all its importance, *ordinatio* is only one part of the story, however. As we will see, the organizational ambitions of *DA* are predicated on a deeper

[5] Building (*aedificatio*) was always central. Galen (*Aff. pecc. dig.* K 5.68.14–69.1) places horology and "all machines" (μηχανημάτων ἁπάντων), including pneumatics, under ἀρχιτεκτονία (cf. Soubiran 1969: ix–xi); and Pappus (8.1022.3–1024.11), reporting οἱ περὶ τὸν Ἥρωνα μηχανικοί, connects architecture and μηχανική but subordinates the former to the latter (Fleury 1994: 187–94).

[6] For *utilitas*, see esp. Novara 1994; also André 1985: 378 (1987a: 274); Fleury 1994: 203–5; Callebat 1997: 109–10; Fögen 2009: 126–8; Oksanish 2011: 69–75.

[7] Ch. 2, n.29, plus Ch. 3, n.11.

[8] *Ordinatio, dispositio, eurhythmia, symmetria, decor, distributio*, for which see Watzinger 1909: 209–23; Scranton 1974; Frézouls 1985, esp. 217–20; Knell 1985: 30–4; Frézouls 1989: 43–5; 1994: 165–9; Callebat 1994: 93–103; Geertman 1994: 17–23; Lefas 2000; Smith 2003: 15–25; Lagopoulos 2009: 196–203; Wilson Jones 2014: 45–58.

[9] For example, 2.1.8, 3.pr.4, 4.pr.1. Critics since the Renaissance have pointed out, however, that Vitruvius' schematizing is not always successful (e.g. Soubiran 1969: xv–xvii; Gros 1975 on Books III–IV; Romano 1987: 7–8; Callebat 1989: 75–6; Gros 1997: l; Wilson Jones 2000: 35); cf. Ch. 1, n.32 (on Varro, *Rust.*).

[10] Cf. Morgan 1906a; Callebat 1982; Romano 1997; Adams 2016: 164–92; Courrént 2019, esp. 87–281.

[11] For example, Fuhrmann 1960: 79–80; Gros 1975: 1986; Novara 1983: 287–8; Romano 1987: 185–7; Callebat 1989: 75; Gros 1992: 42–3; Frézouls 1994: 155; Novara 1994: 57–8; Callebat 1997: 107; Gros 1997: xliii; Callebat 2005: 68–9; Novara 2005: 130–2; Fögen 2009: 125; Saliou 2009: 99–100; Oksanish 2011: 55; Reitz-Joose 2016: 186–8.

conception of the *ars* of architecture that Vitruvius develops throughout his work. This is a view of *architectura* as essentially "architectonic": a discipline whose expertise is relational and depends on a command of the products furnished by other branches of specialized knowledge – what Vitruvius calls variously *artes, disciplinae, doctrinae, eruditiones,* and *scientiae*. This architectonic view of architecture is connected with another fundamental aim of *DA* – that of explanation, or the "unfolding" of an explanatory account (*explicatio*). Vitruvius casts *explicatio* as inseparable from *ordinatio* and sees the *perfecta ordinatio* of architecture as impossible without explanation of the parts (*genera, membra,* or *partes*) in relation to the whole (*corpus* or *totum corpus*). Among many statements to this effect,[12] his envoi at the end of Book X is especially telling (10.16.12):

> in prioribus uero nouem (sc. uoluminibus) (sc. rationes) de singulis generibus et partibus conparaui, uti <u>totum corpus omnia architecturae membra</u> in decem uoluminibus <u>haberet explicata</u>.
>
> In the previous nine (sc. volumes) I furnished (sc. the principles) of each of its divisions and parts, so that <u>the entire body of architecture might be explained in all its limbs</u> in ten volumes.

Explicatio is crucial to Vitruvius' architectonic conception of the *ars* because it is implicated in a process of demonstrative reasoning that relates the architect's own methods and the principles or accounts (*rationes*) of other disciplines. The unity of architecture in *DA* – that is, its textual *ordinatio* – depends on the a priori intellectual unity of architecture vis-à-vis the nature of the things (*res*) with which architectural expertise is concerned. *Explicatio* is therefore no less fundamental than *ordinatio*.

In this chapter, our principal aim will be to unpack Vitruvius' architectonic notion of the *ars* of architecture and the related project of explanation in *DA*. We will begin by studying the critical opening chapter of *DA* (1.1), in which Vitruvius sets out his views on the "architect's knowledge" (*architecti scientia*). *Architectura* is a composite expertise reliant on the principles furnished by many other fields of knowledge. Its epistemic unity is accounted for by the operation of a set of crucial theoretical terms that include "reasoning" (*ratiocinatio*) and its constituent activities "explanation" (*explicatio*) and "demonstration" (*demonstratio*). This conception of architecture is fundamental to Vitruvius' exposition of architectural doctrine in *DA*, as we will see in the remainder of the chapter. We will consider two case studies: first, city-planning; next, selection of building

[12] Cf. 4.pr.1, 5.pr.5 (both quoted in Ch. 2.2), 6.pr.7, 10.pr.4.

materials. Both case studies show how the explanatory project of *DA* aims to establish the methods of architecture on natural first principles. We will also see, however, that Vitruvius' explanations betray a characteristically Roman perspective on the natural world, underscoring the mutual predication of objective, scientific cognition and the Roman Imperial view of the world outlined in Chapter 2.

Vitruvius' conceptualization of architecture and his explanatory project deserve close study because they are broadly important for the Imperial *artes*. His treatise realizes, for the first time in the Roman period, the model of the higher *artes* for a discipline whose subject matter is inextricably connected with *rerum natura*. Cicero had articulated the relevant aims, but only as intellectual ideals for the complete orator; the project of natural explanation itself was left as a theoretical capacity of the best speaker and never substantially worked out. Other plausible candidates, such as Varro's *De re rustica*, only partially anticipate the expansive refashioning of *ars* in *DA*, as we will see further in Chapter 4. The singular virtue of Vitruvius' work in this domain was thus to present a coherent view of architecture that set understanding and explanation of nature at the center of the *ars* at the same time as it presented architecture as only one among many *artes* that shared these principles. Vitruvius' efforts are thus foundational for the scientific culture of the *artes* that would develop over the long century to follow.

3.1 The Architectonic Art

Vitruvius opens *DA* with three successive accounts of *architectura* (1.1–3). Of these definitional sketches, the first and longest (1.1) assumes the highest level of generality, describing the bases for "the architect's knowledge" (*architecti scientia*). In what follows we will concentrate on this section, which also sheds light on Vitruvius' understanding of *ars* as such. Commentators have spilt much ink on the difficult theoretical claims and distinctions that Vitruvius offers in *DA* 1.1.[13] A significant challenge is the abstract terminology, often of rhetorical or philosophical provenance but highly polysemous, that he employs in scaffolding the art.[14] Although we will not bring all of Vitruvius' statements to

[13] Goguey 1978; Sallmann 1984, esp. 15–22; Knell 1985: 20–30; Romano 1987: 47–80, Frézouls 1989: 39–43; Schrijvers 1989; Geertman 1994: 8–14; Anderson 1997: 5–11; Fleury 2003: lxxxv–xc; Bohlin 2009: 140–4; Fögen 2009: 110–19; Courrént 2011: 23–42; Oksanish 2011: 130–59; Callebat 2017b; Oksanish 2019: 126–40.

[14] On the characteristic abstraction of Vitruvius' language, see for example Callebat 1982: 331–2; Romano 1997: lxxxii–lxxxvi; Courrént 2011: 24; Adams 2016: 189.

complete harmony, we will chart a course through them that emphasizes the insight and coherence of his vision of *ars*.[15]

The first chapter of *DA* (1.1) can be understood as an epistemological reflection unfolding in three connected parts. We will briefly sketch the movement of these parts before examining some key moments more closely. In the first part (1.1.1–3), Vitruvius defines the architect's *scientia* and presents the central thesis that architecture is an "architectonic" *ars* – that is, a special kind of expertise both *of* and *through* other disciplines properly distinct from architecture itself. *Ratiocinatio*, *explicatio*, and *demonstratio* are responsible for synthesizing the principles drawn from these disciplines in the formation of the *ars*. The second part (1.1.4–10) is both exemplification and proof of this conception of architecture, as Vitruvius takes the reader through nine *doctrinae* critical to the *ars*. The third part (1.1.11–18) is an apologia of sorts for this ambitious, multidisciplinary view of *architectura*. In arguing that it is both necessary and possible for the *perfectus architectus* to achieve the educational standard he has set, Vitruvius offers the deepest and most general view of *ars* yet. Here, all *artes* are shown to be filiated at the level of their natural principles (*rationes*), a claim which provides Vitruvius with another vantage point from which to unpack the architectonic character of architecture. Altogether, *architectura* emerges as a form of relational expertise dependent on and, within certain limits, exercising judgment and direction over other disciplines; *ratiocinatio*, in the form of *explicatio* and *demonstratio*, names the crucial theoretical operation that establishes the relationship among these disciplines vis-à-vis their shared principles (*rationes*). Although Vitruvius' architectonic approach is shaped by his appropriation of the original notion of *architectura* as "master" (ἀρχι-) discipline, his epistemology for the *artes* is more general.

3.1.1 Defining Architecture

The very first sentence of *DA* proposes the architectonic view of architecture at stake in the first chapter (1.1.1):

> architecti est scientia pluribus disciplinis et uariis eruditionibus ornata cuius iudicio probantur omnia quae ab ceteris artibus perficiuntur opera.[16]

[15] Contrast Oksanish 2019: 126–40, arguing that Vitruvius gives us a "basically irrational account of knowledge" (134) and speaking of his "opacity on the matter of disciplinary knowledge" (137); cf. Kühnert 1961: 140–2.

[16] **H** (Harleianus 2767) and descendants omit *cuius iudicio probantur omnia*. Only Granger follows, taking *opera* awkwardly as the subject of the following sentence (cf. Fleury 2003: 65–6). More

> The knowledge of the architect is equipped with many disciplines and various fields of study; by his judgment are evaluated all the works which are produced from[17] other arts.

Vitruvius' opening words frame our introduction to the discipline. First, we should notice that he speaks not of *architectura* but *architecti scientia*, which is not merely a periphrasis: Vitruvius is interested in the agency of the *architectus* qua expert and in the precise character of his knowledge (*scientia*).[18] Both concerns are significant, because this *scientia* is first and foremost a capacity, namely, for the judgment (cf. *iudicio*) and evaluation (cf. *probantur*) of "all of the works which are produced by other *artes*." The *architecti scientia* is thus defined not by its subject matter but by relational mastery of *opera* of other *artes*. The order invites us to understand that it is the complex character of the architect's own expertise, "equipped with many disciplines and various fields of study," that underlies this capacity for judgment. Vitruvius' presentation develops the semantic kernel of the word *architectus*, which was taken over in Latin directly from Greek *architectôn* (ἀρχιτέκτων). The basic sense of the Greek word is "master-" or "chief-builder," and the Greek sources accordingly emphasize that the *architectôn* proper was no mere craftsman but director or chief of other craftsmen.[19] Vitruvius builds on this etymological proposition in arguing that the essence of *architecti scientia* really is its "architectonic" character, its directive relation to other *artes*. This is not so much the last word as the first, however.

Much of the remaining chapter represents an attempt to work out the epistemological underpinnings of this relational view of the architect's expertise. Vitruvius continues (1.1.1):

> ea (sc. architecti scientia) nascitur e fabrica et ratiocinatione. fabrica est continuata ac trita usus meditatio quae manibus perficitur e materia cuiuscumque generis opus est ad propositum deformationis.[20] ratiocinatio autem

seriously, the omission obscures the logic of Vitruvius' whole argument in 1.1: it is precisely the judgment (*iudicium*) of other disciplines' products which is at stake (see also Ch. 3.1.3).

17 Taking *ab* as cause or source (Adams 2016: 172–7 on Vitruvius' usage of the preposition).

18 See Oksanish 2011: 136–7; 2019: 123–6, comparing Cicero's focus on the (*perfectus*) *orator*. For Vitruvius' *artifex perfectus* (1.1.3), see Ch. 3.1.2; for Columella's *agricola perfectus*, Ch. 4.2.3 and n.63.

19 Cf. Pl. *Plt.* 259e8–260a7 καὶ γὰρ ἀρχιτέκτων γε πᾶς οὐκ αὐτὸς ἐργατικός ἀλλ' ἐργατῶν ἄρχων, κ.τ.λ.; Arist. *Metaph.* 981a30–b2 τοὺς ἀρχιτέκτονας . . . μᾶλλον εἰδέναι νομίζομεν τῶν χειροτεχνῶν καὶ σοφωτέρους, ὅτι τὰς αἰτίας τῶν ποιουμένων ἴσασιν. For the history of the Latin word, see Callebat 1998, esp. 13–19; for the Greek etymology, Beekes 2010: II 1460.

20 The text from *quae* to *est* has troubled editors. Rose and Müller-Strübing (followed by Gros et al.) adopt Buttman's *qua* for *quae*; Krohn changes *perficitur* to *perficit* and deletes *est*; Fensterbusch tacitly transposes *ad propositum deformationis* before *quae*. But the paradosis is defensible if *perficere* means "brought to its highest point of development" or "made perfect" (*OLD* 3a), a sense applied

est quae res fabricatas sollertiae ac rationis pro portione[21] demonstrare atque explicare potest.

This (sc. the architect's knowledge) is born of both construction and reasoning. "Construction" is the continual and familiar practice of experience that is perfected by manual work from whatever type of material is required according to the design of a drawing. "Reasoning" on the other hand is what is able to demonstrate and explain built things in accordance with skill and reason.

Vitruvius' distinction between *fabrica* and *ratiocinatio* is famous but controversial.[22] It seems clear at least that Vitruvius reprises a traditional dichotomy between a practical or experiential component of *ars*, which for architecture he calls *fabrica*, and a theoretical component, which he calls *ratiocinatio*. The language used to describe *fabrica*, especially *meditatio*, *trita* (<*terere*), and *usus*, evokes terms from the Greek *technê*-tradition – *meletê* (μελέτη), *tribê* (τριβή), and *empeiria* (ἐμπειρία), respectively – often used for the experiential element of expertise.[23] Vitruvius specifies by *quae manibus perficitur*, and so forth, that the empirical formation of the architect requires actual handiwork with the materials. *Ratiocinatio* is presented, on the other hand, as a thoroughly intellectual activity, a capacity "to demonstrate and explain" (*demonstrare atque explicare*).[24] It is associated with book-learning, as Vitruvius indicates just afterwards (1.1.2) when he first substitutes *litterae* for *ratiocinatio* and then puts *ratiocinationes et litterae* together.[25]

Vitruvius goes on to underscore that the complete architect requires both *fabrica* and *ratiocinatio*: those who lack the latter cannot endow their work with authority (1.1.2 *auctoritatem*), whereas those who lack the former pursue a mere "shadow of the thing rather than the thing itself" (*umbram*

also to abstractions such as *ratio* or *ars* (*TLL* X 1.1370.51–1371.22 [Reineke]). Retaining *meditatio* as the subject fits the epistemological context (pace Romano 1987: 51n12).

[21] Some editors prefer *proportione* to *pro portione*; Maufras and Rose and Müller-Strübing (followed by Gros et al.) further alter the text to read *sollertia, ratione proportionis*. Schrijvers 1989: 51 and Callebat 2017b: 133–4 cogently defend *rationis pro portione*.

[22] Callebat 2017b with earlier lit. (cf. 2001).

[23] For *meditatio* = μελέτη ("practice" or "experience"), cf. Watzinger 1909: 203; Fensterbusch 1976: 533; see also *OLD* 4a; *TLL* VIII 572.33–46 (Buchwald). For *tritus* and τριβή (etymologically related: de Vaan 2008: 616), see Fleury 2003: 65–6. See also Ch. 1.5 on "balance."

[24] I gloss over *sollertia* here, though it is not unimportant. See Ch. 6.5.3 on *sollertia* and explanation in Book X (10.16.2, 10.16.12); and for *sollertia* generally in *DA*, esp. König 2009: 43–51; also Romano 1987: 166–7; Traina 1988: 335–6, 347–8; Schrijvers 1989: 51; Callebat 2017b: 134–5. Cf. *DA* 2.pr.1, 7.pr.15.

[25] *Litterae* here refers expansively to knowledge or learning (*TLL* VII 2.1523.42 (v. Kamptz), as Romano 2013: 206 emphasizes (cf. Courrént 2011: 32–5). See also n.39.

non rem). This claim does three things. It projects a balanced view of the *ars* that does not privilege either extreme of theory or practice. It reserves architectural expertise for the practitioner (Vitruvius) alone, establishing his authority against his nonexpert readers.[26] And, finally, it identifies a kind of embodied knowledge that cannot be textualized, and which is thus excluded from *DA*, while not diminishing the importance of such knowledge.[27] After this partitioning, *ratiocinatio*, the theoretical part of *architecti scientia*, remains as the subject of Vitruvius' work.

But what, precisely, is *ratiocinatio*? A demonstrative and explanatory form of cognition, yes. But this only defers the problem. There is a certain opacity to Vitruvius' definition that follows from his choice to define it through two abstract and highly polysemous verbs: *demonstrare* and *explicare*. In the rhetorical tradition, *ratiocinatio* appears before Vitruvius' time as a technical term for "reasoning," either in analysis (educing the *ratio* for something), as in the *Rhetorica ad Herennium*, or in deduction (the syllogism), as in Cicero and others.[28] In recent decades, most scholars have leaned toward the former sense – reflection or analysis – and, because Vitruvius says that it operates on things built (1.1.1 *res fabricatas*), maintain that *ratiocinatio* is an essentially a posteriori, rhetorical activity.[29] Accordingly, the function of *ratiocinatio* is to express and justify the architect's practice in terms comprehensible to the audience. Yet this interpretation is certainly wrong, at least in the restrictive form in which it is usually presented: *ratiocinatio* may sustain a communicative function, but it is not the entirely a posteriori activity many commentators understand.[30] To the contrary, it has a constitutive role in the formation of the architect's multidisciplinary expertise. A first clue is given already at *DA* 1.1.1, where Vitruvius says that *ratiocinatio*, along with *fabrica*, is one of the *sources* of the architect's expertise (*ea* [sc. *architecti scientia*] <u>*nascitur e*</u>), not a mere expression of it. More evidence is found throughout 1.1, as he develops the notions of *ratiocinatio*, *explicatio*, and *demonstratio* and expounds the connection between *architectura* and other *artes*.

[26] Cf. König 2009: 51–2; Oksanish 2016: 279. [27] See Ch. 1, n.77.

[28] For the former, cf. *Rhet. Her.* 4.23 (further *TLL* XI 2.205.24–40 [Spoth]); for the latter, Cic. *Inu.* 1.51, 1.57, Quint. *Inst.* 5.10.6 (further *TLL* XI 2.204.65–205.22). For a third technical meaning "de genere status controversiae," see *TLL* XI 2.204.51–65. Cf. also Schrijvers 1989: 52.

[29] Ferri 1960: 33: "per V. la *ratiocinatio* è semplicemente 'commento' o esposizione della '*fabrica*'," followed by Goguey 1978: 101; Gros 1982: 670; Frézouls 1989: 41 (cf. 1985: 215, 221), Schrijvers 1989: 52; Callebat 1994: 93 (with qualifications); Fleury 2003: 68; cf. Gros et al. 1997: 66.

[30] Dissatisfied with an entirely a posteriori role are Gros et al. 1997: 66; Courrént 2004; 2005; 2011: 27–31; Callebat 2017b: 135–8. Courrént still sees a rhetorical dimension in *ratiocinatio* (e.g. 2011: 30) but views it more expansively as a "système intellectuel" (2004: 25–6, 28).

To get to these *artes*, however, we need to tackle one more difficult distinction in the first part of 1.1 that presents an even more general view of architecture (1.1.3):

> cum in omnibus enim rebus tum maxime etiam in architectura haec duo insunt: quod significatur et quod significat. significatur proposita res de qua dicitur; hanc autem significat demonstratio rationibus doctrinarum explicata.
>
> As in all other matters, so certainly too in architecture there are these two things: that which is made known, and that which makes known. What is made known is the established subject of speech; and what makes this known is a demonstration explained on the basis of the principles of fields of learning.

This new dichotomy cannot be mapped directly onto that of *fabrica* and *ratiocinatio*, if only because the drift of the present distinction is cognitive and semantic and hence elides the manual activity of *fabrica*.[31] Yet this dichotomy may be related to the earlier if we see it as an effort to pull us up from the level of architecture qua discipline to that of *ars* in general. At this higher level of abstraction, *architectura* is susceptible to the same bipartite analysis as "all other matters" (*omnibus . . . rebus*). We should observe that Vitruvius' key term *significare* and its double deployment in the active and passive are attested elsewhere in (Stoic) definitions of dialectic.[32] This should not surprise if we recall that dialectic was regarded by Republican authors as the *ars* responsible for the constitution of other *artes*.[33] It would thus be highly appropriate that Vitruvius appeal to the vocabulary of dialectic when describing the epistemological structure of *architectura* in the most general possible way.

If we attempt a looser correlation between this dichotomy and *fabrica* and *ratiocinatio* on this dialectical assumption, there is much to learn. *Fabrica* itself may have dropped out, but the products of *fabrica* – that is, *res fabricatas* (1.1.1) – remain in *quod significatur*, which is defined as *proposita res de qua dicitur*. This would include the subject matter of architectural theory and discourse, not limited however to things actually

31 The dichotomy has received many, divergent interpretations that connect it more or less closely with what precedes: for example, Watzinger 1909: 204–6; Pellati 1927; Ferri 1951: 210–11; 1960: 34 (followed by Gros 1982: 670; Schrijvers 1989: 52; Fleury 2003: 69–70); Fensterbusch 1976: 534; Romano 1987: 55–6; Callebat 1994: 91–3; Gros et al. 1997: 67–8.

32 Cf. esp. Diog. Laert. 7.62 = E–K F188 = *SVF* II 122 (σημαίνειν); also Sen. *Ep.* 89.17, Quint. *Inst.* 3.5.1, Sext. Emp. *Math.* 8.11, Diog. Laert. 7.43. See further Ferri (previous n.); Edelstein and Kidd 1972–99: III 684–6.

33 Ch. 1.2, and n.14.

built but also embracing the architect's plans, blueprints, and so on (hence *proposita*).[34] The other half of the new dichotomy, *quod significat*, is joined more tightly to *ratiocinatio*. The repetition of *demonstrare* (~*demonstratio*), *explicare*, and *ratio* invites us to regard the predication of *quod significat* and *demonstratio rationibus doctrinarum explicata* as another view of *ratiocinatio* (cf. 1.1.1). If we accept this easy equation, we win two conclusions. First, *demonstratio* is apparently distinguished from *explicatio*: the former appears as demonstration in the sense of a communicative "pointing out," while the latter remains "explanation" proper. Second, and more importantly, *rationibus doctrinarum* connects this act of explanation (cf. *explicata*) with a set of principles drawn from, or shared by, other branches of specialized knowledge (cf. *doctrinarum*). Through the dialectical formulation *quod significat*, Vitruvius thus locates the multidisciplinary character of architectural *scientia*, announced in the first line of the work (*pluribus disciplinis et uariis eruditionibus*), in *ratiocinatio*. In other words, *ratiocinatio*, and more specifically *explicatio*, emerges as the crucial site for the architectonic construction of *architectura*, connecting the architect's expertise with the disciplines whose products he "judges" (*iudicare*) and "evaluates" (*probare*).

3.1.2 The artes *of Architecture*

These conclusions might seem to overstate what we can infer from the allusive opening of 1.1, but they are corroborated later in the chapter (not to speak of other books of *DA*). The mention of *doctrinae* (1.1.3) thus prompts Vitruvius to turn to a new subject: the formation of the architect's polymathic *scientia*. The architect who will successfully employ *ratiocinatio* so as to be the *perfectus artifex* (1.1.3) must benefit from both natural ability (*ingenium*) and training (*disciplina*).[35] He then offers a list of nine *doctrinae* in which the architect must be trained (1.1.3):

> et ut litteratus sit, peritus graphidos, eruditus geometria, historias complures nouerit, philosophos diligenter audierit, musicam scierit, medicinae non sit ignarus, responsa iurisconsultorum nouerit, astrologiam caelique rationes cognitas habeat.

[34] Cf. Fensterbusch 1976: 534.

[35] See Gros et al. 1997: 68 on the Ciceronian background to this pairing (esp. *De or.* 1.113–18); cf. Goguey 1978: 101–2.

> And (sc. therefore) he must be acquainted with literature, experienced in drafting, learned in geometry; derive an acquaintance with a great many works of history; assiduously listen to the philosophers; have knowledge of music; be well versed in medicine; familiarize himself with the responses of the jurisconsults; and understand astronomy and the principles of the heavens.

As it turns out, this list is merely a preamble to 1.1.4–10, where Vitruvius returns to each *doctrina* and spells out its relevance for architecture: letters, drafting, and geometry in 1.1.4; history in 1.1.5–6; philosophy (both natural and moral) in 1.1.7; music in 1.1.8–9; and medicine, law, and astronomy in 1.1.10.[36] It is not insignificant that Vitruvius continues to represent these disciplines as connected with the *architectus* himself – that is, as fields in which the architect must be educated – rather than with *architectura* taken *in abstracto*: this perspective retains the focus of the opening of *DA*, in which the architect as master-craftsman is defined through his peculiar capacity to evaluate the products of other disciplines.

This second part of *DA* 1.1 has attracted considerable interest. It is routinely pressed into service to support theses about the "encyclopedic" preoccupations of Roman education in the late Republic.[37] It is also sometimes seen as an aspirational effort to make architecture appear a socially and intellectually worthy discipline.[38] Vitruvius' interdisciplinary approach reflects the intellectual concerns of the *artes* (Chapter 1), but to go further runs the risk of distorting his program in *DA*. His appeal to these disciplines is more than a nod to an educational ideal: it is a serious advertisement for the architectonic conception of architecture which informs the whole of his work. Hence, these disciplines are not ambitiously grafted onto *architectura* but rather set down as its natural roots, such that the architect's *ratiocinatio* as well as Vitruvius' explanatory project in the *DA* are unrealizable without them. We will see as much later, but an easy proof can be adduced here: all the *doctrinae* discussed in 1.1.4–10 are invoked implicitly or explicitly at some point in *DA* for the purpose of explaining architectural method. Consider Table 3.1, which correlates (1) each of the nine *doctrinae* of 1.1.3 with (2) Vitruvius' explanation of their relevance in 1.1.4–10 and (3) a representative example of the use of the *doctrina*.

[36] 1.1.4–10 is introduced by 1.1.4 *quae cur ita sint, haec sunt causae*, which points back to 1.1.3 *et ut*, etc.

[37] Cf. Ch. 1, n.5. For Vitruvius in this light, see for example Watzinger 1909: 208–9; Brown 1963; Goguey 1978: 102–9; Sallmann 1984: 15–21; Romano 1987: 47–80; Geertman 1994: 11–13; Callebat 1997; Romano 2003b: 103–8; Masterson 2004: 392–5; Fögen 2009: 110–19; Romano 2013; Courrént 2019: 134–48, 281–308.

[38] Previous n.; for appropriate skepticism, Oksanish 2019: 120–1, and Ch. 1, n.167.

Table 3.1 *The disciplines of Vitruvian architecture* (DA *1.1.4–10*)

Doctrina	Vitruvius' statement of its relevance in *DA* 1.1	Representative example from elsewhere in *DA*
Litterae (1.1.4)	The use of *commentarii*[39]	7.pr.11–15: Precepts of Greek and Latin architectural authors
Scientia graphidis (1.1.4)	Illustration of architectural plans	Ten figures accompanying *DA* (now lost)[40]
Geometria (1.1.4)	(a) The use of compass (*circinus*) and rule (*euthygrammum*)	(a) 1.6.6–7, 1.6.12–13: Orientation[41]
	(b) Optics (*optice*)	(b) 6.2: Proportion and symmetry in construction
	(c) Arithmetic (*arithmetice*), incl. accounting and measurement	(c) 6.3: Dimensions in construction
Historiae (1.1.5–6)	Justification of design[42]	7.5.5–7: Apaturius and Licymnius on fresco styles
Philosophia (1.1.7)	(a) Moral philosophy (virtue and good reputation)	(a) 3.pr.1–3, 6.pr.5–7, 10.pr.1–4: Ethical standards in profession
	(b) Natural philosophy (φυσιολογία)	(b)
	(i) Varied applications	(i) 2.3–10: Properties of materials from their elements[44]
	(ii) Water supply (*aquarum ductiones*)	(ii) 8.6: Water supply
	(iii) Engineering and mechanics[43]	(iii) 10.3: Motion of machines
Musice (1.1.8–9)	(a) Torsion war engines	(a) 10.12: Tuning war engines
	(b) Resonating vessels (1.1.9 ἠχεῖα)	(b) 5.4–5.5: Resonating vessels in theaters
	(c) Hydraulic machines	(c) 10.8: Ctesibius' water organ
Medicina (1.1.10)	Knowledge of healthful airs, waters, and places	1.4: Healthful site for a city
Iura (1.1.10)	Conforming to law in architectural practice[45]	2.8.17: Legislation concerning thickness of walls
Astrologia (1.1.10)	Construction of clocks (*horologia*)	9.7: Description of analemma

Nota bene that column 3 often offers only one among many possible examples, and again that Vitruvius often employs a given *doctrina* for more purposes than announced in 1.1.4–10 (although these are not addressed in the table).

39 1.1.4 *litteras architectum scire oportet, uti commentariis memoriam firmiorem efficere possit*. Some scholars argue *litteras scire* refers only to literacy, others that it involves higher literary cultivation (Romano 2013, esp. 201–3); here an example of the former, utilitarian purpose is given.

40 These may not be precisely what Vitruvius has in mind with the architect's ability to *operis speciem deformare* (1.1.4) but evidently share a skillset. For the figures in *DA* (none extant), see Haselberger 1989; Fleury 2003: lxii–lxviii; Corso 2016: 15–25; Gros 1996 is classic, but see Cuomo 2016b: 135–6.

41 See Ch. 3.3 on the windrose.

42 1.1.5 *historias autem plures nouisse oportet, quod multa ornamenta saepe in operibus architecti designant, de quibus* (sc. *ornamentis in operibus*) *argumenti rationem cur fecerint quarerentibus reddere debent.*

43 *Machinatio* (1.3.1), here given indirectly as *Ctesibii aut Archimedis et ceterorum qui eiusdem generis praecepta conscripserunt*.

44 See Ch. 3.4 on physical explanation and timber.

45 See Callebat et al. 2003: 144 on legislation concerning architecture.

3.1.3 An Interdisciplinary View of ars

Following the enumeration of the disciplines required by the architect, Vitruvius feels entitled to claim again that "this great discipline is adorned with and abounds in many and various fields of study" (1.1.11 *tanta haec disciplina sit condecorata et abundans eruditionibus uariis ac pluribus*). Accordingly, only those "nourished by a knowledge of much book-learning and many arts" (*scientia plerarumque litterarum et artium nutriti*) can reach the "highest sanctum of architecture" (*summum templum architecturae*).[46] Vitruvius' lofty educational demands set the key problem of the third part of the chapter (1.11–18): a defense of the feasibility of this multidisciplinary view of *architectura*. Is it not too much to expect that the would-be architect acquire knowledge in so many disciplines?[47] Vitruvius' response affords him the occasion for three important moves. First, he unpacks his interdisciplinary view of the *artes*. Second, by appealing to *ratiocinatio* he explains how an expert in one *ars* engages with other *artes*. Third, he underscores pragmatic limitations on the expertise the architect needs to accomplish the aims of his practice. Throughout, Vitruvius uses the opinions of the fourth-century Greek architect Pytheus, builder of the temple of Athena at Priene, as a foil to his own account.[48] We will go through each of these moves in turn.

Vitruvius' concern that the educational requisites for the architect may be off-putting in their steepness and broadness is traditional, echoing worries that Cicero had voiced for oratory and that would be reprised in other Imperial *artes*.[49] It is perhaps fitting, therefore, that Vitruvius also appeals to the notion of a common bond among the *artes*, found first in a Roman setting in Cicero, to defend himself: "all disciplines are joined in their subject matter and mutually connected" (1.1.12 *omnes disciplinas inter se coniunctionem rerum et communicatio habere*).[50] This claim extends of course beyond architecture proper to include all of what we called the "higher *artes*" in Chapter 1. He continues: "encyclic education is composed, like a body, from these as its members" (*encyclios… disciplina uti*

[46] Vitruvius often uses *templum* in the sense of a consecrated (augural) space, perhaps underlying the metaphor here (Morgan 1906b: 12–13; cf. Fleury 2003: 93); McEwen 2003: 28–30, 148, 301 speculates that the actual temple of Capitoline Jupiter is meant. Oksanish 2019: 126–7 compares Varro, *Ling.* 5.8 (cf. 2011: 6, 107–9, 227).

[47] Cf. 1.1.12 *at fortasse mirum uidebitur*, and so forth.

[48] Pytheus is mentioned 4× in *DA* (1.1.15, 4.3.1, 7.pr.12 *bis*); his writings on the Mausoleum at Halicarnassus, of which he was the architect, are mentioned at 7.pr.12. On him see *RE* (1) XXIV 371.21–513.18 (Ziegler); Svenson-Evers 1996: 116–50.

[49] For the Ciceronian background, see Ch. 4.2.3 and n.63 (Columella on the *agricola perfectus*).

[50] See Ch. 1.3 on *Pro Arch.* 2, *De or.* 3.21.

corpus unum ex his membris est composita). Hence, those who are trained in the full range of *eruditiones* since their youth "recognize the same marks in all learning from books and a mutual connection among all disciplines" (*omnibus litteris agnoscunt easdem notas communicationemque omnium disciplinarum*). Vitruvius emphasizes that the interrelations of the *artes* relieve the architect's educational burden: it is "easier" (*facilius*) to acquire knowledge of all these disciplines, since learning in one can be transferred to another. A more detailed justification for this claim will follow. In the comparison of "encyclic education" (whatever that precisely means) to a *corpus* and the subsidiary fields to its *membra*, we may also detect the archetype for Vitruvius' figuration of architecture as a *corpus* and its departments as *membra*.[51] We now also realize that the interdisciplinarity of architecture is not unique to the discipline, but a property of the whole constellation of higher *artes* which are united by the common bond. This point of convergence between architecture and the *artes* is surprising and, in some respects, challenging. After all, Vitruvius introduced other *doctrinae* to explain the architectonic character of architecture – but we now learn that this interdisciplinarity belongs to all the *artes*. What remains for architecture? We will return to this important point shortly.

Vitruvius' second move is prompted by a comparison of his own theory of architecture with that of Pytheus, who (according to Vitruvius) argued that the architect had to be capable of more "in all the arts and disciplines" (1.1.12 *omnibus artibus et doctrinis*) than even their respective experts. Vitruvius rejects this claim as incredible (1.1.13–14). His restatement of his own position is exceedingly interesting, both for how it develops a general conception of *ars* and for the role attributed to *ratiocinatio* (1.1.15):

> igitur in hac re Pytheus errasse uidetur quod non animaduertit ex duabus rebus singulas artes esse compositas, ex opere et eius ratiocinatione. ex his autem unum proprium esse eorum qui singulis rebus sunt exercitati, id est operis effectus, alterum commune cum omnibus doctis, id est ratiocinationem.[52]
>
> Therefore Pytheus seems to have erred in this matter because he does not recognize that each art is constituted by two things, its "work" (*opere*) and the "reasoning" (*ratiocinatione*) about it; nor that, on the basis of these, there

[51] For the disputed meaning of ἐγκύκλιος παιδεία, see Ch. 1, n.5; for *corpus* as implying order, Ch. 2, n.11.

[52] *ratiocinationem* is conjectured for the transmitted *rationem*, which I once defended (Zainaldin 2020b: 59n130). I now believe emendation is required by grammar and sense: *unum* … *alterum* suggest a restatement of the two parts of the *ars* (*ratiocinatio* and *opus*), and Vitruvius is unlikely to have substituted *ratio* as a synonym for *ratiocinatio* when he has hitherto carefully distinguished between them as "principle" and "reasoning (from principles)."

> is the one that belongs to those who are practiced in the particular matters (sc. of an art) – that is, the production of the work (*effectus operis*) – and the other that is held in common with all learned people – that is, the reasoning (*ratiocinationem*).

The *coniunctio* or *communicatio* among disciplines obtains precisely in *ratiocinatio*. Pytheus' mistake was to extend his claim to *opera*, the particular "works" of different disciplines, when he should have spoken only of *ratiocinatio* ("reasoning"), which is "held in common with all learned people." The role now attributed to *ratiocinatio* also squares with his remarks earlier in the chapter, where Vitruvius maintained that at the highest level of dialectical analysis, all *artes* involve *demonstratio rationibus doctrinarum explicata* (1.1.3), an expression that we noted paraphrases *ratiocinatio*. Vitruvius is affirming once more that all *artes* rely on the same cognitive and linguistic process in explaining (*explicare*) and demonstrating (*demonstrare*) their subject matter – that is, on *ratiocinatio* – even where the subject matter is different.

Vitruvius offers three examples to illustrate his claim that *ratiocinatio* is shared while *opera* are not: doctor and musician, musician and astronomer, and astronomer and geometer (1.1.15–16).[53] The first example is the fullest and most illuminating (1.1.15):

> uti medicis et musicis est (sc. ratiocinatio) de uenarum rhythmo et ad pedes motu: at si uulnus mederi aut aegrum eripere de periculo oportuerit, non accedet musicus, sed id opus proprium erit medici; item in organo non medicus sed musicus modulabitur ut aures suauem[54] cantionibus recipiant iucunditatem.

> For example, physicians and musicians share (sc. reasoning) about the rhythm of blood vessels and motion on the feet. But if one must heal a wound or rescue a sick person from danger, it is not the musician who will do, but that work belongs rather to the doctor; likewise, in the case of a musical instrument it is not the physician but the musician who will play it so that the ears receive a pleasant sweetness in song.

Although the text of the first sentence is difficult,[55] the idea is that physicians and musicians share *ratiocinatio* (the natural subject for *est*) as regards the pulse and the metrical foot. This claim alludes to a theory of

[53] The third pair is uncertain (Fleury 2003: 101). For musician and astronomer, cf. (among the land-surveyors) Hyg. *Const.* L 184.14–185.9 = T 148.12–149.4, discussed in Ch. 7.5.3.

[54] Choisy's conjecture (MSS *suae*).

[55] With Gros et al., I follow Rose and Müller-Strübing, rather than Fensterbusch and Fleury (who print *de uenarum rhythmo et [ad] pedum motus*). For attempted solutions, see also Schievenin 2012; none of the plausible emendations affects the general import of the comparison.

sphygmology associated with the third-century Hellenistic physician Herophilus, who borrowed metrical descriptions for the motion of the pulse.[56] On the strength of this theory, Vitruvius can say that both physician and musician reason with the help of a quantitative analysis of natural motion into rhythmical units: thus, *ratiocinatio* is employed by both. But they apply this analysis to the objects peculiar to their own expertise – patients and musical instruments, respectively – and no one would confuse the physician's ability to produce health with the musician's to produce euphonious sound: thus, their *opera* are different. Significantly, Vitruvius also calls the overlap among *artes* that extends through *ratiocinatio* a *disputatio communis*, "shared disputation" (1.1.16), and says that "all branches of specialized knowledge share much or even all of their subject matter so far as disputation goes" (*omnibus doctrinis multae res uel omnes communes sunt dumtaxat ad disputandum*). The notion of *disputatio communis* suggests not only mutual intelligibility but also the potential for influence and exchange, much as the physician (Herophilus) found in music the inspiration for a quantitative analysis appropriate to his own *opera*.

Ratiocinatio as demonstrative and explanatory "reasoning" is in this sense common to all *artes*. In this meaning it is not a countable noun. But complicating the picture is some evidence that Vitruvius believes *ratiocinatio* can be differentiated within and across disciplines, just as *opera* are. A hint for differentiation *within* a discipline is found in the passage quoted earlier in this section, 1.1.15 *ex opere et eius ratiocinatione*: the pronoun *eius* (the genitive is probably objective) picks up *opere* and attaches *ratiocinatione* closely to it. Does this imply that distinct *opera* within the same *ars* have distinct *ratiocinationes*? Vitruvius does not answer this question directly, but it seems probable, given his usage in the plural elsewhere – including at the end of the chapter, when he announces of architecture his intention to present "the power of the art and the *ratiocinationes* which are in it" (1.1.18 *de artis . . . potestate quaeque insunt in ea ratiocinationes*).[57] There is even stronger evidence for *ratiocinationes* distinguished *across* disciplines earlier in 1.1, when Vitruvius contests Pytheus' view of architecture by retorting that "it is not possible to achieve individual excellence in such a great variety of subjects, because it does not fall within a person's power to know and understand their *ratiocinationes*"

[56] Cf. Plin. *HN* 29.6, Cens. 12.4–5; further, von Staden 1989: 276–82.

[57] Further examples of plural *ratiocinationes* in reference to specific subjects within architecture: 1.3.2, 4.8.7, 5.3.8, 6.pr.7, 10.8.1.

(1.1.13 *non ... in tantis rerum uarietatibus elegantias singulares quisquam consequi potest, quod earum ratiocinationes cognoscere et percipere uix cadit in potestatem*). Here, in the plural and attached to individual disciplines, the word obviously means something different than *ratiocinatio* as the unitary "reasoning" underpinning the *communicatio* of the *artes*. The plural cases can be reconciled, however, by the economical supposition that Vitruvius uses the term not only to refer to "reasoning" as a cognitive and linguistic process but also to the worked-out results of reasoning – that is, the specific demonstrations and explanations that furnish the *rationes* for a given *opus*. While all experts share *ratiocinatio* in the former sense, they do not necessarily share specific *ratiocinationes* in the latter sense. To return to the physician and the musician, we might say that while they both employ reasoning with or about (*ratiocinatio*) the same quantitative (metrical) principles (*rationes*), their precise explanations in connection with music and health, their worked-out *ratiocinationes*, are different.

Some such constraint on the sharing of *ratiocinatio* as just suggested is in fact required by Vitruvius' third move, as he both delimits and defines the architect's expertise. He argues (1.1.16):

> ergo satis abunde uidetur fecisse qui ex singulis doctrinis partes et rationes earum mediocriter habet notas eas quae necessariae sunt ad architecturam uti, si quid de his rebus et artibus iudicare et probare opus fuerit, ne deficiatur.
>
> Therefore the architect seems to have done enough who possesses a moderate knowledge of the parts of each discipline as well as the principles of them which are necessary for architecture such that, whenever he is required to judge and evaluate anything pertaining to these subjects and arts, he would not be found wanting.

Vitruvius' statement solves some problems occasioned by his account of *architectura*. In the first place, he puts limitations on the knowledge of other disciplines required by the architect. Pytheus' thesis was rejected because it implied that the architect has control over the *opera* of other disciplines; *ratiocinatio* is rather the bridge between different fields. Even so, the architect's *ratiocinatio* cannot entail complete theoretical mastery of other fields, or Vitruvius would be little better off. Rather, *ratiocinatio* between disciplines manifests itself only as a *disputatio communis* – that is, a capacity for discussion, understanding, and exchange at the level of principles (*rationes*), and the *ratiocinationes* specific to other fields will fall at least in part outside the architect's expertise. That is why Vitruvius says here that the architect need acquire only a "moderate" (cf. *mediocriter*) knowledge of the *partes* and *rationes* of other disciplines. He reiterates this claim just after (1.1.17); the limiting principle is, sensibly, what is "necessary" (1.1.16 *necessariae*) for architecture. Vitruvius further specifies that the "matters and arts" with which the architect must be acquainted are just those covered in 1.1.4–10 (hence 1.1.16 *his rebus et artibus*), meaning only the higher *artes* and their contents relevant for architecture and not absolutely all *artes*. These concessions render plausible the ambitious conception of architecture developed to this point without sacrificing the multidisciplinary character of its expertise.[58]

Vitruvius' admission does double duty, however, because it also restates the special place afforded architecture vis-à-vis other *artes*. We asked earlier in this section what distinguishes *architectura* from other *artes* if they are all in some fashion interdisciplinary. Vitruvius answers this question here, although he is in a sense only fleshing out a thesis we have already met: architecture is not merely *interdisciplinary*, it is also *architectonic* in that it uniquely guides the application of other *artes* to particular ends. Hence, Vitruvius says again that the architect's interdisciplinary knowledge allows him to "judge" (1.1.16 *iudicare*) and "evaluate" (*probare*) things pertaining to other *artes*, echoing the opening sentence of *DA* on the *architecti scientia* (cf. 1.1.1 *iudicio probantur*). Architecture is not unique in its filiation with other *artes* through *ratiocinatio*, but in the evaluative capacity that instrumentalizes *ratiocinatio* in relation to the *opera* of other *artes* relevant for architectural purposes: in other words, the judgment (*iudicium*) of the architect makes *ratiocinatio* actionable by bringing the contents of other disciplines into his own service. This does not destroy the distinction between *ratiocinatio* and *opus* proposed in response to Pytheus or imply, in a Pythean fashion, that the architect can produce the *opera* of other *artes*. The manufacture of these works indeed falls outside the architect's ability; it is just that *ratiocinatio* allows him a grasp of the principles (*rationes*) of the *opera* sufficient to understand how and for what purposes to employ them. This is the architectonic significance of the "master art" of architecture.

From this vantage point, the long first chapter of *DA* is ultimately an attempt to work out the complex definition of architecture offered at the very outset: "The knowledge of the architect is equipped with many disciplines and various fields of study; by his judgment are evaluated all the works which are produced by other arts" (1.1.1). To support this claim, Vitruvius offers an epistemological analysis of *architectura* as well

[58] Compare the position of the Heronian μηχανικοί reported by Pappus (n.5).

as *ars* as such, which brings to light the crucial importance of *ratiocinatio* in the architectonic constitution of the discipline. *Ratiocinatio* is not, as many have supposed, the mere rhetorical communication of the architect's products, but rather a constructive form of reasoning that demonstrates (*demonstrare*) and explains (*explicare*) the subject matter at the level of the principles (*rationes*) shared by all disciplines. Hence, *ratiocinatio* is the basis for the common bond (*coniunctio* or *communicatio*) among the *artes*, through which the architect acquires the capacity to judge (*iudicare*) and evaluate (*probare*) the products of other disciplines. Vitruvius' investigation has consequences for the *artes* more generally: his approach to the discipline indeed posits a whole constellation of higher *artes* which are filiated in their theoretical (explanatory) content. This is ultimately a fresh attempt to synthesize the disparate elements of specialized knowledge that had emerged in the intellectual culture of the late Republic into a coherent notion of *ars* and, more to the point, to work out the implications of this notion for *architectura*. Vitruvius is not only witness but also contributor to developments in the *artes* at the beginning of the Roman Empire.

3.2 Explaining Architecture

Vitruvius' prefatory sketch of the art of architecture is only half of the story of his reconceptualization of the discipline in *DA*, however. To appreciate the project in its totality, we must consider the extent to which he makes good on his ambitious vision. There is indeed a special imperative to do so, because the literary and rhetorical turn in Vitruvian studies, despite all it has done to improve our picture of *DA*, has tended to prioritize the work's prefaces, philosophical excursus, and historical anecdotes to the exclusion of the architectural method and doctrine. A basic contention throughout this book is that we can only understand the *artes* if we read the self-conscious reflections of the prefaces *with* and *against* the technical and didactic contents of the works.[59] This effort is especially rewarding for *DA*. In a narrow way, it addresses some questions left outstanding from the preface. More broadly, it reveals an essentially novel application of *ars* to the understanding of the world: in Vitruvius' hands, the architectonic *ars* of architecture becomes a vehicle for inquiry into the *rationes* that underpin nature and society.

[59] See also the Introduction.

In the remainder of this chapter, we will examine two cases studies from *DA* illustrative of Vitruvius' explanatory mode of discussion. But first I suggest a few problems to motivate these readings. In studying the preface we concluded that *ratiocinatio* and its elements *explicatio* and *demonstratio* are responsible for the interdisciplinary relationship of architecture and other *artes*; that they are not wholly a posteriori activities but are instead sources of the architect's expertise; and that this expertise consists not in absolute mastery of other *artes* – a thing impossible – but in a theoretical acquaintance sufficient to allow the argument to evaluate the products (*opera*) of other *artes* for his own purposes. The general picture is clear, but many questions remain: What do the relevant principles (*rationes*) shared by the *artes* look like? What counts as explanation and demonstration? How do explanation and demonstration inform the architect's judgment and evaluation of other disciplines' products? Where exactly do the limits on the architect's interdisciplinary knowledge stand, and what kind of exchange with other *artes* does *ratiocinatio* as "shared disputation" (*disputatio communis*) sustain?

While Vitruvius uses the terms *ratiocinatio*, *explicatio*, *demonstratio*, and *ratio* in illuminating ways throughout *DA*,[60] we will eschew a strictly lexical approach in favor of a conceptual study of his project. *DA* is full of explanations of architectural method that depend on the *doctrinae* enumerated in his preface. Indeed, the explanations come so densely that commentators have at times charged Vitruvius with inept digressiveness. As we will see, these complaints are off the mark if we discard modern presuppositions about what Vitruvius' architectural treatise should look like and take his own architectonic model of the discipline seriously. The often lengthy theoretical explanations are precisely examples of *ratiocinatio*, the reasoning from principles, with which the architect justifies his selective use of *opera* of other disciplines. Two cases will make this point. First, we will consider Vitruvius' recommendation for the proper orientation of a city in *DA* 1.6; next, we will study his discussion of different woods as building materials in 2.9–10. As we will see, Vitruvius' explanations assert and argue theories drawn from other branches of specialized knowledge and deploy various forms of scientific reasoning: in this way, they communicate an understanding of the natural world as much as they

[60] See Ch. 3.4 on 2.1.9; cf. 4.2.5–6, 7.5.4, with Courrént 2011: 205–32; Oksanish 2019: 85–8; further, Courrént 1999a; 2004: 17–25; 2005; Meyers 2005: 74–5; Callebat 2017b: 132–3, 37; and the index of Courrént 2011. Grammatical connectors, etc., aside, *ratio* is the most common word in Vitruvius' treatise (331×: Courrént 2004: 17n10); its usage may be clarified by inspecting the *TLL* article (esp. XI 2.180.30–51, 183.61–184.19, 189.69–190.45, 197.15–32 [Wick-Beikircher]).

justify architectural practice. We will also see that, despite the universalizing character of Vitruvius' scientific approach, his *ars* remains an *ars* as conceived from the standpoint of a Roman knower.

3.3 Putting the City in Order

After the definitional chapters of Book I (1.1–3), Vitruvius discusses city-planning: the choice of a salubrious location (1.4); the construction of walls and fortifications (1.5); the proper orientation of the city (1.6); and the disposition of certain public buildings and spaces (1.7).[61] These town-planning chapters also afford Vitruvius the occasion to introduce physical principles to which he will appeal throughout *DA* and to model the kind of explanations which are the bread and butter of his work. We will concentrate on his recommendations for the proper orientation of the city (1.6), which illustrate the style of Vitruvius' typical explanation: an argument, or complex of arguments, premised on principles (*rationes*) furnished by other branches of specialized knowledge (*doctrinae*). These principles are taken as a secure and natural basis for the explanation, so that their connection with Vitruvius' method at the same time justifies it. While Vitruvius does not in most cases attempt to establish these principles independently, it is a hallmark of his explanatory strategy to use observation, experiment, and analogy in a *disputatio communis* with other fields to strengthen his account. Such explanations ultimately exhibit the architect's *ratiocinatio* and, along with it, his capacity for the judgment and evaluation of the *opera* of other disciplines.

Vitruvius' subject in 1.6 is "the direction of the streets and alleys with respect to the region of the sky" (1.6.1 *platearum ... et angiportuum ad caeli regionem directiones*). Although he does not directly address the organization of the roads themselves, his discussion is best understood assuming a rectangular grid of thoroughfares or larger "streets" (*plateae*) and narrower cross-streets or "alleys" (*angiportus*) intersecting the *plateae* at right angles.[62] (More on this assumption later.) Vitruvius argues that the streets and alleys, and thus the city as a whole, "will be correctly directed if winds are carefully kept out of the alleys" (1.6.1 *dirigentur haec* [sc. *plateae et angiportus*] ... *recte*

[61] Further Gros et al. 1997: 8–9; Fleury 2003: xciv–ciii.

[62] The standard reconstruction of Vitruvius' city plan (popular in the Renaissance) was radial: Lagopoulos 2009: 207–12. (This plan is still followed by Granger 1931–4: I, pl. A; Plommer 1971: 161.) However, the grid plan better fits the text and is almost certainly correct in its essentials: Hamberg 1965, esp. 106–16; Fensterbusch 1976: 538 and pl. 4; Knell 1985: 38–43; Haselberger 1999: 90–1; Rowland and Howe 1999: 167–8; Fleury 2003: xcix–cii, 146–9, 186–8; Lagopoulos 2009, esp. 212–28. Grid-based reconstructions differ in certain details; I follow Fleury (cf. esp. pl. 4).

si exclusi erunt ex angiportis uenti prudenter). After defending this claim (1.6.1–3), he explains how the orthogonal grid of streets and alleys should be oriented to achieve this effect (1.6.4–5, 8–11): with the help of a windrose, the architect rotates the grid obliquely to the cardinal and ordinal directions, from where the winds arise, so that the winds are blocked by the walls and cannot enter the streets and become trapped in the alleys. Vitruvius offers a procedure (1.6.6–7) for constructing the windrose that his scheme requires (a duplicate procedure is given at 1.6.12–13[63]). These prescriptions and procedures are outlined as in Table 3.2 to reveal the structure of *ratiocinatio* in this chapter of *DA*.

Table 3.2 *The explanatory economy of* DA *1.6, on the orientation of a city*

Argument I The city will be oriented correctly if the winds are excluded from the alleys (*angiportus*) (1.6.1).

Explained by Direct exposure to winds is harmful to health (1.6.1).

Explained by (1) Health is the maintenance of the natural proportion of elements of which the body is constituted.[64]

(2) Contact with wind produces a (harmful) change in the natural proportion of elements in the body (1.6.3).

Explained by Wind is a flowing wave of air with chaotic motion (1.6.2).

Argument II The winds will be excluded from the alleys (*angiportus*) if the streets (*plateae*) and alleys are turned obliquely to the cardinal and ordinal directions (1.6.8).

Explained by There are eight primary winds, whose origins are respectively located in the eight regions of the world centered on the cardinal and ordinal directions (for the architect's determination of the directions, see the **Procedure**).

Explained by (1) The empirical evidence agrees with the most authoritative researchers, who have concluded that there are not four but eight winds, and that their origins are on the compass points (1.6.4–5).

(2) Other winds besides these eight are not primary winds but permutations of the primary winds (1.6.9–10), or else local winds (1.6.10) or temporary breezes (1.6.11).

Procedure The geometric construction of a windrose with the aid of a gnomon will indicate the cardinal and ordinal directions, hence the regions where the winds arise (1.6.6–7, 1.6.12–13).

[63] Fleury 2003: cv–cvi argues the duplicate is authentic but was dislocated in transmission; Plommer 1971, who reconstructs Vitruvius' city plan radially (previous n.), excises 1.6.12–13 after *erit autem* as inept explanation of an inauthentic diagram.

[64] This lemma must be supplied from Vitruvius' remarks earlier in *DA* 1 (see below).

This outline is artificial in that it prioritizes the logical structure of Vitruvius' explanations above the narrative order. It is nonetheless a useful schematization of *ratiocinatio*: of note is how the larger prescriptions for architectural method can be seen to depend on more basic explanations that are grounded eventually in physical theory, as one moves rightward/downward in the outline. Thus, *ratiocinatio* ensures that the architect's design choices are epistemically secure because they are ultimately derived from natural first principles. Let us consider each of the main arguments in greater detail.

3.3.1 Winds and Health: Vitruvian Physics and Medicine

Vitruvius' first order of business (**Argument I**) is to justify his claim that the proper orientation of a city is the one that will keep winds out of the alleys (*angiportus*). The basis for this view is the supposition that direct exposure to winds is inimical to health: "if they are cold, they damage; if hot, they corrupt; if moist, they injure" (1.6.1 *qui si frigidi sunt, laedunt; si calidi, uitiant; si umidi, nocent*). The Greek town of Mytilene in Lesbos exemplifies the harm that results when the winds are not averted, but Vitruvius does not defend his prescription merely by assertion and example; rather, he carefully explains it on the principles that govern the winds (1.6.2) and human health (1.6.3). Let us consider first Vitruvius' theory of wind (1.6.2):

> uentus autem est aeris fluens unda cum incerta motus redundantia. nascitur cum feruor offendit umorem et impetus inflationis[65] exprimit uim spiritus flatus.
>
> Wind is a flowing wave of air with a chaotic surge of motion. It arises when heat strikes moisture and the pressure of the swelling expels a forceful and gusty exhalation.

The definition of wind as a "flowing wave of air" was not the only one in Greco-Roman antiquity but was probably the most widely subscribed, found already in Anaximander and the Hippocratic corpus and recurring to Vitruvius' time and beyond in Aristotle, Varro, Cicero, Seneca, Pliny, and others.[66] Vitruvius' account adds an important further detail,

[65] Wistrand's conjecture (MSS *factionis*); another possibility is *tactionis* (Rose and Müller-Strübing, Rose later *calfactionis*), but *inflationis* is corroborated by *DA* 8.3.3. Cf. *Aetna* 213 (following n.) and n.68.

[66] Similar definitions at Vitr. *DA* 8.2.1, 8.2.2. Cf. Anaximand. D–K A24 = L–M D34, Hippocr. *Flat.* 3 L 6.94, [Arist.] *Mund.* 394b8–9, *Pr.* 882b15–16, Cic. *Diu.* 2.44, Varro *apud* Seru. *Aen.* 5.19, Sen. *QNat.* 5.1.1, Plin. *HN* 2.114; cf. *Aetna* 213. See also *RE* (2) VIII A.2.2229.22–2230.6 (Böker). There

however, in that wind is accompanied by a "chaotic surge of motion" (*incerta motus redundantia*).[67] This property is related to its violent genesis: heat "strikes" (*offendit*) the water; the "pressure" or "force" (*impetus*) of the "swelling" (*inflationis*) "expels" (*exprimit*) the "force" or "strength" (*uim*) of an exhalation. Others believed that wind arises from heat (the sun) or water and similarly appealed to contractionary and expansionary mechanisms, but Vitruvius is unusual in emphasizing the essential instability and irregularity of *uentus*.[68] This view will prove important in explaining the pernicious effects of the winds on health.

Vitruvius claims that it is possible to ascertain that his account is "true" (1.6.2 *uerum*) by experiment with hollow bronze vessels that are laden with water and heated over flame. These vessels, called "aeolipiles,"[69] will allow us "as regards the hidden principles of the sky to form a conception of the truth about divinity by skillfully contrived investigations into nature" (*de latentibus caeli rationibus artificiosis rerum inuentionibus diuinitatis exprimere ueritatem*). The proposition that *ars* (cf. *artificiosis*) can aid in the discovery of the principles (*rationes*) of nature is a central feature of Vitruvius' thought and underlies his frequent appeal to experiments and mechanical contrivances not just to illustrate but to explain phenomena.[70] He describes the experiment as follows (1.6.2):

> fiunt enim aeolipilae aereae cauae. hae habent punctum angustissimum. quae aqua infunduntur conlocanturque ad ignem; et antequam calescant, non habent ullum spiritum, simul autem ut feruere coeperint, efficiunt ad ignem uehementem flatum. ita scire et iudicare licet e paruo breuissimoque spectaculo de magnis et immanibus caeli uentorumque naturae rationibus.
>
> Hollow bronze aeolipiles are prepared. These have a (sc. single) very small hole. They are filled with water and exposed to fire. And before they are

were other theories: for example, Aristotle (*Mete.* 349a16–b2, 360a27–33) rejects the mainstream view, maintaining that wind is a dry and smoky exhalation (cf. 359b27–360a17).

67 *Redundantia* is not "abundance" (Fleury) but "a flowing back (and forth)" (cf. 8.pr.2; so *TLL* XI 2.580.29–35 [Viredaz], Fensterbusch, Gros et al.). *Incertus* may refer to oscillation in intensity (Fleury 2003: 152) but probably picks out changes in direction: cf. Cic. *Nat. D.* 2.101 (Gros et al. 1997: 95; further *TLL* VII 1.879.75–880.42 [Ehlers]).

68 See *RE* (2) VIII A.2.2226.24–2227.41 and esp. 2232.22–2234.8 (Böker); cf. Plin. *HN* 2.114 (with Beaujeu 1950: 194), Sen. *QNat.* 5.6.1. Lucr. 6.685 also presents the generation of air as turbulent.

69 *Aeolipila* or *Aeoli pila* is derived from *Aeolus* (the name of the wind god) and *pila* ("sphere," "ball"), but there are further uncertainties. At 1.6.2, MSS transmit *ex aeolis aereis* and afterwards *aeolipilae aereae cauae*. Most correct the former to *ex aeolipilis aereis* after Rose and Müller-Strübing; Fleury 2003: 153–5 retains the MS reading, however, taking *aeolis* in the first instance to mean "Aeoli" (i.e. bronze effigies of the wind god) and reading *Aeoli pilis* in the second as a restatement (*pila* as "puppet" or "effigy," a rare meaning: *TLL* X 1.2132.40–6 [Ebbeler]).

70 Cf. Romano 1985: 195–200; Courrént 2011: 212–13; Roby 2013 (436–7 on the aeolipile).

> heated, they have no air-currents, but as soon as they begin to grow warm, they produce a violent blast in response to the fire. In this way it is possible from a small and extremely brief spectacle to know and form a judgment about the lofty and immense principles of the climate and of the nature of the winds.

The small-scale demonstration (1.6.2 *paruo breuissimoque spectaculo*) is probative of the earlier definition: the "violent blast" (*uehementum flatum*) produced by the application of heat to water corroborates the claim that winds are violent owing to their generation. Vitruvius' device is paralleled by other such machines described by mechanical authors such as Philo and Hero (Figure 3.1).[71] This small experiment is typical of Vitruvius' explanatory method *DA*. Beginning from first principles, he adopts an account in

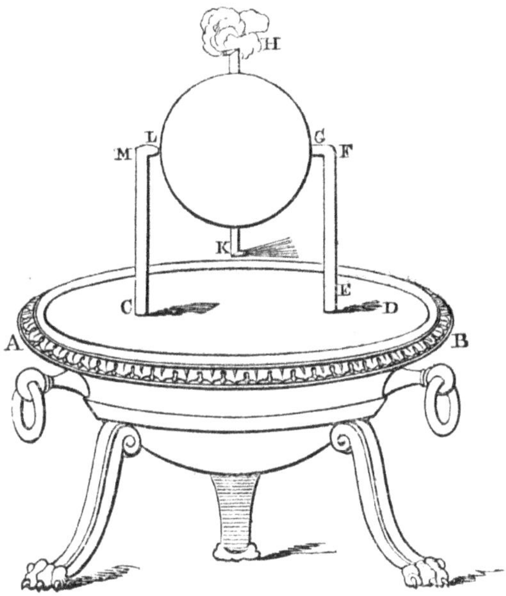

Figure 3.1 In Hero's construction (*Mech.* 2.11), the hollow sphere HK revolves owing to the force of the escaping steam, as water contained in the cauldron AB is heated.

[71] (1) Philo's censer (*Pneum.* 57), a vessel which when heated dispenses vaporized liquid through a nozzle (Carra de Vaux 1903: 196; regarded by Drachmann 1948: 125–6 as posterior to Heron, *Pneum.* 2.34–5); (2) Hero's hollow sphere on a pivot, which rotates when heated owing to steam escaping through small holes (*Pneum.* 2.11; cf. Argoud and Guillaumin 1997: 140–1, and see Figure 3.1).

conformity with the leading theory. While the architect is not strictly responsible for furnishing these principles – only evaluating what follows from them – Vitruvius characteristically goes further and proposes an experiment which makes architecture a helper in confirming the standard account. This is a good example of the "shared disputation" (1.1.16 *disputatio communis*) among the *artes*.

So much for the nature of wind. How do winds affect health? Vitruvius' initial statements (1.6.1, quoted earlier in this section) have various ancient parallels but reveal no specific theory of health and disease.[72] We have more to go on in what follows (1.6.3). Vitruvius says that excluding winds will produce a salubrious environment (1.6.3 *locum salubrem*) for those already in good health (*corporibus ualentibus*). More revealingly, he explains how excluding winds will affect those who are sick by dividing diseases into two classes – call them (a) and (b) – and comparing the effects of climate on therapy. Class (a) comprises diseases which, because they are not caused or exacerbated by airs, could be cured in a wind-exposed locale by ordinary medicine (*morbi ex aliis uitiis ... qui in ceteris salubribus locis habent curationes medicinae contrariae*). The word *ceteris*, "other kinds," contrasts the *locus salubris* of Vitruvius' ideal city shielded from winds with *loci salubres* which are wind-exposed but otherwise healthy. In the ideal city, those suffering from class (a) diseases, who would have recovered anyway, "will heal more quickly by the exclusion of the winds, owing to the temperateness (sc. of the climate)"[73] (*propter temperaturam exclusione*[74] *uentorum expeditius curabuntur*). Class (b) comprises diseases which are more difficult to cure in the aforementioned wind-exposed but otherwise healthy locales (*uitia ... quae difficulter curantur in regionibus quae sunt supra scriptae*[75]), and includes illnesses such as cough (*tussis*), consumption (*phthisis*), and "others which are cured not by removals but by additions" (*cetera quae non detractionibus sed adiectionibus curantur*).

Certain details throw light onto Vitruvius' theory of health. First, *medicinae contrariae* highlights the allopathic principle that underpins his reasoning: opposites cure opposites.[76] For class (a) diseases, *contrariae* specifically takes its point from *ex aliis uitiis*: as long as the source of the

[72] Cf. Gros et al. 1997: 94.

[73] The precise meaning of *temperatura* in Vitruvius' physics is discussed later in this section.

[74] MSS have *exclusiones*, impossibly (Fleury 2003: 156–7); I follow Rose and Müller-Strübing, who adopt the reading of some *recentiores* (Fleury secludes, others transpose).

[75] *Quae ... scriptae* is a circumlocution for the (wind-exposed) *ceteris salubribus locis*.

[76] The principle is common in ancient medicine and often articulated in the Hippocratic corpus: for example, *Aph.* 2.22 L 4.476, *Nat. hom.* 9 L 6.52–6, *VM* 13 L 1.598, and esp. *Flat.* 1 L 6.92 ἑνὶ δὲ συντόμῳ λόγῳ, τὰ ἐναντία τῶν ἐναντίων ἐστὶν ἰήματα. See also Ch. 5.3.1 for allopathy in Cels. *DM*.

injury is not wind (hence *aliis*), it will suffice to offer the treatment appropriate to that malignant defect or excess (*uitium*), regardless of whether the sick person is still exposed to winds. But where the disease is related to airs, as in the second class, we require an allopathic remedy of the appropriate polarity. Vitruvius identifies the relevant opposites in his remarks on class (b) diseases: "addition" (*adiectio*) and "removal" (*detractio*). As regards health and airs, this polarity has a specific precedent in earlier Greek writings such as the Hippocratic *On Winds* (Περὶ φυσῶν, *De flatibus*, probably fifth c. BC),[77] though, as we will see, Vitruvius' aetiology is developed in terms specific to the physical theory of *DA*. He argues that the class (b) diseases cannot be cured without difficulty in wind-exposed locales for two reasons. First, these diseases are caused by cold (*ex frigoribus concipiuntur)*; second, for those already weakened by disease (*defatigatis morbo uiribus eorum*), winds will exacerbate the extenuating *detractio* at the root of their illness (1.6.3):

> aer agitatus e uentorum agitationibus extenuatur unaque a uitiosis corporibus detrahit sucum et efficit ea exiliora.[78]
>
> The air is stirred up and rarefied in consequence of the rushings of the winds and in the same process draws nutrients away from the sick bodies and makes them feebler.

Such people will not recuperate while they are exposed to winds which exacerbate *detractio*. But in the city from which winds have been excluded (1.6.3):

> lenis et crassus aer, qui perflatus non habet neque crebras redundantias, propter inmotam stabilitatem adiciendo ad membra eorum alit eos et reficit qui in his sunt inpliciti morbis.
>
> A gentle and thick air, which does not have strong blasts or excessive surges, on account of its constant stability by adding to their limbs nourishes them and restores those who are involved in these illnesses.

The city shielded from winds offers the appropriate contrary, *adiectio* (cf. *adiciendo*), to allow the sick to convalesce.

It is possible to understand Vitruvius' account of winds and health in 1.6.1–3 by itself. These sections depend on orthodox medical and meteorological views

[77] *Flat.* 1 L 6.92 ἰητρικὴ γάρ ἐστι πρόσθεσις καὶ ἀφαίρεσις, κ.τ.λ.

[78] Printing the text of Rose and Müller-Strübing; with some manuscript support Granger (followed by Fensterbusch, Fleury) prints *est* for *e*, attaching *aer* to the preceding *eorum* (the sick) and punctuating after; Krohn prints *est* but adds <*molestus; namque ubi*>.

to develop a self-consistent explanation for the harmful effects of wind on health. Wind is chaotic and turbulent (1.6.2); these same properties are related to its exacerbation of diseases caused by *detractio* (1.6.3). Yet we can acquire a deeper view of the situation, especially of the "temperateness" (*temperatura*) that characterizes good climate and of the harmful mechanism of *detractio*, if we connect Vitruvius' account with the physical theory of *DA*. This is because his theory of health can be understood simply as a special case of this physical theory, which he expounds most directly in 1.4 but elaborates throughout the whole work. This theory is worth consideration here both for its intrinsic interest and because it is relevant to both case studies considered in this chapter.[79] The premises of his theory are expressed as follows (1.4.5):

> e principiis, quae Graeci στοιχεῖα appellant, ut omnia corpora sunt composita, id est e calore et umore, terreno et aere, et ita mixtionibus naturali temperatura figurantur omnium animalium in mundo generatim qualitates.
>
> According as all bodies are composed of elements (which the Greeks call στοιχεῖα), that is, heat and moisture, earth and air, so by their mixing in a natural proportion are formed the qualities of all the animals in the world, species by species.

Vitruvius adopts a version of the theory common in Greco-Roman antiquity that the physical building blocks are the four elements (typically earth, air, fire, and water).[80] The qualitative properties of things depend on the natural proportion of elements (1.4.5 *naturali temperatura*) that compose them; each thing, according to its type or species (*generatim*), has a characteristic mixture of elements. He illustrates this claim by distinguishing among the natural mixtures of birds, fishes, and terrestrial animals (cf. 1.4.7 *aliam . . . mixtionem habet genus auium, aliam piscum, longe aliter terrestrium natura*): birds have less earth and moisture, moderate heat, and much air; fish a little heat, much air and earth, and almost no moisture; and terrestrial animals moderate air and heat, less earth, and much moisture. Their adaptation to their environments depends on the elemental mixture they possess (1.4.7).[81] Although Vitruvius uses living creatures to illustrate

[79] The framing of Vitruvius' theory that follows is my own, but see usefully also Courrént 1999b; Callebat 2003; Courrént 2011: 121–72; 2016; Callebat 2017c.

[80] Vitruvius' theory is essentially that made famous by Empedocles (D–K A33 = L–M D56, B6 = D57, B71 = D58, B98 = D61), though he substitutes here the qualitative principles *umor* for *aqua* and *calor* for *ignis* (cf. Callebat 1982: 337). Vitruvius himself attributes the four-element theory variously to the Pythagoreans, Empedocles, and Epicharmus (as discussed later in this section). Cf. also Pellati 1951; and on 1.4.5, Courrént 2011: 122–3.

[81] Courrént 2011: 125–32 notes similarities with Aristotle (cf. *Resp.* 477a17–478a10) but observes that Vitruvius is closer to Empedocles in the details.

his theory in 1.4, the same analysis applies to inanimate objects: all things derive their qualitative properties from the natural proportion (*naturalis temperatura*) or mixture (*mixtio*) of elements, including stone, wood, water, and even macroscopic phenomena such as geography and climate.[82]

Vitruvius reprises this elemental theory twice elsewhere in *DA*, at 2.2.1–2 and 8.pr.1–4. At 2.2.1–2, the account is motivated by his argument (2.1.9) that a solid understanding of the architect's building materials can only be attained if explanation of their properties is founded on *rerum natura.*[83] Vitruvius offers a brief history of different physical theories: Thales posited water as the first principle, Heraclitus fire; Democritus and Epicurus believed that atoms (2.2.1 *insecabilia corpora*) were at the root of things; and the Pythagoreans developed the four-element theory proper (*aquam et ignem aera et terrenum*).[84] Vitruvius syncretizes the elemental and atomist theories (2.2.1–2), finding in Democritus' *insecabilia corpora* both an explanation for the permanence of the substance underlying the elements and warrant for a combinatorial account of the emergence of the "infinite varieties" (2.2.2 *infinitis generibus*) of things in nature. In 8.pr.1–4, a similar potted history begins with Thales positing water and Heraclitus fire, but has the Magi (followers of Zoroaster) arguing both water and fire and the playwright Euripides, "student of Anaxagoras" (8.pr.1 *auditor Anaxagorae*), earth and air. We come finally to Pythagoras, Empedocles, Epicharmus, "and other scientists and philosophers" (8.pr.1 *aliique physici et philosophi*) who espoused the four-element theory. Reprising 1.4.5, he submits that on this theory, "the mutual conjunction (sc. of the elements) produces, by their natural formation, qualities based on the differences in types" (8.pr.1 *eorum . . . inter se cohaerentiam naturali figuratione e generum discriminibus efficere qualitates*). Vitruvius works out some biological implications of this theory (8.pr.2–3) before arguing, in rather Stoic fashion, that a "divine mind" (8.pr.3 *diuina mens*) has distributed the elements so as to sustain life.[85]

Vitruvius' four-element theory provides the general framework for physical explanation in *DA*, which he often formulates qualitatively with reference to natural elemental composition. Equally important, or even

[82] For stone, cf. 2.5–7 (with Courrént 2011: 149–59, 165–8); for water, 8.2–3; for wood, Ch. 3.4. For geography and climate, cf. 1.4, 6.1, and of course 1.6 (also Courrént 1999b: 545–51; 2011: 132–49; Callebat 2017c).

[83] On 2.1.9 see Ch. 3.4 (second case study).

[84] For Vitruvius' philosophical and scientific "library," see Courrént 2011: 43–50.

[85] See Ch. 3.3.3 and n.110. Vitruvius observes no contradiction between this providence and the atomist theory of 2.2, probably because his interest in atomism is limited to explaining natural kinds by combination of different elements (cf. also Romano 1987: 125–7; Gros et al. 1997: 179; Novara 1994, esp. 594).

more so in an architectural context, however, is a complementary vocabulary which recasts elemental composition in terms of states of material coherence that range from firm, stable, and solid (*firmus*, *integer*, *solidus*, etc.) to soft, weak, porous, and insubstantial (*exilis*, *inanis*, *inbecillus*, *infirmus*, *mollis*, *uanus*, etc.). When a thing is in its natural (and thus normatively best) state – that is, in possession of the elemental proportion (*temperatura*) that belongs to its natural type – then its physical structure is firm, stable, and solid.[86] When a thing deviates from this elemental state, it weakens and loses its coherence; its physical structure becomes discontinuous, and it dissolves.[87] The element-focused and coherence-focused descriptions provide different but equally valid explanations of a thing's natural properties. On each occasion, Vitruvius chooses which description to prioritize in relation to the *opera* that he hopes to explain.

It is worth observing finally that, despite Vitruvius' interest in the four-element theory, he never attempts an independent proof of it. This fact incidentally reveals the limits of architectural *ratiocinatio*. Two points can be made. First, the four-element theory was the standard theory of matter in Greco-Roman antiquity; as for his theory of wind, Vitruvius probably felt no need to justify its assumption, which would have been unobjectionable to the largest part of his audience. Second, it falls anyway outside the scope of the architect's expertise to know more of other disciplines' theories (their specific *rationes* or *ratiocinationes*) than required for the judgment of their products (*opera*). Vitruvius' syncretizing approach to Greek philosophy underscores that the various theories ultimately converge on a shared strategy of explaining qualitative properties with reference to natural proportion. The *architectus* need not discriminate further if the choice is irrelevant for his practice.

Vitruvius' theory of health, where we began, is a special case of his physical theory in the sense that health is simply the natural (and thus normatively best) state of elemental proportion and coherence of the human body and sickness simply a departure from this mixture and the concomitant dissolution of the body.[88] At 8.pr.2–3, Vitruvius explains why each of the four elements must be present in a healthy creature, pointing

86 At 1.3.2 *firmitas* is one of the three aims of architectural construction (the others are *utilitas* and *uenustas*), a more elaborate concept still dependent, however, upon *firmitas* as physical property (cf. Courrént 2011: 121).

87 Cf. 1.4.6 (where note *interficere*, *dissoluere*, *diluere*, *dissoluere*, *infirmare*, etc.).

88 Vitruvius uses the same word, *uitium*, of defects in elemental composition (1.4.6) and human illness (1.6.3). The idea that health is a balance is fundamental to ancient medical thought (e.g. Alcmaeon, D–K B4 = L–M D30).

out the dangers of a deficit in air, earth, water, or fire. Elsewhere (1.4.6), he considers the alternative problem of an excess in one of the constitutive elements, which is equally harmful. In principle any such imbalance can pose a problem, but in point of fact Vitruvius tends to dilate on one or another as especially injurious.[89] Heat (*calor*) or fire, for example, can be said under the element-focused description to have an attractive power that draws out the moisture (*umor*) of a thing, or it can be seen on the coherence-focused account as cooking and dissolving a thing's physical structure.[90] (In describing the positive effects of cold, *frigus*, Vitruvius prefers the coherence account of solidifying a thing, making it more firm and stable.[91]) The same double explanation holds for another element, air (*aer*), which may be said to change the structure of something either by introducing imbalance into its elemental makeup or else by permeating, extenuating, and thus dissipating its material structure.[92] As health is the correct proportion (*temperatura*) of the elements that constitute the human body, so disease is a physical imbalance whose effects can be analyzed in material terms.

We return to 1.6.3, where Vitruvius' physical theory may elucidate his explanation of the effects of wind on health. First, we can now appreciate his claim that patients "will heal more quickly by the exclusion of the winds *propter temperaturam*." The word *temperatura* here refers, as elsewhere in *DA*, to the elemental proportion of the environment with a view to the elemental proportion of the creatures – humans – that inhabit it.[93] Winds both precipitate a change in temperature (cf. 1.6.1) and bring an excess of air (*aer*) in chaotic and violent motion. Excluding winds thus prevents the unbalancing changes that would yield too much or too little heat, moisture, or air, providing instead well-balanced *temperatura*. "Temperateness" of the environment preserves or promotes *temperatura* (or *naturalis temperatura*, 1.4.5) of human beings, which is by definition good health. We may now also reconsider Vitruvius' statement that only a city protected from winds will allow those who are suffering from diseases marked by agitated air (cf. 1.6.3 *aer agitatus est*) and cured by additions (*adiectiones*) rather than removals (*detractiones*) to recuperate. As we saw, Vitruvius

89 See also Courrént 1999b: 538–45; 2016: 257–9, on the mechanics of physical change in *DA*.

90 For example, 5.3.2 (element-focused description of the effect of heated air on spectators in a theater), 1.4.3 (element- and coherence-focused descriptions of heat-softened iron).

91 For example, 1.4.3 (again of heat-softened iron), 1.4.4.

92 For example, 1.4.6 and esp. 1.6.3 (quoted earlier).

93 *Temperatura* and *mixtio* are quasi-technical and "distinctively Vitruvian" in reference to his elemental theory throughout *DA* (Adams 2016: 180). For environmental *temperatura* and human health, cf. 1.4.8.

holds that rushing winds will "thin out" the air (cf. 1.6.3 *extenuatur*), thus drawing away nutrients (*detrahit sucum*) and rendering the already-sick body (*uitiosum corpus*) "more feeble" (*exilius*). This explanation is presented in terms of both the elemental makeup and the physical coherence of the body. Under the elemental description, excess air effects a removal (*detractio*) of the body's nourishment, *sucus*. Qua bodily nourishment, *sucus* is an earthly element, as remarks later in *DA* prove: "the limbs of the body are nourished by food that consists of the earthy" (8.pr.2 *terrestri cibo membra corporis alantur*); "earthly produce . . . sustains and nourishes animals by continually feeding them" (8.pr.3 *terrenus fructus . . . alit et nutrit animales pascendo continenter*). Under the coherence description, the chaotic dispersal and increase of *aer* dissipates the structure of the body, causing it to become more hollow or feeble (1.6.3 *exilius*). A "gentle and thick air" (*lenis et crassus aer*), however, will restore by adding (*adicere*) and nourishing (*alere*): when *aer* achieves its proper balance in the bodily *temperatura*, so can the other elements, too.

3.3.2 *To Catch the Winds: Geometry, Meteorology, Architecture*

Vitruvius thus explains how winds are harmful to health and, consequently, why the architect must exclude them from the city. This was **Argument I** in the outline presented in Table 3.2. The next task (**Argument II**) is to show how exactly the winds are to be excluded. Based on a grid plan of streets (*plateae*) and alleys (*angiportus*), Vitruvius frames the problem and the solution (1.6.8):

> cum enim plateae contra directos uentos erunt conformatae, ex aperto caeli spatio impetus ac flatus frequens conclusus in faucibus angiportorum uehementioribus uiribus peruagabitur. quas ob res conuertendae sunt ab regionibus uentorum directiones uicorum, uti aduenientes ad angulos insularum frangantur repulsique dissipentur.
>
> For when the streets are laid out directly opposite the winds, then from the open space of the sky the continual rushing blasts will be caught in the narrows of the alleys and roam about with greater force. Hence the directions of the residential ways ought to be turned away from the regions of the winds, so that when they come on toward the angles of the town-blocks they are broken and, repulsed, lose their strength.

The argument is straightforward if we accept Vitruvius' claim about the danger of unmitigated exposure to winds and his orthogonal plan of *plateae* and *angiportus*. The architect's task accordingly becomes to determine, first, what the *regiones uentorum* are, and second, how to "turn away" (cf. *conuertendae*) the streets and alleys (altogether the "residential ways," *uici*) to shield them. Vitruvius contends that the winds will be excluded from the alleys if the streets and alleys are rotated obliquely to the cardinal and ordinal directions. The principles for this argument are both meteorological and geometrical.

Vitruvius says that "there are quite a few who believe that there exist four winds" (1.6.4 *nonnullis placuit esse uentos quattuor*), whose origins are the cardinal directions: *Septentrio* the north wind, *Solanus* east, *Auster* south, and *Fauonius* west. However, he immediately adds that "according to those who have made a more careful study, there are eight" (*qui diligentius perquisierunt tradiderunt eos esse octo*). Chief among this latter group is Andronicus of Cyrrhus (*maxime quidem Andronicus Cyrrestes*), the architect of the so-called "tower of the winds" in Athens, which was constructed sometime in the second or first century BC and still stands today (Figure 3.2).[94] Vitruvius calls the tower an "illustration" or "model" (*exemplum*) of the eight-wind theory, describing it for his readers thus (1.6.4):

> qui etiam exemplum conlocauit Athenis turrem marmoream octagonon et in singulis lateribus octagoni singulorum uentorum imagines excalptas contra suos cuiusque flatus designauit, supraque eam turrim metam marmoream perfecit et insuper Tritonem aereum conlocauit dextra manu uirgam porrigentem et ita est machinatus uti uento circumageretur et semper contra flatum consisteret supraque imaginem flantis uenti indicem uirgam teneret.

> He (sc. Andronicus) also made as an illustration (sc. of the eight-wind theory) an octagonal marble tower in Athens, and on each side of the octagon depicted carved likenesses of each of the winds opposite the blasts that belong to them. Above this tower he made a conical roof and, on top of it, set a bronze Triton holding out a wand in its right hand, which he so engineered as to be turned around by the wind and always stand opposite the blast and hold its wand as a pointer over the statue of the wind which was blowing.

The eight sides of Andronicus' tower face the cardinal and ordinal directions, thus adding another four winds: in Latin, these are *Caurus* in the north-west, *Africus* south-west, *Eurus* south-east, and *Aquilo* north-east (Figure 3.3).[95] The octagonal design of Andronicus' tower assigns a *regio* constituting a 45° arc of the horizon to each of the eight winds (Figure 3.4).

94 On the tower, Kienast 2014; Webb 2017. Varro describes it at *Rust.* 3.5.17; Vitruvius might have learned of it via Varro or might have seen it in person (cf. Corso 1997: 373–7).

95 The eight are Ἀπηλιώτης (Vitr. *Solanus*), Εὖρος (*Caurus*), Νότος (*Auster*), Λίψ (*Africus*), Ζέφυρος (*Fauonius*), Σκίρων (*Eurus*), Βορέας (*Septentrio*), and Καικίας (*Aquilo*). See Kienast 2014: 169–96.

Figure 3.2 The tower of Andronicus of Cyrrhus that still stands in the Roman agora in Athens, from a southeast-facing perspective with the Acropolis in background.

Figure 3.3 Relief of Boreas (the north wind) on Andronicus' tower, one of the eight reliefs illustrating the principal winds.

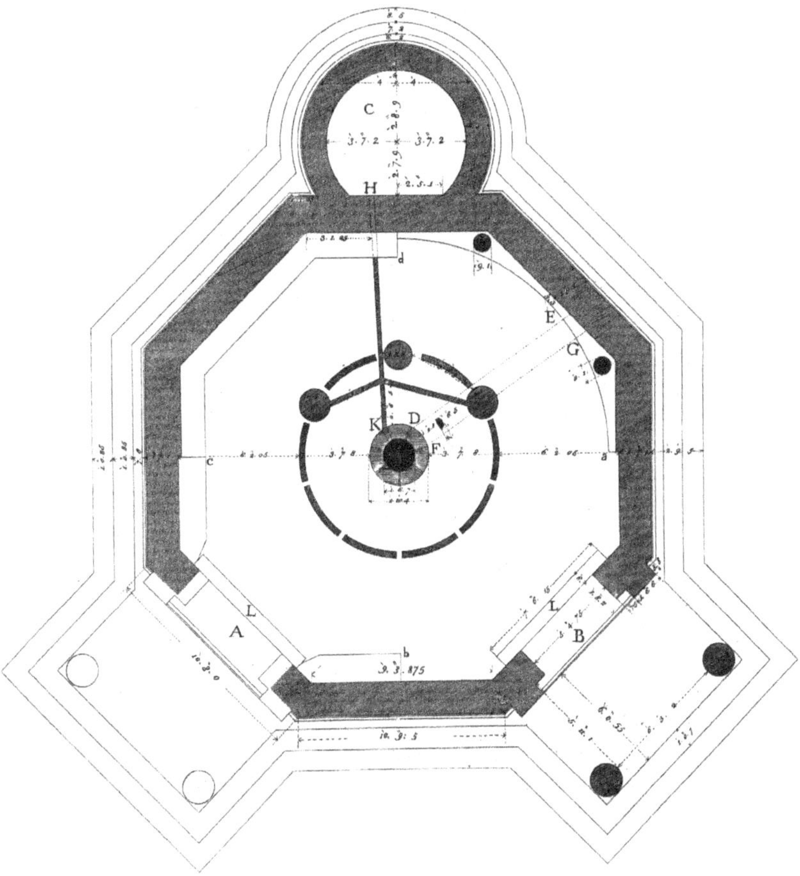

Figure 3.4 Cross-section of Andronicus' tower of the winds, showing the octagonal design that presents the eight faces of the tower to the eight principal winds.

The whole tower thus constructs a windrose, with the Triton weathervane pointing toward the region of the wind (Figure 3.5).

Vitruvius takes over the theory of winds implied by Andronicus' tower: "it seems to have been produced in this way so that it may embrace the numbers, names, and quarters from which the definite blasts of the winds blow" (1.6.5 *hoc modo uidetur esse expressum uti capiat numeros et nomina et partes unde flatus certi uentorum spirent*).[96] Vitruvius' choice to follow Andronicus is notable, especially because his theory was not the only

[96] Vitruvius later describes a twenty-four-windrose (1.6.9–10, n.105) but it is not integral in his city plan, as explained later in this section.

Figure 3.5 Elevation of Andronicus' tower of the winds, showing the Triton weathervane atop that would revolve to point in the direction from which the wind was blowing and identify it by its corresponding relief.

available.[97] While Andronicus stipulated eight winds, others before Vitruvius counted more or fewer: the Hippocratic *On Hebdomads* gives seven winds, Aristotle ten, and Timosthenes (fl. third century BC) twelve, to mention only a few examples.[98] Apart from its number and names of

[97] From [Gal.] *Hippocr. Hum.* K 16.403.1–404.1, 406.6–8, Eratosthenes was once thought to have developed the Hellenistic eight-wind theory (Kaibel 1885, e.g. 587), but since the passage was revealed to be a Renaissance forgery, its history before Andronicus is again uncertain (Rehm 1916: 70–5; *RE* (2) VIII A.2.2364.16–2366.42 [Böker]).

[98] Hippocr. *Hebd.* 3 Roscher 7–8, Arist. *Mete.* 363a21–365a13, Timosthenes *ap.* Agathemerus sec. 7 = Müller 1882: II 473; generally on Greco-Roman windroses, Kaibel 1885; Steinmetz 1907; Rehm 1916; *RE* (2) VIII A.2.2325.20–2381.21 (Böker); Fleury 1991; 2003: cvii–cxiv.

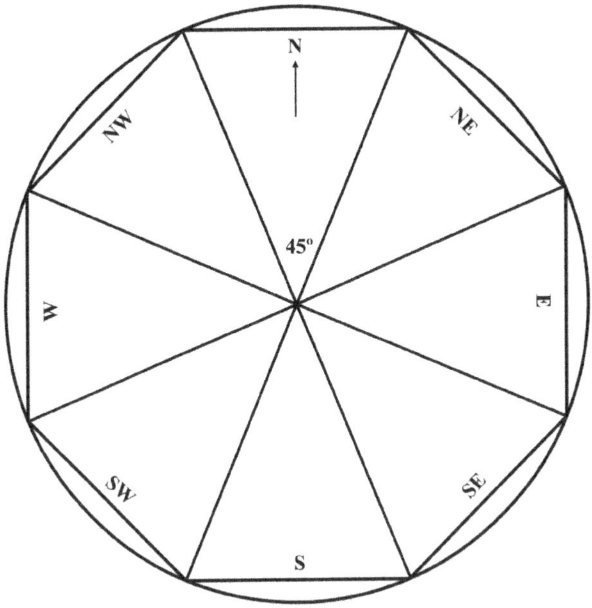

Figure 3.6 Schematic representation of Vitruvius' windrose based on Andronicus' tower of the winds.

winds (*numeros et nomina*), Andronicus' windrose was also importantly different from others in how it delineated the locations of the winds: his tower gives each wind a 45° *regio* constituting a sector of the circle (the horizon) bisected by a radius on the cardinal or ordinal point; consequently, the wind from a particular region is visualized as originating from or being felt in the whole area of the sky 22.5° on either side of the horizon around each cardinal or ordinal point (Figure 3.6). Other ancient windroses, by contrast, usually assign the origin of the winds to points on the horizon, leaving the extent of the wind's force less clearly defined.

Both the number of winds and the geometrically precise division of *regiones* are crucial for Vitruvius' solution to orienting the city. But does he give us independent reasons for adopting Andronicus' theory? This question is significant for the cogency of the whole explanation. In fact, there are three arguments of varying strength. (Vitruvius also comes back later to shore up the theory, as we will see shortly.) First, an appeal to authority (cf. 1.6.4 *qui diligentius perquisierunt*).[99] Vitruvius elsewhere indicates the

[99] Besides *diligentius*, *perquirere*, a verb used of scientific investigation, conveys the authority of Andronicus' view (*TLL* X 1.1660.60–74 (Kruse); cf. *OLD* 3).

propriety of deferring to experts in other fields, as is indeed necessary given the impossibility of attaining mastery in all disciplines.[100] Second, an appeal to empirical observation. Andronicus' tower is called an *exemplum* of the theory, and the Triton's wand serves as a "pointer" (1.6.5 *indicem*) indicating the winds. Third, a *disputatio communis* with other disciplines. Much as Vitruvius' experiment with the aeolipile elucidated the origin of wind, so Andronicus' tower contributes to meteorological discussions about the number and regions of the winds. Andronicus' tower represents a specific application of *ratiocinatio* insofar as it instantiates a rigorous geometrical relation among three crucial terms: the octagon (the tower itself), the circle (windrose), and the rectangle or square (grid of city streets). This geometrical relation is at the heart of Vitruvius' method for orienting the city.

Once Andronicus' theory is accepted, Vitruvius' technique for excluding the winds is straightforward. What follows is a simplified account of the **Procedure** mentioned in the outline presented in Table 3.2, in the version at 1.6.6–7. First, Vitruvius has the architect use a gnomon to ascertain and mark true north as well as the other compass points and to describe a circle carrying these points (Figure 3.7). This circle is then to be divided into eight equal sectors (45° each), each around a cardinal or ordinal point. Because this is equivalent to inscribing an octagon in the circle (again Figure 3.6), Vitruvius is in effect giving us a procedure for constructing Andronicus' windrose.[101] In the next step, the architect overlays his blueprint for the grid of streets (*plateae*) and alleys (*angiportus*) upon the north–south axis of the windrose. From there, he simply rotates the grid so that the streets and alleys have their outlets "through the angles between (sc. each) two regions of winds" (1.6.7 *per angulos inter duas uentorum regiones*) (Figure 3.8). This oblique rotation, 22.5° from the meridian, ensures that the strength of the winds breaks against the walls around the city, since Andronicus' theory implies that the blasts arrive perpendicularly to the faces of the inscribed octagon. Because the outlets are at the interstices of this octagon, they fall between the principal winds, which thus do not enter the streets directly.

While the chief part of Vitruvius' instructions has been given, his explanatory account is not entirely complete. After this procedure he admits, almost by way of postscript, that "perhaps those who know of many names of the winds will be surprised at the fact that only eight winds were said by us

[100] We noted another example in Vitruvius' choice of physical theory; cf. also 7.pr.1–2, 9.pr.1–17 on his attitude toward earlier expert authors (with Novara 1983: 294–300; 2005: 35–109; Callebat 2011; Oksanish 2019: 37–42).

[101] In the second method, Vitruvius does speak of the *octagonus* within the circle (1.6.13 *bis*).

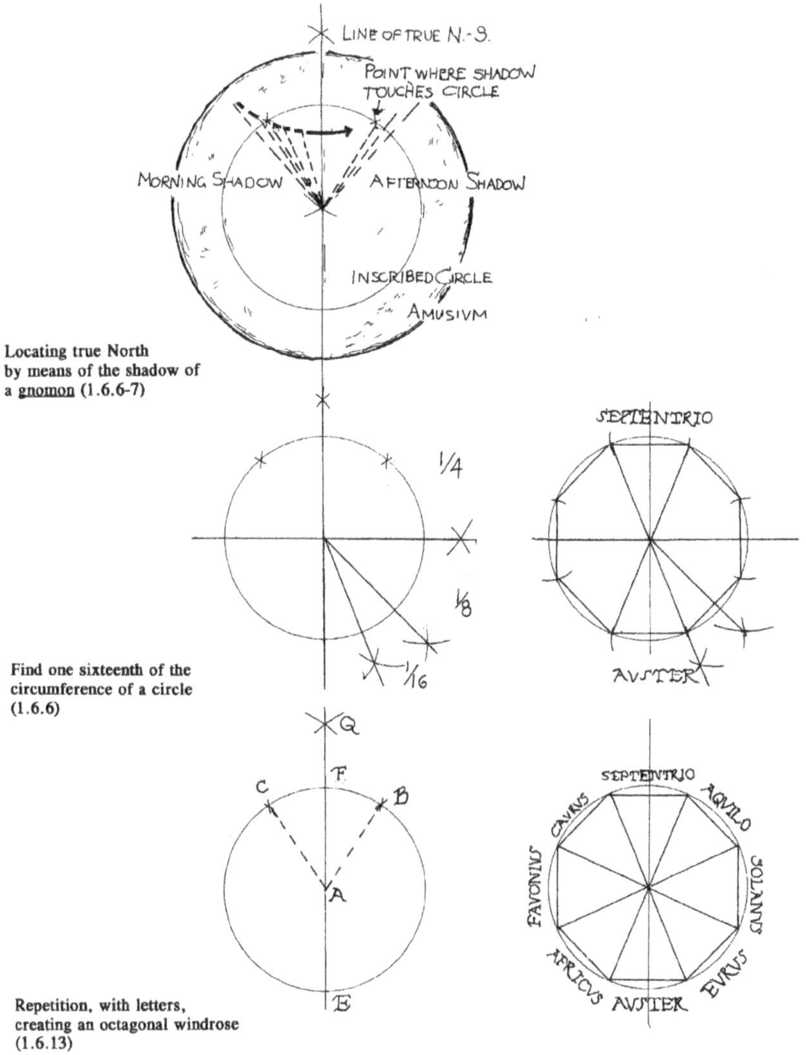

Figure 3.7 Detail of T. N. Howe's illustration of Vitruvius' procedure for constructing, with the help of a sundial, an octagonal windrose to orient the city (method of 1.6.6–7 above, method of 1.6.13 below).

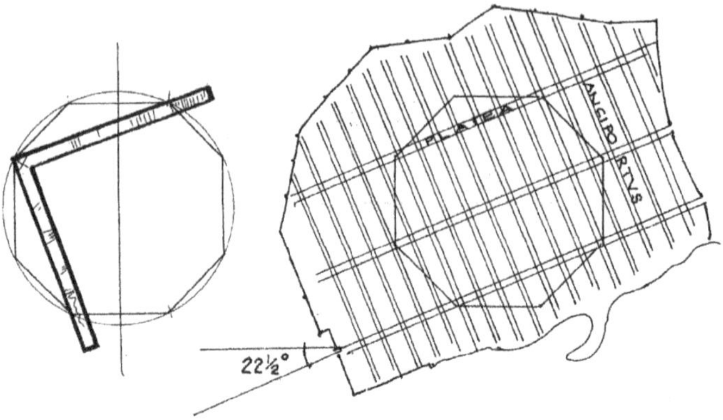

Figure 3.8 Detail of T. N. Howe's illustration of the effect of rotating the grid of streets (*plateae*) and alleys (*angiportus*) based on the octagonal windrose so that their outlets/junctions are located at the vertices of the figure.

to exist" (1.6.9 *fortasse mirabuntur ii qui multa uentorum nomina nouerunt quod a nobis expositi sunt tantum octo esse uenti*).[102] As we remarked, alternative accounts of the winds in antiquity gave different numbers, and there were of course far more than eight names for the winds. Vitruvius recognizes that the existence of additional winds could cast doubt on the completeness of his precepts: if there are other winds originating in other locations, what is to guarantee that they will not enter the *plateae* and *angiportus* and wreak havoc on the citizens' health? In defusing this worry, Vitruvius is thus also buttressing the account he has just offered. He responds not by denying the existence of other winds, but by showing how they can be understood within the confines of his eight-wind theory. His strategy is, in effect, to draw a distinction between the eight principal winds on Andronicus' windrose – what he called the "definite blasts of the winds" (1.6.5 *certi flatus uentorum*) – and other winds which are either derivative of these eight or else are merely local or temporary.

Vitruvius argues that most other winds can be accounted for by attending to some facts about the size of the Earth. He invokes Eratosthenes of Cyrene (c. 240–194 BC), who concluded "by mathematical principles and geometrical methods" (1.6.9 *rationibus mathematicis et geometricis methodis*) that the circumference of the Earth is 252,000 stades.[103] Vitruvius

[102] Fleury 2003: civ–cv believes 1.6.9–12 were added in a second redaction of Book I (cf. lxxx). I do not think the hypothesis necessary; cf. Courrént 2001: 131–4, reading the sections as Vitruvian *disputatio*.

[103] Cf. Plin. *HN* 2.247 for Vitruvius' figures. On the measurement, see Keyser and Irby-Massie 2008: 299; and for the Περὶ τῆς ἀναμετρήσεως τῆς γῆς, Roller 2010: 263–7. Vitruvius acknowledges at 1.6.11 that some criticize Eratosthenes' result but does not believe any such error would overturn the Andronican theory.

converts this figure into paces (= 31,500,000) and then finds the eighth part of that (= 3,937,500), an arc of the great circle which he assigns to each of the eight principal winds. (Although this is not a calculation of area, Vitruvius slips from arcs to sectors under those arcs, which are handled as the *regiones* that belong to the winds.) He continues with an argument that invokes the size of the Earth: those who earlier wondered (cf. 1.6.9 *fortasse mirabuntur*) at his using only eight winds "ought not wonder if in so great a space (sc. an eighth part of the Earth) a single wind by its wandering should, through twistings and reversals, produce other kinds (sc. of winds) by the change of its blasts" (*non debebunt mirari si in tam magno spatio unus uentus uagando inclinationibus et recessionibus uarietates mutatione flatus faciat*). The idea is that each of the eight regions of the world is sufficiently large to allow that a single wind (*unus uentus*) originating in one region could give rise to permutations (cf. *uarietates*). Eratosthenes' exact figure is not critical to the argument but lends rhetorical authority by quantifying the Earth's vastness.[104] Vitruvius goes on to add sixteen subsidiary winds to the eight previously mentioned;[105] because these are derivative of Andronicus' principal winds, in the sense that they emerge from the same region of the sky as their parent winds, their dangers are nullified by the procedure already given. Vitruvius also mentions two other kinds of minor winds: local ones arising from the topography of a place (1.6.10), and temporary breezes (1.6.11), such as the morning wind produced by the rising sun's heat. He does not address the influence of every such wind on the city, but his purpose is obvious: by providing an alternative account of their origins that does not interfere with the *certi flatus uentorum* of Andronicus' windrose, he defuses the potential disturbance to his theory.[106]

3.3.3 *The Roman City as Universal City*

We come to the end of Vitruvius' instructions for the proper orientation of the city. Before we move on, we should say something about the consequences of Vitruvius' urbanism. We have seen that his instructions are ultimately rooted in a set of theories about nature: chiefly meteorological (the origin and properties of wind, the number and direction of winds),

[104] Compare Hyginus' invocation of Archimedes (*Const.* L 184.3–13 = T 147.21–148.11), discussed in Ch. 7.5.3.

[105] Yielding a twenty-four-windrose: *RE* (2) VIII A.2.2372.49–2374.4 (Böker); Fleury 1991: 63–4.

[106] Some minor winds are indeed assimilated to the principal winds: for example, morning breezes caused by the sun "occupy the region of the wind *Eurus*" (1.6.11 *Euri uenti tenent partes*).

but also physical (the four-element theory), medico-biological (human constitution, health, and disease), geographical-astronomical (use of the gnomon, size of Earth), and mathematical (construction of the windrose). Vitruvius furthermore takes for granted that the city will be laid out in an orthogonal grid. This is an important presupposition that we will explore more closely in Chapter 7; for now, we point out simply that it lends geometrical rigor, since the grid plan can be fitted onto the octagon/circle of the windrose, which in turn maps the world's regions and winds. The total effect of Vitruvius' web of explanations is to make his city the "natural city" – that is, the city as conceived in conformity with natural principles. The theoretical underpinnings of this approach are essential to securing this effect: by reasoning appropriately from the right accounts about nature we come to the orientation regarded as maximally salubrious for the inhabitants. This is just the orientation that Vitruvius prescribes. While he casts his plan at times in Roman terms, the deeper justification is not essentially Roman. To the contrary, this natural city is *ipso facto* a "universal city" whose proper axes are determined not by culturally relative concerns or needs but by the axes of the heavens and Earth. The orientational procedure using a gnomon that stands at the heart of Vitruvius' account thus inscribes the streets and boulevards of this universal city into the cosmic order, vis-à-vis the diurnal motion of the sun, and into the terrestrial order, vis-à-vis the geographical *regiones* of the wind.[107]

To that extent, we may say that nature predominates over culture in Vitruvius' town planning, the universal and objective facts of scientific knowledge over the historically specific and subjective imperatives of Roman civilization. Yet while this conclusion has a piece of the truth, this antithesis oversimplifies the relationship between nature and culture in *DA*. A good complement to the universalizing urbanism of 1.6 is Vitruvius' treatment of private buildings (*priuata aedificia*) in 6.1. The thrust of this section is that such buildings can only be "correctly laid out" (6.1.1 *recte disposita*) if they are done so with a view to "what regions or which latitudes they are established in" (*quibus regionibus aut quibus inclinationibus mundi constituantur*): the "principles of the regions and varieties of climate" (*regionum rationes caelique uarietates*) guide the architect. So far we have a close correspondence with 1.6, where nature plays a leading role in design. The situation is afterwards complicated, however, as Vitruvius maintains that regional differences are also reflected in the inhabitants, whose physical characteristics (6.1.3–4), voices (6.1.5–8), and mental qualities (6.1.9–10)

[107] See also Ch. 7.4 on Frontinus' and Hyginus' accounts of the history of centuriation.

depend on their environment. This kind of ethnography has deep Greek roots and was reflected in some quarters of contemporary Roman thought.[108] The salient point here is that Vitruvius' presentation effectively pierces the boundary between nature and culture, since the latter is influenced by physical determinants. Vitruvius goes further as he reflects on the implications of this view for Roman civilization, which he compares with other nations (6.1.10–11). Rome is at the "center of the world" (6.1.11 *medio mundi*): consequently, whereas other peoples' constitutions are imperfectly balanced owing to the climatic excesses of their environment, the Roman peoples' is "the most well-tempered" (*temperatissimae*) because of the climatic equilibrium.[109] Roman *temperatura*, the perfect medium, is authorized as the natural ideal.

Even more significantly, the distribution of nations and national characters is not arbitrary, Vitruvius says. To the contrary (6.1.11):

> ita diuina mens ciuitatem populi Romani egregia temperataque regione conlocauit uti orbis terrarum imperii potiretur.

> In this way the divine mind located the state of the Roman people in an excellent and temperate region so that it might acquire rule over the world.

The natural Roman constitution is thus providentially bent toward securing the global dominance of Roman *imperium*. The *diuina mens* mentioned here is evoked several times in *DA*. At 8.pr.3, it is said to have distributed the elements necessary to sustain life (earth, air, fire, water) in abundance all over the world; at 9.1.1, the fact that the shadow of the gnomon falls differently in different places is a consequence of the design of the *diuina mens* (cf. *ea . . . diuina mente comparata*); and at 9.5.4, along with *natura* it is responsible for the shapes and positions of the constellations. In all these instances, the "divine mind" is an impartial controlling force in nature, most closely Stoic in inspiration.[110] It is the more striking,

108 Classic early sources are Herodotus and the Hippocratic *Airs, Waters, Places* (with e.g. Jouanna 1999: 211–32; Thomas 2000: 42–74, 86–98, 103–14; Chiasson 2001; Borca 2003; Kennedy 2016); see further Pease 1920–3: I 234–6.

109 Vitruvius' argument attributing superiority to those living in the mean condition (μεσότης) is essentially that of Arist. *Pol.* 1327b20–33, save that he substitutes Rome for Hellas. Romano 1987: 26–30 (esp. 27) believes that Vitruvius read a version of Aristotle's argument in Posidonius (who has been claimed a source for 6.1: Reinhardt 1921: 79–87). Cf. also Varro, *Rust.* 1.2.3.–7; Str. 6.4.1; Dion. Hal. *Ant. Rom.* 1.36–7, and for *laudes Italiae*, Ch. 4.4.1 on Col. *RR* 3.8, esp. n.127.

110 The attribution of *diuina mens* to nature or cosmos is not exclusively Stoic (cf. Cic. *Nat. D.* 1.34) but is common in Stoic settings (Cic. *Nat. D.* 1.100, 2.80, 2.88, *Diu.* 1.17 *bis* = *De consulatu suo* Büchner frr. 6.3 and 6.10, 1.110, etc.). What clinches the Stoic connection is *DA* 9.5.4 *siderum in mundo simulacra, natura diuinaque mente designata*: for the collocation of *natura* and *diuina*, cf. Cic. *Fin.* 4.12 *diuina mente atque natura mundum uniuersum et eius maximas partis administrari* (of Zeno and

therefore, that it takes providential care for the Romans. (Nor should we forget that Vitruvius addresses Augustus himself as *diuina tua mens et numen* at the outset of *DA*.[III]) From the perspective of 1.6, the naturalization of Roman civilization in terms of a providentially guaranteed geographical mean collapses the distinction between the "universal city" calibrated to nature and the city of Rome itself. Hence, the art of architecture that takes due consideration of *rerum natura* is in its highest expression a Roman art of architecture – Vitruvius' plan for a city founded on nature is at the same time an expression of the privileged place that Rome occupies in the world.

3.4 The Architect's Materials: The Matter of Wood

We come to the second case study. Book II of *DA* is devoted to a discussion of building materials. Vitruvius advertises his intentions in a passage that draws heavily on the theoretical language introduced in Book I, including *ratiocinatio*, *explicatio*, *demonstratio*, and *ratio* (2.1.9):

> nunc reuertar ad propositum et de copiis quae aptae sunt aedificiorum perfectionibus, quemadmodum uideantur esse ab natura rerum procreatae quibusque mixtionibus principiorum congressus temperentur, ne obscura sed perspicua legentibus sint, ratiocinabor. namque nulla materiarum genera neque corpora neque res sine principiorum coetu nasci neque subici intellectui possunt, neque aliter natura rerum praeceptis physicorum ueras patitur habere explicationes, nisi causae, quae insunt in his rebus, quemadmodum et quid ita sint, subtilibus rationibus habeant demonstrationes.
>
> Now I will return to our subject and provide an account of which materials are appropriate for sound construction, how they seem to have been produced by nature, and in what mixtures the conjunctions of elements (sc. in each) are proportioned, so that these things may not be obscure but transparent to readers. For no kinds of materials nor bodies nor things can come into being or be perceived except by the union of elements, nor can nature receive any true explanations from the teachings of the physicists, unless it is demonstrated with exact principles (*or* precise accounts) how and what are the causes which exist in things.

This view of architectural *ratiocinatio* fits well with Book I. Vitruvius uses *ratiocinari* of unfolding an account (*ratio*) of the creation of things from

the Stoics, following the Peripatetics); and for *diuina . . . mente designata* in particular, Cic. *Diu.* 1.17 = *De consulatu suo* Büchner fr. 6.10 *omnia iam cernes diuina mente notata* (likewise of the heavenly bodies).

[III] 1.pr.1, with Morgan 1909: 152–3.

their elemental beginnings. Both explanation (*explicationes*) and demonstration (*demonstrationes*) are implicated in reasoning to and from the causes (*causae*) of nature (*rerum natura*). This account is motivated by the architect's need to know "which materials are appropriate for sound construction," or, as Vitruvius puts it later, "the varieties and differences in their (sc. materials') use, as well as what qualities they have in buildings" (2.2.2 *de uarietatibus et discriminibus usus earum quasque haberent in aedificiis qualitates*). This is a good example of how theoretical access to other disciplines afforded by *ratiocinatio* serves the architect's judgment (*iudicium*) regarding the *opera* appropriate for a given task.

Having announced the explanatory project of Book II, Vitruvius goes on to account for a wide range of materials. He relies throughout on the four-element theory detailed earlier, explaining the properties of different substances in terms of the "temperament" (*temperatura*) or "mixture" (*mixtio*) of each.[112] He sometimes focuses on the qualitative properties of the element, sometimes on their structural coherence or stability in various combinations. After proceeding through materials including mud brick, sand, lime, and stone (2.3–8), he affords a substantial treatment to timber (2.9–10), which will be our focus here.[113] Because Vitruvius' explanations support the architect's capacity for judgment in design and construction, the inquiry into natural properties is subordinated to well-defined questions regarding the applicability of different materials for building purposes. In the sections on timber, these questions include: When should trees be felled for timber? How can they be cut to maximize their utility as a building material? What qualities do different woods have, and which are best for which purposes? What environments produce the best timber? Vitruvius' answers, as well as the supporting explanations which constitute *ratiocinatio*, can be schematized as in Table 3.3.

Taken individually, the arguments are less elaborate than those of the first case (1.6) but exhibit the same structure: they are based on principles shared with other branches of specialized knowledge and augmented by observation, experiment, and analogy in a *disputatio communis*. We will investigate Vitruvius' *ratiocinatio* for the first and second arguments (2.9.1–4), which are noteworthy for their use of analogy and observation. We will not handle the third and fourth claims (2.9.5–2.10), though we will return briefly to Vitruvius' discussion of the larch (2.9.14–17) in conclusion.

[112] These words are used at, for example, 2.5.2, 2.7.1, 2.7.2, 2.7.3, 2.9.9, 2.9.10, 2.9.11, 2.9.12, 2.9.14; cf. 2.10.3.

[113] For timber in Roman building, see Anderson 1997: 127–39.

Table 3.3 *The explanatory economy of* DA *2.9–10, on the architect's use of wood in building*

Argument I Trees should be felled for timber from the beginning of autumn until the beginning of spring (2.9.1).
- **Explained by** Trees have less robust wood in spring and summer than in autumn and winter (2.9.1–2).
  - **Explained by** (1) When preparing to bear fruit, trees redirect nourishment from their wood to their fruits (2.9.1–2).
    (2) Wood is strengthened by exposure to the cold air of winter months (2.9.2).

Argument II Before trees are felled for timber, they should be drained of their sap (2.9.3).
- **Explained by** Excess sap within the tree weakens its wood and reduces its longevity (2.9.3–4).

Argument III Various woods have different properties that commend them to different uses in building (2.9.5).
- **Explained by** Each tree species is composed of a different proportion of elements in its mixture (2.9.5): fir (2.9.6–7); oak (2.9.8); winter oak, turkey oak, beech, white poplar, black poplar, willow, linden, vitex (2.9.9); alder (2.9.10); elm, ash (2.9.11); hornbeam, cypress, pine (2.9.12); cedar, juniper (2.9.13); larch (2.9.14–17).

Argument IV Trees in sunny regions produce denser and more durable timber (2.10.2).
- **Explained by** The sun draws off the excess moisture from the Earth and trees (2.10.1–2).

3.4.1 Humans and Plants: The Biology of Conception

We look first at **Argument I**. Vitruvius maintains that "timber should be cut from the beginning of autumn until the time before Favonius begins to blow" (2.9.1 *materies caedenda est a primo autumno ad id tempus quod erit antequam flare incipiat fauonius*) – in other words, in autumn and winter, not spring and summer.[114] The basic explanation is straightforward. In spring, trees become "pregnant" (*praegnantes*) and "put the strength of their particular character into leaves and annual fruits" (*suae proprietatis uirtutem efferunt in frondem anniuersariosque fructus*).[115] In autumn, however, "when the fruit have ripened and the leaves have begun to wither" (2.9.2 *maturitate fructuum flaccescente fronde*), their strength is rather directed to their wood. These claims are paralleled in botanical and agricultural writings. Theophrastus maintained that "as a general rule every (sc. wood) is at the best season with regard to its strength not just

[114] The blowing of *Fauonius*, the west wind, marks the beginning of spring (Col. *RR* 8.11.7, Plin. *HN* 18.222).

[115] *Proprietas* refers to natural condition of something qua distinguishing factor (*TLL* X 2.2085.46–7 [Ottink]), a favorite word of Vitruvius in this sense (ibid. 2089.53–64).

when it has stopped putting out leaves, but when it has further finished ripening its fruit" (*Hist. pl.* 5.1.1 ὅλως πᾶν (sc. τὸ ξύλον) πρὸς ἰσχὺν ὡραιότατον οὐ μόνον πεπαυμένον τῆς βλαστήσεως ἀλλ' ἔτι μᾶλλον ἐκπεπᾶναν τὸν καρπόν).[116] The Roman agronomists give advice that relies on a similar principle – namely, that the finite nutrients available to a tree are directed at different times to different aspects of its growth.[117] As in his definition of *uentus* (1.6.2, Ch. 3.3.1), Vitruvius is thus following a mainstream account, but as also with wind, he does not leave the matter there.

Vitruvius provides a deeper account of the "pregnancy" of trees (cf. 2.9.1 *praegnantes*), building up theoretical scaffolding that supports his explanation in terms congenial to his own theory of matter. He offers the following remarkable analogy (2.9.1):

> cum ergo inanes et umidae[118] temporum necessitate eorum fuerint, uanae fiunt et raritatibus inbecillae; uti etiam corpora muliebria, cum conceperint, ad fetus partum[119] non iudicantur integra, neque in uenalibus ea, cum sunt praegnantia, praestantur sana, ideo quod in corpore praeseminatio crescens ex omnibus cibi potestatibus detrahit alimentum in se, et quo firmior efficitur ad maturitatem partus, eo minus patitur esse solidum id ex quo ipsum procreatur. itaque edito fetu, quod prius in aliud genus incrementi detrahebatur, cum ab disparatione procreationis est liberatum, inanibus et patentibus uenis in se recipit et lambendo sucum etiam solidescit et redit in pristinam naturae firmitatem.

> Whenever therefore they (sc. the trees) become hollow and moist under the compulsion of these seasons (sc. spring and summer), they become insubstantial and weak owing to the hollows within them. In like manner women's bodies, after they have conceived, are not considered sound up to the birth of child, nor among those who are enslaved are they (sc. women) advertised as healthy while they are pregnant. This is for the reason that the fetus, when it is growing in the body, draws to itself the nourishment from all the powers of food, and to the same extent that it grows firmer approaching the proper time for birth, so the less solid does it permit that (sc. body) to be from which it is generated. Accordingly, once the child has been born, it

[116] "Round" (στρογγύλον) timber – that is, uncut (*Hist. pl.* 5.5.6) – is exceptionally felled when the leaves come in. Theophrastus is an important source for Vitruvius's remarks on trees in Book II: Callebat et al. 2003: 153 and *passim*; \ 2011: 161–2.

[117] Cf. Garg. Mart. *Pom.* 2.9.2 (with Zainaldin 2020a: 196), 3.7.6, Pallad. 3.25.16, 7.5.1, 8.3.1, *Geop.* 10.13.2. Consider also Columella's discussion of the plant's vascular system (*RR* 3.10.2–3), discussed in Ch. 4.4.3.

[118] The conjecture *tumidae* (Rose and Müller-Strübing after Galiani) is unnecessary; Vitruvius explicitly associates *umor* with porousness (2.10.1–2).

[119] MSS have *ad faetus apartu*, whence *ad fetus partum* (Marini followed by Rose and Müller-Strübing); other editors hazard different solutions but aim at the same meaning.

> (sc. the body) now that it has been freed by being released from its generation takes into itself that which had previously been drawn away for another form of growth and, by absorbing the sap, also grows solid and returns to its former natural firmness.

Comparison of plant and animal life was widespread among ancient biological, botanical, and medical writers.[120] Several writers before Vitruvius' time employed analogies specifically between the growth of the child and the tree, including Empedocles, Democritus, and, at greatest length, the author of the Hippocratic *On the Nature of the Child* (Περὶ φύσιος τοῦ παιδίου, *De natura pueri*).[121] But while Vitruvius' analogy belongs to this tradition, his account differs from earlier treatments, according to which the Earth is like the womb and the tree its embryo; his model operates rather between the womb and wood and the fetus and the fruit.[122] The closest precedent, and probably Vitruvius' inspiration here, turns out again to be Theophrastus, who uses pregnancy (κύησις) and associated metaphors to discuss the seasonal bearing of fruit.[123] It must be stressed, however, that Theophrastus does not offer a full-blown comparison between the pregnant woman and the plant. While Theophrastus reflects on biological processes affecting the tree as an entire organism, Vitruvius' analogy develops a more detailed explanation of how "tree pregnancy" affects the quality of the wood in particular. This shift in focus suits the purposes of architectural *ratiocinatio*: he is not studying plant generation *in abstracto*, but how it changes the material structure of timber, since it is this quality which concerns the architect. It is therefore worth examining the details of Vitruvius' analogy.

We observed earlier that, for human beings, Vitruvius' biological or medical theory is simply a special case of his physical theory: he eschews appeal to specifically organic processes and understands the body in terms of its natural proportions of elements (*mixtio* or *temperatura*) and the physical coherence and solidity that result therefrom. Vitruvius similarly analyzes the pregnant tree and the pregnant woman in almost exclusively physical terms: his language is that of material attraction and withdrawal (*detrahit … in se*, *detrahebatur*, *in se recipiens*); insubstantiality, emptiness, and porosity (*inanes*, *uanae*, *raritatibus*, *non … integra*, *minus … solidum*); and firmness, coherence, and solidity

120 Alcmaeon, Empedocles, Hippocratics, Aristotle, Theophrastus, Galen, et al.: see, for example, Lloyd 1966: 323–4, 335, 347–8, 370–2; Lonie 1981: 211–18; Holmes 2017. See also Ch. 4, *passim*.

121 Emp. D–K A70 = L–M D245, Democr. D–K B148 = L–M D172, Hippocr. *Nat. pue.* 22–7 L 7.514–28, with previous n.

122 More comparable in this respect at least is the Empedoclean snippet οὕτω δ ᾠοτοκεῖ μακρὰ δένδρεα πρῶτον ἐλαίας (D–K B79 = L–M D254).

123 *Passim* in *Caus. pl.* 1.12.8–14.4. Varro's *praegnatio* in reference to plants (*Rust* 1.44.4), the earliest before Vitruvius, follows κύησις at *Caus. pl.* 1.14.1.

(*firmior, solidescit, naturae firmitatem*). There is thus an intense abstraction in his choice to focus on the bodies (*corpora*) of tree and woman qua physical rather than organic systems. This abstraction produces what must be purposeful ambiguity in his discussion of procreator and offspring: although the words following *ideo quod* ("this is for the reason that") refer properly to the human, the language could apply almost equally well to the plant. The ambiguity is thrown into even sharper relief by expressions such as *maturitatem* and *lambendo sucum*, which, though strictly referring to the fetus and female body, seem more appropriate for the tree and its fruit.[124] It could be regarded a defect of Vitruvius' biological theory that he conceives of organic processes largely in terms of his theory of matter,[125] but the same fact may alternatively count as an explanatory virtue: there is an economy in his attempt to reduce disparate phenomena to a very few principles in this interdisciplinary *disputatio*. Even more to the point, his discussion is, as we saw, directed to the specific end of illustrating the *usus* and *qualitates in aedificiis* of the (dead) wood of trees (cf. 2.2.2). The elision of biological processes is unremarkable from this perspective, for the architect is interested chiefly in the properties of bodies so far as they render them *apta* for construction.

Another crucial feature of Vitruvius' analogy is the way that it repurposes and extends a fundamental botanical and biological principle – that of the vital fluid or sap in an animal or plant necessary for growth and reproduction. In Theophrastus' botanical language, this fluid is called *to hugron* (τὸ ὑγρόν) or *hê hugrotês* (ἡ ὑγρότης): literally, "the moist" or "moisture." It is identified, along with fiber, veins, and flesh, as one of the primary substances (*Hist. pl.* 1.2.1 ἀρχαί) or basic principles (τὰς τῶν στοιχείων δυνάμεις) of plant life: "the whole being and nature (sc. of plants) consists in these" (ἡ μὲν οὖν οὐσία καὶ ὅλη φύσις ἐν τούτοις).[126] Vitruvius calls this substance *sucus*, the usual Latin word for it in botanical and agricultural contexts.[127] He shows that he thinks of *sucus* as a nutritive vehicle for growth and reproduction when he mentions *alimentum*,[128] but

[124] For *sucus* = sap, see n.127; *maturus* (*maturitas*) is the *vox propria* for ripe fruit (*OLD* 1; *TLL* VIII 498.24–58 [Brandt]).

[125] So Callebat et al. 2003: 152, 155; Callebat 2003: 152.

[126] At *Hist. pl.* 1.2.3, 1.2.5, Theophrastus discusses alternative terminology connected with τὸ ὑγρόν (e.g. ὁ ὀπός, τὸ δάκρυον, but often given no particular name); for τὸ ὑγρόν as principle of growth (in conjunction with heat, ἡ θερμότης), cf. *Hist. pl.* 1.2.4, 1.12.2, 9.1.1–5, *Caus. pl.* 1.2.4, 1.6.1, 3.15.3, 4.16.3–4, 6.7.1.

[127] *Sucus* = sap (*OLD* 3a) but also sustenance in the soil as potential sap (so 2.9.2, quoted later in this section); for the latter sense, see also Ch. 4, n.102.

[128] For this reason it may not be wholly fair to say that Vitruvius allows for no organic processes in his biology (n.125).

his twist is to analyze its effects in terms of one of his four elements, *umor*: this view is established already in the opening sentence, where he states that the trees become "moist" (*umidae*) in spring and summer. By linking *sucus* and *umor*, Vitruvius brings to bear his coherence-focused account of matter. The problem that the analogy highlights is the siphoning away of food from the pregnant body/tree "for another kind of growth" (*in aliud genus incrementi*, the fetus/fruit). Vitruvius thinks of this redistribution of nutriments in physical terms: as the *sucus* critical for growth is drawn out of one body and into another (the offspring), it leaves "pores" or "gaps" (*raritatibus*) that make the wood of the trees fragile. The flowing *sucus* opens these channels in the main body, but in vacating them creates a material imbalance in the tissue which leaves the substance hollow and weak (cf. *inanes*, *uanae*). Vitruvius' account thus elaborates in his own physical terms what is merely a passing claim in Theophrastus that trees are "emptied in summer because of their sprouting and conceiving fruit" (*Caus. pl.* 1.13.3 κενωθέντα τοῦ θέρους ἐκ τῆς βλαστήσεως καὶ τῆς καρπογονίας).

After the fetus is born or the fruit finishes its ripening, *sucus* may be retained in the body again. This change is conceived, again on the coherence-focused physical account, as a closure of gaps in the material structure (cf. 2.9.1 *lambendo sucum etiam solidescit et redit in pristinam naturae firmitatem*). In completing the analogy, Vitruvius uses nearly the same terms for the tree's revival after it has fruited (2.9.2):

> eadem ratione autumnali tempore maturitate fructuum flaccescente fronde, e terra recipientes radices arborum[129] in se sucum recuperantur et restituuntur in antiquam soliditatem.

> In the same way, when in autumn the fruit are ripe and the leaves have begun to wither, then the roots of the tree, taking into themselves sap from the earth, recover and are restored to their previous solidity.

Like the female body, the tree absorbs and retains *sucus* previously diverted for the offspring (2.9.1 *in se recipiet* ~ 2.9.2 *recipientes . . . in se*), a change which effects a return to an earlier state of material firmness (2.9.1 *pristinam firmitatem* ~ 2.9.2 *antiquam soliditatem*). Finally, Vitruvius clinches his argument for cutting trees in the fall and winter by adding that the "force of the winter air compresses and solidifies them" (2.9.2 *aeris hiberni uis comprimit et consolidat eas*). The wood made dense by the cold yields

[129] Following the reading of the MSS; Rose and Müller-Strübing print Marini's conjecture *recipientes per radices arbores*.

a better timber.[130] Hence, it is autumn and winter that are the best time for obtaining sturdy wood for building.

Vitruvius' discussion of the proper time to fell trees for timber exemplifies his use of analogy to support explanation. The tree–woman analogy allows him to recast a principle fundamental in both botany and biology – vital nourishing fluid (τὸ ὑγρόν or *sucus*) – in terms congenial to his elemental theory of matter. Here is another example of the architect's *disputatio communis*, again bent to practical needs. Linking the descriptions of *sucus*-as-*alimentum* and *sucus*-as-*umor*, Vitruvius prioritizes a physical analysis that highlights the aspects most relevant for the architect: the strength and durability of the material. The abstraction of his account is remarkable for offering the most general possible view of the situation based in *rerum natura*. Yet there is a crucial detail that reminds us, again, that this naturalized architecture is an architecture for the Romans: 2.9.1 *neque in uenalibus ea . . . praestantur sana* appeals to experience of Roman slave auctions.[131] It is appropriate to find a hint of Roman Imperialism here, especially if we recall that Vitruvius likely served on campaign under Caesar, whose Gallic conquests brought large numbers of slaves to Italian land.[132] Perhaps the assumption of non-Roman identity makes it easier for Vitruvius to minimize the humanity of the enslaved pregnant woman in the literally inhuman (because wholly materialistic) analysis that he offers. Again, we are looking at nature through Roman eyes.

3.4.2 *Treating Lumber: Agriculture and Architecture*

We come now to **Argument II**. In 2.9.3–4, Vitruvius discusses how trees may be cut so that the wood "will remain useful in constructions for a very long time" (2.9.4 *habere poterunt magnas in aedificiis ad uetustatem utilitates*). Here is his method (2.9.3):

> caedi autem ita oportet uti incidatur arboris crassitudo ad mediam medullam et relinquatur uti per eam exsiccescat stillando sucus. ita qui inest in his inutilis liquor effluens per torulum non patietur emori in eo saniem nec corrumpi materiae qualitatem. tum autem, cum sicca et sine stillis erit arbor, deiciatur et ita erit optima in usu.

[130] For *frigus* as a solidifying force, see n.91.

[131] For Roman slave auctions, see Harris 1980: 121; Bradley 1994: 51–6; Schumacher 2001: 54–65. Vitruvius' claim that pregnant women are "not advertised as healthy" can be compared with the legal requirement to disclose *quid morbii uitiiue cuique sit* (*Dig.* Vlp. 21.1.1).

[132] For Vitruvius' service, see 1.pr.2 and Ch. 2.2 (n.3), and the siege of Larignum discussed later in this section. For the number enslaved by Caesar, cf. *BGall.* 2.33, Plut. *Vit. Caes.* 15.3, App. *Celt.* 1.6 (with Harris 1980: 122).

> Moreover, the cutting should be done in such a way that a notch be made in the trunk of the tree to the center of its core, and it (sc. the tree) be left so that it may dry by dripping its sap out through this. In this way, the useless liquid in it will flow out through the sapwood and not permit the discharge to decay in it nor the quality of the timber to be spoiled. Then once the tree is dried and no longer dripping, let it be felled and it will be most suitable in use.

Sucus again plays a prominent role in Vitruvius' explanation, though its function is conceived differently here: while earlier the outflow of *sucus* was detrimental, now draining *sucus* or *liquor* will improve the quality of the timber. This difference does not reflect an underlying inconsistency but the different physical contexts of the explanations, as the architect is concerned in each case with the effects of *sucus* on his material. During tree-pregnancy, the veins become especially large and swollen (cf. 2.9.1 *inanibus et patentibus uenis*); as the wood is evacuated of *sucus* flowing to the fruit, it becomes hollow and gappy. If cut in season (autumn and winter), however, the veins of the wood are compact, so that it has its ordinary *firmitas* or *soliditas*. Removing *sucus* at this time will not impair the strength of the timber. To the contrary, it is necessary because the fluid that is retained in the tree after felling will be superfluous (cf. 2.9.3 *inutilis liquor*), or rather, worse than superfluous: *emori in eo saniem*, medical language denoting putrefaction and taint, indicates that the degradation of this organic compound will vitiate the material.[133]

Vitruvius expands on the potentially corrupting aspect of *sucus* afterwards (2.9.4), when he appeals to agricultural practice to support his prescription to drain the trees of sap: "the fact of the matter can indeed be observed also in plantations of trees" (*hoc autem ita esse licet animum aduertere etiam de arbustis*).[134] He explains (2.9.4):

> ea enim cum suo quoque tempore ad imum perforata castrantur, profundunt e medullis quem habent in se superantem et uitiosum per foramina liquorem, et ita siccescendo recipiunt in se diuturnitatem. qui autem non habent ex arboribus exitus umores intra concrescentes putrescunt et efficiunt inanes eas et uitiosas.

> For when they (sc. the trees) are pierced and pruned very low at the proper time, they pour out from their pith the superfluous and unsound liquid which they retain in themselves, and by being dried out in that way take on

[133] *Sanies* is typically of putrefying blood or ichorous discharge (*OLD* 1), hence invoking an analogy with animal life; for *sucus* compared with blood, cf. Thphr. *Hist. pl.* 1.2.5.

[134] As an agricultural term, *arbustum* usually means a plantation of trees for supporting vines (*OLD* 2; *TLL* II 430.34–63 [Bannier]; Bruno 1969: 62; White 1970: 236).

durability. But the moistures which are unable to leave the tree clot inwardly and putrefy, making the trees hollow and unsound.

In looking to the effects of cultivation on trees as evidence for their natural qualities, Vitruvius is following a tradition that stretches back at least to Theophrastus.[135] Versions of the practice described here are indeed found in the agricultural authors.[136] Noteworthy here is how Vitruvius harnesses the observation of agricultural practice in service of a more abstract account of *sucus* supporting his precept: excess liquid clots and putrefies, making trees "empty" (*inanes*) and "unsound" (*uitiosas*). Because he has formulated the effects of putrefying sap in terms of its effects on the material structure of the trees, it is easy to see why fluids retained in the timber are detrimental for the architect's practice. He makes the connection between agriculture and architecture explicit in the conclusion of the passage (2.9.4):

> ergo si stantes et uiuae siccescendo non senescunt, sine dubio, cum eae ad materiam deiciuntur, cum ea ratione curatae fuerint, habere poterunt magnas in aedificiis ad uetustatem utilitates.
>
> Therefore if they (sc. the trees) while standing and alive remain young by being dried out, then without doubt when they are cut down for lumber after receiving the same treatment they will remain useful in buildings for a very long time.

This second argument (2.9.3–4) thus provides another illustration of *ratiocinatio*, which not only draws on the principles of other fields of specialized knowledge (*doctrinae*) for explanation but also bolsters them in a *disputatio communis*.

We conclude our discussion of the architect's use of timber with a brief look at a digression embedded in **Argument III** in our outline (2.9.5–17). The last of the woods that Vitruvius treats is larch (*larix*, 2.9.14), a tree remarkable for, among other things, its resistance to fire (*flammam ex igni non recipit*). After attempting to explain its unusual properties from its elemental makeup, he says that "it is possible to know how it was

[135] In *Caus. pl.* 3, Theophrastus investigates the ἀρχὴ ἐκ τῆς ἐπινοίας καὶ παρασκευῆς (3.1.1) (i.e. under human cultivation), as opposed to the ἀρχὴ τῆς φύσεως (i.e. spontaneous natural growth). In the former respect, however, his investigations proceed on the assumption that human cultivation "aids nature in its ends" (συνεργεῖν τῇ φύσει πρὸς τὸ τέλος), thus complicating the dichotomy.

[136] Compare the tradition that the almond tree should be drained of its noxious sap (variously called τὸ δάκρυον, ἡ ἐπιρροή, *pituita*, *umor noxius*): Thphr. *Hist. pl.* 2.2.11, 2.7.7, *Caus. pl.* 1.17.9, Nic. Dam. *De plantis* 821a39–40, Plin. *HN* 17.252, Garg. Mart. *Pom.* 3.7.2, Pallad. 2.15.11, Zainaldin 2020a: 273–4.

discovered" (2.9.15 *quemadmodum sit inuenta est causam cognoscere*[137]). He relates an anecdote from Julius Caesar's siege of the Italian town Larignum, which he might have witnessed in person.[138] The defenders used a tower of larch as a perch for harassing Caesar's troops; when he attempted to burn it down (2.9.15), he found it remained unscathed (2.9.16). "Wondering" (*admirans*), Caesar sought from the inhabitants the location of the trees after the town surrendered; the local *copiae* of larches were then made available for exportation and use elsewhere in Italy. In this way, Vitruvius connects the discovery and commodification of this new and especially useful kind of timber with the conquests of Caesar. The discovery was in the Romans' own backyard, so to speak, but nonetheless illustrates how Roman *imperium* sets into circulation the products of nature.[139] Indeed, the larch proper is described for the first time here in Greco-Roman writings: Vitruvius' application of his physical theory to its properties is thus a novel contribution to ancient botanical theory.[140] His awareness that he is offering something unknown to his earlier authorities, chiefly Theophrastus, doubtless explains his interest in discussing the origins of the tree. This historical anecdote again recalls us to a Roman point of view, especially because Vitruvius goes on to discuss the availability and usefulness of this wood in the Roman urban center (2.9.16).

3.5 Conclusion

In this chapter, we explored the *ars* of architecture that Vitruvius presents in *DA*. In studying the *architecti scientia* (1.1), we saw how he presents architectural expertise as "architectonic": the architect's peculiar capacity resides in his ability to judge (*iudicare*) and evaluate (*probare*) the products (*opera*) of other branches of specialized knowledge (*doctrinae*) for his own purposes. Vitruvius grounds the architect's capacity for judgment in a painstakingly developed notion of *ratiocinatio* (reasoning), whose constituent activities are demonstration (*demonstratio*) and explanation (*explicatio*). *Ratiocinatio* operates at the level of the principles and accounts (*rationes*) shared by the *artes* and affords a competent albeit limited grasp of

[137] For text and interpretation (*causam* an emendation of *causa*), see Fensterbusch 1976: 182; Adams 2016: 181–2.

[138] For the location see Šašel 1981: 255–6, arguing that the episode occurred probably in 59 or 58 BC; for the possibility that Vitruvius was at the siege, see also Gros et al. 1997: 215.

[139] Cf. Plin. *HN* 14.2, 27.2–3, discussed in Ch. 2.3.

[140] For *larix*, see André 2010: 139. Theophrastus' πεύκη (~ Pliny's *larix*) is not larch but stone pine (ibid. and Meiggs 1982: 24). See also Callebat et al. 2003: 171; cf. Meiggs 1982: 248–50.

their theoretical contents. Architecture is not unique in its connection with other disciplines – all higher *artes* share this *coniunctio* through reasoning – nor again in its "shared disputation" (*disputatio communis*). But it is unique in that its proper *scientia* depends, alone among the *artes*, on the evaluative grasp of the products (*opera*) of other disciplines. As we have seen, Vitruvius does not indulge in idle theorizing in defining architecture in this expansively interdisciplinary fashion: his epistemology is borne out in the precepts of *DA*. Our case studies revealed the central role played by *ratiocinatio* in informing the architect's judgment and highlighted the architect's own contribution to a *disputatio communis* via analogy, experiment, and observation.

This sophisticated approach to *ars* has an obvious benefit for Vitruvius' project in *DA*. His explanations, based as they are on authoritative first principles from nature and leading accounts of other *doctrinae*, are at the same time justifications of his precepts. *Ratiocinatio* thus supports a redoubtable, naturalistic account of the discipline. But his approach has a further payoff that goes beyond the architectural objectives of *DA* narrowly defined. In exhibiting *ratiocinatio* and *disputatio communis*, Vitruvius also presents the fruits of an inquiry into nature (*rerum natura*) which underpins the whole treatment of his subject matter vis-à-vis Roman civilization. *DA* becomes, through the lens of Roman architecture, a vehicle for synthesizing a full-scale account of the natural and human worlds: Vitruvius thus unfolds not only his vision of what *architectura* can and should be, but also an understanding of *rerum natura* that makes that vision possible – a complete "image of expertise."[141]

Vitruvius' *DA* is fundamental for establishing the contours of the scientific culture of the *artes* in the Imperial period. We will conclude this chapter with a few comments on its broader significance in this regard. There is no need to recapitulate how Vitruvius' treatise responds individually to the criteria for the higher *artes* that had developed by the late Republic (systematic, interdisciplinary, explanatory, and balanced: see Chapter 1), since these are the objects of explicit reflection throughout *DA*, as we saw time and again. What does bear emphasizing is how Vitruvius unites these features into a single, well-developed notion of *ars*. Equally significant, the subject matter of *DA* entails the application of this unified notion of *ars* to a discipline concerned with manipulation of the physical world and, hence, with the principles of *rerum natura*. Vitruvius' work thus becomes a model for the scientific character of the *artes*, in that it

[141] See further the Introduction.

shows how an epistemologically self-aware approach to *ars* can also sustain wider-ranging reflection on nature. The universalizing project of *DA* aspires to give a true account of the relevant domains of *rerum natura* that will both secure the status of the *ars* and ground its precepts. Yet we also saw that the same investigation is always related to the practical imperatives set for the *ars* and thus ultimately related to a Roman point of view. Vitruvius' treatise is only the first of several Imperial *artes* which would develop and follow to its natural conclusion this rich and uniquely Roman conception of specialized knowledge. In the following chapter, we will trace an analogue for Vitruvius' novel approach to *architectura* in another of the Imperial *artes*, on agriculture, before we look more widely in subsequent chapters to other problems that emerged as the scientific culture of the *artes* took shape.

CHAPTER 4

Columella and the New Roman Agronomy

The Art of Agriculture and Knowledge of Nature in Res rustica

The appearance of Vitruvius' *De architectura* at the dawn of the Roman Empire heralded the beginning of a new age for the *artes*. This expansive treatment of architecture in the mold of the higher *artes* (Chapter 1) showed how the *artes* could, in presenting an image of disciplinary expertise, also prompt further-reaching reflection on knowledge, nature, and Roman society. The notion of specialized knowledge embodied in Vitruvius' treatise animated the growth of the *artes* as a scientific culture in the early Imperial period. In this chapter, we will jump forward from Vitruvius' treatise to study the most ambitious agricultural work extant from Greco-Roman antiquity: Columella's *Res rustica* (*RR*). This leap is worth the cost of a chronological wrinkle – Celsus, whom we will study in Chapter 5, wrote before and was a source for Columella – because agriculture affords a unique opportunity to consider the influence of the new model of the *artes* on a discipline which already possessed a rich Latin tradition. No pre-Imperial Roman precedents are extant for the other *artes* of this book: architecture, medicine, the art of war, and surveying.[1] The situation is the reverse in agronomy, with two Republican works surviving: Cato's *De agri cultura* (before 149 BC) and Varro's *De re rustica* (37 BC). We also know something about several other full-scale Latin agricultural treatises from Republican times (not to mention didactic poetry) and a famous Punic work translated into Latin. This intense interest was continued into the Imperial age, when numerous agronomic writings, including five books of Celsus' *Artes*, appeared before Columella's own *RR* was published in AD 60/1–65.[2] These earlier works, especially the Republican agronomic writings, serve as a measure for judging the nature and extent of the change in the *artes* in the early Empire.

[1] For Latin architectural writings, see Ch. 3, n.2; for the medical writings of Cato and Varro, Boscherini 1993; for Republican military writings, Ch. 6, n.5. The Republic may have seen no theoretical surveying writings at all: Ch. 7, n.1.

[2] Date: Ch. 2, n.94.

4.1 The Agronomic Background

Columella himself was acutely aware of his many predecessors in agronomy, both Greek and Roman. As to the former – "the great crowd of Greeks who give instructions about agriculture" (1.1.7 *magna . . . Graecorum turba . . . de rusticis rebus praecipiens*) – he follows Varro in producing a lengthy list of names, forty-five in total, that includes the famous (from Hesiod to Xenophon to Theophrastus) and the obscure (1.1.7–11).[3] More original is his account of the Roman agronomists, which figures the growth of Latin-language agricultural literature as the education of a child (1.1.12–13):[4]

> et ut agricolationem Romana tandem ciuitate donemus (nam adhuc istis auctoribus Graecae gentis fuit), iam nunc M. Catonem Censorium illum memoremus, qui eam Latine loqui primus instituit, post hunc duos Sasernas, patrem et filium, qui eam diligentius erudiuerunt, ac deinde Scrofam Tremelium qui etiam eloquentem reddidit et M. Terentium qui expoliuit, mox Vergilium, qui carminum quoque potentem fecit, [13] nec postremo quasi paedagogi eius meminisse dedignemur Iuli Hygini: uerumtamen ut Carthaginiensem Magonem rusticationis parentem maxime ueneremur, nam huius octo et uiginti memorabilia illa uolumina ex senatus consulto in Latinum sermonem conuersa sunt.

> And to at last confer upon farming Roman citizenship (for, up to this point, it belonged to the aforementioned authors among the Greek peoples), let us now recall the famous M. Cato, the Censor, who first taught it (sc. farming) to speak Latin, and then afterwards the two Sasernae, father and son, who more diligently educated it, and next Tremelius Scrofa, who rendered it eloquent as well, and M. Terentius, who gave it polish, and soon thereafter Virgil, who made it capable of song too; [13] nor should we disdain to recall Iulius Hyginus, who was like a teacher to it. Nonetheless, (sc. our recollection of all these should be such) that we honor Mago the Carthaginian above all as the father of agriculture, for his twenty-eight books worthy of remembrance were translated into the Latin language *ex senatus consulto*.

This lineage reconstructs the development of Roman agronomy into an *ars*. To appreciate the nature of this intellectual inheritance, it is worth considering what we know of these figures.

Cato is aptly said to have taught *agricolatio* to speak Latin: his *De agri cultura* (*Agr.*) was not just the oldest Roman agricultural text, but one of the earliest pieces of Latin prose. The work is full of interesting and useful agricultural advice, in style didactic, workmanlike, free of theorizing or

[3] See on these Martin 1971: 53–72; Hentz 1978: I 39–46, III 137–44; 1979: 154–9; Di Lorenzo 1986.
[4] Text of *RR* is Rodgers 2010.

explanation.[5] Each of the 162 chapters develops its own idea, though the forms are extremely varied: sometimes a large topic is developed systematically, but we are frequently thrown directly into procedures, aphorisms, lists, recipes, incantations, observations, notes, or general rules. Certain organizing principles can be inferred – a group of chapters on the management of a country estate, an agricultural calendar, blocks of recipes – but these impose only the loosest structure on the work.[6] As we said in Chapter 1, it is fair to say that, despite traces of Greek learning, the work patently lacks not only the spirit of systematic organization which would later characterize the *artes*, but also the rest of the marks of the higher *artes* at the end of the Republic. Cato's treatise was a landmark for all later agricultural authors, as much owing to the authority of its author as to the basic soundness of its advice. Yet it seems that by Varro's time, *Agr.* was felt in certain respects to be old-fashioned, as were the writings of the Sasernae, father and son, who Columella says "more diligently educated" agriculture (1.1.12). Traces of their lost work, which was probably in multiple books and appeared sometime in the late second or early first century BC, justify Columella's view that it represented an advance beyond Cato in its learning.[7] Nonetheless, their work too seems to have lacked the intellectual sensibilities of the systematic treatises that had begun appearing in the mid-first century BC: hence, both Cato and the Sasernae are criticized by Varro's interlocutors in *Rust.*, including Scrofa, for including irrelevancies in their work – the essential complaint here is that they lacked a clear notion of the *ars* – as well as for insufficient abstraction in their attempt to give general rules, such as in calculating the labor required to cultivate a field.[8]

5 On the style (the oral dimension is frequently stressed), see e.g. Mazzarino 1952: 123–32; White 1970: 19–20; Astin 1978: 190; Dalby 1998: 16–18, 25; Reay 2005: 340–50; Diederich 2007: 156–72; Sciarrino 2011: 141–56; further Suerbaum 2004: 222–3, 257–8

6 Brehaut 1933: xvi–xxiii; White 1970: 44–5 (~ 1973: 447–50); Astin 1978: 191–203; further Suerbaum 2004: 250.

7 The Sasernae are mentioned as father and son by Varro, *Rust.* 1.2.22, Plin. *HN* 17.199. The agronomists also speak of "Saserna" (singular), which has been taken to suggest that the work was co-written, or else that the son supplemented, revised, or edited what his father wrote; an extended period of composition is likely. On the Sasernae's agronomy, see Kolendo 1973 (62–70 on the treatise); also Speranza 1971–2; Martin 1971: 81–5; White 1973: 459–60; Hentz 1978: I 53–4, Richter 1981–3: III 572–3. Testimonia and fragments: Kolendo 1973: 71–80; Speranza 1974: 33–45. It is doubtful the Sasernae knew Hipparchus' discovery of the precession of the equinoxes (as is often claimed) and so this cannot be taken as a mark of their erudition (rightly Kolendo 1973: 42).

8 Cf. Varro, *Rust.* 1.2.22–28 (with Kolendo 1973: 66; Martin 1971: 81–3), 1.18.1–6 (with Kolendo 1973: 64; White 1973: 480). See Ch. 1.2 (and n.26) on early Latin technical writing.

Cn. Tremelius Scrofa, who wrote in the first century BC, rendered agriculture "eloquent" (1.1.12).[9] This should be taken at least to imply a stylistic advancement in his treatment of the discipline, one that would have included, it is natural to assume, a more fluent and systematic approach.[10] Certainly this would be true if his style is reflected in what Varro attributes to the "Scrofa" who appears as an interlocutor in *Rust.* and demonstrates great concern for systematization and the epistemological niceties of the discipline.[11] The significance of some evidence is disputed, but there are hints of the relative sophistication of Scrofa's treatise, some of which will be discussed later in this chapter.[12] With Varro himself we are of course on firmer ground. The dialogue *De re rustica* (*Rust.*) covers in three books agriculture proper, animal husbandry, and *pastio uillatica* (products of the *uilla*, including, for example, domestic fowl, fish, and honey). The work is highly systematic, as we noted in Chapter 1,[13] and full of learning, especially Greek: witness already the fifty-plus Greek sources trotted out in the bibliography of sorts that Varro sketches for his wife Fundania, addressee of Book I.[14] Puns and humor abound; strong (and characteristically Varronian) interest in etymology is frequently evidenced in passing remarks; references to Greek anthropology and philosophy are casually introduced; and scholars have even sought to recover sophisticated social and political commentary from the dialogue.[15] Varro's work is precisely the

[9] Historical Scrofa: Martin 1971: 237–43; Brunt 1972: 304–7; Heurgon 1978–97: I 109–11; Perl 1980; Nelsestuen 2011: 317–21. Testimonia and fragments: Speranza 1974: 46–55.

[10] At *RR* 2.1.1, Scrofa is said to have written *eleganter et scite*; note that *elegantia* refers as much to organization as style: Martin 1971: 244–5; Hentz 1978: I 52–3; Rawson 1985: 136; Nelsestuen 2011: 323–4.

[11] For example, Scrofa's analysis of soil into ninety-nine varieties (*Rust.* 1.9.1–4 with Skydsgaard 1968: 14) and the *scientia pastoralis* into eighty-one subdivisions (2.1.11–12); consider further his portentous claim that *agri cultura* is both *ars* and *scientia* (1.3.1). Reconstruction of Scrofa's work is complicated by different views of the evidentiary value of *Rust.*: did Varro know the agricultural writings of Scrofa, and if so, how did he use them? Four approaches may be distinguished: (1) Varro's Scrofa accurately represents the historical Scrofa's work (Martin 1971: 237–55; cf. White 1973: 460–1; Hentz 1978: I 52); (2) Varro was ignorant of Scrofa's work, and his character Scrofa is a "mouthpiece" for his own ideas (Brunt 1972; Richter 1981–3: III 574); (3) Varro did not use Scrofa's published work but had personal knowledge of his opinions (Heurgon 1978–97: I xliv, Richter 1981–3: III 574); (4) Varro knew Scrofa's work but manipulated the persona and opinions for literary purposes, perhaps parodically (Kronenberg 2009: 77–81; Nelsestuen 2011: 328–47; Sáez Fernández 2013; Doody 2017: 195–6; cf. Skydsgaard 1968: 37; Noè 1977b; Rawson 1985: 136–7).

[12] Uncertainties (previous n.) do not undermine the plausible supposition that Scrofa's contribution to Latin agronomic literature was a rationalizing and systematic approach (cf. White 1973: 460; Rawson 1985: 136; Nelsestuen 2011: 324). See also nn.36,38.

[13] Ch. 1, n.32. [14] See n.3.

[15] On these features see, for example, Tatum 1995: 195–200; Green 1997; Agache 1998; Diederich 2007: 172–209; Kronenberg 2009: 73–129; Nelsestuen 2011: 328–47; Diederich 2013; Nelsestuen 2015, 2016, 2017; Haß 2018; Brown 2019; earlier, for example, Heurgon 1950; Heisterhagen 1952; Cossarini 1976–7; Noè 1977a; Laughton 1978; Traglia 1985.

refined literary treatment of the discipline suggested by Columella's claim that he "gave it polish" (1.1.12).

Virgil with his *Georgics* (c. 29 BC) is said to have made agriculture "capable of song" (1.1.12). This versification stood outside the *artes* proper but was of great significance for later authors of prose treatises, who often handled him as an authority.[16] Naturally, his prestige was more closely related to the literary achievement of his poetry than to advances in the technical treatment of the material, which was, however, in many respects sound.[17] He is the most frequently cited of all of Columella's sources, but this has much to do with the compulsive elegance of Columella's style, which leads him to speak with the mellifluous voice of *noster Vergilius* whenever he can – to say nothing of the didactic poem of Book X that purports to fill a lacuna in the *Georgics*.[18] We know little in turn of the agricultural treatise of C. Iulius Hyginus (Augustus' freedman), although the description *quasi paedagogus* (1.1.13) has stimulated speculation that it was elementary in character.[19] At the end of the list we leap back in time to Mago, who is called the "father of agriculture" (1.1.13 *rusticationis parentem*). Varro has similarly high praise for the Punic agronomist (*Rust.* 1.1.10). We know next to nothing about Mago, but the twenty-eight books under his name ran the gamut of rural business, from general principles of farm management to the cultivation of crops to animal husbandry and apiculture.[20] By Columella's honorific it is likely meant that Mago was responsible for assembling the full range of subjects that

[16] Virgil is the most frequently cited authority in Columella, who has a very positive but not wholly uncritical attitude; Pliny, who also cites the poet frequently, is more polemical. For Virgil in Columella's prose books, see De Saint-Denis 1971; Cossarini 1977; Richter 1981–3: III 640–4; Dumont 2008; Bretin-Chabrol 2010; Mielke 2018, 2023: 455–76 and *passim*; in Pliny, Bruère 1956; Howe 1985: 563, 570–2; Rex 2001: 125–7; and in both, Wilkinson 1969: 270–3; Christmann 1982; Doody 2007; Thibodeau 2011: 222–4. See also n.18.

[17] Spurr 1986a.

[18] Book X has attracted outsized attention (four commentaries: Ash 1930; De Saint-Denis 1969; Boldrer 1996; White 2013); see recently Gowers 2000; Henderson 2002; Diederich 2007: 227–58; Cowan 2009; Fögen 2009: 182–5; Prioux 2013; Gindhart 2017; Reitz 2017; Formisano 2018; Mielke 2023 *passim*; further Fögen 2009: 182n94; Mielke 2023: 115n391.

[19] So, plausibly, Hentz 1978: I 50; cf. Noè 2000: 432; Mielke 2023: 128n471, 162n642. The pronoun *eius* after *paedagogus* has sometimes been taken to refer to Virgil (who precedes) but should certainly point to *agricolatio*. For Iulius Hyginus, see *RE* (1) X 2.628.12–651.67 (Tolkiehn); also Reitzenstein 1884: 18–27.

[20] The only biographical notice is at Plin. *HN* 18.22; Mahaffy 1889: 31–2 thinks "Mago" a mere name attached to Punic agricultural lore (cf. White 1973: 475; Krings and Devillers 1996: 511; Krings 2008: 25); others accept his reality and have him active sometime between the fourth and second centuries BC, when in any event the writings undoubtedly took shape (e.g. Martin 1971: 45n3; Speranza 1974: 77–8; Heurgon 1976: 442; Greene and Kehoe 1995: 110–15; Domínguez Petit 2004: 181–2). Testimonia and fragments: Speranza 1974: 76–119, adding from the *agrimensores* L 348.19–350.16; see also n.22.

belonged to the *res rusticae*, becoming a sort of progenitor for the discipline as conceived in its total extent by the Roman agronomists.[21] Mago's work was translated into Latin *ex senatus consulto* probably not long after Carthage fell, but was apparently used more often in Greek translations and epitomes, which also added material from Greek sources.[22] It is obviously difficult to know what the texture of Mago's treatise was like, but nothing in the fragments contradicts the impression of a pragmatic, un-theoretical approach.

It bears observing, finally, that the education of agriculture figured by Columella took place on a uniquely prominent stage – for agriculture was never merely a discipline but also the way of life for the great majority of Romans and a backdrop against which Roman culture developed. Hence, it offered the bones of an idealized Roman identity connected with the *mos maiorum*, a fact of no little importance for the Roman agronomists, Columella included, as we will see in this chapter.[23]

4.2 Following Nature with Columella

What was Columella's own contribution to be in the face of this venerable tradition – one continued by men of his own time such as Celsus, whose writings embraced the "whole body of the discipline" (1.1.14 *totum corpus disciplinae*), and others such as Iulius Atticus and Iulius Graecinus?[24] What more could be done now that agriculture had, so to speak, been educated, and spoke Latin in prose and verse? Straightforwardly, *RR* meets this problem by surpassing previous Roman writings in scope and detail. In the longest of all *artes* now extant, the reader is guided through the

[21] Hence Varro says of him, *qui res dispersas comprendit* (*Rust.* 1.1.10). For Mago's achievement, see Martin 1971: 44–6; Richter 1981–3: III 621–8; Santini 2000: 370; Domínguez Petit 2004: 189–90; Thibodeau 2018: 467–8.

[22] Latin translation: Plin. *HN* 18.22 (D. Iunius Silanus oversaw the effort); for its possible political motivations, see Heurgon 1976; Krings and Devillers 1996: 498. Greek translation and epitomes: Varro, *Rust.* 1.1.10. On Mago's treatise see further Krings and Devillers 1996; Santini 2000; Cataudella 2002: 41–6; Domínguez Petit 2004; Krings 2008.

[23] For idealized or moralizing Roman views of agriculture, see, for example, Heitland 1921: 437–40; Kier 1933 (including Greek material); Cairns 1975; White 1977 *passim*; Miles 1980: 1–63; Reay 1998 *passim*; Spanier 2010, esp. 33–94; also n.57.

[24] For Celsus' *Artes*, see Ch. 2, n.62; for the agricultural books, Reitzenstein 1884: 30–41; Marx 1915: 5–13; Krenkel 1973: 22; Schulze 2001: 15–16, 85–6. For Celsus' reception in Columella, see Baldwin 1963: 788–9; Hentz 1978: I 47–8; Richter 1981–3: III 586–7, 619–28; Noè 2000: 433–5; Fögen 2009: 178–9; Mielke 2023: 127–86. I. Atticus wrote one volume *de una specie culturae pertinentis ad uitis* (Col. *RR* 1.1.14): *PIR*² I 183; Reitzenstein 1884: 27–30; S–H II 791; Bardon 1952–6: II 140; White 1970: 25. L. Iulius Graecinus (father of I. Agricola) wrote two books *de uineis* (*RR* 1.1.14): *PIR*² I 344; Reitzenstein 1884: 41–4; S–H II 791, 864; Bardon 1952–6: II 140–2; White 1970: 25–6.

principles of rural economy, including the establishment of the farm (Book I); cultivation of cereal crops (II); viticulture and arboriculture (III–V); the husbandry of herd animals (VI–VII); *uillatica pastio*, including birds, fish, and bees (VIII–IX); horticulture (X, a poem); the duties of the *uilicus* (estate manager), the agricultural calendar, and more horticulture (XI); and the duties of the wife of the estate manager (*uilica*) and the preparation and preservation of various farm products (XII). Authorial indices make it easier going to find and apply Columella's instructions.[25] Throughout, the advice is deeply informed by personal experience and delivered in an easy and elegant style. For the breadth and value of its *praecepta* alone, *RR* justifies its value in a crowded field of agricultural writings, and the didactic poem of Book X makes a bold move toward the summation of a discipline both *eloquens* and *carminum potens* (cf. 1.1.12–13).

Beyond this, however, Columella's *RR* is also marked by an approach that sets his work apart from earlier agricultural writings and locates it within the scientific culture of the early Imperial *artes*. At the outset, Columella indeed not only addresses the challenge of the rich heritage of Roman agricultural writings, but also establishes the premises for a deeper and more complete treatment of the discipline. He does so by identifying a sort of crisis of declining yields in Italian farming, which some – following no less an agricultural authority than Tremelius Scrofa – blame on the senescence of Nature (*Natura* or *Rerum Natura*) and Earth (*Tellus*). In refuting this claim, Columella undertakes to vindicate Nature with a true account (*ratio*) according to which it is divine, eternal, and providential. This move ultimately lays the groundwork for a new conception of the *ars*, one dependent on an explanatory grasp of natural principles. The farmer who possesses the requisite interdisciplinary knowledge – the *agricola perfectus*, in Columella's language – will best propagate and cultivate his crops. Columella's effort to connect the *ars agricolationis* with theoretical understanding of the natural world distinguishes his work from earlier Roman efforts.[26] Cato's *Agr.* need not be mentioned. Varro's *Rust.* does anticipate Columella's *RR* in its highly systematic character, and Varro has a penchant for explanation, especially etymological or

[25] These "*argumenta*" are found at the end of Book XI (11.3.65). See further Schröder 1999: 131–42; Henderson 2002: 111–13; Christmann 2003: 135–6; Mielke 2019: 78; 2023: 13, 42, 59, 78–9. Compare Pliny's *summarium* (Ch. 2, n.49) and Frontinus' *species* (Ch. 6.6.1 and n.104).

[26] Thibodeau 2018: 473–4 sees "absence of causal explanations" (474) and, generally, lack of theoretical interest, as basic to ancient agronomy; on the view proposed here, this would not be entirely true of Columella, because he does claim to disagree with others precisely "on the basis of principles" (2018: 473).

linguistic but sometimes also historical or philosophical.[27] But these explanations only throw into sharper relief the novelty of Columella's program: Varro's piecemeal accounts demonstrate the intellectual interest of agriculture and situate it in a broader Roman cultural milieu, but are essentially adventitious and rarely informative of practice; Columella's explanations, by contrast, are constitutive of a larger project that casts explanatory knowledge as integral to good agronomy.[28] It is because this project is presented as crucial to practice – not simply as exhibiting erudition – that Columella also emphasizes the interdisciplinary character of the *ars* vis-à-vis the farmer's education – a concern absent from earlier agricultural writings but familiar to the Imperial *artes*, as we already saw in Vitruvius.

Beginning from Columella's own response to the Latin agricultural tradition, in this chapter we will study his approach to agronomy, focusing on those aspects of *RR* which reveal the place of his work within the scientific culture of the Imperial *artes*. We will consider chiefly the cultivation of crops, what he calls *agricolatio* (1.pr.25), although our findings hold mutatis mutandis for *agri cultura* more broadly. First, we will examine Columella's response to the alleged crisis in the fertility of Nature and Earth in the preface to *RR*. The debate about the cause of declining yields is in fact a debate about the possibility and character of an *ars* of agriculture. In staging this controversy at the beginning of his work, Columella thus affords himself the occasion to unfold his own conception of the discipline. His response leads him to assert an alternative view of Nature that underpins good agronomic practice as well as to outline a theoretically rich and interdisciplinary notion of the *ars*. In the remainder of the chapter, we will study the consequences of Columella's position for agricultural practice in *RR*. In short, Columella holds that explanatory knowledge of Nature conceived as divine, eternal, and providential is necessary for good agronomy. This is because all life bears the stamp of Nature's intelligent design, and if the farmer is to cultivate his crops effectively, he must understand the processes that underpin plant growth. Columella's refutation of the alleged crisis of fertility which threatens in *RR* I and II is a programmatic instantiation of this vision: if the farmer appreciates the mechanisms for the Earth's

[27] For example, etymological/linguistic, *Rust.* 1.29.3 (*porca*), 1.48 (names of grain parts) 2.1.6–7 (*balare*) 3.10.19 (*Melicus* and *Medicus*), 3.16.5 (*alui* and *alimonium*); historical, 1.17.1 (*obaerarii*), 2.5.2–5 (cattle and places), 3.9.16–17 (*gallinae rusticae*); philosophical, 1.4.1 (the four elements, via Ennius), 1.40.1 (seed), 2.1.3–5 (anthropology), 3.12.4 (age of hares).

[28] To be sure, Columella's theorizing in all likelihood does not actually inspire the practices he reports, but the key point is that he *presents* it as doing so, whereas Varro does not. See the Introduction for the value of theorizing vis-à-vis the scientific character of the *artes*.

providential fertility, he will also know how to remedy any temporary barrenness with the aid of manuring (*stercoratio*). Columella does not drop this view of Nature/Earth once he has banished the specter of the crisis, however; to the contrary, he appeals to it to refute other erroneous doctrines held by farmers of his day and to establish the principles for a correct approach to the *ars*. A paradigm example is his advice for selecting and propagating vines in *RR* III, which offers a probative illustration of Nature's divine *ratio* and demonstrates the value of a precise apprehension of her workings. Columella's *RR* in this way not only teaches good agronomy, but also communicates a larger conception of Nature consequential for *agri cultura*, an "image of expertise."[29]

4.2.1 *The Age of the Earth: Theories of Agricultural Decline*

No less than in other *artes*, the preface of *RR* affords Columella a privileged place to present his view of his discipline. He opts for a polemical tack that both sets his work into the tradition of earlier agricultural writing and emphasizes the differences in his own approach. The whole matter turns on an allegation made by the "leading citizens" (1.pr.1 *principes*) that declining yields owe to the exhaustion of Earth and Nature. This claim bears directly on the *ars*, since if yields are limited by the failing soil, the possibility of successful practice is, too. Sensitive to this logic, Columella rejects the explanation suggested by the *principes*: Nature and Earth have suffered no loss in their fertility, which is divinely eternal; it is only that moral decay has made Romans forget the art. Agriculture is in fact an august discipline worthy of engagement by the best Romans and demanding not only extensive experience but also broad learning. The themes established in the preface shape Columella's whole approach throughout *RR*.

Columella announces a crisis of sorts in agriculture in the very first words of *RR*. Whether or not any such crisis really obtained,[30] as a conceit it immediately places Columella into dialogue with a larger group of Romans engaged in the discipline (1.pr.1):

> saepenumero ciuitatis nostrae principes audio culpantis modo agrorum infecunditatem, modo caeli per multa iam tempora noxiam frugibus intemperiem, quosdam etiam praedictas querimonias uelut ratione certa mitigantis, quod existiment ubertate nimia prioris aeui defatigatum et effetum solum nequire pristina benignitate praebere mortalibus alimenta.

[29] See further the Introduction. [30] Scholarly opinions diverge: cf. Mielke 2023: 21n56.

> Time and again I have heard the leading citizens of our nation complaining, sometimes about the barrenness of the fields, sometimes about the intemperateness of the climate which has for many seasons now been harmful to crops. I have heard some also temper the aforementioned complaints as though with a sure account, because they believe that the soil is worn out and exhausted by its excessive fertility in earlier times and cannot produce food for men with its previous bountifulness.

Two complaints are attributed to the *principes*: that the fields are unfruitful, and that the climate has proven unfavorable to crops. The latter is quickly dismissed (1.pr.3), but the former is supported by what some proponents fancy (cf. *uelut*) a *certa ratio*: the soil is "worn out and exhausted." We learn more about this account: *Rerum Natura* (Nature) is said to be "afflicted by infertility as if by some disease" (1.pr.2 *quasi quodam morbo sterilitate adfectam*), and *Tellus* (Earth) to have "grown infirm with age, just like a human being" (1.pr.2 *uelut hominem consenuisse*).

Columella's language is anthropomorphizing and medical but sufficiently vague to square with multiple theories, especially in conjunction with the ascription of "excessive fertility" and "previous bountifulness" to earlier ages. Beliefs in the gradual diminishing of agricultural yields were early and widespread, thanks to cosmogonies and histories of human civilization which posited a bygone "Golden Age" when the Earth spontaneously fed human beings. A *locus classicus* for Golden-Age narratives is Hesiod: while "the wheat-bearing fields used to bear a plenteous and bountiful harvest of their own accord" (*Op.* 117–18 καρπὸν δ' ἔφερε ζείδωρος ἄρουρα / αὐτομάτη πολλόν τε καὶ ἄφθονον), now humans "will never be free from toil and misery" (176–7 οὐδέ . . . / παύσονται καμάτου καὶ ὀιζύος) because of the hard labor required in agriculture. Here the decline in the Earth's fertility is bound up with a myth of races, but in other narratives the aetiology differs. In the *Critias*, for example, Plato explains how the once-rich soil of Attica had declined to the point that "only a spare body of the land remains" (111b6–7 τοῦ λεπτοῦ σώματος τῆς χώρας μόνου λειφθέντος). He blames violent earthquakes for sweeping rich deposits of soil into the ocean (111a6–b6). An explanation of a different order is given by Diodorus, who attributes the waning fecundity of the Earth to the action of the winds and the fiery sun, which have rendered the Earth "more solid" (1.7.7 μᾶλλον στερεουμένην).[31] More generally, declining fertility was implicated in any account of human prehistory which

[31] Diodorus' sources: Cole 1967: 174–92; Sacks 1990: 56–7.

stressed the past productiveness of the Earth, even when expositors did not spell out physical causes.[32] And many more thinkers could be adduced to support the *certa ratio* if we broadened our scope to include theories of natural flux which did not specifically highlight a decline in productivity but would logically lead to that conclusion.[33] Columella's claim to have "time and again" (1.pr.1 *saepenumero*) heard complaints of the Earth's infertility might allude to the wide currency of such narratives.

It is possible, however, to identify a more specific version of the *certa ratio* thanks to the preface of Book II. There, Columella responds to Siluinus, who has asked him why "right away from the beginning" (2.1.1 *a principio confestim*) he set about rebutting the opinion of those who maintain that "the soil has become decrepit, fatigued and exhausted by its age-long deterioration and its long-standing cultivation" (*longo aeui situ longique iam temporis exercitatione fatigatam et effetam humum consenuisse*). This idea, earlier in the mouths of the *principes* in Book I (1.pr.1), is now said to be "the opinion of virtually all of the earlier generation who have written about the cultivation of fields" (2.1.1 *ueterum opinionem fere omnium qui de cultu agrorum locuti sunt*).[34] In shifting the *ratio* from the *principes* to agronomists proper, Siluinus' remarks show that the "crisis" at the beginning of *RR* (1.pr.1–2) already places Columella into conversation with – and distinguishes him from – earlier agriculturalists. An even firmer lead appears in the name of Cn. Tremelius Scrofa, whom Siluinus reveres because "he transmitted a great many precepts regarding *res rusticae* at once elegantly and learnedly" (2.1.1 *plurima rusticarum rerum praecepta simul eleganter et scite memoriae prodiderit*).[35] Apparently, Scrofa "expressed the belief that the parent of everything, the Earth, now weakened by womanish old age like the female sex, is incapable of bearing offspring" (2.1.2 *credidit parentem omnium terram, sicut muliebrem sexum aetate anili iam confectam, progenerandis esse fetibus inhabilem*). Varro's "Scrofa" in *Rust.* never espouses this theory, but that is no reason to doubt the ascription to the Republican agronomist.[36] In any event, the allusion to Scrofa substantiates

[32] For relevant *loci*, see Campbell 2003: 341–3.

[33] For example, in Empedocles' cyclical theory creation and destruction continually alternate: the only constant is change, τῇ μὲν γίγνονταί τε καὶ οὔ σφισιν ἔμπεδος αἰών (D–K B26.12 = L–M D77b.10).

[34] I follow Rodgers in reading *ueterum* (in the *recentiores*) for *ueterem*; Martin 1971: 293–4 attaches no little significance to *ueterem*, but a better point is obtained from *ueterum*: Siluinus accepts Columella's disagreement with his contemporaries (because they are morally corrupt: see Ch. 4.2.2), but how does he explain that earlier generations also endorsed it?

[35] For Scrofa, see nn.9–11.

[36] The theory is germane to the problem of the Varronian vs. historical Scrofa (n.10): see Nelsestuen 2011, esp. 337–47.

the attribution of *ratio* to the older agronomists (2.1.1): not only is Scrofa himself of an earlier generation, but he is said to have been "seduced by his excessive partiality for those of the ancients who discussed similar material" (2.1.2 *inlectus nimio fauore priscorum de simili materia disserentium*). The *certa ratio* of Books I and II thus sets up an intradisciplinary dispute with significant consequences for the *ars*.

Unfortunately, we have no direct evidence outside Columella for Scrofa's version of what we will call the "senescence theory." However, we do find a similar, and highly developed, line of argument in one of Scrofa's contemporaries: Lucretius. In *De rerum natura*, Lucretius often appeals to the idea of *Mater Terra* (Mother Earth) and on several occasions argues that her generative capacities have declined with age.[37] His language is so close to Columella's, or rather to Scrofa's, that it seems certain that Lucretius must be either source or witness to the senescence theory. While we do not know how Scrofa justified himself, or whether he was influenced by Lucretius,[38] Lucretius provides us with perhaps the clearest idea of what a philosophically robust *ratio* underlying the infertility of the Earth might have looked like. (We will return to some details of Scrofa's presentation in *RR* II later.)

Lucretius argues for something resembling Scrofa's theories in two passages, in Books II and V.[39] In the earlier (2.1105–74), Lucretius discusses the growth and demise of the universe (*mundus*, 1105, 1144). This account is not so much biological as mechanistic, but he develops it most closely in relation to organic growth (1122–43). Organisms grow as they "take in more particles than they discharge" (1124 *plura sibi adsumunt quam de se corpora mittunt*). Once an organism reaches the height of its increase (1130 *alescendi summum … cacumen*), however, "now from all over it scatters more (sc. particles) in all directions and emits particles from itself" (1134–5 *in cunctas undique partis / plura modo dispargit et a se corpora mittit*). As the loss of *corpora* exceeds the gain, decay sets in. This process respects the "limit of growth" (1116 *crescendi … finem*) established by Nature (1117 *Rerum Natura creatrix*), which is fixed as the moment "when what is supplied within the vital passages does not exceed that which flows away

[37] For "Mother Earth" in Lucretius, see Campbell 2003: 60–1; Garani 2007: 81–93. The collocation of *mater* and *terra* seems to appear first in Latin at Lucr. 1.251 (*TLL* VIII 443.12–16 (Bulhart)).

[38] Often supposed, but difficult to prove: Bardon 1952–6: I 319–20; Martin 1971: 238; Nelsestuen 2011: 339–40.

[39] For Lucretius' theory of terrestrial decline, see esp. Campbell 2003: 91–8; also, Gale 2000: 81n73; Garani 2007: 92–3; Gale 2013: 31. The most extended discussion of Lucretius' position in relation to Columella is Martin 1971: 292–300; also Gummerus 1909–10: 18; Richter 1981–3: III 575n16; André 1994: 181–2; Mielke 2023: 352–5.

and departs" (1118–9 *ubi nihilo iam plus est quod datur intra / uitalis uenas quam quod fluit atque recedit*). In the last part of this passage (1150–74), Lucretius turns specifically to the senescence of the Earth (*tellus*), which is now "exhausted" (1150 *effeta*), its "life broken" (*fracta . . . aetas*). Though it once spontaneously produced all creatures, the Earth now scarcely generates a little life; Lucretius then pictures the effects of this decline on agriculture (1157–74):

> praeterea nitidas fruges uinetaque laeta
> sponte sua primum mortalibus ipsa creauit,
> ipsa dedit dulcis fetus et pabula laeta;
> quae nunc uix nostro grandescunt aucta labore,
> conterimusque boues et uiris agricolarum,
> conficimus ferrum uix aruis suppeditati:
> usque adeo parcunt fetus augentque laborem.
> iamque caput quassans grandis suspirat arator
> crebrius, in cassum magnum cecidisse laborem,
> et cum tempora temporibus praesentia confert
> praeteritis, laudat fortunas saepe parentis.
> tristis item uetulae uitis sator atque uietae
> temporis incusat momen saeclumque fatigat,
> et crepat antiquum genus ut pietate repletum
> perfacile angustis tolerarit finibus aeuom,
> cum minor esset agri multo modus ante uiritim;
> nec tenet omnia paulatim tabescere et ire
> ad scopulum,[40] spatio aetatis defessa uetusto.

> What is more, (sc. the Earth) herself at first created luxuriant produce and fertile vineyards for men of her own accord; she herself granted sweet produce and rich sustenance. [1160] But as it is now, these things hardly attain great size even aided in growing by our labor, and we wear out cattle and the strength of farmers and blunt the ploughshare but are hardly fed by the fields – so stingy are they with their produce, so much do they increase our toil. And now the old plowman shakes his head and sighs [1165] often that his great labor has fallen in vain, and when he compares present times with times past, often praises the good fortune of his father. Likewise the sad sower of the little old shrunken vine blames the progress of time and finds fault with his own era, and [1170] grumbles at how the generations of yore, full of piety, very easily supported life on a narrow plot, when the size of land assigned man by man was previously much smaller. He does not know that all things decay little by little and come to their destruction, worn out by their old age.

[40] The text is controversial: I translate the paradosis *scopulum* (defended by Possanza 1990), but for renewed support for Vossius' *capulum* ("coffin"), see Galzerano 2015; 2018: 287–90.

Lucretius' querulous farmers nicely illustrate what Columella's *principes* might have argued. Just as Lucretius' plowman and winegrower envy their forefathers for whom the Earth produced lavishly (cf. 1167, 1170), so Columella's *principes* yearn for the *pristina benignitas* of the soil (1.pr.1).[41] The words *spatio aetatis defessa uetusto* (1174), which are the last of Book II and restate Lucretius' claim that all things come to an end (1116–17), are echoed by Columella in specific application to the soil (cf. 2.1.1 *longo aeui situ longique iam temporis exercitatione fatigatam et effetam humum consenuisse*). Finally, Columella's figurative extension of *effetus* to both *Natura* (1.pr.1) and *humus* (2.1.1) recalls Lucretius' usage shortly before the quoted passage (2.1150 *effeta... tellus*).[42] Lucretius' account shows that an atomistic theory could support the grumblings of the *principes*.

Although growth and decay are strictly mechanistic processes in Lucretius' physics, his language also figures the Earth as a mother whose bearing age is past, as in Scrofa's senescence theory. This metaphor, only implicit in Book II, comes to the fore in Book V. Lucretius asserts that "the Earth has time and again acquired the name of mother and rightly keeps it" (5.821–2 *etiam atque etiam maternum nomen adepta / terra tenet merito*) because all creatures spring from her (cf. 5.795–6). He continues (5.826–36):

> sed quia finem aliquam pariendi debet (sc. terra) habere,
> destitit, ut mulier spatio defessa uetusto.
> mutat enim mundi naturam totius aetas,
> ex alioque alius status excipere omnia debet,
> nec manet ulla sui similis res: omnia migrant,
> omnia commutat Natura et uertere cogit.
> namque aliud putrescit et aeuo debile languet,
> porro aliud succrescit et e contemptibus exit.
> sic igitur mundi naturam totius aetas
> mutat, et ex alio terram status excipit alter,
> quod tulit ut nequeat, possit quod non tulit ante.

> But because there must come an end to (sc. the Earth's) bearing, she ceased, like a woman exhausted by old age. For the long passage of time brings change to the nature of the whole world, and all things must pass from one state to another, nor does anything remain the same as it was: all things change, [830] Nature alters all things and compels them to transform. Truly, as one thing decays and dissipates, weakened by the long passage of time,

[41] Pace Mielke 2023: 354–5, Lucretius' *arator* and *sator* are not more admirable than Columella's *principes*, like whom they retail false claims of declining fertility.

[42] For *effetus* of infertile soil (e.g. Verg. *G.* 1.81 [with Servius *ad loc.*], Manil. 4.825, Sen. *Ben.* 1.1.2, *Ep.* 90.44), see *TLL* V 2.156.42–63 (Kapp–Meyer); Linderski 1999: 261–3.

> another grows up and issues forth from contemptible things. So therefore does long age bring change to the nature of the whole world, [835] and the Earth passes from one state into another, so that what bore (sc. the Earth) cannot any longer and what did not bear before can now.

Lucretius aims to explain why the Earth, which once spontaneously produced all the animals that now exist, ceased to do so (5.791–820). The focus is different than in Book II, but the terms for the Earth's senescence are similar. Lucretius again emphasizes that it is "long age" (5.828, 834 *aetas*; cf. 2.1120, 1150, 1174) and "the long passage of time" (5.832 *aeuum*; cf. 2.1114) that have lessened the Earth's fecundity. He also revisits his earlier claim that all things decline with age (2.1174 *spatio aetatis defessa uetusto*), applying the notion figuratively to *terra* "like a woman exhausted by old age" (5.827 *ut mulier spatio defessa uetusto*) who can no longer bear (cf. 836 *quod tulit ut nequeat*).[43] This figurative treatment of Earth as a "woman" responds to the image of *Mater Terra* established just before (5.795–6, 821–2).

Lucretius' account, predicated as it is on atomic theory, is essentially mechanistic and antiteleological. The metaphor of Earth as an aged mother/woman sustains a useful analogy between biological senescence and long-term decline in the Earth's fecundity but is ultimately limited within Lucretius' physics; a more accurate view must appeal to the atomic flux that is responsible for both growth and decay. We do not know how Scrofa himself exploited the metaphor of an aging woman to explain the infertility of the soil – whether as cover for a mechanistic account such as Lucretius', or as the basis for an organic and teleological theory of the Earth or Nature. Indeed, we do not even know whether his figure was anything more than just that. Nonetheless, Lucretius illustrates the redoubtable foundation on which the *ratio* of Columella's opponents could be set. He treats the loss in agricultural productivity alleged by cultivators as an empirical fact requiring natural-philosophical explanation. At the same time, his account substantiates the farmers' complaints. Whatever the details of Scrofa's position, Columella was aware of the larger philosophical implications of a *ratio* that put Nature or Earth into decline.

4.2.2 *Of Nature, Knowledge, and Morals*

The "crisis" that Columella announces forms an important component of the broader justification for his writing in the preface. Scholars have usually foregrounded the social, political, and economic dimensions of the

[43] *tulit ut* is Bentley's conjecture for *potuit*; for the *quod*-clauses as subjects, not objects, see West 1964: 102.

challenge of the *principes*.[44] Here we will rather consider how the crisis supports Columella's construction of an *ars* that differs in key respects from that of "virtually all of the earlier generation who have written about the cultivation of fields" (2.1.1 *ueterum* ... *fere omnium qui de cultu agrorum locuti sunt*).[45] The alleged crisis affords scope for reflection on the status of the discipline precisely because it also presents a challenge to the possibility of productive agriculture: if it were true that the fertility of Nature and Earth were declining, the farmer's success would also be fixed by the decayed state of the soil. Columella's response therefore aims not only to vindicate Nature and Earth, but also to refound the *ars agricolationis* on a proper knowledge of them. We will explore three strands in the preface vital to his project of reconstituting the discipline. First, Columella asserts an alternative conception of Nature and Earth that makes them divine, eternal, and providential. These properties will prove significant for agricultural practice. Second, he argues vigorously that the crisis in agricultural productivity results from moral decay. This line undercuts the proposed senescence of Nature and Earth by suggesting an alternative aetiology for the poor agricultural returns bewailed by his contemporaries. Finally, Columella sketches the capacious knowledge required by the *agricola perfectus*, tempered by practical experience, which substantiates his vision of a discipline based on correct apprehension of nature.

We begin from Nature and Earth. At the same time as he introduces the charges leveled by *nostrae ciuitatis principes*, Columella rejects their view as discordant with both propriety and wisdom. He maintains that it is not right (*fas*) to believe that "Nature, whom the first creator of the universe endowed with perpetual fertility" (1.pr.2 *Rerum Naturam, quam primus ille mundi genitor perpetua fecunditate donauit*), could become infertile, nor the mark of someone wise (*prudentis*) to attribute old age to "Earth, who possesses divine and eternal youth and is called the common parent of all because she has on each occasion borne all things and will bear again" (*Tellurem, quae diuinam et aeternam iuuentam sortita communis omnium parens dicta sit, quia et cuncta pepererit semper et deinceps paritura sit*). Brief as they are, these statements give us to know something about Columella's own cosmology and philosophical orientation. There are three entities: the

44 For example, Martin 1971: 289–310; Cossarini 1980: 98–9; Martin 1985: 1969–71; Richter 1981–3: III 644–5; Noè 2002: 75–7; Diederich 2007: 385.

45 Compare Mielke 2023: 26–7, who sees the moral dimension of the crisis as motivating Columella's turn from mere "Wissensvermittlung" to "Charakterschulung" (for which Mielke 2019: 11; 2023: 17–34).

primus mundi genitor ("first creator of the universe"), *Rerum Natura* (Nature), and *Tellus* (Earth). Of these, *Tellus* is strictly speaking contained by *Rerum Natura*, and we will see that Columella sometimes conflates the two; the present focus on *Tellus* addresses a specific aspect of *Rerum Natura* relevant to agricultural production – that is, Earth as nourishing basis for plant life.[46] His description makes it clear that Nature and Earth – we can collapse the two to say simply "Nature/Earth" – are divine, eternal, and providential.

There is a Stoic coloring to Columella's views,[47] and his thought does often show affinities with Stoicism.[48] At the same time, many of his ideas were the common property of more than one school of thought, and his philosophical terminology is often open to various interpretations.[49] The *primus mundi genitor* could be identified with reason (*ratio*, λόγος) on a Stoic reading, for example, but other schools also allowed for a rational creator (immanent or transcendent).[50] And as regards orthodox Stoic views, there is a wrinkle in Columella's ascription of "eternal youth" (1.pr.2 *aeternam iuuentam*) to Earth.[51] Columella's pluralistic philosophical position bears comparison with that of Vitruvius, who we saw was content to draw on different doctrines useful for physical explanation.[52] Vitruvius was no philosopher but an architect; and Columella is no philosopher but an agronomist. This need not imply carelessness toward truth for the sake of utility, but we should not expect him to adhere to sectarian doctrines when more general versions will do. So, while *RR* often leans toward Stoicism, as Vitruvius' *DA* occasionally does, Columella is no doctrinaire

46 For religious overtones in "*Tellus*," cf. Varro, *Rust.* 1.1.5 *Tellus terra mater* (invocation of agricultural gods), and see Mielke 2023: 291n492.

47 Martin 1971: 298–302; Richter 1981–3: III 646; Martin 1985: 1969–70; 1988: 301–2; Carretero Gálvez 1997: 130; Noè 2000: 438; 2002: 18; cf. André 1987b: 28; 1994: 182.

48 See nn.122, nn. 150–1, 155, plus Mielke 2023: 346n17.

49 The classic study is Gummerus 1909–10; also Carretero Gálvez 1997; Diederich 2007: 135–41 (Columella's zoogony); Mielke 2023: 348–70.

50 The same figure, called *genitor uniuersi*, appears at 3.10.10 and is again distinguished from Nature (Ch. 4.4.5); see Gummerus 1909–10: 8–15 (requiring correction, however, because he wrongly reads *aeternus* at 3.10.10; see n.153). A "first creator" was entertained by many philosophical schools, but Columella's language does bear closest affinity with Stoicism. In Stoic terms, the *genitor* would be *ratio . . . faciens, id est deus* (Sen. *Ep.* 65.12; cf. Diog. Laert. 7.134). But while the *genitor* is distinct from Nature at 1.pr.2 and 3.10.10 (cf. Dumont 1993: 91; Noè 2000: 438; 2002: 18), Gummerus (8) rightly observes that in other instances they are not clearly distinguished (cf. Mielke 2023: 355–6) – possibly an argument for the immanence of divinity.

51 Most Stoics thought the Earth was periodically destroyed in the cosmic conflagration (cf. *SVF* II 592). Gummerus 1909–10: 15–20 finds Columella's position Peripatetic (cf. Phil. *De aeternitate mundi* 61–2, reporting Critolaus); Martin 1971: 301–2, however, has Columella following Panaetius (according to whom the cosmos is ἄφθαρτον: cf. van Straaten frr. 64–6, 68–9).

52 See Ch. 3.3.1 and n.85 (*DA* 2.2.1–2, 8.pr.1–4).

adherent.[53] Questions of philosophical allegiance aside, the properties ascribed to Nature/Earth have programmatic significance for *RR*. For now, we may simply stress that by supplanting the view of the *principes*, Columella's theory stabilizes the discipline: the unchanging fertility of Nature/Earth allows the agriculturalist to develop his discipline without fear that success depends on conditions beyond his control.

Columella's characterization of Nature/Earth frees the soil and weather from blame for poor harvests. Yet he does not dispute that Roman agriculture is in a bad way – only the reason (*ratio*) claimed by the *principes*. It thus falls to him to provide an alternative explanation. This he does by arguing that paltry returns are the result of culpable ignorance of farming matters following the decline of the old agrarian *mores*: because men today have abandoned agriculture for the corrupting modern pursuits of the city, they unsurprisingly reap less from the soil than their forefathers. Hence, Columella argues that the deplorable state of farming "rather occurs by our own failing, for we have handed the *res rustica*, which the best of our forefathers had handled in the best way, over to all the worst slaves as though to the executioner for punishment" (1.pr.3 *nostro potius accidere uitio, qui rem rusticam pessimo cuique seruorum uelut carnifici noxae dedimus quam maiorum nostrorum optimus quisque et optime tractauerat*). Language of capital punishment (*dare noxae, carnifex*) emphasizes the dire state into which agriculture has fallen – it is not merely a base occupation, but something almost criminal or illicit.[54] But the transfer of the occupation from the "best" (*optimus*) men to the "worst" (*pessimo*) slaves is not just a mark of the social degradation; it also explains the decline in agricultural productivity. The operative word is *pessimus*, which activates ethical and social associations (especially in conjunction with *seruus* in antithesis to *maiorum nostrorum optimus quisque*), but demands more specific interpretation here with reference to lack of skill or capacity.[55] Hence, Columella is not rejecting slave labor but making a point about the poor fitness of those in whose hands the best and most capable freeborn men (*optimi*) have left the discipline. The relation between the discipline's social debasement, on the one hand, and its technical and epistemic crisis, on the other, becomes clearer when Columella

[53] See Gummerus 1909–10 generally, and for the limits of Columella's Stoicism, also Weinold 1959: 83; André 1987b: 27–8; Dumont 1993: xiii–xiv; Carretero Gálvez 1997: 130–1; Rex 2001: 50–1; Keßler and Ott 2017: 213; cf. Mielke 2019: 271–93; 2023 *passim*.

[54] *Dare* (or *dedere*) *noxae* as a legal term refers to the surrender of a slave or dependent for punishment for a crime for which the master is liable (Berger 1953: 601), sometimes more loosely simply "to hand over for punishment" (*OLD noxa* 2b); for juridical language in Columella, see Lundström 1928: 117.

[55] For *pessimus* of superlative lack of skill in an art, see *OLD* 1; *TLL* VIII 216.32–47 (Krieg-Rubenbauer). It appears most often with an agent noun in this sense, for example Col. *RR* 2.11.6 *pessimi agricolae*.

describes how *domini* are now accustomed to entrust their farms to the least capable of their slaves or freedmen (1.pr.12):

> nunc et ipsi praedia nostra colere dedignamur et nullius momenti ducimus peritissimum quemque uilicum facere . . . sed siue fundum locuples mercatus est, e turba pedisequorum lecticariorumque defectissimum annis et uiribus in agrum relegat, cum istud opus non solum scientiam sed et uiridem aetatem cum robore corporis ad labores sufferendos desideret; siue mediarum facultatium dominus, ex mercennariis aliquem iam recusantem cotidianum illud tributum, quia uectigali esse non posset, ignarum rei cui praefuturus est magistrum fieri iubet.

> As it is now, we ourselves are unwilling to cultivate our own estates and do not think it of any importance to appoint whichever bailiff is the most experienced . . . On the contrary, if a rich man purchases a farm, from his crowd of footservants and litter-bearers he dispatches to the fields the man who is most infirm in age and strength, although this job (sc. the bailiff's) requires not only knowledge, but also youth with strength of body to endure hard work. Or if the owner is of moderate means, he orders one of his slaves for hire who is now failing to bring in his daily tribute and thus cannot be a source of income to be made superintendent, although this man is ignorant of the business which he will be in charge of.

Romans regard the job of farm management as appropriate only for those whose infirmity renders them incapable of being otherwise useful or profitable – precisely the *pessimi serui*. The rich man's decrepit footservant or litter-bearer lacks not only the *scientia* but also the *robur* required for agricultural work, and the middle-class man's *mercennarius* presumably does, too, which is why he can no longer find other kinds of gainful employment.[56] Small wonder that farms are faring so badly when the least competent are in charge. What has brought agriculture so low that it should suffer this indignity?

Most of Columella's preface (1.pr.3–20) is devoted to an account of moral decline at Rome that explains the low worth in which *res rustica* is held. We mentioned earlier (Ch. 4.1) the special place agriculture held in Roman culture: this part of the preface constitutes a masterly synthesis and updating of moralizing traditions both Greek and Roman, as Columella expands on themes in earlier authors ranging from Xenophon and Cato to Cicero and Varro.[57] Inter alia, Columella weighs agriculture against other professions

[56] For the meaning of *mercennarius*, see Beare 1978; Scheidel 1990; Hawkins 2016: 145–6. This *mercennarius* is too old to undertake work that would earn a wage for his master (*cotidianum illud tributum*), so he is transferred to the farm; cf. Hentz 1978: III 113–14. On *mercennarius*, see also Bürge 1990; Möller 1993.

[57] See Zainaldin 2023: 78–84 (also 2020a: 348–60).

popular at Rome (1.pr.3–10) and vaunts the simple virtue of Rome's ancient agrarian life, adducing such legendary farmer-generals as L. Quinctius Cincinnatus, M'. Curius Dentatus, and C. Fabricius Luscinus (1.pr.13–14, 17–19). This is all intended to demonstrate the essential nobility of farming and thus to give the lie to the contemporary opprobrium attached to it. Having proven by appeal to the *mos maiorum* that farming is no unworthy business, Columella shows that it is rather declining Roman morals which have brought the discipline into ill repute. He follows Varro's narrative of the development of early Roman society and the antagonism between city (*urbs*) and country (*rus*) (*Rust.* 2.pr.2–5).[58] Varro deplored the ignorance of rural matters that followed the Romans' move to the city;[59] Columella says that Varro "made his complaint already in the time of our grandfathers" (1.pr.15 *iam temporibus auorum conquestus est*) and claims that the situation has become much worse.[60] Apart from the emasculating effects of urban life, the temptations of the city are especially grave because, as Columella will later argue, nothing is so important to the success of an agricultural operation as the "presence of the owner" (1.1.18 *praesentia domini*) at his villa.[61]

Significantly, the moral decay which has disinclined Romans to agriculture has also had the effect of suggesting that the *ars* as such requires virtually no instruction or training. Columella does not know of any who profess to be teachers of farming (1.pr.5 *agricolationis … doctores*) or students (*discipulos*). Men are completely ignorant of the *praecepta* of agriculture (1.pr.10). Romans must now buy imported foods and wine, because they have forgotten how to farm "here in Latium and in this land of Saturn, where the gods had taught their offspring about the fruit of the fields" (1.pr.20 *in hoc Latio et Saturnia terra, ubi di fructus agrorum progeniem suam docuerant*).[62] This is indeed "no

[58] For *urbs* vs. *rus* in Varro, *Rust.*, see Janson 1964: 88–9; Martin 1971: 275–7; Flach 1996–2002: II 177–8; Cardauns 2001: 21–5; Diederich 2007: 335–6; Wiseman 2009: 97–8. Varro's complaints about city life are read ironically by Kronenberg 2009: 100–2; Nelsestuen 2015: 147–8.

[59] *Rust.* 2.pr.4. Varro follows Dicaearchus of Messana (cf. 1.2.16, 2.1.3–5), according to whom there are three *gradus* of civilization: first the "natural" mode (gathering but not cultivating), second pastoralism, third agriculture proper. (For Dicaearchus' theory, see Reischl 1976: 90–6; Saunders 2001; for his influence on Varro, Nelsestuen 2017; Piras 2017, esp. 11–13.) Varro assigns the founders of Rome to the second stage and their children to the third; Romans now ignorantly confound the practices of the *gradus* (2.pr.4).

[60] *RR* 1.pr.15–17, a passage evoking Varro but full of elements common in Imperial moralizers. Some elements are also older: Columella's *nec orientem solem uidemus nec occidentem* (1.pr.16), for example, is proverbial (Otto 1890: 326; Castro Gasalla 1993: 68–9) and perhaps originally Catonian (cf. Cic. *Fin.* 2.23, Sen. *Ep.* 122.2).

[61] See esp. 1.1.18–2.1, a theme connected with Mago (Speranza 1974: 80); further, Mielke 2023: 29n91.

[62] Since Lundstrom, *hoc Latio et Saturnia terra* has been taken as citation or paraphrase of Ennius (cf. Varro, *Ling.* 5.42; Weinold 1959: 90; Hentz 1978: III 124), though the whole sentence is in some respects modeled on Varro, *Rust.* 2.pr.4 (n.59).

wonder, since there has been generally conceived and indeed already grown strong a popular belief that the *res rustica* is low work, and that it is the sort of business which does not require any teaching or instruction" (*nec mirum, cum sit publice concepta et confirmata iam uulgaris existimatio rem rusticam sordidum opus et id esse negotium quod nullius egeat magisterio praeceptoue*). The present dearth of agricultural training of course highlights a gap into which Columella will step as *doctor agricolationis*. But it also serves as a foil to the far more demanding agricultural education that he next sketches (1.pr.21–33). The "crisis" is then also connected with a dispute about the intellectual profile of the discipline.

4.2.3 What the Farmer Knows

Columella's aim in the last part of the preface is not only to refute the idea that agriculture is a trivial business, but also to present a new take on the discipline that brings it closer to an ambitious, interdisciplinary view of *ars* such as we saw in Vitruvius' *DA*. He is concerned chiefly with what "the person who will wish to declare himself perfect in this branch of knowledge" (1.pr.22 *qui se in hac scientia perfectum uolet profiteri*) must know. The *agricola perfectus* (cf. 1.pr.28, 1.pr.32) can be compared with Vitruvius' *artifex perfectus* (*DA* 1.1.3), both figures inspired by Cicero's *orator perfectus* – a title no speaker can earn "unless he has acquired knowledge of all important subjects and arts" (*De or.* 1.20 *nisi erit omnium rerum magnarum atque artium scientiam consecutus*) – and serving as the implicit subject of the educational ideals in their respective disciplines.[63] Also perhaps with a nod to Vitruvius, Columella uses the *corpus* figure to pictorialize the difficulty of the discipline, speaking of the "great extent of the whole subject matter, like a certain vastness of body" and the "fineness of its parts, as of individual limbs" (1.pr.21 *magnitudinem totius rei quasi quandam uastitatem corporis … partium eius uelut singulorum membrorum subtilitatem*).[64] He then surveys the farmer's knowledge (1.pr.22–7), illustrating both its extent and subtlety. Heading the list is the claim that the *agricola perfectus* must be "extremely discerning of nature" (1.pr.22 *rerum naturae sagacissimus*); he must also know the geography and regional climates of the world, recall the motions of the celestial bodies and their relations to the weather, read the signs of seasonal change (1.pr.23),

[63] On the *orator perfectus*, see Ch. 1, n.38; for the *agricola perfectus* in this vein, see Mielke 2023: 412–20, 452–5; also Diederich 2007: 216–17; Fögen 2009: 164, 199–200.

[64] See Ch. 2, n.29 and esp. Ch. 3, n.11.

apprehend the many varieties of soil and land (1.pr.24–5) and know what they will bear, appreciate the habits of propagation and care for countless species of tree and vine (1.pr.25), and know the modes of life of many beasts and birds (1.pr.25–7). Then there is grafting, pruning, cultivation of fruits and vegetables, and so on (1.pr.27).

Columella's elaborate treatment of "so vast and such varied knowledge" (1.pr. 28 *tam uariae tamque uastae scientiae*) serves several purposes. It confutes the "popular opinion" (1.pr.20 *uulgaris existimatio*) that farming requires no instruction, a belief partly responsible for the crisis deplored by the *principes*. But insofar as it serves as an index for the theoretical investment of Columella's approach to agronomy, it also rebuts more respectable views that played down the intellectual demands of the discipline. In the *Oeconomicus*, Xenophon has Socrates' interlocutor Ischomachus repeatedly claim that the work of farming is "the easiest (sc. τέχνη or ἐπιστήμη) to learn" (6.9 μαθεῖν ... ῥᾴστη; cf. also 15.4, 15.10, 18.10), something presented as a virtue of the discipline. Moreover, Cicero and Celsus in different contexts report the idea that agriculture does not require *ratio* or *disputatio*, only observation and experience (*usus*).[65] (Columella's own appreciation of experience is discussed later in this section.) Columella appears to address this tradition directly when he argues that "the widely held belief that agriculture is very easy and does not require any intelligence is far from the truth" (1.pr.33 *illud procul uero est quod plerique crediderunt, facillimam esse nec ullius acuminis rusticationem*). Cato and Varro had not belittled the knowledge required for practice – indeed, Varro affirms that agronomy has a theoretical component[66] – but neither author had unpacked its epistemic principles, in particular the comprehension of nature (*rerum natura*), that underpinned the *ars*. Columella's self-conscious treatment of the *scientia agricolae*, like Vitruvius' inquiry into the *scientia architecti*, reflects an attitude toward specialized knowledge that first began to hold sway in the Imperial *artes*.

Also like Vitruvius, Columella is sensitive to the objection that his educational requirements are unattainably high. "After this announcement of so many and such complex subjects" (1.pr.28 *post hanc tam multarum tamque multiplicium rerum praedicationem*), he acknowledges that the pendulum may have swung in the other direction: some now might be

[65] Cic. *De or.* 1.248–9, Cels. *DM* 1.pr.32 (though in neither of these passages does Cicero or Celsus speak frankly in his own voice). Cf. Cic. *Verr.* 2.3.227 (chance dominates farming rather than *ratio* or *labor*). See also Martin 1971: 95–7; 1995; Noè 2000: 409; 2002: 151–2; Blandenet 2009, esp. 4–9; Fögen 2009: 163; Thibodeau 2011: 34; 2018: 474.

[66] It is both *ars* and *scientia* (*Rust.* 1.3.1; cf. n.11).

discouraged from agriculture because, far from believing the discipline to be easy, they fear the impossibility of mastering all its prerequisites. Here again he is following Cicero, whose *Orator* he invokes to justify his curriculum.[67] It is appropriate for those studying "subjects most useful for the human race" (1.pr.29 *generi humano res utilissimas*) – now not just oratory, but also agriculture – to "attempt all things" (*cuncta temptare*),[68] and in what follows (1.pr.29–31) he adopts Cicero's argument (*Or.* 3–9) that one ought not be discouraged from pursuing perfection, because even if one falls short, one will still have accomplished much. We know the names of many poets besides Homer and Vergil, orators besides Demosthenes and Cicero, sculptors besides Phidias. As Cicero puts it, the reward for those who fail to become the best is to be great (cf. *Or.* 8 *magna sunt ea quae sunt optumis proxima*).

Unlike Vitruvius, Columella does not invoke other *artes* by name in his curriculum, instead merely gesturing at the varied learning of the ideal agriculturalist. This material is interdisciplinary in spirit if not letter, however. In Chapter 3 we emphasized that, for Vitruvius, the measure of the knowledge of other *artes* required by the architect is the judgment and use of their *opera* for his own purposes.[69] While Columella sets lofty theoretical standards for the *ars* (cf. 1.pr.32), he similarly establishes a pragmatic check against this requirement in the form of first-hand experience. A lacuna obscures the reasoning,[70] but he seems to balance knowledge of nature against the demands of practice (*usus*), for whom the models are Roman agriculturalists such as Scrofa, the Sasernae, and C. Licinius Stolo.[71] Hence the balanced conclusion: "country work cannot be handled by those who possess neither the 'most acute' nor again 'a dull Minerva,' as they say" (1.pr.33 *potest ... nec subtilissima nec rursus, quod aiunt, pingui Minerua res agrestis administrari*).[72] The *agricola perfectus*

[67] Cic. *Or.* 3 ~ Col. *RR* 1.pr.28.

[68] Cic. *Or.* 4 ~ Col. *RR* 1.pr.29. Columella notably changes Cicero's statement to refer to those aiming not just to learn but also to communicate the discipline (*memoriae tradere*); this appears self-referential (cf. Mielke 2023: 416–18), though the topic remains the *agricola perfectus* "whom we desire and will describe" (1.pr.28 *quem desideramus agricolam quemque describemus*).

[69] See Ch. 3.1.1, 3.1.3, on *DA* 1.1.1, 1.1.16.

[70] *Tamen* at 1.pr.32 introducing the clause with *usus* looks adversative given the dangers of excessive book-learning described thereafter, but the transition is abrupt; Holford-Strevens reasonably posits a lacuna (adopted by Rodgers): [32] ... *et* (sc. *fuerit consecutus*) *in agrorum solique molitione Triptolemi aut Aristaei prudentiam* <***>, *multum tamen profecerit, si* <*quis*> *usu Tremelios Sasernasque et Stolones nostros aequauerit.* [33] *potest enim*, and so forth. Rodgers gives the following diagnostic supplement in the *app. crit.*: <*multo plus temporis in discendo consumpserit quam ad agros colendos reliquum habebit*>.

[71] Stolo is an interlocutor in Varro, *Rust.*, and is praised for his diligence (1.2.9).

[72] For the idiom, see Otto 1890: 224–5; Castro Gasalla 1993: 74.

must acquire some knowledge of other fields, but he cannot feasibly match the respective experts in their own realms of inquiry – Vitruvius makes the same point, of course.[73] Columella later continues to emphasize the importance of experience, claiming that *monumenta scriptorum* – including his own work – "more prepare (*or* teach) than make the practitioner" (1.1.15 *magis instruunt quam faciunt artificem*).[74] "Experience and practice are supreme in the *artes*, and there is no discipline in which learning does not proceed from making mistakes" (1.1.16 *usus et experientia dominantur in artibus, neque est ulla disciplina in qua non peccando discatur*). Experience is the final arbiter of *praecepta*, which can only "aid" (1.1.17 *adiuuare*), not "complete" (*consummare*), the art.[75]

Columella's attitude toward experience (*usus*) further establishes the limits of *ars* in *RR*. Thus, when he reflects elsewhere on the relationship between agriculture and other disciplines, the principle bounding the *scientia agricolae* is the farmer's practice. Three such fields peripheral to agronomy crop up: land-surveying, biology, and astronomy. Land-surveying is the subject of an important set piece in Book V, where M. Trebellius has claimed that it is appropriate for the same person demonstrating (5.1.2 *demonstranti*) how to prepare the soil for a vineyard also to teach (*praecipere*) how to measure it (*metiri*).[76] Columella acknowledges that the farmer requires quantified knowledge of the area of his fields but argues that instruction in this area "is the duty not of the farmer but of the surveyor" (5.1.3 *non agricolae sed mensoris officium esse*).[77] His hesitation turns on the question of what "belongs to the profession" (5.1.2 *propria professionis*): agriculture (5.1.3 *nostra disciplina*) must only "declare in what way each thing must be done, not how much is that which it has accomplished" (*dicat qua quidque ratione faciendum, non quantum id sit quod effecerit*). To buttress this criterion, he appeals to architecture as a parallel case: though architects must know the principles of measurement (5.1.3 *mensurarum … rationem*), even they leave calculations of the completed

[73] Hence, I understand Columella's claim that the person of consummate art (1.pr.32 *artis consummatae*) will "attain" or even "outstrip" (*fuerit consecutus*) the *sagacitas* of Democritus and Pythagoras and the *prouidentia* of Meton and Eudoxus as referring to intellectual capacity (not theoretical achievement). For Vitruvius' position, see Ch. 3.1.3 on *DA* 1.1.15.

[74] Cf. Cels. *DM* 1.pr.47, 74 (Ch. 5.1).

[75] By contrast, Columella uses *consummare* at 1.pr.32 to stress the theoretical knowledge required of the *agricola perfectus*; the difference is one of perspective (cf. Cels. *DM* 1.pr.47).

[76] For Trebellius, see Ch. 2, n.98.

[77] The distinction between *teaching* and *using* is important, for Columella rejects only the former; the *use* of surveying data is vital, and Columella's advice thus often quantifies the area under cultivation (of cereal crops, cf. *RR* 2.5, 2.9–10, 2.12 with Spurr 1986b: 136–40; of vines, Duncan-Jones 1982: 39–59).

work to others.[78] Columella's brief excursus on surveying (5.1.4–5.3.9) comes only after he has thus disclaimed expertise.[79]

Biology – or, more generally, natural philosophy – features in the discussion of apiculture in Book IX, where Columella mentions that Hyginus, Virgil, and Celsus all handled the subject in different but outstanding ways (9.2.1–2).[80] This preamble marks the limit of Columella's own discussion, as he indicates that he will not spend any time weighing questions such as the origin or generation of bees or their manner of producing honey: "for it belongs more to those searching into the recesses of *rerum natura* than to countrymen to investigate these and like things . . . since they are of no use at all either in (sc. agricultural) work or in household management" (9.2.5 *haec enim et his similia magis scrutantium rerum naturae latebras quam rusticorum est inquirere . . . quoniam neque in opere neque in re familiari quicquam iuuant*). Finally, astronomy. In Book XI, Columella provides an astronomical calendar (παράπηγμα) that will equip the *uilicus* to determine the seasons and weather (11.2).[81] Thus, the *uilicus* will ensure that agricultural operations are conducted always at the right moment (cf. 11.1.28–2.3). Columella rejects the more subtle correlations between climate and stars that some have established (11.1.32), maintaining that less intricate prognostication is sufficient for the *uilicus*.[82] All three instances reveal Columella's awareness of the need to define the relationship between *res rustica* and related disciplines. Like Vitruvius, he makes full use throughout his treatise of the interdisciplinary learning stipulated in the preface to his work, but he too carefully limits the required knowledge to what serves the practitioner's purpose.[83]

4.3 Farming, Naturally

Columella's response to the *principes* establishes several themes of enduring importance for his approach to *agricolatio*: he redefines Nature/Earth as

[78] 5.1.3. Vitruvius in fact stipulates that the architect must know *arithmetice* (*DA* 1.1.4), whereby *sumptus aedificiorum consummantur, mensurarum rationes explicantur*. But no contradiction with Columella need be found if we recall that Vitruvius makes the architect's knowledge extend only as far as to judge the suitability of the products (*opera*) of other disciplines for his own purposes; this accords with Columella's view.

[79] 5.1.4, on which Scivoletto 1992: 803; Noè 2002: 46–7; Fögen 2009: 169. For the surveying material, see Guillaumin 2009a: 189; Bertoni 2017: 537–51.

[80] Hyginus *diligenter*, Virgil *ornate*, Celsus *eleganter* (cf. 9.2.1); for Columella's comparison of these with his own treatment, see Noè 2000: 418, 432–3; Mayer 2005: 231–2; Doody 2007: 196; Fögen 2009: 180–1; Bertoni 2017: 533–4; Mielke 2019: 122–8; 2023: 128–9.

[81] Lehoux 2007: 244–52. [82] See Zainaldin 2022.

[83] Mielke 2023: 420–52 tracks Columella's use of the knowledge advertised in 1.pr.22–7.

divine, eternal, and providential; diagnoses moral decay in Roman society that explains the current woes of agriculture; and proposes an interdisciplinary conception of the *scientia agricolae* based in knowledge of *rerum natura.* In the remainder of this chapter, we will explore how these threads are woven together into the *ars* presented in *RR.* The weft is Columella's conception of nature – or, rather, Nature – whose providential fertility provides the blueprint for the farmer's work. Nature is teleological on the small scale of plant parts and physiology as well on the large scale of ecological self-regulation. The warp is human reason (*ratio*), for it is only by understanding the intelligence of Nature's designs that agriculture can be placed on a secure foundation. The farmer's broad knowledge of *rerum natura*, buttressed by experience, allows him insight into the principles of plant life relevant to this purpose. Columella's moralizing critique of his contemporaries furthermore recurs as a useful means of explaining (erroneous) variation in practice.

We will investigate several moments in *RR* in which Columella argues from Nature to best agricultural practice. First, we will consider his theory of manure, which concretely interprets the *perpetua fecunditas* of Nature/Earth asserted in the preface: observation of wild, uncultivated growth reveals that decaying organic material is the nourishment (*alimentum*) by which Nature restores itself. This insight demonstrates the importance of manure (*stercus*) in cultivation and decisively refutes Scrofa's "senescence theory" by revealing the mechanisms that contribute to the enrichment of the Earth. Next, we will leave behind theories of soil fertility and explore further consequences of Columella's view of Nature in his instructions for selecting and propagating vine cultivars. Italian cultivars are said to produce excellent wine but in less copious amounts than foreign varieties. However, an appreciation of the benevolence of Nature's designs will persuade the farmer that the productiveness of Italian vines can be improved, allowing him to maximize the flavor and quantity of his vintage. As the eternal creatress (*aeterna opifex*) of all living beings, Nature has intelligently shaped the anatomy and physiology of vine plants: by appreciating her designs the farmer will grasp the correct methods of propagation.

4.3.1 Quid loquar de utilitate stercorandi? *The Senescence Theory Revisited*

In Book II, Columella returns to the senescence theory of the Earth and offers a more closely argued rebuttal of the account. In so doing, he rejects the specific interpretation of the Mother Earth analogy espoused by Scrofa (and with various nuances by the *principes*, Lucretius, and others) and explains how the Earth remains perpetually fertile. This rebuttal depends on the role that organic detritus plays in renewing the productivity of wild land: Columella describes such detritus as Earth's nourishment (*alimentum*), by which soil is fattened and made vigorous for plant growth. An appreciation of the natural alimentary cycle of soil and plants suggests a "nutritive theory" of manure, in which manuring (*stercoratio*) is envisioned as a process of feeding and restoring soil that has depleted itself on growth. The nutritive theory provides Columella with a conceptual foundation for reconsidering the sources and uses of *stercus* in Book II and elsewhere in *RR*. Ultimately, *stercoratio* becomes a potent illustration of the cooperation between Nature and human skill that makes the *ars agricolationis* possible: Nature's fecundity teaches the farmer how to cultivate his own land so that it may remain continually productive.

We saw earlier that Columella attributes to Scrofa a version of the senescence theory that draws out the implications of figuring Earth as a "mother" and understands declining soil fertility as a sort of menopause (Ch. 4.2.1). To refute Scrofa's theory, Columella tries to show that the interpretation of the Mother Earth analogy on which it is based is incoherent. He begins by examining the "mother" element of the analogy more closely to define the infertility that results from old age (2.1.3):

> hominis tum demum declaratur sterile senium non cum desinit mulier trigeminos aut geminos parere, sed cum omnino nullum conceptum edere ualet, itaque transactis iuuentae temporibus, etiam si longa uita superest, partus tamen annis denegatus non restituitur.

> Infertile old age is attributed to a human being not when a woman ceases to bear triplets or twins, but only when she is completely incapable of conceiving and consequently, since the period of her youth has past, even if a long period of life remains to her does not recover the capacity for bearing[84] which is denied by her years.

The infertility which results from old age has two essential features: it is (1) complete (cf. *omnino nullum*) and (2) irreversible (cf. *denegatus non restituitur*). We might be able to accept that Earth had, like a (human) mother, become infertile through age "if crops were failing entirely" (2.1.2 *si in totum nullae fruges prouenirent*). But that is manifestly false: no one thinks fruits have failed completely, only perhaps that they have become smaller or total

[84] For *partus* as "facultatem vel potestatem pariendi, fecunditatem" (not recognized by L&S or *OLD*), see *TLL* X 1.539.35–44 (Kruse).

yields have declined. Think again of fallowed land: soil abandoned (2.1.3 *derelicta humus*) for its infertility should remain infertile if senescence is the cause; yet no such thing happens, and old plots brought under cultivation produce bountiful returns (2.1.3). Columella does not yet attempt to explain how this happens (cf. 2.1.3 *seu sponte seu quolibet casu*), but no explanation is required for the present argument, which must only show that land does not remain permanently infertile. Hence (2.1.4):

> non ergo est exiguarum frugum causa terrae uetustas si modo, cum semel inuasit senectus, regressum non habet, nec reuirescere aut repubescere potest, sed ne lassitudo quidem soli minuit agricolae fructum.
>
> The old age of the Earth is not therefore the cause of poor crops if it is the case that, as soon as old age has set in, it cannot be reversed, nor can it be revitalized or rejuvenated; but as it is, not even the soil's fatigue diminishes the farmer's harvest.

The "menopause"-version of the senescence theory must be rejected because, as even ordinary experience of fallowing shows, *lassitudo* does not prevent the soil from bearing again – something incompatible with the total and irreversible effects of *senium sterile*. Since Scrofa's version of the Mother Earth analogy is falsified in this key respect, its broader explanatory appeal also crumbles. Even if Scrofa's analogy is given up as mere metaphor, so that the biological exactness of Columella's criticism is less pertinent, the increased fertility of fallowed land still poses a strong empirical objection to claims of a decline in the Earth's fecundity.

Dispatching Scrofa's first argument, Columella turns to a more qualified version of the senescence theory which we might deem simply the "exhaustion theory." This account retains the image of an aging Earth but deemphasizes the analogy with female childbearing, maintaining rather that "just as in humans exhaustion is consequent to excessive bodily exercise or the heavy load of some burden, so too is it to the cultivation and working of fields" (2.1.4 *tamquam in hominibus nimiae corporis exercitationi aut oneris alicuius ponderi, sic cultibus et agitationibus agrorum fatigationem succedere*).[85] In proof, Scrofa points to the fact that when wild territory (2.1.5 *intacta et siluestria loca*) is brought under cultivation, growth is at first luxuriant (*exuberare*), but that "soon after it fails to respond in the same way to the husbandmen's efforts" (*mox deinde non ita respondere labori colonorum*). Columella responds that Scrofa "no doubt

[85] So already 2.1.1 *longo aeui situ longique iam temporis exercitatione fatigatam et effetam humum consenuisse*, which shows that the notion of physical exhaustion does not supplant that of aging.

sees what happens, but he does not perceive why it occurs" (*uidet sine dubio quid eueniat, sed cur id accidat non peruidet*). This punning line names a charge with which Columella again and again reproaches farmers whose beliefs or methods are erroneous:[86] observation alone, without a more searching consideration of the causes, will lead the farmer astray. We will see other examples in the case of the propagation of grapevines.[87]

Columella offers an alternative explanation for the sharp decline in productivity of wild land brought under cultivation (2.1.5–6):

> neque enim idcirco rudis et modo ex siluestri habitu in aruum transducta fecundior haberi terra debet quod sit requietior et iunior, sed quod multorum annorum frondibus et herbis quas suapte natura progenerabat uelut saginata largioribus pabulis, facilius edendis educandisque frugibus sufficit. [6] at cum perruptae rastris et aratris radices herbarum ferroque succisa nemora frondibus suis desierunt alere matrem, quaeque temporibus autumni frutectis et arboribus delapsa folia superiaciebantur, mox conuersa uomeribus et inferiori solo, quod plerumque est exilius, permixta atque absumpta sunt, sequitur ut destituta pristinis alimentis macrescat humus.

> This does not prove that land which is uncultivated and has just been converted into farmland from its wild state ought to be regarded as more fertile because it is less disturbed and younger; rather, because the foliage and vegetation of many years which it naturally engendered have as it were fattened it with more abundant food, it is more readily capable of bearing and rearing up produce. [6] However, once the roots of the vegetation which have been broken up by hoes and plows and the trees which have been cut down with iron cease nourishing their mother with their foliage, and the leaves which in autumn fell from shrubs and trees and lay atop the earth are later turned over by plows and mixed into and depleted by the lower soil, which is for the most part thinner, it results that the soil is deprived of its earlier nourishment and grows meager.

In this extraordinary passage Columella argues that the Earth is naturally (cf. *suapte natura*) sustained by the very plants which it produces. The conceptual crux is the notion of feeding: the soil is "fattened" (*saginata*) by the vegetation it has produced as though by "food" (*pabula*); trees and shrubs "nourish" (*alere*) the land; the soil goes hungry and "grows meager" or "wastes away" (*macrescat*) without the dead leaves and boughs which are

[86] Scrofa *sees* what happens, he does not *see through* it to the deeper truth; for the force of *per-* in *peruideo* thus, see *TLL* X 1.1870.3–4, 14 (Banta).

[87] As explained later (Ch. 4.4.3–4), farmers know that the "head" (*caput*) of the vine is the most fruitful part, but their failure to understand why undermines their methods for selecting *semina*. Cf. also 3.8.1 (Ch. 4.4.1), and see Mielke 2023: 87–107, on perception in *RR*.

its "nourishment" or "sustenance" (*alimenta*).[88] The idea is that, so long as wild plants flourish, the Earth is sustained by the matter they return to it; when this detrital "food" (*pabula, alimenta*) vanishes with their extirpation, however, the soil begins to fail.

This "nutritive theory," as we may call it, supplants Scrofa's "exhaustion theory." Earth does not become less fertile because it grows tired or old as a human being does, but because the *alimenta* once available to it are removed. The notions of sustenance and feeding conspicuous in the passage are crucial to Columella's argument; not merely figurative, the language describing the soil's changes represents an attempt to explain the biological and ecological processes which underpin fluctuations in its fertility by appeal to analogous processes of animal nutrition.[89] Moreover, the appearance of *mater* (2.1.6) in reference to the Earth reminds us that one of the principal aims of this account is to reinterpret the Mother Earth analogy on which the senescence and (albeit less directly) exhaustion theories depend. For Columella does not discard the appellation "mother" altogether, only his opponents' use of it: in his own cosmology, Earth is a creatress divine, eternal, and providential. These properties are largely borne out in his account of the Earth's nutrition, which reveal some of the mechanisms that enable the soil to remain fertile: when left to its own devices, the Earth *suapte natura* bears the plants which nourish its soil, creating a self-sustaining process of renewal.

As in the preface, Columella shifts blame for declining harvests on cultivated land from Nature/Earth onto human beings. Failure to understand the importance of the nutritive processes of the soil breeds indolence, as farmers cart away fruit without leaving any food for the Earth: "in sum, it is neither exhaustion (the belief most have expressed) nor old age which causes our fields to repay us less generously, but manifestly our own idleness" (2.1.7 *non igitur fatigatione, quemadmodum plurimi crediderunt, nec senio sed nostra scilicet inertia minus benigne nobis arua respondent*). *Inertia* ("idleness") is difficult to translate but refers to culpable inaction or neglect.[90] Hence, we have it in our own power to stem the loss of fertility in soil: "it is permitted to realize a greater harvest, if Earth (*or* the earth[91]) were to be restored by

[88] For *saginare* and *macrescere*, see n.89 (cf. n.96); for *alimentum* and *pabulum*, n.96.

[89] *Saginare* is typically of animals (L&S I; *OLD* 1), although once of a plant at Plin. *HN* 19.54. *Macrescere* is rare and elsewhere only of animals (*TLL* VIII 20.49–56 [Hofmann]), but unlike *saginare* derives from an adjective, *macer*, which has currency in agricultural authors ("lean" or "meager" soil, White 1970: 99–100; the opposite is *pinguis*, n.95.) See further n.96.

[90] *OLD* 2a; *TLL* VII 1.1314.35–6 (Rehm). Relevant is the meaning "want of skill/art" (*OLD* 1; *TLL* VII 1.1315.55), suggesting that epistemic failure lies behind neglect; cf. Mielke 2023: 27n85.

[91] It is unclear whether Columella still thinks primarily of (*Mater*) *Terra* or now merely *terra* (earth = soil); certainly, the latter soon becomes the key meaning. Keßler and Ott 2017: 201–2; Mielke 2023:

frequent, timely, and moderate manuring" (2.1.7 *licet enim maiorem fructum percipere, si frequenti et tempestiua et modica stercoratione terra refoueatur*). Here we finally pivot from the Earth's own processes for regulating fertility to the contribution that agriculture makes to plant growth. Manuring (*stercoratio*) would allow the cleared land to be "restored" or "revived" (*refoueatur*) if it were applied often, at the correct times, and in the appropriate amounts by farmers. The effect of introducing *stercoratio* after the account of soil fertility is to construe this technique in terms of the Earth's own means of nourishing itself. Columella does not say much about manure here, but this suggestion is developed throughout *RR* and especially in the discussion of *stercoratio* in Book II. Before we consider his remarks there, however, we should clear up a small issue of terminology: "manure" in English and "*stercus*" in Latin both refer chiefly to animal dungs, but they may also be used generally of any fertilizing materials, such as compost or green manure (for the latter, see Ch. 4.3.2); in Latin, *stercus* is found of vegetative substances already from Cato's time.[92] In what follows, I will often use "*stercus*," "manure," and related words in both English and Latin in their broader significances, in keeping with Columella's practice. Departures from ordinary Latin usage will be noted for their significance to Columella's theory where relevant.

4.3.2 Feeding the Fields: Ecology, Biology, Agriculture

After discussing various aspects of the cultivation of grains and legumes (2.2–13.3),[93] Columella comes to manuring (2.13.3–4):

> sed omni solo quod praedictorum leguminum segetibus fatiscit una praesens medicina est, ut stercore adiuues et absumptas uires hoc uelut pabulo refoueas, [4] nec tantum propter semina quae sulcis aratri committuntur, uerum etiam propter arbores ac uirgulta quae maiorem in modum laetantur eiusmodi alimento.
>
> But there is one ready medicine for every soil which becomes exhausted by the growth of the aforementioned legumes: that is, that you aid it with manure and restore its depleted strength with this food as it were, [4] doing

290–2 stress a shift in viewpoint from Earth as living to non-living, but I attach less significance to it: Columella simply turns from a higher order of explanation (religious/cosmological) to lower (physical/biological), without repudiating the former.

92 *OLD* 2b. For words used only of animal dung (e.g. *fimus*/*fimum*), see Zainaldin 2020a: 224–5.

93 Soil and preparation for sowing (2.2–5); varieties of grains (2.6) and legumes (2.7); sowing of grains (2.8–9.19) and legumes (2.10); cultivation (2.11); reckoning of manhours (2.12); green manuring (2.13.1–3, quoted later in this section).

> so not only with a view to the seeds (sc. of grains or legumes) which are sown in the furrows of the plow, but also the trees and shrubs which to an even greater degree flourish by such nourishment.

This language draws together threads interwoven at the beginning of Book II: soil (*solum*), exhaustion and fatigue (*absumere, fatiscere*[94]), food and nourishment (*pabulum, alimentum*), and restoration or renewal (*refouere*). Columella thus indicates that his approach to *stercoratio* should be regarded as an extension of the nutritive theory articulated earlier: therefore, whereas *frondes* and *folia* were called *pabula* or *alimenta* in the setting of *siluestria loca* (2.1.5–6), on the farm it is *stercus*, especially animal dung but also vegetative compost and other fertilizers, which fulfills a similar role. While this loaded transition establishes the connection with the nutritive theory, it is not the only place where Columella takes seriously the idea of manure serving as food for the Earth. The language of nutrition recurs continually throughout *RR* and is already found scattered throughout Book II before the passage just quoted. For example, Columella enjoins that "before we plow thin soil again, it will be appropriate to manure it, for it fills out with this food, as it were" (2.5.1 *prius . . . quam exilem terram iteremus, stercorare conueniet, nam eo quasi pabulo gliscit*). He recommends mixing dung (*fimus*) into the earth so that "the soil, thoroughly mixed with the aforementioned nourishment, may grow fat" (2.5.2 *ut permixta humus praedicto alimento pinguescat*).[95] Elsewhere he speaks of the effects of "satiating" the soil "with manure" (2.9.15 *stercore saturare*; cf. 2.10.23 *largo . . . stercore saturatum*).

To appreciate the significance of Columella's nutritive theory of manure, it helps to observe that there is no real precedent for what might seem the natural analogizing of *stercus* as food for the Earth among earlier agronomists or writers on agricultural themes. The few isolated examples cannot be knitted together into anything like the sentiment expressed in 2.13.3 or even into a consistent pattern of figurative language.[96] Columella's own

[94] *Fatiscere*, of soil only here (*TLL* VI 1.353.80–6 [Pflugbeil]), picks up *fatigare* from 2.1.1, 2.1.4 (this verb also never elsewhere of soil).

[95] *Pinguescere* picks up *saginare*; the former has somewhat wider currency of plants and soil (*TLL* X 1.2163.67–78, 2164.5–19 [Nikitinski]), although the only author to use it of soil before Columella is Virgil (*G.* 1.498; but cf. Mynors 1990: 95). *Pinguis* is a standard word for "fat" or "rich" soil in the agronomists (*OLD* 3b; White 1970: 99–101; Fruyt 2014 *passim*).

[96] Regarding the vocabulary of nutrition: (1) *alimentum* is so used only by Columella (2.5.2, 2.17.6, 11.3.13), although it is not uncommon of food for plants (*TLL* I 1586.34–54 [v. Mess]; *OLD* 1b, most frequently in Columella; cf. Vitr. *DA* 2.9.1, in Ch. 3.4.1); (2) *gliscere* is so used only by Columella (2.5.1, 2.13.2); (3) *pabulum* is so used once by Virgil (*G.* 1.86–7) but otherwise exclusively by Columella thus (2.1.5, 2.5.1, 2.5.2, 10.84) (it appears at Cato, *Agr.* 7.3, of fruit trees, but the authenticity of the sentence is doubted: Richter 1978: 176–8); (4) *pinguescere* is so used once by Virgil before Columella (2.5.2, 2.10.7) and then after him sporadically (previous n.); (5) *saturare* is so

characterization of manure as food should not, of course, be read merely as a literary flourish or figurative motif.[97] His analogizing of manure as food follows the biological/ecological account propounded at the start of Book II; this broader theoretical context helps us to understand that if the soil is to remain fertile when it cannot renew itself, the farmer must intervene to complete the process of restoration (cf. *refouere*). In this sense, *stercoratio* represents a cooperative regulation of the soil between Nature and human beings, with agricultural art and effort completing the ends that Earth cannot fulfill once her self-activity has been interrupted.[98] Columella's refutation of the senescence theory thus shows itself to have broader significance than might first appear: his solution to the crisis underpins a model for agronomy founded on knowledge of Nature.[99]

The nutritive theory serves as a conceptual nucleus for reflection on the varieties and uses of manure in agricultural practice. It is worth considering some of Columella's *praecepta* for the light they shed on his Nature-based approach to agronomy. We begin from green manuring. Some crops, called "cover crops," are planted chiefly to protect and enrich the soil for the principal crop; these crops are eventually plowed or dug under to fertilize the soil and are thus called "green manures." Although the expression is modern, the practice is ancient and was observed by all the Roman agronomists.[100] Columella's main treatment comes immediately after his discussion of leguminous plants, where he follows the Sasernae in dividing legumes into those which serve as green manures and those which do not. He gives lupin (*lupinum*) and vetch (*uicia*) his special stamp of approval, and in so doing details the correct method of using green manure (2.13.1–2):

> de lupino nihil dubito atque etiam de pabulari uicia, si tamen eam uiridem desectam confestim aratrum subsequatur et quod falx reliquerit priusquam inarescat uomis rescindat atque obruat; id enim cedit pro stercore. [2] nam si

used only by Columella (2.9.15, 2.10.23); (6) *saginare* is so used only by Columella (once of a plant by Pliny: n.95). Columella's vocabulary of exhaustion is similarly scarce in earlier agronomists (nn.89, 94).

97 Despite occasional qualifications such as *uelut* and *quasi*: Columella invents the vocabulary to support his novel account, so these words simply apologize for the extension of words typically applied to animal life (for *quasi*, *OLD* 9a; *uelut*, *OLD* 4). Compare *uidetur* at 2.13.2 (quoted later in this section).

98 The idea that art completes what nature cannot is old (cf. Arist. *Ph.* 199a15–17 ὅλως δὲ ἡ τέχνη τὰ μὲν ἐπιτελεῖ ἃ ἡ φύσις ἀδυνατεῖ ἀπεργάσασθαι, τὰ δὲ μιμεῖται); in an agricultural context, cf. Thphr. *Caus. pl.* 3.1.1 (Ch. 3, n.135). See also Richter 1981–3: III 647 (Virgil's influence). Finally, cf. Col. *RR* 4.28.2 (a report from Celsus and Atticus possibly not endorsed by Columella; see Ch. 5.5).

99 See Keßler and Ott 2017: 199–20 for an ecocritical reading of 1.pr.1–3, 2.1.

100 White 1970: 135–7; Spurr 1986b: 127. Columella often simply says a plant "manures" (*stercorare*) the land (e.g. 2.10.25, 2.13.1, 11.2.44, 11.2.80, 11.2.81).

> radices eius desecto pabulo relictae inaruerint, sucum omnem solo auferent uimque terrae absument, quod etiam in faba ceterisque leguminibus quibus terra gliscere uidetur ueri simile est accidere, ut nisi protinus sublata messe eorum proscinditur, nihil îs segetibus quae deinceps in eo loco seminari debent profuturum sit.
>
> I am in full agreement as regards lupin and also pabular vetch, provided that it (sc. vetch) be cut down while still green and immediately plowed under, and that the leavings from the sickle are cut again and buried by the plowshare before they wither; for this serves as manure. [2] For if the roots that remain once the fodder has been cut off wither, they will remove all the sap from the soil and deplete the strength of the earth – something which apparently happens also in the case of beans and other legumes by which the earth seems to fill out – so that unless they are plowed under directly after the harvest has been collected, they will provide no benefit to the crops which are afterwards to be planted in that place.

Columella assumes the legumes are grown as cover crops and that the parts which can be used for animal fodder (*pabulum*) will be harvested.[101] The details of the account are expressed in terms complementary to the nutritive theory. The action of the green manure itself resembles the self-nutrition of the Earth, whose growth returns to the soil to nourish it. The earlier notion of fattening (*saginare*) is thus recalled here by *gliscere*. Additionally, the claim that the plowed-in plants "serve as manure" is more than mere periphrasis for the fertilizing potential of the crops if we recall that it is plant detritus which inspires *stercoratio* in the first place. Of course, the cultivated leguminous crops do not directly serve the natural cycle of *intacta et siluestria loca*, because the farmer reaps the stem, leaves, fruit, and flowers – hence the necessity of breaking up and plowing in the roots and stubble of the legumes after the harvest. If the roots that remain behind wither, they will diminish the soil (*uim . . . terrae absument*): here the appearance of *absumere* reminds us of the nutritive theory, for it has been twice used to refer to the depletion of the Earth (2.1.6, 2.13.3). To understand such depletion, recall that *sucus*, "sap," means not only the vital fluid within plants but also the nutritive moisture within soil.[102] When Columella says that the remnants of the legumes "will remove all the sap (*sucum*) from the soil," he seems to mean that the *sucus* taken up into their roots will be lost by the parching sun and hence cannot return to enrich the

[101] For this relatively uncommon use of *pabulum*, "respicitur quodammodo pars herbae," see *TLL* X 1.6.80–3 (Kruse).

[102] *OLD* 4; cf., for example, Cic. *Nat. D.* 2.120 with Pease 1955–8: II 852. See also Ch. 3.4 on Vitr. *DA.* 2.9–10.

earth. To prevent this, the farmer must plow the remnants of the crop back into the soil. Columella's discussion of green manuring thus builds on the nutritive theory by suggesting that the specific mechanism for the recycling of nutrients back into the soil is *sucus*.

After green manures, Columella turns to a general discussion of manure (2.14–15), whose preamble (2.13.3–4) we discussed at the outset of this section. He begins by ranking animal dungs and commenting on their qualities (2.14.1–4). Next, he turns to alternative sources of manure for those who cannot keep animals (2.14.5–6), a subject that he reprises (2.15.4–6) after discussing storage (2.14.7–9) and application (2.15.1–3). His declaration (2.13.4) that manuring has not been given sufficient attention by earlier generations of writers seems exaggerated when one considers that the basic framework of his ranking system – bird dung is best, next human, last animal[103] – is taken over from earlier agronomists, along with certain other recommendations.[104] Indeed, the traditional character of Columella's discussion may explain why it shows fewer overt traces of the nutritive theory.[105] Yet some remarks hint at a deeper view of the situation. For example, when he says that ass dung is best among *pecudes*, he explains that "this animal chews most slowly, and therefore digests easily and produces dung that is nearly prepared and ready for the field" (2.14.4 *id animal lentissime mandit ideoque facilius concoquit et paene*[106] *confectum atque idoneum protinus aruo fimum reddit*).[107] This statement implies that dung of herbivorous animals is "concocted" or "digested" from plant matter in its diet. When Columella elsewhere discusses the preparation of plant material for compost, he says that it must "rot" (*putrescere*, 2.14.7, 2.14.8) so that it can become "ready for the fields" (2.14.8 *aruis idoneum*). On this evidence, admittedly slight, the analogy between mastication/digestion (*mandere*, *concoquere*) and decay (*putrescere*) could indicate that varieties of manure can ultimately be explained on their vegetable components, in accordance with the nutritive theory.

Columella's broader concerns are evinced somewhat differently in his discussion of alternative manures for farmers who do not keep birds or animals. Earlier he asserted that it was "our own idleness" (2.1.7 *nostra* . . .

103 Varro, *Rust.* 1.38.1, attributes this ranking to Cassius Dionysius, the Greek translator and epitomizer of Mago; it is probably Magonian (Zainaldin 2020a, Appendix II).

104 For example, Col. *RR* 2.14.6 ~ Varro, *Rust.* 1.38.3.

105 To be sure, it is unclear how much is traditional in Columella's account: the general structure resembles Varro's (n.103), but his account is longer and contains important differences that may be his own additions, details culled from other sources, or the fuller report of Mago/Cassius Dionysius.

106 A *uaria lectio* in the MSS for *paene* is *bene*.

107 Cf. also 11.3.13 (ass dung is *optimum* for gardens *quia minimum herbarum creat*).

inertia) which caused the fertility of the land to fail. This moralizing backdrop recurs now when Columella appeals to it in passing from one subject to another (2.14.5):

> nec ignoro quoddam esse ruris genus in quo neque pecora neque auis haberi possint; ac tamen inertis est rustici eo quoque loco defici stercore.
>
> I am well aware that there are certain kinds of country in which it is not possible to keep either beasts or birds. Nonetheless, it is the mark of an idle farmer to lack manure even in such a place.

He lists four other sources (2.14.6), from assorted vegetation to refuse accumulated among brambles or along roads, composted ferns, and so forth. This is not to mention green manures: "if the farmer should lack all of these things (sc. other sources of manure), certainly he does not lack the aid of lupin, which is very easy to procure" (2.15.5 *si deficiatur omnibus rebus agricola, lupini certe praesidium expeditissimum non deest*). If we have kept the thread from Books I and II in our hands, we can identify this *iners rusticus* as one of those who have failed to appreciate the causes of the Earth's fertility: it is unsurprising, therefore, that he does not see how plant detritus may remedy a lack of bird and animal dung. The resourceful and industrious farmer, however, will replenish his soil if he recalls all the materials that can serve *stercoratio*; because he knows how important fertilizing is if the Earth is to repay his investment, he will strive not to be without.

We will conclude our discussion of *stercoratio* by remarking an aspect of Columella's treatment that elaborates and complicates the nutritive theory: that the effects of manure are described at times not with the language of nutrition but of heat – a "thermodynamic theory." Manure had long been attributed the capacity to warm plants or soil.[108] This description is appropriate on at least two grounds: first, because decaying organic matter in compost heaps gives off palpable heat; second, because fresh or potent manures can cause plants to wither or brown, giving the impression that the plant has been "burned."[109] The thermodynamic view emerges most clearly in the following passage (2.15.2–3):

> nec ignorare colonos oportere reor, sicuti refrigescere agrum qui non stercoretur, ita peruri si nimium stercoretur,[110] magisque conducere agricolae frequenter id potius quam immodice facere. [3] nec dubium quin aquosus

[108] For example, Thphr. *Hist. pl.* 8.7.7, *Caus. pl.* 3.6.2.

[109] "Burn" is still used of the effects of excessive manuring today (*OED* 13d).

[110] Rodgers (*app. crit.*) proposes *stercore adiuuetur* for *stercoretur*, comparing 2.13.3, 3.11.9, 4.32.5.

> ager maiorem eius copiam, siccus minorem desideret, alter quod adsiduis umoribus rigens hoc adhibito regelatur, alter quod per se tepens siccitatibus hoc adsumpto largioribus torretur; propter quod nec deesse ei talem materiem nec superesse oportet.
>
> I also think that farmers should know full well that, just as the field which is not manured grows cold, so the one manured excessively is burned, and that it more behooves the farmer to perform (sc. manuring) frequently than extravagantly. [3] Nor is there any doubt that a watery field needs a greater amount of it, a dry one less: the former because (sc. the field) that is chilled by continuous moisture is thawed when (sc. manure) is applied, the latter because (sc. the field) that is warm by itself is scorched by additional dryness once (sc. the manure) is absorbed. This is why there ought neither be too little nor too much of such material.

Columella argues from a basic property of manure – its capacity to generate heat – to a doctrine of the mean in its application: to achieve a happily fertilized plot, the farmer must take care not to apply too much (which will burn the field, *perurere*, *torrere*) or too little (which will leave it chilled, *refrigescere*, *rigere*). The farmer must reckon with the fact that moisture (*umor*, *aqua*) chills, so that wet fields require more manure for the same heating effect and dry ones less. The principles that Columella articulates in this passage inform *praecepta* for manuring elsewhere in *RR*.[111]

It is not obvious how the less prominent thermodynamic theory is to be related to the nutritive theory, but the complication is instructive. The theories should not be supposed incompatible just because Columella does not explain their equivalence: life/growth and heat were closely associated in popular as well as philosophical thought,[112] and each theory provides a useful perspective on the effects of manure: the nutritive from what we might consider a biological perspective and the thermodynamic from a physical one. Recall that Vitruvius could toggle between element- and coherence-focused accounts of physical structure depending on explanatory need. What is more, the addition of the thermodynamic account illustrates the general theory-building potential of the nutritive account. As the conceptual nucleus for reconsidering manure, this theory becomes a peg from which to hang related phenomena and advice. These descriptive or prescriptive elements gain by association with the nutritive account even where they appeal to explanatory factors *prima facie* unrelated to feeding (such as on the thermodynamic theory), because they are set into an

[111] For example, 4.8.3 (application of manure conditional on the cold environment).
[112] For example, Arist. *Resp.* 474a25–8.

explanatory matrix that relates *stercoratio* to the fertilizing processes of Nature. Analysis of green manure, rankings of animal dungs, maintenance of the manure heap, heating effects of fertilizers – these topics are unified not only practically but also theoretically, because they represent efforts to explain and teach the restoration of the soil, which the nutritive theory reveals as the essential aim of manuring.

In sum, Columella's more detailed rebuttal of the senescence theory in Book II allows him to present an alternative explanation for declining agricultural productivity bewailed by the *principes* in Book I. As Columella holds, the Earth replenishes itself with food in the form of detritus from the plants growing upon it. This account buttresses his earlier assertion (Book I) that the Earth is divine, eternal, and providential by accounting for the mechanisms for its perpetual fertility. When the natural cycle is disrupted and *alimenta* removed by cultivation, however, the soil is depleted. It is our laziness and ignorance, the result of fallen *mores*, which blind us to the problem: if we recall that the Earth requires that its growth be returned to it, we will know to supply this nourishment ourselves through *stercoratio*. On this view of manuring, the farmer learns from Nature herself how to cultivate his land. This theory not only motivates systematic discussion but also provides a vocabulary and conceptual apparatus to discuss manuring throughout *RR*. *Stercoratio* thus becomes a conspicuous example of the need to understand Nature to understand farming, and hence of the new *ars* of agronomy proposed in *RR*.

4.4 Viticulture, Naturally

Nature functions teleologically at a large scale, that of the ecological system of forest or cultivated field. But it also functions teleologically at a small scale, that of the individual plant, and the farmer must no less grasp the relevant biological mechanisms here. After cereals and legumes (Book II), Columella turns in Books III, IV, and V to the grapevine (*uitis*), addressing all aspects of its cultivation.[113] In what follows we will focus on Book III, examining his *praecepta* on the choice of vine cultivar and the selection of vine slips for propagation. These two subjects are connected, as Columella first undertakes to defend the worthiness of Italian vine cultivars, especially the Aminnean *genus* (3.7–3.9.1), and next explains the methods of propagation enabling these cultivars to be productive (*fecundus* or *ferax*; 3.9.2–3.10). Although the senescence theory is no longer the relevant backdrop,

[113] See 11.arg.3–5 for an outline.

Columella's doctrines are still motivated in part by the moralizing view of agricultural decline presented as the stimulus for his undertaking in the preface to *RR* and in Book II on *stercoratio*. Moreover, he relies on the same conception of Nature to explain the proper approach to viticulture; indeed, his intricately argued instructions substantially develop its significance. Those farmers who fail to maximize the returns of grapevines do so because they lack knowledge of Nature, which is divine, eternal, and providential.

4.4.1 Par lex fecunditatis*: In Defense of the Aminnean Vine*

The central question in Columella's remarks on vine cultivars (*genera*) in Book III is which cultivars can be grown with the greatest profit. Vines may be roughly sorted into three grades (*gradus*, 3.2):[114] the first grade (3.2.7–19), comprising native Italian varieties such as the Aminnean, produce the best wines;[115] the second (3.2.19–24) and third (3.2.25–7) grades, mostly foreign types, produce lesser wines, but are commended, inter alia, for their copious yields.[116] The foreign cultivars of the latter two grades are especially preferred by Roman viticulturalists, who condemn Italian vines for their alleged lack of productiveness. Columella rejects the opinion of these farmers and blames their erroneous views on a failure to understand Nature. In particular, he reveals the flaws in their view by demonstrating the benevolence (*benignitas*) inherent in Nature's providential design of creatures: while Nature endows different beings with different properties, she has decreed an "equal law of fertility" (3.8.1 *parem legem fecunditatis*) for all, and has not deprived any one organism of what is granted more abundantly to others. The upshot is that, even if the Aminnean and other Italian vines appear less productive, they have the same potential for fertility apparent in foreign cultivars; it requires only the proper skill and knowledge to express this *fecunditas*. A correct understanding of Nature's workings thus makes attainable the ideal of a vine that produces excellent flavor and an abundant yield.

As with his discussion of the principles of soil fertility, Columella's foray into the biology of viticulture is prompted by Siluinus.[117] Just as Columella

[114] Applicable to vines grown for wine; Columella dismisses vines grown *ad escam* (3.2.1–2), which are less profitable.

[115] Cf. 3.2.19 *hae pretiosi gustus celeberrimae*.

[116] Those of the second grade *possunt ... prouentu et ubertate commendari* (3.2.19); of the third, *fecunditate sola commendantur* (3.2.25). Some cultivars are of uncertain grade (3.2.28–9). On the identification of ancient cultivars, see Thurmond 2017: 57–61 (earlier literature 2017: 58n11); also nn.118,119.

[117] It is significant that *studiosi agricolationis* are frequently responsible for Columella's deeper musings: see Ch. 2.5, and for Siluinus, n.96.

has begun to discuss propagation and care (3.4–6), he stops at an imagined question from his addressee: "What variety is this fruitful vine which we are at such pains to describe? Or are we talking about one of those which are now generally held to be the most productive?" (3.7.1 *cuius generis sit ista fecunda uitis, quam nos tam adcurate describimus, anne de îs aliqua significetur quae uulgo nunc habentur feracissimae*). The query is reasonable, since Columella has slipped into discussing cultivation without telling us which variety he is speaking of. Given the high yields promised, Siluinus assumes that he is talking about one of the cultivars reputed to be *feracissimae*. Perhaps it is the Bituric, which "most extol" (3.7.1 *plurimi . . . Bituricam . . . laudibus efferent*)? Or maybe the Spionan, Baliscan, or Arcelacan, each of which has its proponents (3.7.1)? None of these, in fact, although Columella will not slight such cultivars: they produce "the most bountiful supply of wine" (3.7.2 *largissimi uini*), he admits, and we will recall that they belong to the second and third grades praised for *prouentus*, *ubertas*, and *fecunditas*.[118] But his own ambitions are different (3.7.2):

> proposuimus docere uineas eiusmodi conserere, quae nec minus uberes fructus praedictis generibus adferant et sint pretiosi saporis, uel Aminnei uel certe non procul ab eo gustu.
>
> We have set out to teach how to sow such vines as bear no less abundant fruit than the aforementioned varieties yet are also of excellent flavor, such as of the Aminnean or certainly not far from it.

Columella is unwilling to compromise: he wants a vine which yields as much wine as the productive cultivars, but has the taste of the Aminnean, which "surpasses all others in flavor" (3.2.7 *ceteras omnis sapore praecedere*) and is distinguished for the "nobility of its wine" (3.2.14 *uini nobilitate*; cf. 3.2.28 *generositate uini Aminneae*).[119]

Yet the mere mention of the Aminnean grape has Columella on the defensive. He acknowledges that his opinion differs from that of nearly all other cultivators (3.7.2 *paene omnium agricolarum*), who have long believed that "Aminnean vines suffer from a natural and inborn barrenness" (3.7.2 *Aminneis . . . naturali et ingenita sterilitate laborantibus*). So is his ideal, the

[118] The Bituric and Baliscan are second grade (3.2.19, n.116) and excel the first grade in *fecunditas* (3.2.20). The Spionan and Arcelacan are third grade (3.2.27), the *Arcelaca maior* said by Columella to be *omnium quas cognouimus copiosissima*. The Arcelacan, Baliscan, and Bituric are later adduced as exemplars of *fecunditas* (3.2.28). For these important varieties, see on the (1) Arcelacan, André 1952: 148; Santon 1998: 95; (2) Baliscan, André 1952: 148–9; Santon 1998: 76; Thurmond 2017: 46; (3) Bituric, André 1952: 129; Dion 1959: 121–6; Santon 1998: 74–6; Berthault 2009; Thurmond 2017: 59; (4) Spionan, André 1952: 151–2; Piccoli 2004: 21–4; Thurmond 2017: 59–60.

[119] On the Aminnean vine, see André 1952: 128–9, 132–3, 153–4; Johnston 1999: 216; Santon 1998: 52–4.

ferax Aminnean vine, a figment of his imagination? No, he argues: the disbelief of other viticulturalists has only taken root because "reason has been deprived of the light of truth, condemned by the laziness no less than the want of knowledge of our husbandmen, or rather blinded as it were by the darkness of our ignorance" (3.7.3 *ratio . . . desidia nec minus inprudentia colonorum damnata aut uelut ignorantiae tenebris occaecata luce ueritatis caruit*). Columella's complaint connects viticulturalists' failure to appreciate the Aminnean vine with the moral decline crippling other parts of agriculture. Just as *inertia* led farmers to neglect *stercoratio*, so too have *desidia*, *inprudentia*, and *ignorantia* obscured the "reason" (*ratio*) which would allow correct assessment of the Aminnean vine. Having identified the root problem, Columella says that *ratio* must be "strengthened by many examples" (3.7.3 *conpluribus exemplis firmanda*), and that we should consider "those things which seem capable of correcting this widespread error" (*ea . . . quae uidentur hunc publicum errorem corrigere posse*).[120]

We soon learn what "those things" capable of rectifying the widespread misapprehension of agronomists consist in: a proper understanding of Nature. Columella continues (3.8.1):

> igitur si Rerum Naturam, P. Siluine, uelut acrioribus mentis oculis intueri uelimus, reperiamus parem legem fecunditatis eam dixisse uirentibus atque hominibus ceterisque animalibus, nec sic aliis nationibus regionibusue proprias tribuisse dotes, ut aliis in totum similia denegaret.
>
> So, Siluinus, if we should examine Nature with as it were the keener eyes of the mind, we would find that she has decreed an equal law of fertility for plants and humans and animals, and that she has not endowed some nations and places with their own qualities in such a way as to wholly deny similar things to others.

Scrofa's erroneous theory of soil fertility resulted from his inability to "see through" (*peruidere*) the phenomena to the true causes;[121] here too we must attend not to what appears to sight but what reason and "the keener eyes of the mind" reveal. Nature has established a *par lex fecunditatis* for all living beings – that is, has providentially guaranteed that all organisms are capable of productiveness (*fecunditas*). The universality of this rule is why it is called the "same" or "equal law" (*parem legem*), a Stoicizing expression.[122] The *lex* does not preclude that some beings could be more productive than others, for, as Columella

[120] For methodological error, see also Ch. 7.4.4 on *errores* in centuriation.
[121] See 2.1.5 (Ch. 4.3.1) with nn.86, 87.
[122] See Gummerus 1909–10: 23–4; also Noè 2000: 437; Diederich 2007: 135. Cf. *lex* at 3.9.1, 3.10.11.

observes, Nature has endowed different peoples and places with their own qualities (*proprias ... dotes*), among which could be a special fecundity. Yet – and this is key – Nature has not shared out such qualities in such a way that she has not given other beings "similar things" (*similia*): even if certain organisms have an extraordinary reproductive proclivity, *similia* are by no means forbidden to others.

Columella elaborates this argument before applying the conclusions to the Aminnean vine. He first turns, as promised, to the *conplura exempla* corroborating his account of Nature, proceeding through each of the three categories of living beings he has mentioned (3.8.1 *uirentibus atque hominibus ceterisque animalibus*).[123] The examples center on geographic difference (cf. 3.8.1 *aliis nationibus regionibusue*), with Italy forming one pole and various foreign countries the other; we will return momentarily to the significance of this dichotomous treatment. (1) First, people. Nature may have given to Egyptians and Africans abundant offspring (3.8.1 *numerosam generandi subolem dedit*), such that twins are commonplace among them, but among the Italians can be found the Alban twins who bore triplets (the Horatii and Curiatii) because Nature "wanted them to be of remarkable fertility" (*uoluit eximiae fecunditatis*).[124] Germany may have the tallest people in general, but Nature "has not wholly cheated" (3.8.2 *non in totum fraudauit*) others of lofty stature, as exemplified by the Italian giant Naeuius Pollio in Cicero's time or a Jewish man seen recently, both taller than the tallest German.[125] (2) Second, animals (3.8.3 *pecudes*). The Umbrian city of Mevania is reputed to produce large cattle, Liguria small, yet small Mevanian cattle and large Ligurian are not uncommonly seen.[126] And while the largest beasts, elephants, are not native to Italy, nothing prevents them from reproducing there, as shown by the birth of some at Rome. (3) Finally plants, in particular crops (3.8.4 *genera frugum*). Mysia and Africa are believed to abound with grain, yet Apulian and Campanian fields "do not lack rich fields" (3.8.4 *nec ... opimis defici segetibus*). Mt. Tmolus and Mt. Corycus may yield saffron, and Judaea and Arabia costly

[123] The order of presentation is an inverted *scala naturae*, beginning with humans and ending with plants – appropriately, since we are concerned ultimately with the vine.

[124] Pliny says multiple births are unremarkable in Egypt, *ubi fetifer potu Nilus amnis* (*HN* 7.33 with Beagon 2005: 164). For the "Alban women of the Sicinian line" (*Albanas Siciniae familiae*), see Dion. Hal. *Ant. Rom.* 3.13.4; cf. Dumont 1993: 32n2.

[125] For Pollio, cf. Plin. *HN* 7.74; he may have appeared in Cicero's *de admirandis* mentioned elsewhere by Pliny (Beagon 2005: 257). As for the second figure, cf. Joseph. *AJ* 18.103–4 with Melfi 1972: 185–7; Bloch 2010: 213n31; possibly the same man appears at Plin. *HN* 7.74 (Melfi 1972: 187; Beagon 2005: 257).

[126] For Mevanian cattle, Mynors 1990: 121 on Verg. *G.* 2.146–8; Dumont 1993: 33n5.

perfumes, but Rome lacks not even these: such fragrant foreign plants grow in gardens around the city.

Columella draws two conclusions from these *exempla* calculated to remove doubts about the potential for fertility possessed by the Aminnean vine and other Italian cultivars. The first conclusion does not depend specifically on Nature's distribution of qualities to organisms but a *laus Italiae* asserting the centrality of Italy for (Roman) agricultural practice.[127] The geographic slant of the examples tends to show that, while certain creatures and plants might flourish outside Italy, outstanding examples may also be found at Rome. Hence, Columella surmises that "Italy is most tractable to the attention of mortals" (3.8.5 *curae mortalium obsequentissimam esse Italiam*) and argues that, if Italian soil is capable of learning to support foreign fruit, we should harbor no doubts "about that fruit which is as it were indigenous and belongs natively to this soil" (*de eo fructu qui uelut indigena peculiarisque et uernaculus est huius soli*). The loosely reasoned argument seems to be that because Italian soil is excellent (sc. because foreign species transplanted in it will flourish), its native plants must be especially excellent. Columella feels, at any rate, that his argument is bolstered by the observation that Italian cultivars of the grapevine are "foremost in the excellence of their wine" (3.8.5 *nobilitate uini principes*).[128] Italy had always been the epicenter of the Roman agricultural imagination, and Columella relates this perspective to the *ars*. (We saw also in Vitruvius the special place accorded to Rome vis-à-vis its people and, by extension, its architecture.[129]) Nonetheless, Columella concedes that "these (sc. Italian vines) could perhaps be found wanting in <u>productiveness</u>" (3.9.1 <u>*fecunditas*</u> *ab his forsitan desideretur*), echoing his contemporaries' complaints.

To disarm the apparently more credible charge that Italian vines are insufficiently fruitful, Columella offers a second conclusion drawn from the *exempla*, which is more closely related to the view of Nature set out just before. Asserting that unproductiveness "can be aided by the industry of the cultivator" (3.9.1 *adiuuari potest cultoris industria*), he continues (3.9.1):

[127] For *laudes Italiae* see Canter 1938; 1939; Thomas 1982: 39; 1988: I 180; Gärtner 2002. Especially famous is Verg. *G.* 2.136–76 (Thomas 1982: 36–51; Ross 1987: 116–28); earlier, Varro, *Rust.* begins with a *laus Italiae* (1.2.6–7, Boscherini 1986; Nelsestuen 2015: 86–92). See also Dench 2005: 189–93. For Columella's idealization of Italy, see also Santon 1998: 161; Diederich 2007: 224; Fögen 2009: 197–9; Mielke 2023: 398–406.

[128] *nobilitate* recalls earlier descriptions of the Aminnean vine (3.2.14, 3.2.28, quoted earlier in this section).

[129] See Ch. 3.3.3 on *DA* 6.1.

> nam si, ut paulo ante rettuli, benignissima rerum omnium parens Natura quasque gentis atque terras ita muneribus propriis ditauit, ut tamen ceteras non in uniuersum similibus dotibus fraudaret, cur eam dubitemus etiam in uitibus praedictam legem seruasse, ut quamuis earum genus aliquod praecipue fecundum esse uoluerit, tamquam Bituricum aut baliscum, non tamen sic Aminneum sterile reddiderit, ut ex multis milibus eius ne paucissimae quidem uites fecundae, tamquam in Italicis hominibus Albanae illae sorores, reperiri possint?
>
> For if, as I said a little before, Nature the most benevolent parent of everything has endowed particular peoples and parts of the Earth with their proper characteristics in such a way that she has not completely deprived others of similar qualities, why should we doubt that she has also preserved the aforementioned law in the case of vine? So while she may have wished certain varieties of them to be especially fruitful, such as the Bituric or Baliscan, yet she did not make the Aminnean so infertile that among its many thousands not even a very few vines could be found to be fertile, just like the Alban sisters among the Italian peoples.

Calling Nature *benignissima parens*, Columella restates the "law" according to which she has not granted unique characteristics (3.9.1 *muneribus propriis* ~ 3.8.1 *proprias . . . dotes*) to some creatures such that others are deprived of similar qualities (3.9.1 *similibus dotibus* ~ 3.8.1 *similia*). This is just what the *exempla* have shown, as the appearance of the Alban sisters reiterates. The foreign Bituric or Baliscan might be "especially productive" (*praecipue fecundum*), but Nature has not made the native Aminnean so infertile that there cannot be found productive (*fecundae*) specimens.[130] In this way, viticulturalists' false belief that the Aminnean vine suffers from a "natural and inborn barrenness" (3.7.2 *naturali et ingenita sterilitate*) results from the same failing we met before: ignorance of the true workings of Nature, exacerbated by incurious laziness.[131] Once more, Romans have failed Earth and Nature, not the other way around.

4.4.2 *Function, End, Explanation: Propagating the Vine*

Columella's defense of the Aminnean vine is premised on the claim that it can be made more productive through cultivation (3.9.1, Ch. 4.4.1). While he has proven that it is not impossible for them to be *ferax* according to

[130] Columella underscores the foreignness of the Bituric and Baliscan vines at 3.9.4 (cf. n.118).

[131] The absenteeism of villa owners lamented in the preface (n.61) is germane, since "experience" (*experimentum*) would prove the existence of fruitful plots – not to mention the copious yields of Italian vines reported by Cato and Varro (*RR* 3.9.2–3).

Nature's "equal law of fertility," he has not shown us how to make them so. Hence, he goes on to argue that "if we are willing to apply our effort and attention, we will easily make vines of the Aminnean stock as fruitful as those of the Bituric or Baliscan" (3.10.7 *si* . . . *laborem et curam uelimus adhibere, facile* . . . *tam feraces Aminnei generis uineas constituemus quam Biturici aut balisci*). The *labor et cura* include selective breeding of the most fertile strains of the cultivar and a good understanding of general transplanting techniques (3.9.4–9). The most important factor, however, is an exact appreciation of how to select the most fertile and productive slips for propagation (*semina*), because choosing more fruitful *semina* will allow the farmer to raise more fruitful plants.[132] We will turn directly to this problem, which is the subject of a long and extraordinarily rich discussion in 3.10. In this section, Columella addresses two connected issues: (1) from where on a vine the *semina* must be gathered and (2) which *semina* must be selected. He submits these questions to a dialectical inquiry that pits *ratio*, the "reason" or "intelligence" which allows insight into the principles of Nature, against *consuetudo*, agricultural "custom" or received practice. Application of *ratio* reveals that *consuetudo* errs in certain crucial respects and must be corrected by a proper understanding of Nature; here is Columella's polemic against the ignorance and laziness of contemporary Roman farmers. Yet before all is said and done, we see that *consuetudo* is not wholly mistaken, but preserves some agricultural experience that is mutually corroborative with *ratio*; here are *usus* and *experientia* crystallized in the form of standard practice.

The dialectic between *ratio* and *consuetudo* structures 3.10, which divides into two main parts. In the first, Columella addresses the supposition of agricultural authorities that the "head" (*caput*) of the vine is the best part for gathering slips. *Ratio* shows that this belief is flawed and results from a failure to understand plant nutrition; rather, the "shoulders" (*umeri*) of the plant are the best source for *semina*, because fruitfulness in these parts is a true indicator of fertility. Columella turns to a moralizing digression that connects contemporary ignorance of viticulture with the general decline of agricultural knowledge: the complaints echo the now-familiar moral narrative. In the second part of 3.10, Columella again appeals to *ratio*, now to explain the providential design with which Nature has created animals and plants. This teleological view of Nature supports a functional description of organisms, according to which each member of the body has been

[132] *Semina* can mean any plant part for propagation, not just seeds proper (*OLD* 3); the common thread is reproductive capacity.

designed by Nature for some proper purpose or function (*proprium officium*). This insight both supports Columella's answer to the preceding question of where to select *semina* from and lays the groundwork for the following argument that the cultivator must take care to select only the slip whose fruitfulness shows that it has been endowed with an innate generative capacity (*uis generandi* or *uis fecunditatis*). Certain objections afford him the opportunity to revisit the *consuetudo agricolarum*, which he presents as reinforcing in some respects the view of the situation articulated by *ratio*. In both parts of 3.10, Nature plays the leading role: only by understanding its botanical designs can we correct or confirm *consuetudo* and establish a rigorous *ars*. We come to know Nature both with and by *ratio*, with which it is divinely instinct (3.10.10 *diuina praedita ratione*).

4.4.3 The Mechanics of Vine Nutrition

Columella begins by addressing the question of from which part of the vine the most fruitful slips for propagation (3.9.9 *feracissima semina*) are to be gathered. According to the "old authors" (3.10.1 *ueteres auctores*), they should be taken from "the tip, what they call the 'head' of the vine, that is, the outermost and most extended shoot" (*extrema pars . . . quod caput uitis appellant, id est ultimum et productissimum flagellum*). The reason (*causa*) for choosing this place is "the first impression, that is, the number of grapes which very often appear on the most extended branch" (*prima species et numerus uuarum qui plerumque conspicitur in productissimo sarmento*). But the phenomenon is deceptive (3.10.1):

> id enim accidit non palmitis ingenita fertilitate sed loci opportunitate, quia reliquas trunci partes umor omnis et alimentum, quod a solo ministratur, transcurrit dum ad ultimum perueniat.
>
> For this occurs owing not to the innate fertility of the vine-shoot but to its advantageous location, because all the moisture and nourishment, which is supplied by the soil, passes by the other parts of the stem until it reaches the outermost (sc. part).

The *ueteres auctores* are hence mistaken in inferring from their observation that the abundance of fruit is necessarily owed to *ingenita fertilitas*.[133] Here is yet another example of the ambiguous character of phenomena absent appreciation of natural causes.[134] It is, rather, the advantageous position of

[133] For misconceptions propagated by the *ueteres auctores*, see also Ch. 4.2.1 on *RR* 2.1.1.
[134] Cf. 2.1.5 (with nn.86, 87), 3.8.1, Ch. 4.3.1, 4.4.1.

the branch that ensures its fruitfulness, since moisture or sap (*umor*) and nourishment (*alimentum*) are channeled to the furthest growth. The difference is significant for propagation because we are of course looking for just that part of the plant whose "innate fertility" promises a happy result from planting or grafting. The dichotomy set up by *ingenita fertilitas* and *loci opportunitas* thus suggests a problem: Which part of the vine is essentially productive, in virtue of its nature, rather than only incidentally productive, in virtue of its location? The problem is not fully developed until later but already motivates Columella's investigation.

In refuting the claim that the tip of the vine is the best place for gathering the *feracissima semina*, Columella elaborates on the suggestion that the sap and nourishment (*umor*, *alimenta*) are channeled to this part of the plant (3.10.2–3):

> naturali enim spiritu omne alimentum uirentis quasi quaedam anima per medullam trunci ueluti siphone, quem diabeten uocant mechanici, trahitur in summum; quo cum peruenit, ibi consistit atque consumitur. unde etiam materiae uehementissimae reperiuntur aut in capite uitis aut in crure uicino radicibus. [3] sed et hae steriles, quae e duro citantur, ac duplici ex causa robustae sunt, quod a fetu uacant quodque ex proximo terrae integro atque inlibato suco aluntur; et illae fertiles ac firmae, quia e tenero prorepunt, et quicquid, ut supra dixi, ad eas alimenti peruenit indiuiduum est. mediae sunt macerrimae, quas transcurrit hinc parte aliqua interceptus, illinc umor adtractus.[135]

> For by natural respiration, all the nourishment of the living plant like a kind of breath is drawn through the core of the stem, as though through a siphon (which engineers call a διαβήτης), to the top; once it arrives there, it remains in that spot and is consumed. This is why the most vigorous material for growth is found either in the head of the vine or in its shin hard by the roots. [3] But (sc. the shoots of) the latter (sc. of the shin), those which are put out by hard (sc. wood), are infertile, yet are vigorous for two reasons, both because they lack offspring and because they are nourished from the whole and undiminished sap closest to the earth; whereas (sc. the shoots of) the former are fruitful and strong, because they are put forth from soft (sc. wood) and because, as I said above, whatever nourishment reaches them is not shared (sc. with other parts). (sc. Those shoots) are the leanest which are in the middle and which the moisture passes by, either being diverted in some part on the one side or drawn along to the other.

[135] *adtractus* is Rodgers' attractive conjecture for *ad se tractus umor* of **SA** and the *recentiores* (in which *a se* is also found).

Columella asserts that plants have a "natural respiration" which draws nutriments "like a kind of breath" to the top, where they are consumed. If conceived as an actual biological process, this doctrine of plant "respiration" seems obscure, and indeed it is not found elsewhere.[136] However, it is more easily understood if we consider that the notion of inspiration is meant chiefly to explain the movements of nutriments to the site of growth: just as animal *spiritus* draws vital breath into the body from without, so plant *spiritus* carries nourishing moisture from the soil to its greening parts. The doctrine of "respiration" is therefore chiefly analogical, underlining the fundamentally mechanical character of plant nutrition; as much is corroborated by Columella's comparison between the "core" (*medulla*) of the trunk or stem and a *sipho* (Gk. σίφων), a "small pipe" through which air or liquid travels under pressure that is indeed frequently handled in mechanical texts (Figure 4.1).[137] The gloss *diabetes* (Gk. διαβήτης), which appeals to διαβαίνειν, helps the reader to understand the relationship between the *sipho* and movement "through" or "across" (*transcurrere*).[138]

The mechanical analogy of the siphon affords the explanatory basis for Columella's account of plant nutrition, according to which sap collects at one or the other end of the plant and passes swiftly through the middle part. Some of the sap is "diverted" (*interceptus*) to the "shin" or stock (*crus*), which is close to the ground; the remainder, entering the siphon-like *medulla*, is "attracted" (*adtractus*) under mechanical force to the "head" or top (*caput*).[139] The middle parts are relatively deprived, because the sap "passes" them "by" in its movement up the stem (3.10.1 *reliquas trunci partes ... transcurrit*; cf. 3.10.3 *mediae* (sc. *partes*), *quas transcurrit*). The verb that Columella uses to describe this action, *transcurrere*, calques διαβαίνειν, reminding us of the physics of the siphon which underpin this account. The unequal distribution of sap has the consequence that the

[136] As Gummerus 1909–10: 26 observes, *anima* here has nothing to do with soul. For *medulla* and *anima*, he compares Plin. *HN* 17.153. Dumont 1993: 39n2 finds in *spiritus ... anima* a reminiscence of Cic. *Nat. D.* 2.136 (with Debru 2001: 46–52; for the significance of *Nat. D.* in *RR*, n.150). See also the following n.

[137] *OLD* 1. When Columella later returns to the passage of nutriments through the *medulla*, the notion of *spiritus* is deemphasized: cf. 4.20.2 (for the Theophrastan background of which, see *Caus. pl.* 3.5.2 with Richter 1981–3: III 625).

[138] διαβήτης most often means "compass" (*DGE* 1), where the relevant sense of διαβαίνειν is "spread the legs" (Beekes 2010: I 327), but Columella clearly does not explain the siphon as διαβήτης "because of the formal similarity" (pace Beekes 2010); the salience of διαβαίνειν is "go through" (δια + βαίνειν), as the calque *transcurrere* shows (3.10.1, 3.10.3, quoted in this section); cf. Strömberg 1944: 89 on διαβήτης as disease.

[139] For ancient explanations of the attractive forces of siphons, see Berryman 2009: 75, 161, 167–8.

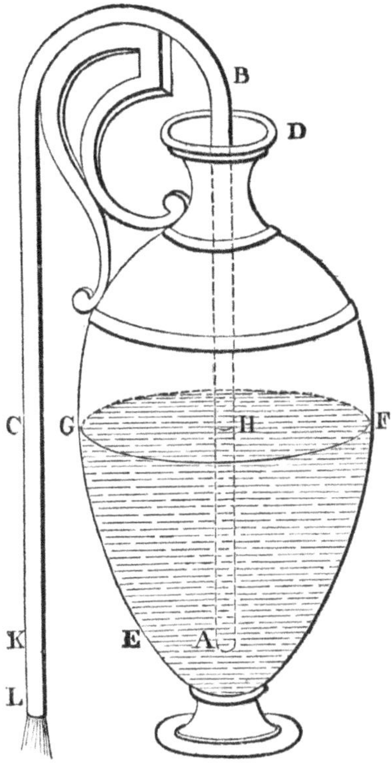

Figure 4.1 A simple bent siphon (καμπύλος σίφων), ABC, described by Hero (*Mech.* 2.1), which can be used to move liquid. Hero discusses the principles of the siphon's action in the same chapter and handles varied applications of the siphon in later chapters.

head and stock grow vigorously, while the parts in the middle are the "leanest" (*macerrimae*). This account substantiates Columella's claim that the fertility of the leading branch is, at least partially, a result of its "advantageous location" (3.10.1 *loci opportunitate*). But how does it bear on the question of where the vine shoots should be collected? Columella argues as follows (3.10.4–5):

> non debet igitur ultimum flagellum quasi fecundum obseruari, etiam cum plurimum adferat, siquidem loci ubertate in fructum cogitur, sed id sarmentum quod media uite situm ne inportuna quidem parte deficit, ac numeroso fetu benignitatem suam ostendit. hic surculus translatus rarius degenerat, quoniam ex deteriore statu meliorem sortitur; siue enim pastinato deponitur siue trunco inseritur, largioribus satiatur alimentis quam

prius, cum esset in egeno. [5] itaque custodiemus ut e praedictis locis, quos umeros rustici uocant, semina legamus.

Therefore, we should regard as fertile not the outermost shoot, even though it bears the most, since it is compelled to yield fruit by the richness of its position, but rather that branch which, situated in the middle of the vine, does not fail even because of its disadvantageous position, and in fact demonstrates its own bountifulness by its abundant produce. This slip less often degenerates after it is transplanted, since it goes from worse to better conditions – for whether planted down into trenched ground or grafted onto a stock, it is satiated with more abundant nutrition than it was before, when it was in a poor (sc. spot). [5] Hence, we will take care to select slips for propagation from the aforementioned locations, which countrymen call the "shoulders."

Columella does not commit himself to the claim that *semina* taken from the *ultimum flagellum* are necessarily unproductive; the argument is weaker, establishing only that the "richness of position" of the leading branch makes it impossible to know whether its abundant production betokens innate fertility or is only a byproduct of extra nourishment. Some slips from this branch could be fertile, but, rather than risk obtaining infertile ones, the cultivator would do better to take them from the middle of the vine, the "shoulders" (*umeri*),[140] where the poor nutrition guarantees that abundant fruit can be a sign only of the natural fertility of a branch (cf. *benignitatem suam*). The addition *hic surculus*, and so forth, simply makes the implication explicit: the productiveness of such a slip is not likely to deceive, because its flourishing in spite of poor conditions suggests that when it acquires better ones, it should thrive even more.[141] Choosing *semina* from the *umeri* of the vine is therefore a pragmatic strategy, calculated to ensure that the farmer does not waste time gathering slips which may produce poor yields.

Columella's answer to the first of the two questions addressed in 3.10 – where on a vine the *semina* ought to be gathered – already shows how knowledge of nature guides agricultural practice. But this mechanistic explanation is not Columella's final word on the reasons for the productiveness of different parts of the vine: it is, rather, one stop on the way to

[140] For *umeri*, see *OLD* 2d. It appears again of vines at *Arb.* 3.1 (where *lumbi* produce the best slips, *umeri* second best!), of fruit trees at *Arb.* 20.1 and, by plausible conjecture from there, Col. 5.10.6; cf. Plin. *HN* 17.105 (pointing to Verg. *G.* 2.299–300, with Mynors 1990: 137). See also Santon 1998: 185–6.

[141] Compare the analogous principle that plants should be transplanted *ex deteriore terra potius in meliorem quam ex meliore in deteriorem* (3.5.2; cf. 3.9.9, and Zainaldin 2020a: 176 on Garg. Mart. *Pom.* 2.5.6).

a deeper explanatory account in which the loose conception of "nature" relevant for the physical explanation of fluid movement is replaced by a more specific conception of "Nature" as intelligent designer of terrestrial phenomena. This development in explanatory strategy comes when Columella turns to the second question addressed in 3.10.

4.4.4 Ratio *versus* consuetudo

Hard on the heels of his advice to take slips from the shoulders of the vine, Columella qualifies his instructions by saying only those which we have seen bear fruit should be gathered: "for if they are bare of produce, then we judge this part of the vine, excellent though it may be, to make no contribution to the fertility of the mallet-shoot" (3.10.5 *nam si fetu uacua sint, quamuis laudabilem partem uitis nihil censemus ad feracitatem conferre malleolo*).[142] This is, in a nutshell, the answer to the second question with which Columella is concerned in 3.10, which *semina* are to be gathered – yet the *disputatio* continues. For we now learn that there are some viticulturalists "who do not believe that it matters at all how many bunches of grapes a branch has borne, provided that it be chosen from a fertile vine and not be grown out of hard wood, which they call *pampinarius*" (3.10.5 *qui minimum referre credunt, quot uuas sarmentum habuerit, dum et ex uite fertili legatur et non ex duro trunco enatum, quod pampinarium uocant*).[143] This *agricolarum opinio* is posed ostensibly in response to the second of Columella's questions (which *semina* should be gathered?), but in fact also offers a challenge to the first (where should the *semina* be gathered?). These farmers not only neglect to use the productiveness of particular shoots as an index for fertility, as Columella has just asserted should be done, but seem also not to discriminate between the *umeri* and other parts of the plant: the only criterion for selection is that the vine itself be fertile and the slip not be grown from a woody part. This "extremely faulty" (*uitiosissima*) view initiates a new turn in 3.10.

Columella's response leads into a moralizing digression of sorts that reframes the dispute in terms of the narrative of agricultural decline repeatedly invoked in *RR*. Arguing that his opponents' belief stems "from ignorance about selecting slips for propagation" (3.10.6 *ex inscientia*

[142] *Malleolus* is a "hammer-shaped slip of a tree for planting" (*OLD* 2): cf. *RR* 3.6.3 and see Bruno 1969: 72; *TLL* VIII 191.37–70 (Bachmann).

[143] *Pampinus* may mean the whole vine shoot or branch or, more narrowly, bring into focus the vine foliage (*TLL* X 1.185.36–65 [Somazzi]); the latter meaning is foremost in the adjective *pampinarius* (so 5.6.29). See Bruno 1969: 82–3; Santon 1998: 186, 190–1.

seminum eligendorum), he warns emphatically that this advice will produce "vineyards first not sufficiently fruitful, and then, quite barren" (*primum parum fecundas uineas, deinde etiam nimis steriles*). How is it that a *uitiosissima opinio* which risks destruction of the whole vineyard is so common? Here is Columella's stern report (3.10.6–7):

> quis enim omnino iam per tam longam seriem annorum agricola malleolum legentibus praecepit ea, quae paulo ante rettulimus? immo quis non inprudentissimum quemque et eum qui nihil aliud operis facere ualeat, huic negotio delegat? itaque ex hac consuetudine ueniunt inprudentissimi ad rem maxime necessariam, deinde etiam infirmissimi; nam inutilissimus quisque, ut dixi, qui nullum alium laborem ferre queat, huic officio adplicatur. [7] is porro, etiam si quam scientiam eligendi malleoli habet, eam propter infirmitatem dissimulat ac supprimit,[144] et ut numerum quem uilicus imperauit explere possit, nihil curiose, nihil religiose administrat; unumque est ei propositum peragere laboris sui pensum, cum tamen, ut et sciat, et quod sciat exsequatur, hoc solum praeceptum a magistris accipit, ne pampinariam uirgam deplantet, cetera omnia ut seminibus contribuat.
>
> For is there any farmer at all who in the many years till now has taught to those selecting a mallet-shoot the things which I wrote about a moment before? Or even any (sc. farmer) who does not assign to this duty whoever is least knowledgeable and the kind of person who is incapable of doing any other work? And it is in consequence of this custom that the least knowledgeable, and indeed also the weakest, take up a matter which is of the highest importance; for it is whoever is the most useless, as I said, so as to able to bear no other job, who is assigned to this duty. [7] This fellow, even if he does know something about selecting a mallet-shoot, hides and conceals it owing to his weakness, and in order that he be able to fulfill the number (sc. of slip-transplants) which the bailiff assigns, accomplishes nothing in a painstaking or scrupulous way. His only intention is to complete his allotted work when, although he may be knowledgeable and strive to carry out what he knows, he nonetheless receives only this instruction from his masters: that he not plant a leaf-branch, but that he take everything else as slips for propagation.

As in the case of declining soil fertility, Columella's attack on viticultural *consuetudo* (3.10.6) unites two themes: critique of tradition and diagnosis of contemporary moral failure. Columella earlier lamented that Romans are unwilling to cultivate their own land and pay little attention to who tends it for them: the rich person chooses for his *uilicus* the one who is "most

[144] MSS transmit *superponit*, but the word is difficult; *OLD* 4 cites only this passage for the meaning "(app.) to mask, overlay (a quality)." I print *supprimit*, Rodgers' conjecture *apud* Lucarini 2014: 651, who also suggests *seponit*.

infirm in age and strength" (1.pr.12 *defectissimum annis et uiribus*), while the middle-class man sends an unprofitable slave to the farm though he be "ignorant of the business which he will be in charge of" (*ignarum rei cui praefuturus est*). The present passage echoes this situation (3.10.6 *infirmissimi* ~ 1.pr.12 *defectissimum*, 3.10.6 *inprudentissimi* ~ 1.pr.12 *ignarum*), locating the erroneous methods of vine propagation in a wider malaise besetting agriculture. Short-sighted interest in profit, compounded by ignorance of the complexity and importance of slip selection, has caused erroneous practices to proliferate.[145] Because earlier generations of agronomists seem not to have communicated proper techniques (cf. 3.10.6 *quis . . . rettulimus?*), tradition offers no bulwark against the common error.

With contemporary and historical *consuetudo* in error, Columella once more turns to *ratio* (3.10.8):

> nos autem primo rationem secuti, nunc etiam longi temporis experimentum, non aliud semen eligimus nec frugiferum esse ducimus, nisi quod in parte genitali fructum attulerit. nam illud quidem quod loco sterili laetum robustumque sine fetu processit, fallacem fecunditatis imaginem praefert, nec ullam progenerandi uim possidet.
>
> At first following reason, and now also experience over a long period, we neither select nor regard as fruitful any other slip for propagation except the one which bears fruit on the reproductive part of the plant. For the slip which in a sterile spot has grown out strong and hearty without produce offers a false appearance of fertility and does not have any capacity for procreation.

Foregrounding reason (*ratio*) confirmed by experience (*experimentum*) as the guide for practice, Columella takes a wider view of the two major concerns in selecting vine slips previously addressed.[146] As regards the second question (which *semina* to gather), fruitfulness (cf. *fructum attulerit*) remains the criterion; but as regards the first (where from), the "shoulders" (*umeri*) are replaced by the "reproductive part" (*parte genitali*). The latter change is significant, because it points to a reframing of the examination of fertility in *functional* – hence biological – and not merely *local* terms. The earlier mechanistic analysis convinced us that *umeri* are the safest place from which to draw the slips but demonstrated only the

[145] Note especially the false "productivity principle" which forces workers to complete a requisite number of tasks (3.10.7 *numerum*) at the expense of a properly careful approach.

[146] *Ratio* here has the same meaning as at 3.7.3 (above), referring to the faculty of reasoning (rather than "account," as often). Below, we discuss the positive contribution of *consuetudo*, which in part covers *longi temporis experimentum* here; but *experimentum* surely includes Columella's own practice or experiment, too.

incidental cause for this conclusion (the local distribution of nutrients). When Columella says that we will now examine the part of the vine qua reproductive (that is, as *pars genitalis*), he indicates that we will be concerned with the essential causes of fertility – the innate biological capacity for reproduction, *uis progenerandi.*[147] We can go further: the functional analysis led by *ratio* hints also at a collapse of the *where* and *which* questions previously distinguished, for an account of vine parts with respect to their productive capacity should ipso facto also tell us where and which slips to gather. This would hold also of the converse: just as Columella links *in parte genitali* with *fructum attulerit* for the productive slip, so he joins *loco sterili* to *sine fetu* for the unproductive, showing that in both cases the phenomena are related to the function of the plant part.

4.4.5 *The Teleology of the Vine*

Columella does not leave us in suspense as to what it means to "follow reason" (3.10.8) in pursuit of this deeper, unifying view of vine propagation. He sets out his central thesis as follows (3.10.9):

> id procul dubio uerum esse ratio nos admonet, si modo ut in corporibus nostris propria sunt officia cuiusque membri, sic et frugiferarum stirpium partibus propria munia.
>
> It cannot be doubted that reason teaches us the truth of this (sc. the correct approach to selecting vine shoots), provided that just as each of the members of our bodies has its own proper function, so too do the parts of fruit-bearing plants have their own proper activities.

Ratio persuades us of a biological analogy between animals and plants: given that in our own bodies each part has a "function" (*officium*) that is "its own" (*proprium*), so we would expect that in plants each part would likewise have a proper "activity" (*munus*).[148] We have already remarked on the prevalence of plant–animal analogies in biological explanation in Chapter 3, where we considered Vitruvius' comparison of trees and pregnant women.[149] Columella similarly appeals to analogy in the present instance, although for a different purpose. As we will see, his analogy serves

[147] For *uis* = "capacity" (cf. Gk. δύναμις), see *OLD* 14, 16. *Progenerare* of plants is nearly exclusive to Columella, who uses it not infrequently: *TLL* X 2.1756.66–75 (Banta).

[148] For *officium* of the function of a body part, see *OLD* 4b; *TLL* IX 2.524.9–22 (Oomes); *munus* is less common in this sense (cf. *OLD* 1b; *TLL* VIII 1663.54–77 [Lumpe]) and seems to be *variatio* (Nyström 1926).

[149] See Ch. 3.4.1 on *DA* 2.9.1 and n.120.

(1) to conceptually clarify the idea of physiological function (*officium* or *munus*) and (2) to relate each function to one and only one plant/animal organ or part (hence *proprium*). In arguing the points, he invokes a grander vision of natural design based on the notion of Nature as a divine and providential creator, thus making the analogy philosophically appropriate.[150] Once plant life and reproduction are comprehended from this teleological perspective, the wisdom of Columella's advice for choosing vine slips becomes clear.

First I quote in full the crucial passage in which Columella develops his analogy (3.10.9–11):

> uidemus hominibus inspiratam uelut aurigam rectricemque membrorum animam, sensus iniectos ad ea discernenda quae tactu quaeque naribus auribusque et oculis indagantur, pedes ad gressum conpositos, bracchia ad conplexum, ac ne per omnes uices ministeriorum uagetur insolenter oratio, nihil aures agere ualent quod est oculorum, nihil oculi quod est aurium. [10] ne generandi quidem data est facultas manibus aut plantis, sed quod hominibus ignotum uoluit esse genitor uniuersi, uentre protexit, ut diuina praedita ratione rerum aeterna opifex quasi quibusdam secretis corporis in arcano atque operto sacra illa spiritus elementa cum terrenis primordiis misceret, atque hanc animantis machinae speciem effingeret. [11] hac lege pecudes, hac uirgulta progenuit, hac uitium genera figurauit, quibus eadem ipsa mater ac parens primum radices uelut quaedam fundamenta iecit, ut îs quasi pedibus insisterent, truncum deinde superposuit uelut quandam staturam corporis et habitus, mox ramis diffudit quasi brachiis, tum pampinos elicuit uelut palmas, eorumque alios fructu donauit, alios fronde sola uestiuit ad protegendos tutandosque partus.

> We see that humans have been infused with a soul, as it were the driver and directress of our limbs; that senses have been instilled for the perception of things which are ascertained by touch as well as those things ascertained by the nose, ears, and eyes; that feet have been formed for walking, arms for grasping; and, in order not to let my discussion wander without restraint through all of the relations of the bodily functions, that the ears cannot do anything which belongs to the eyes, and the eyes nothing which belongs to the ears. [10] Nor indeed was the capacity for generation granted to hands or the soles of feet: rather, because the father of the universe wanted this to

[150] As Gummerus 1909–10: 25 observes, Columella's basically Stoic approach should be compared with *Nat. D.* 2.133–45, where Balbus outlines the *incredibilis fabrica naturae* (2.138). Of course, teleological explanation was by no means exclusively Stoic (e.g. Pease 1955 *ad locc.*) At *Nat. D.* 2.120, Balbus addresses the *ratio naturae intelligentis* evident in plant design (cf. *Tusc.* 5.37). For the philosophical backdrop of Columella's design argument, see further Gummerus 1909–10: 21–31; Dumont 1993: 42nn11–12, 90–1; André 1994: 195; Noè 2000: 438–9; 2002: 108–10; Diederich 2007: 135; Bretin-Chabrol 2012: 101–2.

> remain unknown to humans, he covered it with the belly, so that the eternal creatress of things, endowed with divine reason, would in certain hidden parts of the body, as it were, in a secret and concealed place mingle those sacred elements of the vital breath with the earthly principles and thus fashion this sort of living machine. [11] According to this law she formed beasts, by this (sc. law) shrubs, by this (sc. law) the varieties of vines, to which the very same mother and parent first gave roots, as a certain foundation, so that they might stand on these as if on feet; then she placed above them a stalk as a certain upright bulk of body and bearing; next she enlarged this with branches as arms, after which she teased out tendrils as hands, endowing some of these with fruit and garbing others only with leaf for the purpose of protecting and defending the offspring.

Columella begins by pointing out that each function or activity of the body has been assigned to one organ. It is evident (cf. *uidemus*) that the soul (*anima*) guides the limbs,[151] the senses perceive according to their specific capabilities, the feet are fashioned for walking, the arms for grasping, and so on. Ears cannot see, eyes cannot hear, and if we are to generalize as Columella wishes us to, we must conclude that each organ has its own proper (*proprium*) function and can fulfill no other function or functions except those which belong to it. These correspondences work out the implications of the earlier claim (3.10.9) that "each of the members of our bodies has its own proper function (*propria … officia*)." After adumbrating this rule of proper functional correspondence, he turns to the reproductive capacity (*generandi … facultas*), which is obviously of the highest concern in the present context, and investigates the human organ associated with it. The womb prompts him to expound upon the foundations of the seemingly purposive anatomical and physiological organization of living creatures.

To explain the womb, Columella appeals to the intelligence of the *genitor uniuersi* and the *rerum aeterna opifex* (3.10.10). This philosophical perspective is important for understanding Columella's conception of Nature and, ultimately, the relation of such knowledge to agricultural practice. To begin, it bears emphasizing that Columella's invocation of the two designers is consistent with the cosmological principles asserted in the preface to *RR*: as we will recall, he spoke there of *Rerum Naturam, quam primus ille mundi genitor perpetua fecunditate donauit* (1.pr.2). While the identity of the first mover, the *genitor mundi* or *uniuersi*, remains uncertain, the status of the *opifex* is more secure: it refers to Nature as the

[151] See Gummerus 1909–10: 27–9 on the Stoic coloring to *uidemus … animam* (though not all details are exclusively Stoic; cf. Mielke 2023: 152n592).

creatress and nourisher of living beings.[152] So much is clear not only from the position of *Rerum Natura* in the quotation in 1.pr.2, but also from the epithets *aeterna* and *mater ac parens*, words which Columella has applied to Nature in the guise of *Tellus* or *Terra* elsewhere.[153] Apart from these descriptions, he offers a significant admission about the *opifex* – that is, Nature/Earth: she is "endowed with divine *ratio*." We have already seen that the providential character of Nature is implied in the self-sustenance of Earth as well as in the benevolence (*benignitas*) with which she distributes qualities among living creatures. Columella now makes it explicit that Nature is penetrated by a divine intelligence or rationality, and because Nature/Earth is "mother and parent" of all living creatures (3.10.11 *mater ac parens*), both plant and animal life bear the stamp of her intelligent design.

Nature understood as intelligent designer plays a key role in Columella's argument, because the providential organization of animal and plant parts underpins the proper correspondence between organs and functions. While he does not relate the specific form of particular parts to their activities in detail, the notion of design sets up a teleological approach to explanation, for it imports the idea of natural ends or purposes. This teleological approach is evident in the discussion of the womb, which is covered by the belly (cf. 3.10.10 *uentre protexit*) so that the beginnings of life might be concealed from humans (cf. *ignotum*); despite the religious mystique,[154] the intentionality of the anatomical structure is clear (thus the purpose clause: *uentre protexit, ut*, etc.).[155] Also revealing is Columella's conclusion that Nature has shaped the human as "a sort of living machine" (*animantis machinae speciem*).[156] This view of the living being as an animated (ensouled) "contrivance" or "machine" (*machina*) recalls the mechanistic explanation adopted earlier (3.10.2–3), where Columella described

[152] For the *genitor*, see n.50.

[153] 1.pr.2, 2.1.2. As stated earlier, Earth can be subsumed under Nature because Earth is contained by Nature; Columella's focus on *Tellus*/*Terra* earlier was necessary to rebut the senescence theory, which concerned the soil. Gummerus 1909–10: 8, 10, 22n2 identifies the *opifex* with the *genitor primus*, defending an emendation *aeternus* (and hence taking *diuina praedita ratione* in reference to the *elementa*); that is both unnecessary and implausible. For *opifex aeterna* = *Natura*, see also Richter 1981–3: I 659; Dumont 1993: 91; Santon 1998: 190; Noè 2000: 439; 2002: 18; Diederich 2007: 135; Mielke 2023: 356. For *opifex rerum*, vel sim., in reference to the "creator" or "fabricator" of the world, see Bömer 1969–86: I 43 on Ou. *Met.* 1.79; *TLL* IX 2.704.63–705.12 (Beikircher).

[154] For the mystery, besides 3.10.10 *ignotum uoluit esse genitor uniuersi*, cf. also *quibusdam secretis corporis in arcano atque operto*, and so forth.

[155] The origins of life described by Columella (3.10.10) are not, pace André 1994: 195, Epicurean. *Spiritus* is πνεῦμα, which mixes with the (passive, terrestrial) elements in the Stoic account of generation: see Gummerus 1909–10: 28; Dumont 1993: 91; Diederich 2007: 135.

[156] For *machina* extended analogically, see *TLL* VIII 13.52–62 (Dietzfelbinger). Berryman 2009: 205–15 considers machinery used analogically to explain organisms but does not discuss Columella.

the vine stem as a kind of διαβήτης/*sipho*. This explanation is not repudiated by the teleological account but rather reinterpreted and deepened: the structural homology between the vine stalk and the διαβήτης is a trace of the *ratio* of Nature rather than fortuitous similarity.

The discussion of Nature's intelligent design comes full circle when Columella turns his teleological lens to the functional explanation of vine parts (cf. 3.10.9). With the help of the analogy of the human body – which we now recognize not merely as a heuristic or explanatory comparison, but a reflection of purposive design – Columella proceeds through the plant parts, explaining how Nature, *mater ac parens*, has crafted each part for some end.[157] Thus, vines have *radices* "so that they might stand on these as if on feet" (3.10.11), and a *truncus* to give them upright posture.[158] The account culminates in the tendrils (*pampini*)[159] – the "hands" (*palmae*), as it were – some of which the *rerum aeterna opifex* has made fruit-bearing (cf. *alios fructu donauit*), others of which she has made leafy to shield the produce (cf. *ad protegendos tutandosque partus*).[160] The significance of the account is obvious: while the mechanistic account could show only that and why some tendrils (*pampini*) are fruitful *incidentally*, the teleological account teaches us that and why some tendrils are productive *essentially* – because Nature has fashioned them as the reproductive organs of the plants, just as the womb is the reproductive organ in humans.

4.4.6 *Agricultural Empiricism:* Ratio *versus* consuetudo, *Revisited*

What is the upshot of Columella's teleological account for practice? Apart from corroborating the *praeceptum* to gather slips from the shoulders (*umeri*), the argument substantiates his advice to gather slips only from branches which have previously borne fruit (3.10.8). We saw that other

157 The details of this comparison are not precisely paralleled in other ancient accounts (Gummerus 1909–10: 25). Schneider 1794–7: II.2 154 thinks the particulars might be inspired by rustic usage (cf. 3.10.1, 3.10.5, 3.18.5, *Arb.* 3.1).

158 *Statura* sits oddly in the comparison because it is an abstraction, whereas the other items are concrete body parts; perhaps it is meant concretely (hence my translation "upright bulk"), although this sense is not distinguished by L&S or *OLD*.

159 For *pampinus*, see n.143.

160 Earlier (3.10.10) Columella remarked that for human beings, *ne generandi quidem data est facultas manibus aut plantis*; that he now identifies the generative organ in plants with their equivalent of *palmae* (i.e. their *pampini*) underscores that what plants and animals share is the fact of proper correspondence in structure and function, not specific functional homology (pace Turra 2017: 218). Still, an organ may have more than one proper function, and vine tendrils (*clauiculae*) are elsewhere imagined by Columella *quasi quibusdam manibus* by which the plant grasps a prop (4.6.2, after Cic. *Nat. D.* 2.120; cf. Zainaldin 2020a: 144–5), so a partial functional analogy is not wholly inappropriate.

agricolae maintain that any vine branch can serve as a slip provided it is not a "leaf-branch" (3.10.7 *pampinariam uirgam*) – that is, a leafy growth from hard wood (cf. 3.10.6). While these farmers are right to avoid leaf branches, their criterion is inadequate, since only some branches that are not leaf branches possess innate fertility. But even the viticulturalist who apprehends the intelligent designs of Nature will face a difficult choice in selecting *semina*, because different kinds of *pampini* are structurally identical, at least in some respects. Columella's advice to choose only the fruitful branches is meant to resolve the issue: the phenomena are pressed into service as an index for natural endowment of the *uis generandi.*[161] Proven fruitfulness is a guarantee of reproductive capacity (although we will not forget earlier caveats about the unequal distribution of nourishment), so observation can help the diligent cultivator determine where to gather *semina* from.

This empirical link explains why Columella is at great pains afterwards to stress the essential relation between fertility and the visible production of fruit. Thus, if the farmer gathers from parts "which are barren of fruit" (3.10.12 *quae fructibus uidua*), he is said to be "toil[ing] for shade" (*umbrae … laborauerimus*), because these are destined merely to serve "so to speak as coverings and parasols" (*tamquam tegumina et umbracula*) for fruit-producing parts. But if he takes from the "reproductive members themselves, heavy with offspring and produce" (*ipsa membra genitalia conceptu atque fetu grauida*),[162] his labor happily accrues "to the vintage" (*uindemiae*). Columella uses a possible objection to his account to buttress the epistemic significance of observed fruit production. Since Nature has granted each part of the plant "its own function which is appropriate for it" (3.10.13 *proprium … officium quod scilicet ei conuenit*), would not the reproductive parts, which are essentially fertile, retain their *uis fecunditatis* regardless of whether they have put out fruit? The conclusion could endanger the indexical value of the phenomena. Appealing again to the human–plant analogy, Columella observes that it is not impossible for an anatomically complete person to be unable to reproduce – that is, to be infertile; in such instances, it is just the failure to produce offspring which demonstrates the infertility. Likewise, failure to produce fruit proves the infertility of a branch: even if it is "born in a fertile area" (3.10.14 *genitali loco … nata*), if it lacks fruit, ipso facto it must lack reproductive capacity.

161 Consider Columella's suggestion that, even for those who understand the human body, the womb which Nature has hidden in the belly remains somewhat mysterious (Ch. 4.4.5 and n.154).

162 The basic sense of *grauidus* is "pregnant" (*OLD* 1), so the word is especially apt given the tenor of the passage and especially the collocation with *membra genitalia*, *conceptu*, and *fetu*; cf. 3.6.2, 3.11.5.

This conclusion effectively collapses two distinct causes – (1) essential sterility which results from the absence of the reproductive *officium* and (2) infertility which befalls parts that do possess this *officium* (analogous to human infertility despite intact reproductive organs) – and makes the phenomena the arbiter of the issue. If a branch lacks fruit, it does not really matter what the underlying cause is, because its infertility is assured; conversely, presence of fruit promises future fruit, at least if the branch is growing on the *umeri*.

Columella's discussion of vine fertility presents an opportunity to return to agricultural *consuetudo*. Until now we have seen him adopt a polemical attitude toward the opinions of both earlier and contemporary practitioners, which he aims to correct with *ratio*. Yet the dialectic between *ratio* and *consuetudo* is not so one sided. Columella adduces agricultural custom to corroborate the argument that vines sprouting in a fertile part of the plant cannot be trusted (3.10.15):

> itaque, ut ad consuetudinem agricolarum reuertar, eiusmodi surculos qui nihil tulerint spadones appellant; quod non facerent, nisi eos suspicarentur inhabiles frugibus. quae et ipsa appellatio rationem mihi subiecit non eligendi malleolos quamuis probabili parte uitis enatos, si fructum non tulissent.

> Hence, to return to the custom of farmers, they call these kinds of shoots which bear nothing "eunuchs," something that they would not do unless they suspected that they were incapable of bearing fruit. And it is indeed this name which suggested to me the reason not to select mallet-shoots even though they had sprouted from a good part of the vine, unless they had borne fruit.

Farmers call branches that are barren but grow in fertile spots *spadones*: although the word may be translated as "eunuch," its meaning is wide enough in Latin to refer to males who are impotent not only by castration but also by nature; indeed, the word was sometimes distinguished as the *vox propria* for those who are naturally infertile.[163] It is therefore appropriately transferred to a vine branch which ought to be productive, if one were judging merely from its structure and position, but which lacks generative capacity. Strikingly, Columella says that the name *spado* in fact suggested to him his *ratio*: farmers "would not (sc. have called such vines 'eunuchs') unless they suspected that they were incapable of bearing fruit." Agricultural *consuetudo* thus preserves an insight into the human–vine

[163] See *RE* (2) III A.1.1258.29–44 (Pfaff); Stevenson 1995: 497; cf. Maas 1925: 439–41.

analogy which sets the student of Nature onto the path to right explanation. Yet it is significant that the word he uses of the farmers' insight is *suspicari*: though they might have "suspected" or "formed an idea," this does not mean they had complete knowledge or understanding.[164] *Consuetudo* thus may point toward the truth, but it does not yield the proof that *ratio* does. (We will study the rationalizing approach to etymology exemplified in Columella's analysis of *spado* at greater length in Chapter 7, on the *agrimensores*.) In any case, what deserves emphasis is the cooperative potential of *consuetudo* and *ratio*: *ratio* not only revises and corrects *consuetudo* but can also be strengthened or even guided by longstanding agricultural practice. Here we recall Columella's earlier remark that "long experience" (3.10.8 *longi temporis experimentum*) confirms his account.

This cooperative dialectic between *ratio* and *consuetudo* is developed further as Columella elaborates his account in response to various objections (3.10.15–22). It falls beyond the present need to study these passages in detail,[165] but through all of them past fruitfulness remains the most important indication of suitability for propagation: thus we must not only gather shoots "from the productive part of the vine" (3.10.17 *e fructuosa parte uitis*), but also select only those "which promise productiveness in the future <u>by the fruit that they have already given</u>" (*eos qui futuram fecunditatem <u>iam dato fructu</u> promittunt*) – and the more fruit the better (3.10.17 *qui <u>numerosissimis fetibus</u> conspiciuntur*). While the emphasis on the observation of productiveness has an empirical and common-sense dimension, Columella never allows us to forget the earlier teleological explanation of the vine's parts, reminding us that "the <u>capacity for fruitfulness</u> belongs to <u>certain so to speak members</u> (sc. of the vine)" (3.10.21 *<u>uim fecunditatis</u> certis <u>quasi membris</u> inesse*). It is this deeper view of plant structure and function which underwrites his advice: both biologically, in the anatomical and physiological design of the plant, and epistemologically, in the connection between observed fruit production and fertility.

To conclude, Columella's complex justification of *praecepta* for vine propagation in 3.10 further demonstrates the importance of understanding Nature for right agricultural practice. Using both *ratio* and *consuetudo*, he

164 *OLD* 1. *Suspicari* appears also at 9.8.8 (of an inference); cf. *suspicio* (noun) at 7.12.5.

165 Columella rejects *pampini* grown out of hard wood (*e duro*), which might possess some adventitious fertility (3.10.15–17); argues that the tendency for plants to revert to wild types makes careful selection of slips even more critical (3.10.18–20, where Verg. *G.* 1.197–200 is quoted); and suggests again that *consuetudo* preserves an implicit recognition that only certain cuttings have generative capacity (3.10.20–22).

undertakes a penetrating investigation into the causes of plant biology which will equip the *agricola* to propagate fruitful vines. In addressing one key question – where to gather *semina* from – Columella provides a mechanistic explanation of vine nutrition that supports his advice to gather slips from the shoulders (*umeri*) of the plant, not the head (*caput*) as many believe. But *consuetudo* is stubborn, even where it is mistaken, and the moral decline which has hobbled the *res rustica* generally makes its effects felt in widespread misconceptions in viticulture. In investigating the second key question – which *semina* are to be chosen – we gain a deeper knowledge of Nature with the aid of reason that allows us to ascertain the proper methods for careful vine propagation. To support his view that slips should be drawn only from branches observed to be fruitful, Columella develops a functional account of vine anatomy according to which certain parts are uniquely endowed with the capacity for reproduction (*uis generandi* or *uis fecunditatis*). This account is underpinned by a biological analogy which invokes Nature as a divine, eternal, and intelligent *opifex* of plant and animal life. As providential designer, Nature ensures that the parts and functions of animals are harmoniously and properly matched. This conception of Nature, recognizable in its essentials from earlier in *RR*, thus plays a vital role in securing Columella's advice for propagation, shoring up his arguments that only certain parts of the vine are intrinsically fertile and that the index of their fertility is their fruitfulness. *Ratio* again corrects *consuetudo*. While we observed in our discussion of the agricultural term "*spado*" that reason and custom are not always at odds, opposing them as sharply and prominently as he does allows Columella to make vine propagation an object lesson in good agronomy: proper understanding of Nature is crucial to evaluating the truth of (received) agricultural practice. Columella said it best: "and while perhaps more was said here than an account of the truth required, yet it was far less than what the badly distorted and entrenched opinion of farmers required" (3.10.22 *haec et forsitan pluribus dicta sint quam exigebat ratio ueritatis, minus tamen multis quam postulabat praue detorta et inueterata opinio rusticorum*).

4.5 Conclusion

In this chapter we examined how Columella's conception of nature (or rather "Nature") and his reevaluation of the farmer's *scientia* are vital to the project of *RR*. He outlines his approach to the *ars* in the preface before developing it through the *praecepta* and explanations in his treatise. We focused on several key moments in the first three books, which belong to the subdivision of *agri cultura* that Columella calls *agricolatio* (1.pr.25). To be sure, not all parts of *RR* evince the sort of sustained exposition of *Rerum Natura* prominent in the chosen passages, which are among the most sublime in the entire work. Yet the line of thought we have traced – implicating Nature (*Rerum Natura*); reason (*ratio*); and practice, customs, and morals (*mores, consuetudo*) – puts front and center an approach to *agri cultura* predicated on an understanding of the physical world and made programmatic by its place in the early books of the *ars*. Although there were already highly sophisticated and intellectualized Latin treatments of the discipline by Columella's time, *RR* reflects an interdisciplinary awareness and concern with a unified program of explanation for method that only became central to the *artes* after Vitruvius' *DA*. Of course, Columella's *ars* is not identical to Vitruvius' *ars*: Columella's conceptualization of nature (or "Nature"), the prominence of his own moralizing narrative, the treatment of the interdisciplinary contents of the field, and the style of argumentation are often strikingly different. But the sensibility of their works is very similar. Like the *scientia architecti* of Vitruvius' focus, the *scientia agricolae* of Columella's project necessarily implicates in the *ars* of agriculture a larger view of the human and natural worlds – a view unfolded in *RR* itself. *RR* is thus no less than *DA* an expression of the scientific character of the early Imperial *artes*, which in their construction of disciplinary expertise also become vehicles for the Roman understanding of society and nature.

PART III

CHAPTER 5

Making a Roman ars *of Medicine*

Observation, Explanation, and Judgment in Celsus' De medicina

Columella's *Res rustica* allows us to glimpse the transformation of agriculture in the crucible of the scientific culture of the *artes* in the early Roman Empire. But Columella was not the first Imperial author on the subject: C. Iulius Hyginus, Cornelius Celsus, Iulius Atticus, and Iulius Graecinus all find their way into his register of earlier writers on *res rusticae*.[1] Of these, Celsus is singled out for having embraced "the whole body of the discipline" (*RR* 1.1.14 *totum corpus disciplinae*) in five books.[2] Several fragments confirm the extent of Celsus' coverage,[3] and his reception in later agricultural authors shows the high regard in which these books were held. Columella praises him as a well-informed author who wrote in an elegant style.[4] Indeed, he is the most frequently named prose author in *RR* by a wide margin, and Columella's criticism no less than his praise reveals the significance he attaches to Celsus' opinions.[5] Pliny, who lists Celsus as a source, credits Graecinus with having followed Celsus closely; if Pliny is not mistaken, this would be further proof of Celsus' influence.[6] Finally, no name appears more frequently than Celsus' in the agricultural fragments of the third-century author Gargilius Martialis, who calls him "a man very experienced in the Italian practice" (*Pom.* 4.1.3 *Italicae disciplinae peritissimum*).[7]

[1] See Ch. 4 (nn.19, 24) on *RR* 1.1.13–14. [2] For *corpus*, see Ch. 2, n.34.

[3] Marx 1915: 5–13 sorts the fragments into five categories: *ager*, *arbores*, *pecora*, *aues*, *apes*; see further Ch. 4, n.24.

[4] Columella puts him among *aetatis nostrae celeberrimi auctores* (*RR* 3.17.4; cf. 4.8.1) and says no one wrote *elegantius* on bees (9.2.1). Cf. *RR* 2.2.15 (quoted in Ch. 5.1).

[5] Named 34× in *RR* (whence 29 fragments in Marx). By comparison, Atticus is named 23×, Mago (or *Poenus*) 23×, Cato 19×, Varro 13×, Tremelius Scrofa 12×, and Saserna/Sasernae 9×. For Columella's response to Celsus, see further Ch. 4, n.24, plus Ch. 5.5 (nn.171–173).

[6] Plin. *HN.* 14.33 *Graecinus, qui alioqui Cornelium Celsum transcripsit.* André 1958: 86–7 argues that Pliny is wrong, however, and that his error owes to a hasty reading of *RR* 1.1.14.

[7] Zainaldin 2020a: 15–16, 284.

5.1 Celsus' *Artes* and the *artes*

Celsus did not write solely on agriculture, of course, as Columella pointedly acknowledges when speaking of him as a "man knowledgeable not only of agriculture but of nature in its entirety" (2.2.15 *non solum agricolationis sed uniuersae naturae prudentem uirum*). The agricultural books were only the first five of a larger work called *Artes*, whose next part comprised the still-extant eight books on medicine (= Books VI–XIII of the *Artes*), conventionally called *De medicina* (*DM*). From here the precise contents, number, and order of books of the *Artes* are debated, but there were certainly books on rhetoric (seven?), philosophy (six?), and the art of war (in an unknown number), and possibly also jurisprudence.[8] Regardless of the uncertainties, our knowledge of the *Artes* is sufficient to justify Quintilian's remark that Celsus was "worthy, simply in virtue of his intention, of our credence that he knew all of these things" (*Inst.* 12.11.24 *dignus uel ipso proposito ut eum scisse omnia illa credamus*).[9] What exactly Celsus "knew" (*scisse*) will be a guiding concern of this chapter.

As a whole, Celsus' *Artes* is a fundamental witness to the unity of the intellectual culture of the *artes* studied in this book. Yet even as it confirms some of the themes explored thus far, it also presents an approach to specialized knowledge strikingly different in certain respects. One such area is in the interdisciplinarity that we posited as a premise of the (higher) *artes* in Chapter 1. In Chapters 3 and 4, we saw that Vitruvius and Columella are both sensitive to the relationship of their own disciplines with other branches of specialized knowledge – for Vitruvius this was indeed essential to his architectonic conception of architecture. It is the more notable, therefore, that we find little theorization of the relationship among the *artes* in Celsus' extant books or fragments. It is possible – indeed probable – that he addressed his choice of disciplines in a preface or introduction to the *Artes* now lost; even so, the eight books on medicine show scant interest in drawing connections with other *artes*, or even other

[8] Ch. 2, n.62.

[9] In the same passage Quintilian calls Celsus *mediocri uir ingenio*; the evaluation, which looks depreciatory, has attracted much discussion. Wellmann 1913: 4 and Marx 1915: lxxv found support in it for their view that Celsus' *DM* was merely a translation of a Greek text, but since their hypothesis has been discarded (n.30) and scholars began to esteem Celsus' authorial contribution more highly, the problem has become rather to explain a pejorative judgment which apparently does not square with the quality of *DM* and (inter alia) the warm regard for Celsus' agricultural writings: see Kappelmacher 1918: 38–40; Scarborough 1969: 169n47; Daube 1974; Highet 1975; Capitani 1980; Contino 1988: 20n13.

parts of Celsus' work.[10] There are two exceptions.[11] First and more briefly, at the very beginning of *DM*, Celsus effects a brisk transition to his new subject: "just as agriculture promises sustenance for healthy bodies, so does medicine promise health for those that are sick" (1.pr.1 *ut alimenta sanis corporibus agricultura, sic sanitatem aegris medicina promittit*). The verb *promittere* is regularly used to refer to the special aim (*finis*/τέλος) expected of any *ars* or *technê*.[12] Here it establishes the parallel purposes of these two parts of the *Artes*, adumbrating (however faintly) a larger conception of specialized knowledge.

More interesting is the second exception. As we will discuss, Celsus excludes natural philosophy (1.pr.47 *naturae rerum contemplatio*) from the art of medicine proper. At the same time, however, he asserts that "although there are many things which do not belong properly to the *artes* themselves, yet they aid them by exercising the intelligence of the practitioner" (*quamquam igitur multa sint ad ipsas artes proprie non pertinentia, tamen eas adiuuant excitando artificis ingenium*). *Contemplatio rerum naturae* is among these: "though it does not make the physician, yet it makes him more capable at medicine and renders him complete" (*quamuis non faciat medicum, aptiorem tamen medicinae reddit perfectumque*).[13] The claim is echoed later when he says that causal inquiry into nature "is excluded from the art itself, though not from the practitioner's thought" (1.pr.74 *non ab cogitatione artificis sed ab ipsa arte reiectis*).[14] There are two ways that we can relate this position to other Imperial *artes*. On the one hand, we might emphasize the discrepancy between Celsus' view and Vitruvius' and Columella's, for the latter both regard reasoning with natural first principles as central to their expansive conceptions of their disciplines. Thus, Celsus would be rejecting the kind of interdisciplinarity proposed in other *artes*. On the other hand, we could stress the second part of Celsus' statement – that the inquiry into nature is indispensable to the formation of the complete (*perfectus*)

10 The only cross-reference is *DM* 5.28.16c, referring to a veterinary remedy in the agricultural books (*sicut in pecoribus proposui*).

11 I set aside 1.pr.32, where Celsus is repeating a conventional Empiricist argument while expounding their views (cf. Gal. *Med. exp.* 12 W–F 61–2, with Mudry 1982: 123–4; Deuse 1993: 822–3).

12 Like Greek ἐπαγγέλεσθαι in this sense; see Heinimann 1961: 123, and on *DM* 1.pr.1, Mudry 1982: 51; Zurli 1990: 306; Mudry 2010: 342; cf. *TLL* X 2.1869.1–35, 1873.53–63 (Delhey).

13 Mudry and Serbat omit *perfectumque* (with J), but wrongly. Celsus is appealing to the notion of the *perfectus artifex* central to the *artes* since Cicero's time: cf. Vitr. *DA* 1.1.3, Col. *RR* 1.pr.22, 1.pr.28, 1.pr.32, and Ch. 1, n.38, Ch. 3, n.18, Ch. 4, n.63; also Romano 1985: 140; Zurli 1990: 303.

14 With *reiectis* is understood *obscuris omnibus* (sc. *causis*), referring to natural-philosophical investigation (see Ch. 5.3.2).

practitioner – and view him as affirming a version of the interdisciplinary notion of the *artes* propounded since the late Republic.[15] In this case we would see rapprochement rather than distance, though we would still be compelled to admit that the exclusion of *contemplatio rerum naturae* from the discipline lends Celsus' *ars* a very different cast from those of Vitruvius and Columella.

Celsus' own remarks aside, we can also see in the *Artes* a manifestation of the interdisciplinary character of the *artes* that is embodied rather than abstract: in its very textual form, joining multiple disciplines, his work instantiates the proposition that different branches of specialized knowledge were in some way related. This point raises a larger question about the inspiration of his *Artes*, both set against earlier Roman literature and in his own age. Scholars have traditionally located the *Artes* in a lineage of Roman "encyclopediae" running back through Varro's *Disciplinae* to the so-called *libri ad filium* of Cato the Elder.[16] While it is appropriate to emphasize Roman interest in the polymathic ideal reflected in these figures' writings, the "encyclopedia" narrative obscures as much as enlightens: Celsus' *Artes* are at once more distinctive in their approach to specialized knowledge and more typical of the attitude that prevailed in the early Imperial period than the narrative implies.[17] Cato's technical writings were not much like the more sophisticated *artes* that emerged in the late Republic, let alone those of the Empire, and his *libri ad filium*, even if it be supposed that they handled various disciplines, are hardly comparable to Varro's and Celsus' systematic gatherings.[18] While Varro's *Disciplinae* are a more likely candidate for the inspiration to join multiple disciplines in

[15] So, for example, Mudry 1982: 142–3; Romano 1985: 138–40; Mudry 1990c: 145; Zurli 1990: 319.

[16] So Jahn 1850 (also n.18); this thesis is frequently repeated in scholarship on Roman "encyclopedic literature" vis-à-vis Celsus: for example, Probst 1905 *passim*; Scheller and Frieboes 1906: v–vi; Ilberg 1907: 379, 382; Wellmann 1913: 2–3; Marx 1915: vii–viii; Castiglioni 1940: 859; Krenkel 1959: 112–13; Barwick 1960: 246–7; Krenkel 1973: 2; Harig 1976 *passim*; Romano 1985: 133–4; Zurli 1990: 297; Mudry 1993a: 793–5; Stok 1993: 421–7; cf. Schulze 1999: 104–8; Gautherie 2017: 32–42.

[17] For the problematic appellation "encyclopedia," see Ch. 2, n.64; for appropriate skepticism regarding a "canon" of Roman "encyclopedists," Doody (n.64); König and Woolf 2013c: 37–49. König and Woolf 2013a seeks to rescue "encyclopedism" for antiquity (if not "encyclopedia" as such), for which Doody prefers "polymathy."

[18] That Cato's so-called *libri ad filium* constituted an "encyclopedia" of *artes* including oratory, agriculture, medicine, law, the art of war, and Roman customs is owed to Jahn 1850: 263–72 and reflected (more cautiously) in Jordan 1860: 77–83. It is difficult to maintain belief in this "encyclopedia" despite frequent and continued repetition in scholarship of the alleged lineage Cato ➔ Varro ➔ Celsus: see Astin 1978: 332–40 (cf. Mazzarino 1952: 23–6). Doody 2010: 53–4 (not citing Astin) is also skeptical and reports others' doubts. Yet, it is certain that besides *De agri cultura* Cato also wrote on *res militaris*, law, and medicine (Astin 1978: 184–5 with Boscherini 1993: 730n8), so that he seemed a polymath to later authors (cf. Quint. *Inst.* 12.11.23).

a single work, what we know of it does not much elucidate Celsus' plan. As observed in Chapter 1, Varro's work probably embraced, in one book each, grammar, dialectic, rhetoric, geometry, arithmetic, astronomy, music, medicine, and architecture.[19] Celsus' work covers only two of these subjects (rhetoric and medicine), adding agriculture, the art of war, philosophy, and possibly jurisprudence; it was also far larger: at least twenty-seven books long to Varro's nine.[20] Again, Varro's disciplines seem inspired by the Hellenistic culture of education, with the more idiosyncratic inclusion of medicine and architecture.[21] Celsus' choices look rather like a return to Cato, whose wide-ranging authorship upon such subjects as oratory, agriculture, medicine, law, and the art of war was exemplary of Roman polymathy regardless of whether they were united as an "encyclopedia."[22] While Varro's *Disciplinae* savors of the schoolroom, Celsus' *Artes* seems to spring rather from the life of the elite Roman man, for whom managing an estate, waging war, speaking in public, and engaging the services of educated physicians were all areas appropriate to cultivate interest in. No purely theoretical education, the *Artes* are best seen as an image of Roman practical expertise, on a scale appropriate to the extent of Roman *imperium* itself. In its magnitude Celsus' work indeed anticipates Pliny's *HN* as a monument of Roman Empire, though in its conception it belongs much more closely than does Pliny's to the scientific culture of the *artes* represented by works like Vitruvius' *DA* or Columella's *RR*, since the *Artes* too is ultimately an attempt to carve nature along the joints of Roman know-how.[23]

5.2 *Ars* According to Celsus: Making Roman *medicina*

Celsus' *Artes* represents a turning point, in more ways than one, for our study of the *artes*. At the end of Chapter 4, we made the easy observation that, while Vitruvius' and Columella's treatises share certain preoccupations, they are not wholly alike in their approaches. The point can be generalized further. The *artes* were not univocal, since their intellectual unity derived basically from a set of shared questions or concerns, and only secondarily from shared answers. The scientific culture of the *artes* was a diverse one.[24] In particular, authors handled different practical and

[19] Ch. 1, n.6.

[20] For the *Artes*, see Ch. 2, n.62. Varro wrote on agriculture in *Rust.*, but not in *Disciplinae.*

[21] Ch. 1, nn.5–8; Greek inspiration is certain despite reasonable doubts about a fixed "curriculum" of the ἐγκύκλιος παιδεία or *artes liberales* in Varro's age.

[22] Cf. Marx 1915: vii–viii.

[23] For Pliny and the *artes*, see Ch. 2, n.38.

[24] See the Introduction.

theoretical problems in ways that responded to the exigencies of their own disciplines. Hence, the composite character of architecture lent Vitruvius the armature for his highly interdisciplinary theorization of practice, while Columella's agronomy began from a morally laden philosophical dispute about nature conceived as a systematic whole (capital-N "Nature"). The common thread in both authors is an interest in justifying method by means of explanation from natural principles. But, as we will see in this and subsequent chapters, even this was only one way of establishing a foundation for disciplinary expertise in the *artes*. Thus, in *DM*, Celsus rejects the approach to explanation shared by Vitruvius and Columella while no less critically examining the knowledge required for expert practice. It is not that he repudiates reasoned explanation as such, the examination of causes, or the need for knowledge of nature; rather, he construes these and related concepts differently, and imposes different limitations upon them, than do Vitruvius or Columella.

Celsus' more aetiologically reserved approach to *ars* doubtless reflects his own intellectual sensibilities, but in the case of *DM* may also partly reflect his engagement with the earlier medical tradition. His aim in this part of the *Artes* was to provide for a Roman audience an overview of the entirety of the practice of medicine – and, more specifically, medicine as developed by the Greeks down to his own day, for the overwhelmingly Greek nature of the doctrines presented in *DM* is indisputable.[25] In a highly systematic fashion, he sets out the discipline in three main parts: regimen or dietetics (Books I–IV), drugs (V–VI), and surgery (VII–VIII).[26] These parts are presented as exhausting *medicina*, which Celsus stresses is an interconnected whole.[27] It was possible to subdivide the art in other ways, but this Hellenistic schematization in terms of therapeutic modalities reflects a focus on the practical dimension of medicine – the art as instantiated in the practitioner's judgment and actions.[28] Like Celsus' status as practitioner

[25] See, for example, von Staden 1996: 394–9; the occasional Roman "folk" remedies (3.23.7, 4.7.5, 4.13.3, 5.28.7b) accentuate the Greekness of the remainder (cf. Mudry 1990c: 139–41; Scarborough 1993: 20; Mazzini 1994: 126–7).

[26] Overviews of *DM*: Fuhrmann 1960: 88–91; Serbat 1995: xiv–xx; Mazzini 1999: 18–21; Schulze 2001: 28–31; also, tables in Scheller and Frieboes 1906: xxvii–xl; Marx 1915: 423–30; Gautherie 2017: 116–23.

[27] Cf. 1.pr.9, 1.pr.12, 5.pr.1, 7.pr.1; and for the interconnected whole, 5.pr.2. Cf. Scrib. Larg. 200 and, generally, Mudry 1985.

[28] This subdivision was accepted by the end of the fourth century BC (Mudry 1982: 67–8; von Staden 1989: 99). According to Galen, Empiricists regarded these three parts as departments of "therapy" (θεραπευτικόν); the two other main parts of medicine were "semiotics" (σημειοτικόν) and "hygienics" (ὑγιεινόν): cf. *Subf. emp.* 5 D 52.5–18, *Part. art. med.* 6.3 Lyons 41.3–6 = Kollesch, Nickel, and Strohmaier 125.3–6, and Deichgräber 1965: 288–91. For other subdivisions, see Galen *Part. art. med.* and Mudry 1982: 66–9; von Staden 1989: 89–114; Flemming 2000: 90–122.

or laymen, his precise audience remains uncertain. It is clear, however, that, even if experts were among his readers, so were educated Romans who were not professional physicians – the same group who might have perused the books on agriculture that preceded or those on the art of war or rhetoric that followed in the *Artes*.[29] It was difficult enough to render the whole of medicine systematically and economically for a nonspecialist audience, for whom little knowledge about the body, health, and disease could be presupposed. But Celsus' task was made more challenging by the history of Greek medicine itself, which greeted the would-be author of an *ars* with a theoretical inheritance far richer than that of architecture, agriculture, or, indeed, any other discipline in Greco-Roman antiquity. This was an inheritance marked from the earliest stages by enormous variety, and often forthright disagreement, in the principles and methods claimed to be requisite for expertise.

Celsus could mitigate some of these difficulties by narrowing the scope of his work to the practical dimensions of diagnosis, prognosis, and treatment, organized along the therapeutic modalities that structure the work as a whole. But this was not an automatic solution to the multiplicity of Greek medical theories, since these theories were contested precisely because they were thought consequential for practice. A more direct response was required. Hence the second, and crucial, move for Celsus was to articulate a medical epistemology of his own, one predicated on his intermediateness as recipient of the divergent Greek traditions. In this role, which we will characterize further throughout this chapter, Celsus was a partisan of no particular theory of medicine, and, hence, (at least potentially) a beneficiary of all. The raw material on which his independent judgment would operate were the precepts contained in the Greek texts transmitted to his own time – from the remote age of the *antiqui* (the Hippocratic authors and others) down through the Hellenistic period and late Republic to the early Empire – and sources from all these periods are indeed used in *DM*.[30] Without having Celsus' library before us or knowing

[29] Many commentators assume a mixed audience, hence falling between the extremes of, for example, Jocelyn 1985: 304 (a lay, "purely intellectual interest in the art") and Schulze 1999 (professional education: Ch. 2, n.87). Gautherie 2017: 167–8 compares Vitruvius' double audience of *aedificantes* and *sapientes* (*DA* 1.1.18; see Ch. 2, n.90), suggesting that Celsus addressed both *curantes* (not exclusively *medici*) and *sapientes* (generally Gautherie 2017: 148–77). Cf. Mudry 1990b on *DM* as "littérature de vulgarisation" (30).

[30] Early attempts (Wellmann 1913; 1917; 1925; Marx 1915: lxxiv–xcv; Kappelmacher 1918: 33–58) to find a single source from which Celsus translated his work have long since given way to the view that Celsus employed many sources directly or indirectly: see Ilberg 1907, esp. 391–2; Temkin 1935: 254–62; Meinecke 1941; Pazzini et al. 1953; Contino 1988: 42n35; Deuse 1993, esp. 829–31; Mudry

the extent of his personal experience, it is difficult to account for the criteria by which he selected, adapted, and synthesized this Greek material.[31] Important factors seem to have included the authority of a physician-author, the agreement or convergence of the medical tradition in a doctrine across time, the internal consistency of the doctrine, and the moderate and cautious (rather than daring) nature of therapy.[32] Whatever the case may be, the general point remains that, in virtue of his self-fashioned role as neutral and rational arbiter of the Greek tradition, Celsus could achieve – or, rather, *create* – a consensus that might stand to his Roman readers as an image of the *ars* of medicine. As we discussed in Chapter 2, Celsus' mediating position thus models a distinctively Roman attitude toward the reception of Greek knowledge: the Roman particularity of his *medicina* derives not from supplanting Greek medical precepts with autochthonous alternatives, but from his own selection, alteration, and organization of the material in fashioning an "image of expertise" for the discipline.

In this chapter, we will study *DM* to unpack Celsus' approach to specialized knowledge in the *Artes*. For Celsus no less than for Vitruvius and Columella, a mere grasp of rules is inadequate for *ars*, but natural-philosophical explanation has little role to play in grounding expertise. Rather, it is the development of a cautious and critical judgment vis-à-vis the empirical elements of health and disease that undergirds the physician's skill. To grasp the character of this critical faculty, we will first examine Celsus' own medical epistemology both as proposed in the main preface and as developed throughout the doctrines of the books. This epistemology conditions concepts vital to the physician's decision-making, including nature, cause, reason or account, and experience, and hence establishes a framework for medical thought and practice. Crucially, Celsus restricts medical theorizing to a purely empirical domain, eschewing appeal to principles that are not actually or potentially accessible to the senses. Next, we will consider three examples of Celsus' doctrines in *DM* – regarding bloodletting, critical days, and head injuries – that exhibit different aspects of medical

1993b; Mazzini 1994: 125–31; von Staden 1994a; Serbat 1995: liii–lxviii; Schulze 2001: 73–5; von Staden 2010; Gautherie 2017: 50–80.

[31] For his sources, previous n.; for personal experience, Marx 1915: 451–2; Spencer 1935–8: I xi–xii; Krenkel 1973: 26; and esp. von Staden 1994; Gautherie 2017: 59–61, 63–5.

[32] Authority: for example, Hippocrates (n.165); agreement or convergence: *passim*, this chapter, on medical experience; internal consistency: for example, negative judgment on the doctrine of critical days (Ch. 5.4.2 on 3.4.11–15); moderate therapies: for example, avoidance of *temeraria medicina* (n.127).

decision-making in practice. These examples will show how Celsus models the good physician's practice within the bounds of his medical epistemology.

5.3 Celsus' "Middle Ground": The Medical Epistemology of *De medicina*

In Chapter 2, we briefly examined Celsus' preface as a window onto Roman attitudes toward Greek specialized knowledge (τέχνη). We return now to come to closer grips with the premises for *medicina*. The main preface to *DM* is supremely important for the history of Greco-Roman medicine and is much studied.[33] As observed in Chapter 2, it divides into two parts: a shorter account of the emergence and diversification of the *technê* or *ars* (1.pr.1–11) and a longer account of disputes about the proper theory of medicine (1.pr.12–75).[34] The first part prefaces *DM* in its entirety, the second the books on regimen (I–IV), which Celsus calls the "most difficult but also most illustrious" (1.pr.12 *ut difficillima, sic etiam clarissima*) part of medicine. The arguments of this part center on the epistemologies of the Greek medical sects that emerged in the Hellenistic period and hold a fortiori for the rest of *medicina*.[35] Celsus summarizes the doctrines of the "Rationalists" (1.pr.13–26) and "Empiricists" (1.pr.27–44) before setting out his own position (1.pr.45–53). In a sort of digression he responds to Erasistratus as well as the so-called Methodists (1.pr.54–73), recapitulating and critically examining their views. He concludes by restating his own approach (1.pr.74–5).

In Chapter 1, we briefly characterized the dispute between the "Dogmatist" or "Rationalist" and "Empiricist" sects as a response to earlier philosophical discussions about *technê*. Celsus' preface is a fundamental witness to the schism, and scholars have investigated it to reconstruct this history as well as to investigate Celsus' own doctrinal allegiances.[36] Its evidence for the sects is problematic, however, not just because the account as a whole seems to derive from an Empiricist source,[37] but more generally

[33] See esp. Mudry 1982; also, for example, Pigeaud 1972; Romano 1985; Stok 1985a; Zurli 1990, esp. 304–20; Deuse 1993; Stok 1993: 427–39; Serbat 1995: xxxviii–liii; von Staden 1999; Schulze 2001: 31–7.

[34] Ch. 2, n.70.

[35] On the organization and contents of this part, see Mudry 1982: 77–9, 83–5, 113–14, 139–40, 153–5, 174–5, 201. For the difficulty of different parts of medicine, cf. 7.pr.1–2.

[36] See Ch. 2, nn.68–69; on Celsus' allegiances, for example, Deuse 1993; Mudry 1993b; Stok 1993: 421–3; 1994; von Staden 1994a; Serbat 1995: xl–liii; also, n.30.

[37] And hence misleads on what "Rationalists" might commit themselves to; see Mudry 1982: 83–5; Stok 1994: 63–4; von Staden 1994a: 78; 1999: 280.

because Celsus' presentation is calculated to produce a symmetrical opposition useful for formulating his own position, which is avowedly "somehow midway between (sc. these) divergent views" (1.pr.45 *media quodammodo inter diuersas sententias*). Celsus is indeed clear that the account of the sects merely prepares the ground for his own, moderate viewpoint.[38] This design may complicate the evidentiary value of the preface, but it also underpins the nonsectarian stance that allows him to avail himself of Greek medicine in its branchless historical entirety throughout *DM*. In refusing to affiliate himself with any particular Greek physician or school to the exclusion of others, Celsus leaves himself the latitude to adopt precepts originating with any practitioner without further commitment.

Celsus' epistemic neutrality does not diminish the extent to which his approach to the *ars* was shaped by Greek debate. Most significantly, his medical epistemology centers around theoretically loaded concepts with long histories in the Greek medical tradition and which were at the heart of the Rationalist–Empiricist debate. Celsus takes a stand on four key concepts inherited from the dispute, which also form the armature for his own medical epistemology: nature (*natura* = φύσις), cause (*causa* = αἰτία), reason or account (*ratio* = λόγος), and experience (*experientia*, *experimentum/a*, or *usus* = ἐμπειρία, κ.τ.λ.[39]). His moderating reaction to the diametrically opposed viewpoints of the sects might suggest that finding his position is just a matter of triangulating the mean for each term. And, up to a point, this is true – Celsus does consciously articulate his tenets as a compromise between the two sects. However, a closer look at Celsus' professed middle way (1.pr.45–53, 1.pr.74) within the preface and *DM* as a whole reveals a richer, more nuanced, and more independent picture of his epistemology. Celsus' true "middle ground," as we will call it, does represent an essentially moderate position between the Rationalists and the Empiricists, but it is a position more flexible, robust, and expansive than the terms of their dispute would imply.

In what follows, we will study Celsus' middle ground through the fundamental concepts just mentioned: nature, cause, reason or account, and experience. Their relationships and place in the medical art can be set out briefly in advance. ***Natura*** (nature) is a necessary object of knowledge for the physician but is almost entirely restricted to phenomena actually or

[38] This is consequently characterized as a *disputatio in utramque partem* by Mudry 1982: 77–9; 1990a: 93–4; Serbat 1995: xli–xlii; cf. Capitani 1991: 111. Cf. *DM* 1.pr.12, 39, 45. Mudry (1982 *passim*; 1990a; 1993b: 802–3) sees the influence of the New Academy's probabilism on Celsus, but cf. von Staden 1994a: 80n5; 2007: 367n73.

[39] See Ch. 5.3.4 and n.110 on various terms for "experience."

potentially observable. Inquiry into natural principles that are inaccessible to the senses has no role to play in medicine (which is not to say that natural philosophy cannot enhance the physician's skill). Even given these empirical strictures, *natura* retains several important functions in the physician's decision-making. Indeed, Celsus' handling of *causa* and *ratio* are conditioned by his view of nature. ***Causae*** (causes) are relevant and indeed necessary for medicine, but only as empirical descriptions of the relationship between observable states of affairs. ***Ratio*** operates with *natura* and *causa* to inform medical practice in several ways. Understood as "**reason**," it is a form of causal deliberation that aids the physician's judgment; moreover, it allows for productive innovation and equips the physician to adjust general rules of therapy to the individual patient. Understood as "**account**," it crystallizes systematic observations of *natura* and *causae* and structures the precepts and methods of medicine. Finally, **experience** broadly underpins successful practice: as the aggregate of historical practice, it is the inventor or witness of medical method; as the habituated practice of the physician, it guides the application of generalized rules and precepts. With these concepts, Celsus' epistemology establishes the parameters of the physician's expertise.

5.3.1 Natura *in the Sects and in Celsus' Middle Ground*

Sharply differing views on the role that nature (*natura*) should play in medicine are at the root of fundamental disagreements between the Rationalists and the Empiricists. (By "Rationalists" and "Empiricists," we mean Rationalists and Empiricists as represented by Celsus, except as noted.) The Rationalists believe that medical practice requires an explanatory theory or account of the human body, health, disease, and things related to these. Because this theory meaningfully informs the physician's decision-making, a corollary is that medical practice will vary as a function of the specific theory adopted.[40] The Rationalist's attitude toward nature involves at least three things. First, he needs an account of the material principles (*principia*) of the human body (1.pr.14–16).[41] Rationalist physicians may disagree about what these principles are and how disease derives from them – whether as humoral or elemental imbalance, an excess of moisture or air, the clogging of microscopic pores, or something else – but

[40] Cf. 1.pr.21–2, with 1.pr.27 (Ch. 5.3.1) for Empiricist criticism.

[41] For *principia* = Gk. στοιχεῖα (to which the term is not entirely restricted in Celsus' usage), cf. Vitr. *DA* 1.4.5 in Ch. 3.3.1; see also Mudry 1982: 88–9.

they agree that "it is he who does not lack awareness of the first origin of the disease who will apply the right treatment" (1.pr.16 *eum uero recte curaturum, quem prima origo causae non fefellerit*).[42] Second, the Rationalist requires an account of the "the natural actions of the body" (1.pr.19 *naturales ... corporis actiones*), such as respiration, sleep, and digestion (1.pr.19–20). As in the case of physical first principles, his medical practice will differ according to his theories – for example, whether digestion is mechanical grinding or concoction (21–22).[43] Third, the Rationalist must know the "internal parts" (1.pr.23 *interioribus partibus*) of the body, including their position, color, shape, size, order, hardness, softness, smoothness, and points of contact; their outgrowths and involutions; and their connections with other parts (24–5). Knowledge in these three domains – (1) material principles (*principia*), (2) natural actions (*naturales actiones*), and (3) internal parts (*interiores partes*) – purportedly affords the Rationalist an understanding of the body, health, and disease indispensable to practice.

The Empiricist, by contrast, rejects the Rationalists' knowledge of nature as both impossible to acquire on its own terms and irrelevant to practice. In a word, "nature cannot be apprehended" (1.pr.27 *non conprehensibilis natura sit*). This axiom excludes all apodeictic knowledge-claims about *natura*,[44] but the Empiricist especially takes umbrage against the Rationalists' appeal to things not directly accessible to the senses. Rationalists claim to infer their principles and natural actions from the phenomena, but their theories cannot be verified save with the mind's eye; the Empiricist produces several arguments to show that there is no sure footing here (1.pr.28–30).[45] In place of the Rationalist's account of nature, the Empiricist finds the observed consequences of phenomena the only reliable guide to practice (1.pr.31–5): such and such a disease suffered for so many days is seen to be followed by this or that result; such and such a cure applied in such and such a circumstance is seen to be followed by health; and so on. Medicine rests on the sum of these observations – that is, on

[42] For *causa* = *morbus*, relatively uncommon in Celsus, see *TLL* III 680.82–681.35 (Meister); Capitani 1967: 139–40; Mudry 1982: 94–5; Svandrlik Civilia 1993: 200–1.

[43] For variation in practice tracking theoretical variation, cf. 1.pr.21 *illud uero conuenit, alium dandum cibum laborantibus, si hoc, alium, si illud uerum est*, etc.

[44] Cf. 1.pr.32 *cum igitur illa incerta, inconprehensibilis sit*, etc. As Mudry 1982: 116; 1990: 88 observes, *conprehendibilis* (later *conprehens-*) is Cicero's coinage for Gk. καταληπτός (*Ac.* 1.41; see *TLL* III 2154.44–75 [Hey]), the Stoic term for what can be apprehended with certainty (κατάληψις). Hence, Celsus is calling *natura* ἀκατάληπτος.

[45] Including disagreement among theoreticians (*discordia*, 1.pr.28); lack of trustworthy criteria for deciding among theories (28–9); the fact that knowledge of nature does not guarantee medical expertise (29); and local variation in practice (30). See Mudry 1982: 117–21; 1990a; also Deichgräber 1965: 281–8.

"experience" (*experientia*; see Ch. 5.3.4) – and, where practice varies, it does so as a function not of theory but of difference in experience. Hence, knowledge of nature makes no appreciable difference to medical practice: this is why the Empiricists say it is irrelevant or superfluous (1.pr.27 *superuacuam*) besides impossible. The only kind of knowledge professed by the Rationalists which the Empiricists also allow is that of the internal parts (cf. 1.pr.43) – understandably, since such parts are always potentially and sometimes actually accessible to the senses (as when treating someone with a grievous wound), and hence the legitimate subject of description.[46]

Celsus' stated opinion concerning natural principles and actions makes some concessions in each direction (1.pr.46–7). In general, he is sympathetic to the Empiricist claim that an explanatory account of nature is unnecessary for practice (1.pr.46). Although expressing himself more moderately than the Empiricists, he agrees that nature cannot be the object of definite knowledge, and hence concludes that theories about nature cannot furnish the basis for definite therapy.[47] Even so, he nods to the Rationalists in allowing that the "inquiry into nature" (1.pr.47 *naturae rerum contemplatio*) is of considerable value, rendering the physician "more capable at medicine and complete" (*aptiorem ... medicinae ... perfectumque*). We discussed this claim in Ch. 5.2, which cuts against the Empiricists' view that philosophical investigation is useless for the practitioner. Here we might only add that Celsus adduces Hippocrates and Herophilus in proof of the salutary effects of the study of *natura* (1.pr.47). Given the high esteem in which Hippocrates is held in *DM*, this is a strong endorsement, tantamount to prescribing natural philosophy for the ambitious physician (even if it is excluded from the art proper).[48] This devaluation of explanatory knowledge of nature joined with the endorsement of extradisciplinary study may seem to be Celsus' "middle ground." If that were so, the Empiricists would surely believe themselves to have fared the better, in that Celsus has uncoupled practice from the kind of knowledge the Rationalists think essential. Yet Celsus' reception of this dispute is more

[46] They cannot be the basis for an explanatory account, however. The Empiricists argue against use of dissection and vivisection (1.pr.40–5, against 1.pr.23–6; cf. 1.pr.74–5), maintaining that the Empiricist will learn what he needs through treatment or chance observation. For Celsus' representation of this dispute, see Edelstein 1932; Pigeaud 1972; Scarborough 1976; Mudry 1982: 107–11, 133–8; Deuse 1993: 831–7; Mudry 1997: 216–19; 1999: 171–4.

[47] Of principles and natural actions, he says *ne sapientiae quidem professores scientia conprehendunt, sed coniectura persecuntur*, adding, *cuius autem rei non est certa notitia, eius opinio certum reperire remedium non potest* (1.pr.46).

[48] For Celsus and Hippocrates, cf. 1.pr.8, and see Mudry 1977; Mazzini 1992b; Serbat 1995: liii–lviii; Ch. 5.4.2–3 on 3.4.11–15, 8.4.3–4.

complex than might first appear, and we have not in fact come to his true middle ground – only, perhaps, to the edge of it.

A fuller look at *DM* confirms Celsus' basic commitments in the preface but reveals an attitude toward *natura* that is more expansive than, and strikingly independent of, the sectarian views foregrounded in the preface. Celsus does maintain his commitment throughout *DM* to the proposition that the physician does not require explanatory knowledge of nature, and accordingly never makes precepts dependent upon any account of health, disease, or the human body that requires causal inference to or from what is remote from observation. The upshot is that he not only foregoes natural-philosophical investigation into first principles, but also largely abstains from physiology beyond observable properties of structures or substances.[49] All the same, Celsus does not hesitate to speak of *natura* as something objectively real and knowable – in fact, as the essence of a thing. *Natura* so understood is crucial both in epistemological and normative roles for the physician: on the one hand, knowledge of *natura* guides medical decision-making, and, on the other, establishes the goals of medicine. Celsus' enlarged conception of *natura* does not respond to or fit entirely within the opinions of either sect, but it is well suited to the systematic exposition of medicine undertaken in *DM*. To get a better sense for its significance, we examine some key uses of *natura* here.[50]

Natura names the natural – that is, healthy and fully functional – state of the body. When the body is in accordance with *natura*, it has neither lack nor surfeit in its constituents and qualities, its parts and substances are in their proper locations, and it displays no localized or generalized dysfunction. Nature in this sense is the normative ideal of medicine. Thus, Celsus often invokes *natura* explicitly to evaluate a part or process of the body, such as the pulse of blood vessels (3.6.7 *uenae naturaliter ordinatae*); the localized collection of fluid from dropsy (3.21.15 *quae* [sc. *aqua*] *contra naturam ibi substitit*); the feel (5.28.1a *corpus aridum et durius quam naturaliter*) or color (5.28.15b *aliter naturali colore mutato*) of the body; the passage of urine (7.26.1c *urina naturaliter profluat*); and so on.[51] More often it is only implicit, however, as the background against which medical precepts are to be understood. Celsus' allopathy, for example, assumes a norm from which the body has deviated: when he says that "every remedy" (2.9.2 *omne . . . auxilium*) effects one or another contrary (heating

49 For a slight exception, see n.53.

50 See von Staden 2007: 348–60; Gautherie 2017: 254–65 (also Mudry 2001) for some other aspects of *natura* not highlighted in this section.

51 Further, for example, 3.4.4, 5.26.27b, 5.26.29, 5.28.2b, 6.6.14, 7.26.1a, 8.10.1e, 8.10.3b, 8.15.1.

or cooling, adding or subtracting), the result is understood to be curative because it corrects the deviation of the body from its natural state.[52] Again, when Celsus sorts pathologies into larger genera (5.26.1, five altogether: traumatic injury, corruption, retention of foreign body, abnormal growth, abnormal absence), he is describing categories of dysfunction compared to the natural state. Or, again, surgical treatments aim to restore through physical manipulation what illness or injury has destroyed: for example, a traumatic injury to the leg induces contraction of tendons and muscles, which the physician must counter by forcefully (8.10.1c *per uim*) stretching them into place to reproduce the natural extension. Many further examples could be given, especially if we consider the ubiquity in *DM* of words such as *nimis* ("too, excessively"), which imply a natural standard.

In a similar sense, *natura* names the natural state or properties of a specific organ, substance, or part. For example, the spleen (*lien*) is "soft and porous in nature" (4.1.5 *natura mollis et rarus*), and the "nature of blood" (4.6.2 *natura sanguinis*) is such as to be rapidly heated or cooled by other bodily substances.[53] There is a "natural order of the eyelashes" (7.7.8a *ordine naturali pilorum* [sc. *qui in palpebra sunt*]) as well as a "natural posture of the arm" (8.10.3b *bracchii naturalis habitus*).[54] As for the body as a whole, the *natura* of a part often serves as a normative ideal: the uvula can be abnormally long (7.12.3b *longior uua . . . quam esse naturaliter debet*), or the pupil (*nigra pars oculi*) may pathologically lose its natural form (7.7.14a *amissa naturali figura*).[55] Sometimes, Celsus uses *natura* in an exceptionally rich sense to refer to the total description of the parts and conformation of an organ: hence, when he speaks of "the nature of the eye" (7.7.13a *oculi natura*), he means a descriptive account of it, including its component parts, as also of the "nature of this region" (7.18.1 *ipsius loci natura*) in reference to the reproductive organs.[56]

Indeed, an important manifestation of *natura* in *DM* is as anatomical account.[57] The physician needs generalized knowledge of the *natura* of the

[52] See *DM* 1.3.7; also Ch. 2.3.1 (and n.76) on Vitruvius' allopathy.

[53] The observation about blood is a rare foray into physiology to support Asclepiades' prescription of bloodletting for tetanus (4.6.2); cf. Temkin 1935: 253; von Staden 2007: 352. Marx 1915 compares Hippocr. *Cord.* 12, but Celsus unlike the Hippocratic author does not speculate about the substance of blood.

[54] Cf. 4.1.4, 8.4.2, 8.4.5, 7.25.1a–b, 8.4.15, and von Staden 2007: 351–2. Compare *natura* in the qualitative taxonomy of fluids issuing from wounds and ulcers (5.26.20) and in other passages describing things not invariably healthy but natural qua bodily products.

[55] Cf. 7.7.8e, plus n.51 *passim*.

[56] Cf. 7.29.1 *uuluae natura mirabilis* (with von Staden 1991: 272, 281–4, 292–3).

[57] On Celsus' anatomical knowledge, see Manetti and Roselli 1994: 116–17; Schulze 2001: 48–9; von Staden 2010, esp. 10–11.

body because he must be prepared to identify its parts in their normal conditions by their qualities, locations, and relations to other parts. In the preface, Celsus rejects vivisection but maintains that dissection is "necessary" (*necessarium*) for students of medicine to learn the position and order of the parts, with other qualities (color, feel, and so forth) becoming known through practice (1.pr.74–5). Besides close descriptions of individual body parts, he offers two larger anatomical discussions that exemplify knowledge of *natura* of the body. At the beginning of Book IV (4.1), Celsus surveys internal anatomy from head to groin, "offering an overview as it were of these things so far as one must know them to administer care" (4.1.2 *his uelut in conspectum quendam, quatenus scire curanti necessarium est, adductis*).[58] At the outset of VIII (8.1), he provides an even more detailed description of the skeleton (*ossa*) from head to foot as a propaedeutic for the surgical precepts.[59]

Another use of *natura* picks out the nature of the individual as a complex of real but contingent facts. The individual's nature informs treatment and establishes norms for health. In the preface, Celsus asserts that the physician must know "what the nature of the sick person is" (1.pr.53 *quae sit aegri natura*), giving as examples the individual's particular moistness (*umidum magis an magis siccum*), robustness (*ualidi nerui an infirmi*), predisposition to illness (*frequens aduersa ualetudo an rara*), and more. In Book I, on regimen, he says from the perspective of the layman that "before all else each person must know the nature of his own body . . . since some are thin, others are fat, some hot, others colder, some moist, some dry; some are troubled by constricted bowels, others by loosened" (1.3.13 *ante omnia . . . norit quisque naturam sui corporis, quoniam alii graciles, alii obessi sunt, alii calidi, alii frigidiores, alii umidi, alii sicci; alios adstricta, alios resoluta aluus exercet*). Celsus constantly reminds us that the physician must attend closely to the individual patient, since age, sex, habitus, and physical disposition, as well as qualities such as temperature and moisture, affect medical decision-making.[60] Finally, *natura* is applied to things other than bodies and parts, from diseases (e.g. 7.7.7a *naturam carcinomatis*) to food and drugs (for example, water at 2.18.11–13). In these cases, the idea is similarly that *natura* picks out the bundle of properties by which a thing can be reliably known. *Natura* has further applications in Celsus, but these are sufficient for our purposes.[61]

58 The *quatenus*-proviso is only partially true (von Staden in n.57) but captures Celsus' intentions.

59 See Scheller and Frieboes 1906: 758–65; Spencer 1935–8: III 593–601.

60 For example 1.1.4, 1.2.7, 3.6.5, 3.18.10, and von Staden 2007: 352–3; Gautherie 2017: 262–3; see also Stok 1997.

61 Consider, for example, nature as creator or agent: cf. 1.pr.24, 3.1.4, 4.7.4, 6.7.1a, with von Staden 2007: 349–50, Gautherie 2017: 254–8.

Celsus' multifarious use of *natura* as a description of the essence of various objects of medicine – above all the body – defines his true middle ground, which is more capacious than the preface suggests. His views owe something to both the Empiricists and the Rationalists, of course, but the sectarian dispute hardly anticipates the conception of *natura* richly developed throughout *DM*. That is not accidental. While Celsus' introduction to the sects is intended to produce a set of antitheses instrumental to articulating his own views, finding his middle ground is not as easy as simply locating the mean between the opposed beliefs. This is because the sectarian debate is conducted in terms too narrow to accommodate the *ars* of medicine Celsus wishes to set out. Nature is a case in point: *natura* imputed to the Rationalists always implies an explanatory natural-philosophical account, a sense the Empiricists accept (if only to criticize it).[62] This polarized presentation – *natura* is an aetiological principle either to be outright accepted (Rationalists) or rejected (Empiricists) – prompts Celsus' more expansive, moderating appropriation of the concept, not mere acceptance of the sects' univocal understanding of the term. Hence, Celsus elaborates a conceptual framework for *natura* independent of both Empiricists and Rationalists and aptly suited to his own project. Specifically, his complex and descriptive use of *natura* defines the domain in which the physician's judgment operates in *DM*. Epistemologically, *natura* marks out a set of facts, both universal and particular, relevant for medical decision-making. Normatively, *natura* suggests a reference point for therapy which supports broader methodological assumptions, such as the aim to restore homeostasis or the appeal to allopathic therapies. In the final analysis, Celsus makes *natura* vital for medical expertise, but *natura* as constructed for his own purposes and to be understood empirically only "so far as one must know (sc. it) to administer care" (4.1.2, quoted earlier in this section). The key applications of *natura* discussed here are represented schematically in Figure 5.1.

5.3.2 Causae *and* ratio *in the Sects and in Celsus' Middle Ground*

In the preface, *causa* and *ratio* are contested chiefly in connection with the Rationalists' and the Empiricists' differing views of *natura*. As Celsus has it, both sects allow that the appraisal of *causae* is crucial to medical practice (cf. 1.pr.13, 17),[63] but differ sharply on the relevant kinds of causes. The key distinction is between "evident" (*euidentes*) and "hidden" or "obscure"

[62] Cf. 1.pr.9, 13, 19, 27, 32, 47, 48, 59.

[63] Contrast the Methodists (1.pr.54), Ch. 5.3.3.

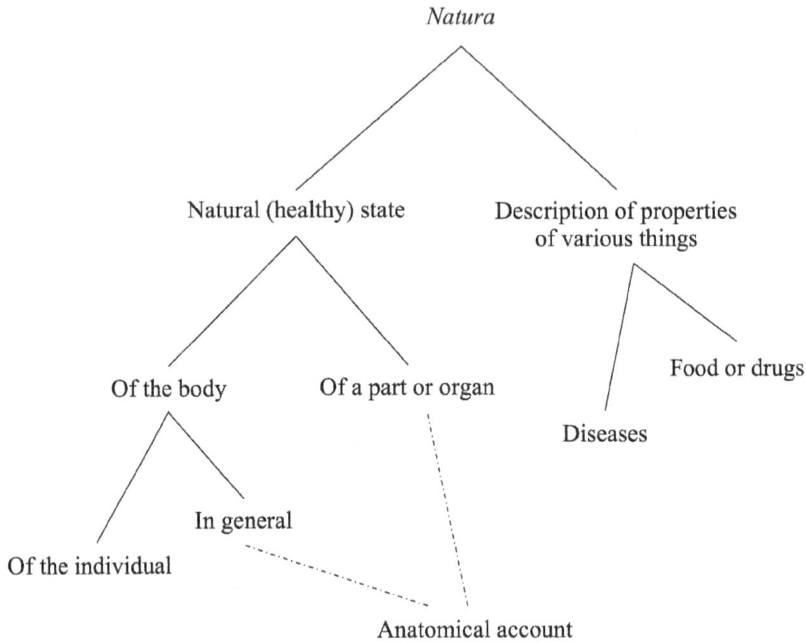

Figure 5.1 Concept map illustrating significances of *natura* in Cels. *DM* discussed in this chapter (not exhaustive of Celsus' use of the word).

(*abditae* or *obscurae*) causes;[64] the terms again reflect an Empiricist framing but are used by Celsus in a relatively neutral way throughout the preface.[65] The Rationalists maintain that both hidden and evident causes are necessary, while the Empiricists permit only the evident. In Celsus' account, hidden causes are connected chiefly with the first of the three kinds of knowledge of *natura* required by the Rationalists: the material principles (*principia*) of the human body.[66] More is at stake, however, since an account of natural actions (*naturales actiones*) or the qualities, powers, or functions of internal parts (*interiores partes*) presupposes an account of bodily principles; hence, the question of hidden causes bears on the whole dispute about *natura*. The causes are said to be "hidden" or "obscure" because they are not evident to the senses: they are reached by inference and include various natural principles, powers, substances, and so forth

[64] *Euidentes* at 1.pr.13, 18, 27, 52, 74 (cf. 3.3.6); *abditae* at 1.pr.13, 14 (cf. 3.3.6); *obscurae* at 1.pr.27, 32, 48, 74. These map onto Gk. φαινόμενος (vel sim.) and ἄδηλος.

[65] Cf. n.37. [66] So 1.pr.14; cf. 1.pr.27, 46.

held to be causally responsible for the phenomena (1.pr.14–16).[67] Celsus thus also says that hidden causes "contain diseases" (1.pr.12 *morbos continentium causarum*), by which he means that they represent a ground-level statement about things in nature that are causally sufficient of the pathology.[68] Not only when a Rationalist appeals to material first principles, but also when he speaks of natural actions or the function of internal parts to explain a state of health or disease, he is using "hidden causes."

"Evident causes" are defined as precipitating causes – that is, those which name the "origin of the disease" (1.pr.18 *initium morbi*). These causes do not "contain" the disease because they make no claim about a natural state of affairs responsible for the continued existence of the pathology.[69] An evident cause is simply that which occasions the onset of the disease – for example, "heat or cold, hunger or satiety, and things similar to these" (1.pr.18 *calor ... an frigus, fames an satietas, et quae similia sunt*).[70] These causes are called "evident" because they are ostensibly patent to observation: they are claims about the succession of different events or states and do not involve any deeper explanatory account grounding health or illness in certain natural principles. When a physician considers that a person drank to excess the evening before he fell ill, or that a boy was bitten by a dog prior to a bout of fever, he is noting the evident cause. The Rationalist uses evident causes because he believes it important to know the beginning or origin (1.pr.18 *originem*) of the disease. But evident causes are also acceptable to the Empiricist, since they do not commit him to any further view of the body, health, or disease. Indeed, they are arguably even more important for the Empiricist because his whole medical practice depends on knowledge acquired from observing successions of states manifest to the senses, which make up *experientia* (Ch. 5.3.4). This knowledge could be structured in terms of evident causes (cf. 1.pr.33–5).

The key difference between Rationalists and Empiricists on *ratio* relates to their contrasting views on *natura* and hence also *causae*. The primary sense in which *ratio* is contested between the sects is *ratio* as "account," where

67 See Mudry 1982: 88–94 for Celsus' examples.

68 *Continentes* (sc. *causae*) translates Gk. συνεκτικαὶ (sc. αἰτίαι), which are joined with hidden causes at Gal. *SI* 7 K 1.85.15–16 = H 17.3–4. "Containing causes" are routinely distinguished by medical authors from two others, προκαταρκτικαί ("antecedent") and προηγούμεναι ("preceding"); Celsus, however, does not refer to this classification, emphasizing only "evident" versus "obscure" (cf. Mudry 1982: 86). On containing causes (and medicine), see Frede 1987: 242–8; Hankinson 1987b: 80–6; 2003: 36–41 and *passim*; Johnston 2006: 34–5, 106–8.

69 See previous n.

70 Hence, these would be προκαταρκτικαὶ αἰτίαι (Mudry 1982: 87–8), for which see Gal. *CP* with Hankinson 1998 and n.68.

"account" is understood to involve an explanatory claim dependent upon hidden causes. Thus, when introducing the Rationalists, Celsus says they believe experience (*usus*) insufficient unless "an account of bodies and nature has been ascertained" (1.pr.12 *corporum rerumque ratione comperta*). Throughout the summary of Rationalist doctrine, *ratio* is used exclusively in this sense;[71] it does not appear with the more general significance "reason" or "reasoning," even if the Rationalist "account" implies the use of reason.[72] It is plain to see why the Empiricists should reject the need for *ratio* (or *rationes*) if construed as the causal account of nature touted by the Rationalists, and indeed the Empiricists' attacks presuppose this conceptually loaded sense of the word.[73] In Celsus' presentation the Empiricists go even further, however, to attack reasoning itself (*ratiocinatio*): "those pursuing philosophy would be the greatest doctors, if reasoning were able to accomplish this: as it is, they overflow with words but lack knowledge of healing" (1.pr.29 *sapientiae studiosos maximos medicos esse, si ratiocinatio hoc faceret: nunc illis uerba superesse, deesse medendi scientiam*).[74] As the connection with philosophy (*sapientia*) shows, *ratiocinatio* is understood as causal reasoning or inquiry which might yield an account (*ratio*).[75] The Empiricists allow that the doctor requires "deliberation" or "judgment" (1.pr.38 *consilio*) and that an "irrational animal" (*inrationale animal*) could not practice the art,[76] but they are much more concerned to establish limits for the operation of reason. Hence, they proscribe *disputatio* (1.pr.32), evidently discursive argumentation about *rationes*; *quaestiones* (1.pr.33), "inquiries" into nature; and *latentium rerum coniecturae* (cf. 1.pr.38), inferences to hidden causes.[77]

[71] 1.pr.17, 19; cf. 1.pr.26. At 1.pr.22 *eadem ... ratione*, it means merely "in the same way," as often in *DM*.

[72] Accounts of *principia*, *naturales actiones*, and so forth, are underpinned by reasoned arguments, but these arguments (inevitably various) are elided in Celsus' doxography (cf. 1.pr.15–16, 20). Reasoning is also implied in the deductive approach to medical decision-making based on a *ratio* (account): cf. 1.pr.21–2.

[73] Cf. 1.pr.29, 36; 1.pr.34 *eadem ... ratione* is irrelevant (n.71). In 1.pr.36 *nec post rationem medicinam esse inuentam, sed post inuentam medicinam rationem esse quaesitam*, the second clause shows that "account" is meant; cf. also 1.pr.12. In 1.pr.36 *requirere etiam se, ratio idem doceat quod experientia an aliud*, there is nothing to prevent *ratio* meaning "reason," but "account" works just as well and should be preferred given the usual sense of the word in the preface: "they also ask whether an account (sc. such as the Rationalists offer) teaches the same thing that experience does or something different."

[74] For *sapientia* = Gk. φιλοσοφία, see Marx 1915: xcvi; Mudry 1982: 59–60; Toninato 1993, esp. 202–4; von Staden 1999: 262–3.

[75] For *ratiocinatio*, see Ch. 3.1 (esp. on Vitr. *DA* 1.1.1, 1.1.15); the comparison is apt even if Celsus' *ratiocinatio* = Gk. λογισμός.

[76] For *inrationale animal* = Gk. ἄλογον ζῷον, see Sconocchia 1993: 224; Svandrlik Civilia 1993: 198–200.

[77] For *disputatio*, see Ch. 3.1.

As with *natura*, Celsus articulates an intermediate position on both *causae* and *ratio* in the preface: "I do think that the art of medicine ought to be rational, but it should be informed by evident causes; all obscure (sc. causes) are excluded from the art itself, though not from the practitioner's thought" (1.pr.74 *rationalem quidem puto medicinam esse debere, instrui uero ab euidentibus causis, obscuris omnibus non ab cogitatione artificis sed ab ipsa arte reiectis*). The statement is calculated to evoke elements of both Rationalist and Empiricist positions. To call medicine *rationalis* is plainly to signal a rapprochement with the Rationalists (cf. 1.pr.13 *ii, qui rationalem medicinam profitentur*, etc.);[78] but the stipulation that follows – to employ only evident causes and reject the hidden or obscure – follows the Empiricists.[79] With this compromise it might appear again that we have found Celsus' "middle ground."[80] But again there are wrinkles. We saw that in Celsus' presentation of the sectarian debate *ratio* was understood chiefly as an (explanatory) account of nature, a sense which the adjective *rationalis* in its earlier appearances also usually carries.[81] But if this kind of account is no part of medicine – as Celsus maintains in consequence of his standpoint on *natura* and *euidentes causae* – then what would it mean to say that medicine is *rationalis*? I do not mean that *rationalis medicina* cannot be translated or given a sense, but that it is surprising that Celsus should deploy the word in his final statements in a way apparently inconsistent with its meaning throughout the preface. In fact, Celsus' composite definition of *medicina* conceals as much as it reveals: it does establish epistemological limits that are not elsewhere controverted, but also gives an inadequate impression of his true "middle ground." As with *natura*, Celsus has a more original and independent notion of *ratio* than indicated by the sectarian debate.

There is one exceptional passage in the preface which broadens the meaning of *ratio* beyond the sense assumed by the Rationalists and the Empiricists and helps us extricate ourselves from the strictures of the debate to appreciate Celsus' own view. After setting aside *contemplatio rerum naturae* (1.pr.47, Ch. 5.1), Celsus continues: "in truth medicine itself *does* require *ratio*, even if not among obscure causes or among natural actions,

[78] Indeed, Marx 1915: lxxvii; Mudry 1993b: 816–17 (modifying 1982: 201–2) take Celsus to be professing himself a Rationalist, but see the discussion in this section.

[79] See Ch. 5.2 on the qualification *non ab cogitatione artificis.*

[80] *Quidem* and *uero*, balancing the opposed positions (like Gk. μέν . . . δέ), reinforce the impression of accommodation between the Rationalist and Empiricist positions.

[81] Cf. 1.pr.10, 11, 57, and esp. 1.pr.62–3, where Celsus tendentiously argues that Methodists might be counted as *rationales* if their own method "is drawn from an account (*ratione*)." An exception is 1.pr.52, discussed later in this section.

yet often . . ." (1.pr.48 *ratione uero opus est ipsi medicinae, etsi non inter obscuras causas neque inter naturales actiones, tamen saepe****). The sentence is truncated by a lacuna, but it is clear from Celsus' summary statements (1.pr.74) as well as later in this same passage (1.pr.52, quoted later in this section) that he would have mentioned *euidentes causae* as the sphere in which *ratio* can legitimately operate. Whatever the specific supplement (or even if there is no lacuna),[82] Celsus' statement is striking for how it alters the sense of *ratio*: for the first time in this part of the preface, the word unambiguously means "reason" or "reasoning" rather than "account." The shift in sense is signaled by the adversative *uero*, which establishes a contrast with the Rationalists' inquiry into nature – and, by implication, their use of *ratio* – that immediately preceded.

Celsus puts this larger notion of *ratio* as "reason" or "reasoning" to good use when handling the problem of how the physician is to respond to a novel or unknown disease (1.pr.49–53). As we learn earlier in the preface, the sects disagree on how often new diseases appear (1.pr.17). The Rationalists think it happens "often" (1.pr.17 *saepe*), and use this to support their position, since they believe that their aetiological inquiries better position them to deduce the cause and hence also the remedy of novel illnesses.[83] The Empiricists, by contrast, do not think new diseases are found at all (cf. 1.pr.36 *neque . . . reperiri*), although they allow that a disease might be "unknown" (*ignotum*). For the individual practitioner, this is almost the same problem as if the disease were new (*nouum*), of course, so the Empiricists must still offer a solution – obviously one that does not depend on an account of the nature of the disease. They argue that the physician "would look to what disease it (sc. the unknown disease) was closest to and would attempt remedies similar to those which often helped the neighboring illness and thus find recourse by the use of similarities" (1.pr.37 *uisurum cui morbo id proximum sit temptaturumque remedia similia illis quae uicino malo saepe succurrerint, et per eius similitudines opem reperturum*).[84]

In returning to the problem, Celsus stakes an intermediate position on the frequency of the occurrence of new diseases, which appear

[82] Marx plausibly suggests the sense should be *in eorum qui euidentibus causis nati sunt morborum curatione*. Less persuasively, Capitani 1967: 117–18; Mudry 1982: 144–5; Serbat 1995: 128 argue that the sentence is complete as it stands.

[83] Rationalist arguments that experience is insufficient to address new diseases are discussed later in this section.

[84] This alludes to the analogical process that Empiricists called ἡ τοῦ ὁμοίου μετάβασις or ἡ κατὰ τὸ ὅμοιον μετάβασις and which they regarded as a "method for discovery" (Gal. *SI* 2 K 1.68.14 = H 4.6–7 ὁδὸς . . . ἐπὶ τὴν εὕρεσιν). On ἡ τοῦ ὁμοίου μετάβασις, see further Gal. *SI* 2 K 1.68.1–69.2 = H 3.21–4.13, *Subf. emp.* 3–4 D 49.16–51.11, 9 D 69.29–74.23, with Deichgräber 1965: 301–5, and following n.

"quite rarely but still from time to time" (1.pr.49 *rarius sed aliquando*). He then attacks the principle of "similarity" (1.pr.51 *similitudo*) with which the Empiricists hope to address the problem of unknown diseases, leveling two complaints. First, he maintains that "similarity" is sometimes simply inadequate for the innovation of medical treatments (cf. 1.pr.51 *ad quod medicinae genus neque semper similitudo aliquid confert*). Second, he argues that "if it (sc. *similitudo*) ever does make a contribution, nonetheless to consider from among similar classes of diseases and remedies which therapy might be used most effectively does in fact involve *ratio*" (*si quando confert, tamen id ipsum rationale est inter similia genera et morborum et remediorum cogitare, quo potissimum medicamento sit utendum*). The second prong suggests that the Empiricists regarded *similitudo* as exclusive of *ratio* (note *tamen*); Celsus simply denies that this could be true of the deliberative process they envision.[85] As just before (1.pr.48), *rationale* here thus seems to mean "involving reason." *Ratio* is not named again in 1.pr.51–3 but remains the implicit topic, as Celsus argues that reason plays a valuable heuristic role in treating new or unknown pathologies. This is not, however, the demonstrative reason desired by the Rationalists, which supplies the physician's judgment (1.pr.52 *consilium*) "from things hidden" (*ab rebus latentibus*), but rather reason which begins "from those things which are capable of investigation, that is, evident causes" (*ab iis quae explorari possunt, id est euidentibus causis*): "for it is relevant whether exhaustion or thirst, heat or cold, wakefulness or hunger, has produced the disease, and whether an abundance of food and wine or excessive sex" (1.pr.52 *interest enim fatigatio morbum an sitis, an frigus an calor, an uigilia an fames fecerit, an cibi uinique abundantia, an intemperantia libidinis*). Celsus thus stakes out an empirically limited causal reasoning (*ratio*) necessary for medical practice. His position differs from that of the Rationalists, for whom *ratio* means an explanatory account based in nature, and of the Empiricists, who reject *ratio* both in the Rationalists' sense and in the more general significance of "reason."[86]

[85] Empiricists apparently did regard ἡ τοῦ ὁμοίου μετάβασις as a purely practical principle, derived from experience (ἐμπειρία), that did not involve reason (λόγος): cf. Gal. *SI* 2 K 1.69.1–2 = H 4.12–13, *Subf. emp.* 9 D 70.10–71.11, with Deichgräber 1965: 302–4; Hankinson 1987a: 333–4; Frede 1987: 250–1; 1988: 87. Exceptional was Theodas, who called transition to the similar "rational experience" (*Subf. emp.* 4 D 50.2–3 *rationabilem experientiam*).

[86] See the discussion of *ratiocinatio* (1.pr.29) earlier in this section; the Empiricist exclusion of *ratio* as "reason" is also active in their implied claim that *similitudo* does not involve reason (previous n.). See however Ch. 5.3.3 and n.109 on ἐπιλογισμός, not mentioned by Celsus.

5.3.3 Rationalis medicina*: Reason and Account*

We have seen one sense in which Celsus believes that medicine must be *rationalis* (1.pr.74) – namely, because *ratio* as "reason" serves heuristically to determine the appropriate response to new or unknown diseases. This is not the only way that Celsus uses *ratio*, however, and we can better understand its place in his *ars* if we consider how it appears outside the preface as well. Throughout *DM*, Celsus elaborates *ratio* in the sense of both "reason," one aspect of which we have met already, and "account," though with a different significance than in the Rationalist use. Although these meanings inevitably shade into one another, they are sufficiently distinct to be considered on their own terms. We will begin with *ratio* as "reason" before turning to *ratio* as "account." It is worth pointing out first, however, that *DM* continuously presupposes a kind of low-grade, algorithmic reasoning in virtue of Celsus' systematic presentation of ailments and remedies. For example: If a disease of the gut presents with X (e.g. pain in the cecum, 4.21.1), do Y (poultices and rubbing); if Y does not work, do Z (cupping). Or: For all varieties (*species*) of disease A (e.g. dropsy, 3.21.1–2), follow procedure B (3.21.2–8); if variety C (e.g. λευκοφλεγματία, 3.21.11), add procedure D (3.21.11–13). We can plausibly regard this sort of rule-following faculty as a form of *ratio*. If so, we must admit, however, that Celsus does not much thematize it – presumably because this basic level of rationality is baked into *ars* as such.[87] There are other, higher-grade sorts of reasoning which more consequentially belong to *ratio* throughout *DM* that we will consider now.

One kind of *ratio* very important for Celsus is a general-purpose reasoning from *euidentes causae*. We saw him defend the value of *ratio* in this application for treating new diseases (1.pr.49–53). But, in fact, the same sense of *ratio* has a much broader scope, and we constantly see the physician in *DM* employing it to make decisions about diagnosis, prognosis, and treatment. Celsus approaches this function of *ratio* by way of the Methodists (1.pr.54–7), who argue that "no cognizance of any cause is relevant to therapy" (1.pr.54 *nullius causae notitiam quicquam ad curationes pertinere*).[88] They argue that the physician need only recognize each disease as pertaining to one of three common states (1.pr.54 *communia morborum*) – corporeal stricture, looseness, or a combination of the two

[87] Even if they would not call it "*ratio*," Empiricists and Methodists could not deny the necessity of such rule-following – hence, for example, the Empiricist claim that the physician requires *consilium* and that medicine cannot be practiced by an *inrationale animal* (1.pr.38, discussed in Ch. 5.3.2).

[88] On the Methodists, see n.74.

(1.pr.55) – and to apply the appropriate contrary (1.pr.56).[89] Celsus levels an extensive critique against the "method" (1.pr.57 *uiam*) of the Methodists (1.pr.62–73), allowing him also to develop his own medical epistemology further. Among other complaints, Celsus attacks the Methodist disregard for causes, arguing rather that "appraisal of the cause ... often resolves illness" (1.pr.69 *causae ... aestimatio saepe morbum soluit*).[90] He presents an instructive anecdote in which the renowned physician Cassius cured a patient simply by administering cold water at the appropriate moment because he recognized that the symptoms followed a bout of drunkenness (1.pr.69–70):[91] "the physician provided this remedy in a timely fashion not following from the fact that the body was constricted or loose, but from the cause which had preceded" (1.pr.70 *quod auxilium medicus opportune prouidit non ex eo, quod aut adstrictum corpus erat aut fluebat, sed ex ea causa quae ante praecesserat*). By the "cause which had preceded" Celsus of course means the *euidens causa*,[92] which does not employ explanation from natural principles but merely indicates a succession of empirically verifiable states. This example illustrates the relevance of reasoning (*ratio*) about causes (*causae*) for the physician's judgment even in ordinary circumstances, where there is no question of a novel illness; the same connection between reason and evident causes is represented in Celsus' considered opinion about *medicina* (1.pr.74).

The medical precepts of *DM* bear out the significance of reasoning with evident causes advertised in the preface. Where Celsus uses *causa* as the "cause" of a disease, he almost always means evident cause.[93] This is made clear sometimes by the addition of an adjective such as *manifestus* ("clear," 2.7.29, 6.6.37b) or a verb such as *apparere* ("be evident," 2.7.10), but is more often simply inferred from the function that *causa* carries as observable source for the illness.[94] Recognition of an evident cause is frequently

[89] In the mixed state, *uehementius malum* is to be treated (1.pr.56). Celsus adds that the Methodists recognized further categorizations of disease (localized or general; acute or chronic; waxing, stable, or waning), according to which treatment varied (1.pr.55–6).

[90] For *aestimatio*, see the case studies in Ch. 5.4, esp. 5.4.1 on 2.10.4–5.

[91] For the Empiricist Cassius, whom Celsus calls *ingeniosissimus saeculi nostri medicus* (1.pr.69; cf. Plin. *HN* 29.7), see *RE* (1) III 2.1678.64–1679.17 (Wellmann); Deichgräber 1965: 210–12; Mudry 1982: 192.

[92] See on 1.pr.18 in Ch. 5.3.2 and n.70.

[93] This is not the only sense of *causa* in *DM*, of course: besides phrasal expressions such as genitive + *causa* (~30× in *DM*), *causa* may also be used, for example, of the "cause" for administering treatment (e.g. 1.3.18, 2.14.6, 6.6.1e, 8.20.7) or the "cause" for the physician's fear on behalf of his patient (7.3.2).

[94] For example, 2.6.2, 2.7.21, 2.10.6, 3.22.1, 4.2.6–7, 4.2.8, 7.7.5, 7.7.8a, 7.7.9a, 7.14.1, plus n.98; cf. 4.2.5, 5.28.2d, 5.28.3a, 6.6.14. At 3.3.6, Celsus distinguishes *abditae causae* from *euidentes* but makes no use of them; at 2.14.2–3 he contemplates reducing various kinds of rubbing (*frictio*) to

significant for therapy. For example, *causae* are relevant to the treatment of wasting (*tabes*), of which Celsus details three kinds (*species* or *genera*, 3.22.1–3).[95] The physician must "first take note what it is that the person suffers from; then, if it is evident only that the body is not nourished, to consider the cause of this" (3.22.4 *animaduertere primum oportet, quid sit qui laboretur; deinde, si tantum non ali corpus apparet, causam eius attendere*). Earlier, Celsus said that wasting resulting from lack of nourishment (*non ali corpus*) is called *atrophia* (3.22.1 ἀτροφία), which "occurs as a rule from one of two causes" (*ea duabus fere de causis incidere consueuit*): either under- or overeating (3.22.1). Different therapies are recommended in each case (3.22.4–6).[96] Sometimes evident causes are handled rather more briefly, as with pain in the flanks: "the onset of this is from cold, trauma, excessive running, or disease" (4.13.1 *initium uel ex frigore uel ex ictu uel ex nimio cursu uel ex morbo est*).[97] Elsewhere the cause is not thematized as such but is obviously crucial, as for animal bites, where different instructions are provided for rabid dogs, various snakes, and scorpions and venomous spiders (5.27.2–9). Often the absence of an evident cause is as important as its presence. For instance, "if a fever along with a swelling in the groin arises for a woman, and no cause is evident, then there is an ulcer in the womb" (2.7.10 *si mulieri inguen et febricula orta est, neque causa apparet, ulcus in uulua est*). The lack of an externally visible cause warrants the inference to another, internal cause, which subsequently guides treatment.[98] (This inference to something unseen is legitimate because the ulcer is potentially observable, and the inference does not depend on any explanatory hypothesis about *natura*.[99]) There are many examples of reasoning from and to evident causes in *DM*, but the instances discussed here are enough to demonstrate the importance of *ratio* in this sense to the physician's critical faculty.

A third function of *ratio* that Celsus makes crucial for medical decision-making pertains to the adjustment of therapy in response to the individual's nature (*natura*). Again, he first outlines this use indirectly, by defending it against the Methodists. We remarked that the Methodists' practice is based

a single cause (*una causa*), but discounts it as something *quod iam ad medicum non pertinet* (anyway, neither *natura* nor *abditae causae* figure there).

95 Ἀτροφία (3.22.1), κακεξία (3.22.2), φθίσις (3.22.3): Scheller and Frieboes 1906: 518–21.

96 For another adjustment of therapy based on precipitating cause, see 4.2.6–7 (discussed in Ch. 5.3.4 in connection with *experientia*).

97 For *euidens causa* = *initium morbi*, see 1.pr.18 (Ch. 5.3.2) and cf. 1.pr.16 *prima origo causae* (where *causa* = *morbus*, n.42).

98 For example, 2.7.2, 2.7.29, 2.8.36, 4.2.7, 6.6.37b; also 3.8.22.

99 For such legitimate inference, see esp. 5.26.7–19; also later in this section on ἐπιλογισμός (n.109).

on recognition of the appropriate "commonalities" (*communia*); the "particulars" (*propria* or *proprietates*) of the patient's nature are held to be irrelevant.[100] Celsus argues that the Methodists are actually inconsistent on this point, since they allow consideration of certain particulars (*locus*, *caelum*, *tempus*) in advising healthy individuals but not in treating the sick (1.pr.70–1). More fundamentally, however, he believes it is simply wrongheaded to ignore the particulars with which the doctor is confronted. Even in the same individual, diseases manifest themselves quite differently (cf. 1.pr.71 *etiam morborum in isdem hominibus aliae atque aliae proprietates sunt*) and may thus require different treatments at different times. Or, again, the same kind of treatment – giving or withholding food (*cibus*), for example – is administered differently based on relevant particulars, including age, habit, fitness, season, and so on (1.pr.72–3). Celsus agrees that *communia* are important – chiefly where the physician does not know or cannot acquire knowledge of *propria* – but that *propria* should take precedent whenever possible (cf. 1.pr.73 *his* [sc. *propriis*] *insistere*).[101] This maxim underpins Celsus' famous claim that the best *medicus* is one who is also an *amicus* (1.pr.73).[102] More generally, prioritization of *propria* means that Celsus frequently introduces therapeutic adjustments or alternatives depending on the particulars of the patient. This concern is related to his development of *natura* in the sense of the *natura aegri* (the nature of the individual sick person) discussed earlier (Ch. 5.3.1). Countless examples of medical precepts calibrated to *propria* can be found throughout *DM*.

We turn now from *ratio* as "reason" to *ratio* as "account." The translation "account" is vague, but intentionally so: Celsus' usage of the word is varied and captures both descriptions and prescriptions. We have already seen that in his summary of Rationalist doctrine, *ratio* means "account" as an explanation or theory grounded in natural principles. This is not the sense of "account" employed by Celsus elsewhere, although both he and the Rationalists share the assumption that *ratio* as "account" is reasoned or systematic. Beyond that, his views differ as a function of his conceptions of *natura* and *causae*, so that "account" in its various senses remains limited to evident causes and things actually or potentially observable. While *ratio* is

[100] Though not all particulars of the case are ignored (n.89); for the Methodists' rejection of *propria*, see 1.pr.65–8, 70–3.

[101] *Communia* are prioritized by those *qui ampla ualetudinaria nutriunt . . . quia singulis summa cura consulere non sustinent* (1.pr.65): see Harig 1971, esp. 189–91; Mudry 1982: 184–5; Contino 1988: 43–6.

[102] On the *medicus amicus*, cf. 3.4.9, and see Mudry 1980; 1982: 199–200; Pisi 1983; Berrey 2017; Gautherie 2017: 310–24.

thus more epistemologically limited for Celsus than for Vitruvius or Columella (both of whom might be lumped in with Rationalists were they physicians), it plays a similar role in the constitution of *ars* in that it crystallizes discursive elements of expertise in *DM*. In this capacity it ranges from an "account" as record of observation to somewhat more developed meanings such as "precept" or "method" (a prescriptive account of practice). Considering some patterns of its usage in *DM* will illustrate the importance of *ratio* in Celsus' construction of *medicina*.

Ratio appears frequently as a descriptive, empirical accounting of the body, health, disease, or other things relevant to the art. In this sense its applications are varied. Celsus suggests that "an account of the strength of the patient" (*ratio uirium hominis*) should guide treatment decisions, such as for the duration of hot baths (2.17.8) or the appropriate time to administer food (3.6.11).[103] Similarly, when discussing the appropriate frequency of sex (*concubitus*), he maintains that "the frequency (sc. of sex) should be judged not by the number of times but by nature, taking account of age and bodily condition" (1.1.4 *frequens non numero sit sed natura <aestimandus, habita> ratione aetatis et corporis*).[104] *Ratio* is also applied to diseases in this way: "the *ratio* of the onset of fever and its remission prescribes the appropriate times for food and drink" (3.6.5 *tempora cibo potionique febris et remissionis ratio det*). Here it means the observed intervals of the fever, which may be of various lengths and are fixed by close attention to the individual case; hence, Celsus later says that "individual (sc. fevers) must be distinguished insofar as they each possess a different account" (3.13.1 *discernendae . . . singulae* (sc. *febres*) *sunt, sicut rationem habent dissimilem*).[105] In such cases *ratio* can almost mean *natura*, because it constitutes a description of the relevant thing. Thus, when advising how to dress a wound with a linen bandage, Celsus recommends that the edges of the torn flesh be drawn closed; "if the *ratio* of the wound does not permit that, then (sc. the bandage) should be placed in the middle first, so that it may then be drawn toward each side" (5.26.24a *si id uulneris ratio non patitur, media primum inicienda est, ut tum in utramque partem ducatur*). The common thread in these instances is that *ratio* is fundamentally descriptive and presents the raw material, so to speak, for the physician's judgment.

[103] For observation of the patient's *uires*, see also Marx 1915: lxxx–lxxxviii; von Staden 1998: 124–5; and on 2.10.2 (bloodletting) in Ch. 5.4.1.

[104] The supplement is Marx's, who posits a lacuna after *natura* (which is also omitted in J). Serbat 1995: 24n2 defends the transmitted text (with *natura*). The different readings do not much affect the point: even if *natura* were omitted with J, "nature" would remain relevant in the sense of the individual's "nature" vis-à-vis age and bodily condition.

[105] For example, 2.10.10, 2.12.2f, 3.4.1, 3.18.1, 4.13.5.

Far the most common usage of *ratio* as "account" is in reference to the precepts or methods of medicine. It would be superfluous to examine all the instances where Celsus applies the word *ratio* to a therapeutic procedure.[106] It is enough to stress that in this application *ratio* has a more strongly prescriptive force. Here it means not just what the physician has observed, but the appropriate – and appropriately systematic – method of healing. This is what Celsus calls in general terms the "*ratio* of therapy" (*ratio curandi* or *ratio curationis*).[107] His medical epistemology brings the descriptive and prescriptive senses of *ratio* close together, however, because medical precepts are ultimately founded on the observation of successful treatment in terms of evident causes. Nonetheless, there remains an important difference in the way that each kind of *ratio* is systematic: a descriptive account is systematic qua structured observation, whereas a prescriptive account is systematic qua goal-directed formula. A "*ratio* of nutrition" (*ratio uictus*) may refer in the former sense simply to an observed pattern of taking food and drink, but in the latter sense is a methodical pattern, informed by the doctor's expertise, conducive to health.[108] A prescriptive *ratio* thus isolates those elements which are actually therapeutic.

So much for *ratio* and *causa* in the epistemology of *DM*. *Causa* is straightforward enough: Celsus rejects the "hidden" or "obscure" causes of the Rationalists and admits only the "evident" or "clear" causes accepted by both sects. *Ratio*, with which the use of *euidentes causae* is connected, is more complex. Celsus' considered opinion is that *medicina* is *rationalis* (1.pr.74). While this view plainly signals sympathy with the Rationalists, he rejects *ratio* in the Rationalist meaning – that of an explanatory account based in *natura*. The tension in his definition points toward a conception of *ratio* different from that fueling dissension between the sects. Casting our net more widely, we found several ways Celsus might say that the art involves *ratio*. *Ratio* as "reason" has a heuristic function in the discovery of treatments for new or unknown diseases; supports all-purpose reasoning from *euidentes causae* in diagnosis, prognosis, and therapy; and aids the physician in adjusting general precepts to the individual patient's illness. *Ratio* as "account"

106 For example, 3.5.2, 3.20.3, 3.21.5, 4.2.4, 4.2.5, 4.5.7, 4.6.2, 4.27.1d (Ollero Granados 104.10 = Capitani 171.52), 5.26.28b, 6.18.2b, 6.18.2d, 6.19.2, 7.5.2c, 7.5.4b, 7.11.1, 7.16.5, 7.20.5, 7.31.1, 7.31.2, 8.3.1, 8.3.10, 8.4.12, 8.4.19, 8.10.7g, 8.16.2, 8.16.4, 8.21.2, and following nn.

107 *Ratio curandi*: 1.pr.47, 1.pr.53, 3.3.6, 3.21.9, 5.28.7b, 7.9.4; *curationis*: 7.2.2, 7.4.4b, 7.7.1a, 7.9.2, 7.27.4 (by conjecture).

108 *Ratio uictus*: 2.7.33, 2.33.1, 4.25.2, 5.pr.1–3, 5.26.24c, 5.26.34c, 5.28.3c, 5.28.16b, 6.6.1h, 6.6.11, 6.6.27a, 6.7.8c, 7.pr.1, 7.3.4, 7.7.15b.

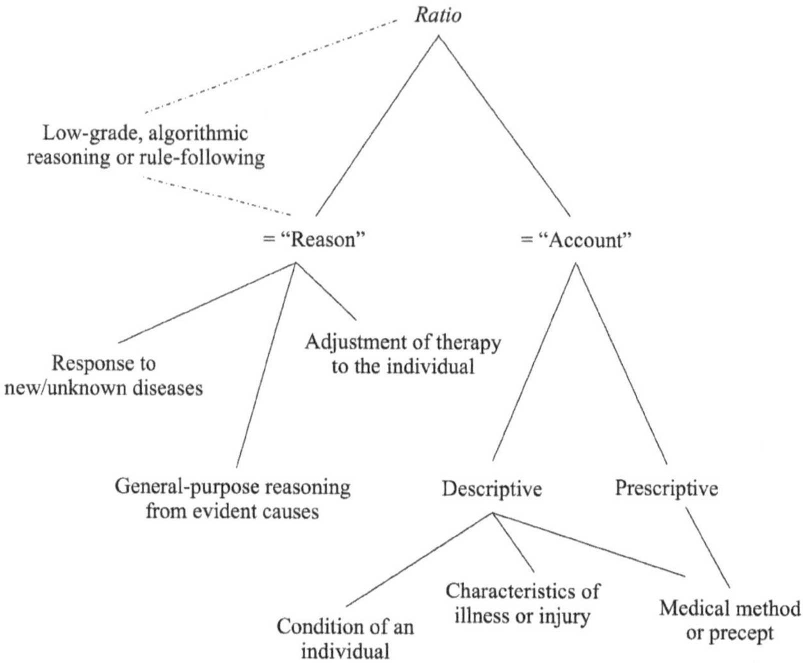

Figure 5.2 Concept map illustrating significances of *ratio* in Cels. *DM* discussed in this chapter (not exhaustive of Celsus' use of the word).

encompasses a variety of systematic statements about phenomena relevant to medicine, both descriptive and prescriptive. Both kinds of "account" are relevant to Celsus' notion of *ratio*, which equally structures the *ars* and informs the physician's judgment. Figure 5.2 schematizes these important functions of *ratio* in *DM*.

Before we move to the last of the concepts we will examine – "experience" – it is worthwhile considering another example of how Celsus presents the dispute between the sects so as to portray his own position as a moderate compromise. We pointed out that he sees *ratio* in one important sense as a kind of reasoning to and from evident causes indispensable to medicine. Against the background of his prefatory summaries of the Rationalist, Empiricist, and Methodist doctrines, this application seems unheralded. For the Rationalists *ratio* means only "account," and their reasoning proceeds to and from natural first principles; the Empiricists not only reject the Rationalists' "account," but also oust "reasoning" as such, substituting a nonrational judgment of similarity (*similitudo*); and the Methodists reject reasoned use of causes altogether (including *euidentes*). Celsus' empirical form of causal reasoning thus seems to have no clear precedent. It is interesting to observe, therefore, that evidence beyond *DM* shows that at least some of the Empiricists accepted a kind of reasoning close to Celsus' conception of *ratio*. They called it *epilogismos* (ἐπιλογισμός), describing it as a reasoning from things evident to conclusions that were themselves evident or observable; it was distinguished from the sign-inference of the Rationalists, called *analogismos* (ἀναλογισμός), which moved from things evident to things that were not.[109] Celsus may have known and suppressed this distinction for rhetorical purposes, but even if he did not, his survey of Empiricist doctrine implies the empirically limited sort of *ratio* that he eventually defends for medical practice; after all, he attributes to them the belief that evident causes are "necessary" (1.pr.27 *necessarias*). Although such speculation can be treacherous, the conclusion stands that by omitting finer-grained consideration of the meanings of *ratio* for the Rationalists and Empiricists, Celsus also allows himself greater latitude to take the word and concept in senses not conditioned by their dispute and strengthens the impression of his own independence.

5.3.4 Experience in the Sects and in Celsus' Middle Ground

We come finally to experience. As we have seen, the debate between the Rationalists and the Empiricists can be understood as a disagreement about the role that explanatory accounts ought to play in medicine. This debate could have been equally well framed as a disagreement over the sufficiency of experience to ground the art – the lens through which we examined the Hellenistic sects in Chapter 1, in fact. As discussed there, the role of experience in *technê* or *ars* could be construed in different senses, either more narrowly as the individual's personal know-how acquired through practice, or more broadly as a body of technical knowledge acquired from repeated observation. In the latter sense, experience could be represented as the epistemic basis for an art, insofar as it preserves a record of the successes or failures that have followed upon certain actions – "what works," to put it simply – and might thus serve as a principle for methodical practice. This latter sense is the one chiefly contested between the Rationalists and the

109 Ἐπιλογισμός aids in "the discovery of things temporarily non-apparent" (*SI* 5 K 1.78.2–3 = H 11.9–10 εὕρεσιν τῶν προσκαίρων ἀδήλων). For ἐπιλογισμός versus ἀναλογισμός, see Gal. *SI* 5 K 1.76.18–78.13 = H 10.14–11.20, *Subf. emp.* 7 D 62.23–63.1, *Med. exp.* 24–7 W–F 89–99.

Empiricists, since their positions are framed in the first instance as competing claims over the constitution of the art itself. Celsus employs several words for experience, which at times correspond to specific Greek words: *usus* can translate as *tribê* (τριβή, "practice"), *experimenta* can translate as *peirai* (πεῖραι, "experience" as repeated observation, "trial and error"), and *experientia* can translate as *empeiria* (ἐμπειρία, "experience").[110] These Greek terms could be technical in Empiricist usage, to which Celsus seems to have been sensitive: *experientia* appears only in the preface and almost always means the Empiricists' technical notion of experience as knowledge sufficient to ground medical precepts; hence, it is never put in the mouths of the Rationalists, who otherwise acknowledge that *experimenta* are necessary (so 1.pr.16).[111] These distinctions are not always preserved, however,[112] and Celsus' simplified presentation of the sectarian debate does not turn on any more demanding notions of "experience" than the two broadest ones distinguished in Chapter 1.

After discussing the Rationalists' hidden causes (1.pr.14–16), Celsus says that they acknowledged that "experience too is necessary" (1.pr.16 *experimenta quoque esse necessaria*), but that such experience had originally to be structured a priori by some account (cf. *ne ad haec quidem aditum fieri potuisse nisi ab aliqua ratione*): "for men of older times did not foist just anything upon those who were sick, but they gave thought to what would be most beneficial, and then put to the proof in their practice that to which some conjecture had led beforehand" (1.pr.17 *non enim quidlibet antiquiores uiros aegris inculcasse, sed cogitasse quid maxime conueniret, et id usu explorasse ad quod ante coniectura aliqua duxisset*). The Rationalists thus have *experimenta* establishing a check on therapy reached from an account (*ratio*). To the extent that *experimenta* confirm what *ratio* suggests – and thus functions as a record of "what works" – the Rationalists are prepared to admit that it preserves many true precepts.[113] Their broader commitments mean that *experimenta* can never be sufficient, however. We discussed the appearance of new diseases, for example, "in which practice has

[110] See Mudry 1982: 96, 115, 123–5; von Staden 1999: 273–4.

[111] *Experientia* = Empiricist ἐμπειρία at 1.pr.27 (*qui se enpiricos ab experientia nominant*), 31, 36, 64; see also following n. See also Ch. 1, n.23.

[112] At 1.pr.47–8, *experientia* picks up Empiricist terminology but more loosely expresses Celsus' own position. For *usus* = "experience" without Empiricist coloring, cf. 1.pr.75, 2.10.2, 2.10.14, 5.26.35c, 5.27.4, 7.4.3a, 8.15.5. Finally, while *experimenta* appears only in the plural in the preface, elsewhere it appears in the singular without technical coloring: cf. 5.28.7b, 6.9.7, with von Staden 1994a: 81n6; 1999: 283n106.

[113] The Rationalists do not believe this concession compromises their position (1.pr.17, with Capitani 1967: 115; Mudry 1982: 98).

not yet brought anything to light, and for that reason it is necessary to consider what the origins of the diseases are" (1.pr.17 *in quibus nihil adhuc usus ostenderit et ideo necessarium sit animaduertere, unde ea coeperint*), such origins residing in knowledge of (hidden) causes (*in obscuro positas causas*).

The Empiricists, by contrast, hold that experience alone is sufficient to establish the medical *ars* – indeed, that it is the only sound basis. We saw them argue that *natura* is not to be comprehended;[114] and "since it (sc. nature) is uncertain and incomprehensible, recourse must rather be sought from things that are certain and that have been put to the proof, that is, from what experience has taught in the actual administration of therapies" (1.pr.31 *cum . . . illa incerta, inconprehensibilis sit, a certis potius et exploratis petendum esse praesidium, id est is, quae experientia in ipsis curationibus docuerit*). Against the Rationalists, the Empiricists argue that experience is effective for practice without direction from *ratio*. In the Empiricist prehistory of the discipline, sick people tried one or another thing as their whims or chance dictated (1.pr.33–4).[115] In the course of things, "diligent people made note of which things often produced a better result, and then began to enjoin these things to those who were ill" (1.pr.35 *diligentes homines notasse quae plerumque melius responderent; deinde aegrotantibus ea praecipere coepisse*). This was the beginning of experience qua record of proven therapies, and, at the same time, the beginning of medicine as such (cf. 1.pr.35 *sic medicinam ortam*). Only after this body of precepts was assembled were *rationes* (accounts) sought for medicine (1.pr.36), reversing the order of discovery claimed by the Rationalists.

Celsus shows little interest, relatively speaking, in polemical claims about the sufficiency of *experientia*. "It is true that nothing contributes more to therapeutic method itself than does experience" (1.pr.47 *uerum . . . est ad ipsam curandi rationem nihil plus conferre quam experientiam*), he says, but this is far from reducing the art to *experientia*. Indeed, he underscores the fallibility of *experientia*: "it is not only conjecture which often fails to produce a result, but also experience" (1.pr.48 *neque respondet . . . plerumque non solum coniectura sed etiam experientia*). And we have already noted that he doubts nonrational *similitudo* is sufficient for

114 1.pr.17, 32 (Ch. 5.3.1 and n.44).

115 The Empiricists distinguished two sorts of experience at the origin of the art: "fortuitous" (περιπτωτικόν) and "improvisatory" (αὐτοσχέδιον) (e.g. Gal. *SI* 2 K 1.66.1–67.2 = H 2.12–3.4), but as Mudry 1982: 126 notes, Celsus' account includes only the fortuitous. (A third kind, μιμητικόν, followed the other two.) For an example of αὐτοσχέδιον εἶδος τῆς ἐμπειρίας, see *DM* 5.27.4 discussed later in this section.

new diseases. Apart from these moderating hints, however, the whole problem simply lacks urgency for Celsus: having appropriated and redefined *natura*, *causa*, and *ratio*, he can put experience – removed from the sects' dispute – into a productive rather than antagonistic relationship with other elements of expertise. Experience is indispensable for the art, but that does not make *natura*, *causa*, or *ratio* any less useful. This is perhaps the more striking because, in a certain sense, medical experience qua body of time-tested methods transmitted through writing holds for Celsus a sovereign place in the art not unlike that claimed by the Empiricists: at least, this is a natural conclusion to draw from Celsus' wide-ranging use of Greek medical texts as an historical archive of medical precepts.[116] He makes little of this from an Empiricist perspective, however, and his middle ground rather emerges between the high implicit valuation he places upon *experientia* and his reluctance to aggrandize it into the whole of the art. Throughout *DM*, Celsus addresses the importance of experience in several senses for medical practice, as a few examples will illustrate.

Granted that experience as the aggregate of historical practice is the source of all Celsus' precepts, occasionally the notion comes to the fore explicitly. Something like the Empiricists' story of the origin of medicine crops up in his discussion of snakebites, where Celsus remarks that "experience itself has taught that the one whom the asp has bitten should consume vinegar" (5.27.4 *usus* ... *ipse docuit eum quem aspis percussit acetum* ... *bibere*).[117] In explanation he offers the story of a boy who, when bitten by an asp, was consumed with thirst; since he could find no water, he drank vinegar that he had at hand and was cured.[118] This is a singular episode, however, and the recounting seems motivated in part by the unusual character of the therapy, which contrasts with other remedies offered for snakebites. When Celsus adduces experience in the broad sense otherwise,

116 The Empiricists distinguished αὐτοψία (experience personally acquired) from ἱστορία, the report (ἐπαγγελία) of autoptic experience. The latter came to be regarded as the more significant, granting the long expanse of time in which medical experience had been aggregated and refined (so Gal. *Subf. emp.* 6 D 54.13–23), although αὐτοψία remained indispensable. Galen allows that "everything which is written in books be called history, because most doctors are so accustomed to call it" (*Subf. emp.* 8 D 67.5–8 *uocare omne quod scriptum est in libris ystoriam, eo quod pluribus medicorum ita consuetum est uocare*). On ἱστορία, see further Gal. *SI* 2 K 1.67.13–18 = H 3.15–20, *Subf. emp.* 3 D 49.14–16, and esp. 8 D 65.23–69.28.

117 Marx follows **J** in printing *debere* after *bibere*, but the auxiliary is unnecessary (*OLD* 4b; *TLL* V 1.1745.3–6 (Bulhart)).

118 An example of the "fortuitous" (περιπτωτικόν) experience the Empiricists put at the root of the art (n.115); cf. *casus cuiusdam pueri* and, of the vinegar, *quod forte secum habebat*. See also Touwaide 1994: 233–4.

it is often because a therapy is contested or novel. Asclepiades is said to have rejected Erasistratus' advice to bind the limbs against hemorrhagic diseases; Celsus retorts that "experience testifies however that this (sc. treatment) often has a good result" (5.11.7 *sed id saepe commode respondere experimenta testantur*), thus invoking past observation as a decisive criterion.[119] Experience is later recruited to Asclepiades' side on a different subject, when he is credited with "a general remedy against all ear conditions that is now confirmed by experience" (6.7.3a *commune ... auxilium aduersus omnes aurium casus iamque usu conprobatum*).[120] The appeal to *usus* probably takes its point from the ambitious claim of the medication – that is, its efficacy against all ear ailments – as well as perhaps the relative recency of the innovation.[121] Experience thus serves to assess the validity of medical precepts; we will see another example later.[122]

Besides understanding experience as the time-tested body of precepts, Celsus also focuses on the physician's personal experience in the sense of skill habituated through practice. He contrasts the "inexperienced" physician (*inperitus*) with the "experienced" physician (*peritus* or *usum habens*), who is also called simply the "good" physician (*bonus*).[123] The inexperienced physician is prone to being "deceived" (*falli, decipi*) by subtle or ambiguous medical phenomena.[124] Celsus is of course alluding to the need for habituated know-how; this is the notion of experience most important for Vitruvius and Columella, inflected in *DM* by the programmatically deceptive character of medical phenomena.[125] Finally, Celsus sometimes employs the language of experience to refer to trial-and-error methods supporting the doctor's decision-making. *Experimentum* thus is complimentary to the physician's reasoning from *natura* and *euidentes causae*. In discussing methods to treat headache (κεφαλαία), Celsus urges consideration in the first instance of "what cause stimulated the pain" (4.2.6 *quae causa dolorem excitarit*); if the precipitating cause was heat (*calor*), the physician is to sponge cold water on the patient's head, but if cold (4.2.7 *frigus*), then warm water.[126] However,

[119] Binding was contentious: cf. Cael. Aur. *Chron.* 2.13.186–7.

[120] See Scheller and Frieboes 1906: 335–6.

[121] Asclepiades lived a century or perhaps somewhat more before Celsus, so the drug was relatively new compared to Hellenistic and Hippocratic remedies. For Asclepiades, see Vallance 1993 (more fully 1990); Polito 1999; he is significant in *DM*: Mudry 1993b; Serbat 1995: lix–lxviii.

[122] 2.10.1, discussed in Ch. 5.4.1; cf. 8.15.5.

[123] *Peritus*: 2.10.16, 3.6.6 (cf. 7.26.2c); *inperitus*: 1.pr.63, 2.6.15, 2.10.3 (cf. 7.29.8); *usum habens*: 2.10.15; *bonus medicus*: 2.6.15, 2.10.7. The instances in 2.10 are discussed in Ch. 5.4.1. See also Gautherie 2017: 187–8.

[124] See Van Schaik 2018; also Ch. 5.4.3 on 8.4.1–6 and n.128.

[125] Cf. Vitruvius' *fabrica* (Ch. 3.1.1 on *DA* 1.1.1) and Columella's *usus* (Ch. 4.2.3 on *RR* 1.pr.32, 1.1.16).

[126] This is an example of reasoning (*ratio*) with evident causes discussed in Ch. 5.3.3.

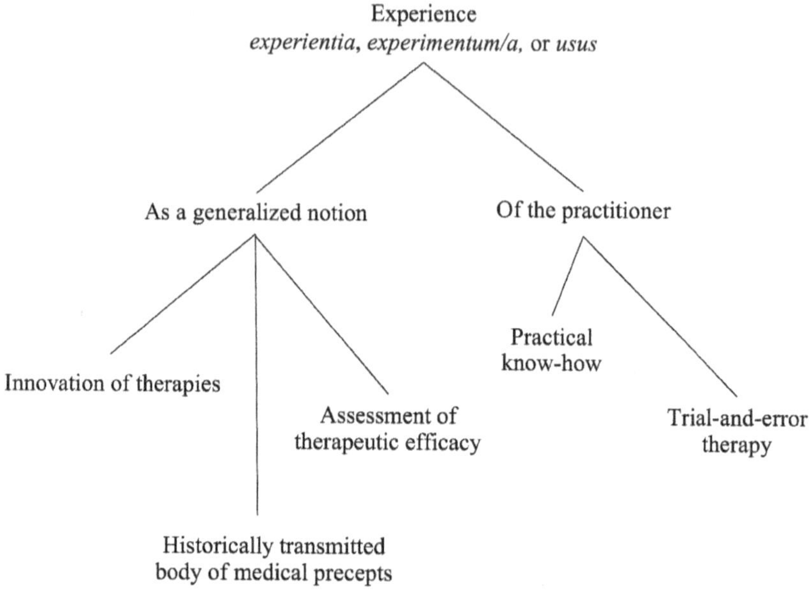

Figure 5.3 Concept map illustrating significances of experience (*experientia, experimentum/a*, or *usus*) in Cels. *DM* discussed in this chapter (not exhaustive of Celsus' use of these words).

"when the cause is unknown, (sc. the physician) ought to see whether things cooling or heating bring more relief, and to employ whichever trial and error confirms" (4.2.7 *ubi causa incognita est, uidere oportet, refrigerantia magis an calfacientia leniant, et is uti quae experimentum adprobarit*). Celsus cautions against "bold medicine" (*temeraria medicina*) and discourages trial of potentially dangerous solutions except where the situation is dire, but a milder form of experiment that tests remedies to determine their efficacy plays a valuable role in calibrating medical precepts to the *natura* of the individual patient.[127] The various significances of experience discussed in this section are schematized in Figure 5.3.

5.3.5 *Celsus' Moderate* ars

This brings us to the end of our examination of Celsus' medical epistemology through nature, cause, reason or account, and experience. In part to

[127] For *temeraria medicina*, see 3.9.10, 7.26.2b, and cf. 2.10.7 *temeraria . . . uia* (discussed in Ch. 5.4.1). See further Gautherie 2017: 269–74 for Celsus' attitude toward risky therapies in *DM*.

underscore the difference between Celsus' approach and that of Vitruvius and Columella, we began with *natura*. All the core terms in Celsus' epistemology are tightly linked, however, and we might have started from any to reach much the same conclusions. From his engagement with the Greek tradition emerges what we called his "middle ground" – that is, his own opinion staged as a moderating compromise between the competing sects. But, as we also argued, his true middle ground defies interpretation as mere appropriation: he adopts the elements of his epistemology from the Rationalists and Empiricists but ultimately redefines them in service of an independent and self-consistent theory of the *ars*. His middle ground is shaped by a rigorously empirical approach to *natura*. *Contemplatio rerum naturae* does not belong to medicine, and he accordingly rejects the use of natural principles that would ground claims about health and disease. This is not the same as arguing that the physician does not require knowledge of nature, however. To the contrary, the physician must grasp the nature of the body and its parts, of individual patients, and of much else; understood as the empirical essence of a thing, *natura* is epistemologically and normatively significant for *medicina*. The boundaries of *ratio* and *causa* are also fixed by Celsus' conception of *natura*. The only *causae* with which the physician will deal are "evident" (*euidentes*) or "clear" (*manifestae*) causes, these being statements about observable conditions that precede illness or health. *Ratio* is a complex and polysemous notion: in the sense of "reason," it encompasses algorithmic decision-making, causal reasoning, and deliberation about particulars; in the sense of "account," it is a descriptive or prescriptive statement about nature or evident causes. Finally, experience has several important functions. In the larger sense of the historical record of practice, it is the basis for the Greek-derived *medicina* communicated in *DM*; in this capacity it is sometimes invoked as a criterion in medical decision-making. In narrower senses, it may refer to the individual practitioner's habituated skill or to trial-and-error methods employed to determine the most effective therapies for the patient.

The functions of nature, reason, and cause in the constitution of disciplinary expertise are very different in Celsus' thought than in Columella's and Vitruvius'. The latter are far more willing to admit natural philosophy into *ars* and are indeed concerned to justify their *praecepta* with explanatory accounts derived from putatively secure natural principles. Celsus' epistemology is no less sophisticated – it is only that the empirical limitations he places upon the physician's decision-making color the whole of *DM*. His approach establishes the parameters for a fundamentally moderate conception of *medicina* that sets great store by the physician's knowledge and experience but allows ample room for chanciness and uncertainty in practice. His is not an art of medicine predicated on absolutes, either the apodeictic certainty of the Rationalists or the skeptical particularization of the Empiricist: "I should submit rather more moderately the following: that the art of medicine is conjectural, and that the nature of conjecture is such that, while more often than not it produces its effect, yet sometimes it disappoints" (2.6.16 *illa tamen moderatius subiciam, coniecturalem artem esse medicinam, rationemque coniecturae talem esse ut, cum saepius aliquando responderit, interdum tamen fallat*). There are general rules, but also individual exceptions; medicine alleviates the burden of disease but does not eradicate it; the practitioner can help his patients, but he will also find himself frustrated in his work, either by his own error or from circumstances beyond his control.[128] Celsus' epistemology is well suited to a key aim of *DM*: to offer his Roman readership a view of the *ars* in its entirety. He pleads for no extravagant understanding of nature, cause, reason, or experience; represents the physician's thought in terms of those things only that his audience might observe or experience; and constructs a systematic body of precepts which illustrate, step by step, how *ars* can make life better by preserving health or combating disease. If one or another of the quarrelsome Greeks might disapprove of Celsus' Roman "midway" (1.pr.45 *media*) in theory, they were unlikely to disapprove of it in practice – for it was almost always their own practice, too.

5.4 Medical Decision-Making in Theory and Practice

Celsus' medical epistemology does not so much downplay the physician's challenges as bring them front and center. It is precisely the inherent variability and uncertainty of nature, the treacherous and changeable character of disease, and the fallibility of the art that require of the physician a finely honed critical faculty in his decision-making. In the remainder of this chapter, we will consider three cases – on bloodletting, critical days, and skull fractures – that illustrate Celsus' epistemology in action and characterize the judgment required of the physician. Our goal is to appreciate how Celsus describes and, in some cases, models the thought and behavior of the *bonus medicus*. Three broader lessons emerge from these cases. First, Celsus' negotiation of the Greek medical tradition

[128] For Celsus' moderate view of medicine, see esp. von Staden 1998 (and 1994a: 80n5; 1994b: 112–13, 1996: 407); also Mudry 1982: 150, 177, 179; 2001, esp. 65–9; 2003, esp. 31–3; 2010, esp. 340–6; Gautherie 2017: 234–75.

sheds further light on how both reason (*ratio*) and experience (*usus*) inform the selection of appropriate therapies. Second, the good physician's investigation (e.g. *considerare, explorare, obseruationes adhibere*), appraisal (e.g. *aestimare*), and inference (e.g. *colligere, coniectare*) regarding the patient's condition reflect an attitude founded on the correct appreciation of nature, cause, reason, and experience in decision-making. Finally, the pragmatic character of Celsus' precepts and his acknowledgment of the limitations and risks of therapy underscore the fundamentally moderate character of his medicine.

5.4.1 *Letting Blood: The Physician's Judgment Between Rule and Exception*

After reviewing the signs and symptoms (*indicia*) of illnesses generally affecting the body (2.1–8), Celsus turns in the second part of Book II to general remedies, treating first those which act by "removing matter" (cf. 2.9.2 *demitur materia*).[129] These include bloodletting, cupping, purging, vomiting, and so on (2.10–17). Here we will consider some important themes in his lengthy discussion of bloodletting (2.10).[130] Celsus considers when the physician should let blood (2.10.1–11), then how (12–19). He begins by foregrounding a dispute in the medical sources, the parties to which go unnamed but are divided into *antiqui* and more recent physicians (2.10.1).[131] With wry humor, Celsus reports that "it is not new to let blood by opening a vessel – but it is new that there is practically no sort of disease for which blood is not let" (2.10.1 *sanguinem incisa uena mitti nouum non est: sed nullum paene esse morbum, in quo non mittatur, nouum est*). The *antiqui* had strict rules for bloodletting, believing that it was unsafe for children, the elderly, and pregnant women (2.10.1). Celsus sets *usus* against this position: "experience later showed however that there is nothing that holds universally in these matters" (2.10.2 *postea uero usus ostendit nihil in his esse perpetuum*). Here we see something of his moderating stance. Historical experience allows for the assessment of precepts handed down from the *antiqui*,

[129] For the structure of Book II, see Serbat 1995: xxvi–xxxviii. For Celsus' allopathy, see Ch. 5.3.1 and n.52; cf. Ch. 3.3.1 on *adiectio* and *detractio* in Vitr. *DA* 1.6.3 (with n.77); also Serbat 1995: 163–4.

[130] For bloodletting in Greco-Roman antiquity, see Brain 1986, esp. 112–57.

[131] For *antiqui* opposed to *recentiores medici* (Celsus often plays the two groups off one another), see 2. pr.1 (where Hippocrates is named as a key source for Book II). It can be inferred from Celsus' usage that all physicians before Asclepiades are *antiqui*, Asclepiades and later being *recentiores*: Temkin 1935: 259n58; Mudry 1982: 185.

including Hippocrates, one of whose *Aphorisms* pronounces against bloodletting for pregnant women.[132] The key insight in this case is a denial of the universality of a rule, a reserve characteristic of Celsus' method: medicine is not devoid of general rules, but absolutes are often found inadequate in decision-making when the situation is scrutinized more closely.[133]

What replaces the opinions of the *antiqui*? Celsus says that "other observations must be brought into consideration, according to which the judgment of the one administering care should be directed" (2.10.2 *alias ... obseruationes adhibendas esse, ad quas derigi curantis consilium debeat*). It is not the patient's age or (in women) pregnancy that matters as such, but his or her *uires* ("strength").[134] A person in good condition (*firmus, robustus, ualens*) can stand bloodletting regardless of other factors, but the weak person (*inbecillus, parum ualens*) cannot. Two features of this advice are characteristic of Celsus' approach: first, the insistence on calibrating therapy to the individual patient (*natura aegri*); second, the reliance on observation (*obseruatio*) to guide the physician's judgment. The same features also raise another thematic problem, however: the difficulty of evaluating the patient's nature from empirical data. "The inexperienced physician is especially apt to be deceived" in challenging cases (2.10.3 *maxime ... medicus inperitus falli potest*) – precisely, for example, when the patient is old or young, or a women is pregnant – because generalizations may obscure his assessment of the individual. What is needed is careful deliberation (2.10.4):

> praecipua in hoc ars sit, quae non annos numeret, neque conceptionem solam uideat, sed uires aestimet, et eo colligat, possit necne superesse quod uel puerum uel senem uel in una muliere duo corpora simul sustineat.
>
> The chief part of skill here is not to reckon the years or take note just of pregnancy, but to appraise (sc. the person's) strength and determine from this whether or not there is enough to sustain the child, elderly person, or together the two bodies in one woman.

The physician's experience helps to size up a situation and determine whether treatment may be safely applied (contrast 2.10.3 *inperitus*[135]), but the nub of the issue is careful examination that will allow inference to the appropriate conclusion (cf. 2.10.4 *colligat*). Celsus lays much weight on

[132] Hippocr. *Aph.* 5.31 L 4.542.
[133] See also von Staden 1994b: 112; 1998: 115; Mudry 2010: 343.
[134] Cf. n.103.
[135] Quoted earlier in this section; see also n.123.

skillful observation: The good physician will not stop at visual inspection (old or young, strong or weak, fat or thin), but will also check the pulse, since "the strength of a body is better appraised from the vessels than from the appearance" (2.10.5 *uis corporis melius ex uenis quam ex ipsa specie aestimatur*).[136] There are other things too which "must be examined" (*consideranda*), including the kind of disease, balance of bodily substances, and more (2.10.5–6). The physician who assesses the situation will sometimes find that although the ailment calls for bloodletting, a patient is not physically suited to it (2.10.7); but where the patient is bound to die "unless aided by a bold method" (2.10.7 *nisi temeraria ... uia fuerit adiutus*) and bloodletting offers the only hope (*spes*), it is the role of a good doctor (*boni medici*) to enjoin this remedy despite the danger (*metus*).[137] This exchange between physician and patient (2.10.7–8) that frankly acknowledges the limitations of medicine is again representative of Celsus' moderate presentation of the art.[138]

The question how to administer bloodletting is almost as fraught as the question of when. A significant issue is from where blood should be let. Celsus proposes a general rule: if the ailment afflicts the whole body, from the arm; if the ailment affects only a part, from the affected part, or as close to the part as possible (2.10.12). He acknowledges that other physicians maintain exactly the opposite, however, holding that bloodletting at the affected part exacerbates the problem by drawing additional blood to the site, so that the best method is to let blood as far from the injury as possible (2.10.13). Celsus calls this view, again attested in a Hippocratic text,[139] "false" (2.10.13 *falsum*), claiming that blood flows only to the affected part so long as the bloodletting is administered (*eatenus sanguis sequitur, quatenus emittitur*); when bloodletting is stopped, the flow stops (*ubi is suppressus est, quia non trahitur, ne uenit quidem*). This argument might not satisfy all, but to the extent that it convinces, it is because it rests on an observable state of affairs (blood is flowing from the wound, or it is not) and appeals to the common-sense intuition that flow is altered only so long as the physician actively intervenes. At any rate, such empirical reasoning is

[136] But elsewhere Celsus says *uenae* are *fallacissima res* (3.6.5, with Van Schaik 2018: 169–71); very few signs are trustworthy in isolation (cf. Ch. 5.4.3 on 8.4.6–10).

[137] *Metus* and *spes* along with related words (for *metus* also *periculum*, *desperatio*, and verbs such as *timere* and *terrere*) regularly name prognostic polarities in *DM*: cf. 2.5.1, 2.6.1, 2.6.17, 2.7.2, 2.7.4, 2.7.8, 2.7.11, 2.8.1, 2.8.9, 2.8.26, 2.8.29, 2.9.1 (programmatic, of prognostics), 2.10.11, 2.12.2f, 3.2.6, 3.5.1, 3.9.2, 3.11.1, 3.21.8, 3.27.1b, 4.9.1, 4.11.6, 4.12.11, 4.18.4, 4.31.9, 5.26.1c, 5.26.21b, 5.26.7, 5.26.20d, 5.27.2c, 6.6.1b, 6.7.8c, 7.3.1, 7.7.14a, 7.8.2, 7.16.1, 7.26.5i, 7.27.1, 7.27.8, 8.4.1, 8.4.9, 8.9.1f, 8.4.20, 8.25.2.

[138] Cf. n.127. [139] Hippocr. *Nat. hom.* 11 L 6.60.

clearly intended as a descriptive "account" (*ratio*) that supports the proposed method. The exception that follows confirms its general applicability (2.10.14):

> uidetur tamen usus ipse docuisse, si caput fractum est, ex brachio potius sanguinem esse mittendum; si quod in umero uitium est, ex altero brachio, etc.
>
> But experience itself seems to have taught that, if the head is injured, blood should rather be let from the arm; (sc. or) if there is any ailment in the shoulder, that it should be let from the opposite arm, etc.

The departure from the *ratio* appears arbitrary because no reason is given, but deference to experience (*usus*) against the imperative of strict explanatory consistency is typical of Celsus' approach. Since he has offered no demonstrative account of bloodletting, only an empirical argument against the Hippocratic precept, he is free to depart from his own generalization where experience recommends. Indeed, because *rationes* for Celsus are always formulations of empirical states of affairs, they must ipso facto be revised or qualified according to observation. This is a good example of his negotiation of the rule and the exception; of the give-and-take between a reasoned account (*ratio*) based on evident phenomena and experience (*usus*); and of the pragmatic attitude that places therapeutic efficacy above theoretical consistency.

There is, finally, the challenge of bloodletting itself: "while letting blood is extremely easy for one who has experience, it is extremely difficult for the unskilled" (2.10.15 *mittere … sanguinem cum sit expeditissimum usum habenti, tum ignaro difficillimum est*). Blood vessels are connected with structures which, if damaged, will cripple the patient; and, again, too much or too little force with the scalpel will fail to accomplish the goal (2.10.15–16). The meaning of experience has shifted back to the practitioner's habituated skill, as Celsus stresses that "there are many things which make what is extremely easy for an experienced (sc. healer) difficult for the ignorant" (2.10.16 *multae res id difficile inscio faciunt, quod perito facillimum est*). Once the physician has cut to the middle of the vein, "he should pay attention to the color and condition (sc. of the blood)" (2.10.16 *colorem eius* (sc. *sanguis*) *habitumque oportet attendere*): the appearance will indicate whether the blood is corrupted, in which case it should be let, or healthy, in which case staunched (2.10.17). Of the latter possibility, Celsus remarks that "this cannot happen under the care of that physician who knows what kind of body to let blood from" (2.10.17 *id euenire non potest sub eo medico qui scit ex quali corpore sanguis mittendus est*). This confident rider points back to the diagnostic assessment of the patient's suitability for bloodletting (2.10.2–6). Yet it can also be seen as a concession to the fallibility of the physician, who may not always correctly judge when or where to let – hence, it serves as an empirical check on his *aestimatio*.

5.4.2 *Fevers and Critical Days: A Failure of* ratio

Throughout his discussion of bloodletting, Celsus emphasizes the fluid and variable nature of medical decision-making, which is marked as much by exceptions as by rules. Each patient must be evaluated individually to determine the appropriate course of therapy. Celsus' own adjudication among alternative therapies, using experience as well as *ratio* as his touchstones, models a moderate and pragmatic approach to medicine. We can fill in this picture by looking at how he handles another contested area of medicine: regimen in fevers. Celsus' account of fevers (*ratio febrium*; cf. 3.18.1) is among the longest and most complicated topics in *DM* (3.3–17), systematically expounding precepts for fevers in general and in particular (tertian, quartan, etc.), as well as attendant pathologies (delirium, headache, etc.).[140] Near the outset, Celsus asserts that "the best medicine for it (sc. fever) is food given at the right time" (3.4.6 *optimum . . . medicamentum eius est oportune cibus datus*). Physicians differ sharply as to when the "right time" is, however, laying down different rules.[141] Celsus rejects all such generalizations, arguing that "none of these (sc. precepts) <u>holds universally without exception</u>" (3.4.7 *nihil . . . horum <u>utique perpetuum est</u>*). Food may be given on the first day, or the second, third, fourth, or fifth; or after the first attack of fever, the second, or more (3.4.7):

> refert enim qualis morbus sit, quale corpus, quale caelum, quae aetas, quod tempus anni; minimeque in rebus inter se multum differentibus <u>perpetuum</u> esse praeceptum temporis potest.
>
> For the kind of illness, the climate, the age (sc. of the patient), and the time of year make a difference: there can be no rule for the time (sc. to administer food) that <u>holds universally</u> for situations that differ so much from one another.

Celsus again eschews absolutes in decision-making, expecting that the physician will observe the patient's *natura*, plus relevant external particulars, to tailor his judgment.

[140] See Scheller and Frieboes 1906: 506–12.

[141] Celsus reports the opinions of *plerique ex antiquis*, Asclepiades, and Themison (3.4.6); for *antiqui* versus *recentiores*, see n.131.

After discussing contingencies that affect the administration of food (3.4.8–10), Celsus turns to the so-called "critical days" (3.4.11–15) observed by the *antiqui*. These days are called "critical" or "decisive" (3.4.11 κρισίμους) "on the grounds that the course of things concerning sick people were decided at such times" (*tamquam tum de aegris iudicaretur*).[142] In Celsus' account these days are handled chiefly as those on which one expects onsets or attacks (*accessiones*) to occur, and hence on which the course and severity of the disease might be assessed.[143] Celsus repudiates the doctrine. His lengthy broadside not only brings his engagement with the Greek medical tradition to the surface of *DM*, but also models the critical application of reason (*ratio*) that must inform the physician's practice. Celsus mainly attacks the theory of critical days as propounded in Hippocratic texts; his strategy is to play various Hippocratic precepts off one another to demonstrate the discordance of the whole doctrine.[144] The best way to grasp his critique is to first lay out the precepts (= P#) of which he takes note:

P1 (3.4.11): The 3rd, 5th, 7th, 9th, 11th, 14th, and 21st days are critical.[145]

P2 (3.4.11): Of the critical days given in **P1**, the 7th, 14th, and 21st days are especially consequential (*summa potentia . . . daretur*).[146]

P3 (3.4.12): Odd-numbered days are especially critical in therapy, because it is on these that onset or abatement of fever is significant.[147]

P4 (3.4.13): The 8th day has the "nature of the 1st day" (*primi die naturam*), since it begins a second 7-day series.[148]

P5 (3.4.14): The 4th day is the most severe (*grauissimum*) for those whose illness will break on the 7th day.[149]

P6 (3.4.14): Every 4th day is especially significant, so that the 4th, 7th, 11th, 14th, and 17th days are critical.[150]

142 For a Hippocratic definition, cf. *Aff.* 8 L 6.216 κρίνεσθαι δέ ἐστιν ἐν τῇσι νούσοισιν, ὅταν αὔξωνται αἱ νοῦσοι, ἢ μαραίνωνται, ἢ μεταπίπτωσιν ἐς ἕτερον νούσημα, ἢ τελευτῶσιν; see also Jouanna 1999: 337–41. Cf. Galen's *On Critical Days*, with Miller 2018.

143 For this medical use of *accessio* (cf. *decessio* "abatement," *remissio* "remission"), first attested in Celsus, see Langslow 2000: 196 (and *OLD* 2; *TLL* I 284.84–85.33 [Bickel]).

144 The major Hippocratic sources for Book II are *Prognostic*, *Aphorisms*, and *Prorrhetic* II (*index nominum propriorum* of Marx s.v. "Hippocrates," plus nn.30, 48).

145 Cf. Hippocr. *Aph.* 4.36 L 4.514–16, where 17, 27, 31, and 34 are also critical; *Aph.* 2.23 L 4.476 mentions only day 14 as critical.

146 Although not spelled out in *Aph.*, *Prog.*, or *Prorrh.* 2, this extrapolation is perhaps warranted by the hebdomad's function as a basic unit in Hippocratic sequences. Cf. **P4–6**, noting that the sequence 4-7-11-14 can be understood as fractional hebdomads, 5-1-1.5-2 (cf. *Carn.* 19 L 8.612–14); in **P6** the final 17 suggests, however, that the tetrad is a base (Jouanna 1999: 338–9). For the hebdomad and Hippocratic critical days, see Roscher 1906: 55–86.

147 Cf. Hippocr. *Aph.* 4.61 L 4.524.

148 Cf. *Aph.* 2.24 L 4.476.

149 Cf. *Aph.* 4.71 L 4.526 (with *Coac.* 145 L 5.614); also *Aph.* 2.24 L 4.476.

150 Cf. Hippocr. *Prog.* 20 L 2.168, counting up to 20 (= 6th in the sequence).

Celsus points out several inconsistencies (= C#) among this patchwork of doctrines:

C1 (3.4.13): Critical days are odd-numbered (from **P3**; cf. **P1**);[151] but the 14th day, which is even-numbered, has "great power" (*magnam uim*) (cf. also **P2**).[152]

C2 (3.4.13): The 8th day has the *natura* of the first and begins a fresh 7-day sequence (from **P4**); but Hippocrates does not regard the 8th (virtually 1st), 10th (virtually 3rd), or 12th (virtually 5th) days, and so forth, as critical, privileging instead the 9th and 11th (from **P1**).[153]

C3 (3.4.14): Critical days are odd-numbered (from **P3**; cf. **P1**); but after the 9th and 11th days (from **P1**; cf. **C2**) the next critical day is not the 13th but 14th (from **P1**), which is even-numbered; the 14th day is also arbitrarily further from the last number in sequence.[154]

C4 (3.4.14): The 4th day can be critical (from **P5**; cf. **P6**); yet the 4th day is even-numbered (an inconsistency: cf. **C1** and **C3**) and not included among other lists of critical days (**P1**, **P2**).

C5 (3.4.15): Every 4th day is especially significant (from **P6**; cf. **P5**); this principle is inconsistent with odd-numbered sequences of critical days (**P3**; cf. **P1**); the sequence 4–7–11–14–17, itself purportedly built on the 4th (also from **P6**), is internally inconsistent because the 11th is the 5th, not 4th, day from the 7th.[155]

Besides all this, Celsus thinks that the determinacy of critical days is simply incompatible with the variability of disease: "often even in the course of the illness the accounting (*ratio*) of days changes, and (sc. a day) becomes more severe, having earlier been freer of fever" (3.4.12 *nonnumquam etiam in ipso morbo dierum ratio mutatur, fitque grauior, qui remissior esse consuerat*). No one sequence of critical days can account for even a single disease, and the various sequences when taken together clash.

Celsus' fundamental complaint is that Hippocrates' doctrine lacks *ratio*. He complains of one sequence that it is constructed "without any

[151] Cf. *DM* 3.4.11 *antiqui potissimum impares sequebantur.* [152] Cf. esp. *Aph.* 2.23 L 4.476 (n.145).

[153] *Aph.* 2.24 L 4.476, the basis for **P4** (n.148), makes the 11th day significant. Celsus omits the Hippocratic explanation (that the 11th day is the 4th day, i.e. half, of the new hebdomad); his inclusion of 9 (not mentioned in 2.24) suggests the dissonant premise comes from *Aph.* 4.36 L 4.514–16, the basis for **P1**. The absence of the 8th day is not conclusive, because the 1st day was not regarded as critical (cf. von Staden 2007: 355n36).

[154] Since Celsus aims to bring out inconsistencies, he does not bother to relate the 14th and 11th via the hebdomad (n.146).

[155] Reckoning inclusively. For the jump to the 14th, cf. n.146.

plausible *ratio*" (3.4.14 *sine ulla probabili ratione*), and concludes his whole examination with the remark that "it appears indeed that, by whatever account we take heed of number (sc. in assessing the course fevers), no method (*or* reason) at all can be found under the authority of this author" (3.4.15 *adeo apparet, quacumque ratione ad numerum respeximus, nihil rationis sub illo quidem auctore reperiri*). In its two appearances, *ratio* shades from "account" or "principle" into "method" or "reason," sharpening the charge of incoherence.[156] Celsus' dispute with Hippocrates indeed centers on medical *ratio*. His doctrine of critical days is not only riddled with internal contradictions or inconsistencies, and hence insufficiently rational and systematic, but also lacks an empirical foundation. The variable progression of diseases thwarts the attempt to fix certain universally valid days, since onset of fever may occur on any day, "whether odd- or even-numbered" (3.4.12 *par inparue*), and no particular day intrinsically possesses "greater or lesser danger" (*maius uel minus periculum*).[157] Celsus turns Hippocrates' error into a lesson about the appropriate sources of medicine (3.4.15):

> uerum in his quidem antiquos tum celebres admodum Pythagorici numeri fefellerunt, cum hic quoque medicus non numerare dies debeat, sed ipsas accessiones intueri et ex his coniectare, quando dandus cibus sit.
>
> The Pythagorean numbers excessively popular at the time deceived the ancients in these matters, however, since here too the physician should not count days, but observe the onsets of fever themselves and from these infer when food ought to be given.

It is not merely that Hippocrates based his doctrine on the sort of natural-philosophical speculation that Celsus excludes from the *ars*:[158] it is that he positively disregards the phenomena in "counting" (*numerare*) days instead of "observing" (*intueri*) the signs of the illness itself.[159] The words *hic quoque* allude to Celsus' earlier claim that none of the doctor's prescribed days for administering foods "holds universally (*perpetuum*) without exception" (3.4.7). The whole episode highlights again the indispensable role of

[156] Inconsistency *in practice* is not necessarily problematic for Celsus, if empirically justified (see Ch. 5.4.1 on 2.10.14), but *logical* inconsistency is a different matter.

[157] These statements come in a denial of the Hippocratic doctrine attributed to Asclepiades (3.4.12).

[158] Pythagoras is invoked thanks to his association with numerology, though Hippocratic critical-day systems cannot actually be reduced to a Pythagorean source (Burkert 1972: 294, 475; Jouanna 1999: 341).

[159] For counting versus observing/appraising, cf. also 2.4.10 *non annos numeret ... sed uires aestimet* (Ch. 5.4.1).

the physician's critical faculty in medical decision-making. General rules are of limited utility in the abstract: the physician must attend carefully to the phenomena and form a unique judgment from them (3.4.15 *ex his coniectare*) tailored to the individual patient.

5.4.3 *Deceptive Breaks: Skull Fractures*

A third and final example – Celsus' discussion of skull fractures – further illustrates the kind of observation and investigation crucial to medicine. Book VIII focuses on surgical interventions for different injuries in bone: after an anatomical overview of the skeleton (8.1), he discusses bones suffering from pathological processes (8.2) before turning in the remainder of the book (8.4–25) to a systematic treatment *a capite ad calcem* of traumatic injuries. He first devotes a lengthy discussion to the skull (*caluaria*, 8.4); we will consider just the opening sections on diagnosis and prognosis of skull fractures (8.4.1–7). First, the doctor assesses whether the patient has any secondary symptoms, such as vomiting, impaired vision, sluggishness, or paralysis. The specific symptoms suggest the prospects for treatment (8.4.1).[160] Where none of these symptoms appears, closer examination is required to determine whether the bone is broken. A guess can be made if the details of the blow are known (material of object, weight, size, speed, etc.), "but nothing is better than to investigate it by means of a more certain indication" (8.4.2 *sed nihil tamen melius est quam certiore id nota explorare*).

The physician's investigation of the skull reveals the interplay between the observational skills and experience of the physician, the potentially deceptive phenomena, and the systematic procedures for investigation intended to minimize the risks of error. The basic technique is to insert a probe (8.4.2 *specillum*) into the wound (*plaga*) and ascertain the damage by the feel of the instrument on the bone: if the bone feels "smooth and slippery" (8.4.3 *leue et lubricum*), "it may appear to be whole" (*integrum id uideri potest*); if there is any roughness (*si quid asperi est*), "it gives evidence that the bone is fractured" (*fractum os esse testatur*). The nature of the skull throws up several obstacles which can impair the physician's judgment, however. If the probe is too thin (8.4.2 *tenue*) or pointed (*acutum*), it may slip into "certain natural hollows" (*quosdam naturales sinus*), thus "producing a false opinion that the bone is fractured"

160 As Celsus later says (8.13), prognosis is important partly so that the physician is not blamed for incurable injuries. A *locus classicus* for the idea is Hippocr. *Prog.* 1 L 2.112, with von Staden 1990 (Hippocratic corpus) and Mudry 1994 (similar themes in Celsus).

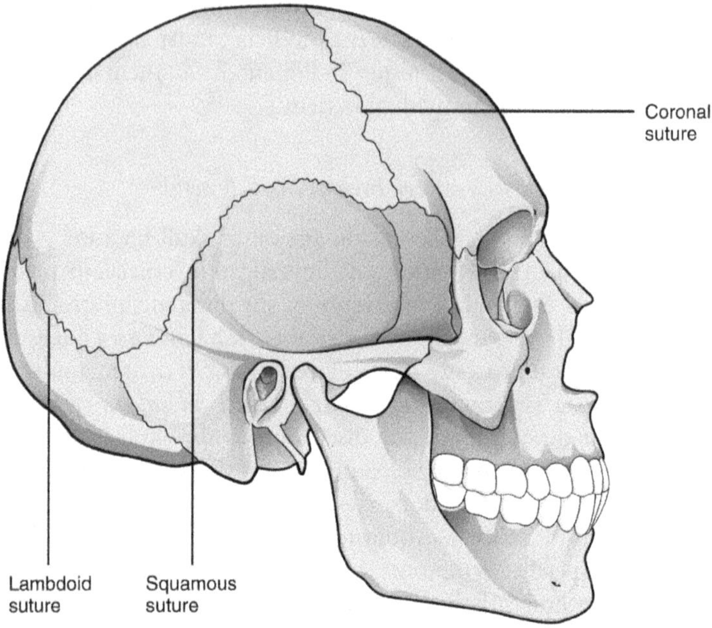

Figure 5.4 Principal sutures of the human skull.

(*opinionem fracti ossis frustra faciat*).[161] If the probe is too blunt (*plenum*), on the other hand, "small cracks may escape the notice (sc. of the physician)" (*paruolae rimulae fallant*).

The most significant stumbling block, however, are the sutures (*suturae*) of the skull, the fibrous bands of tissue connecting the bones (Figure 5.4).[162] To illustrate the danger, Celsus cites Hippocrates, who "recorded that he was deceived by the sutures" (8.4.3 *a suturis se deceptum esse ... memoriae prodidit*).[163] He praises Hippocrates for his "frank acknowledgment of his being mistaken as to the truth"(8.4.4 *simplex ueri erroris confessio*),[164] taking his admission as a sign of the magnanimity of the great doctor, who handed this

[161] For the "natural hollows," see 8.1.7; for ancient medical probes (*specillum*, Gk. μήλη), see Bliquez 2014: 108–13.

[162] Various sutures are described at 8.4.2–4 (with Scheller and Frieboes 1906: 758–9).

[163] Hippocr. *Epid.* 5.27 L 5.226; the admission became an anecdote illustrative of Hippocrates' moral integrity: cf. Quint. *Inst.* 3.6.64, Plut. *Mor. De prof. uirt.* 82d6–e4, Iul. *Ep.* 50 444d, with Mazzini 1999: 314–15.

[164] Marx deletes *erroris*, unnecessarily; I follow Englund 1935: 40; Contino 1988: 198–9 in taking *ueri erroris* to mean "ignorance of the truth" (cf. Tac. *Hist.* 2.72 *errore ueri*, further *TLL* V 2.820.25–38 [Hey]), rather than "true" or "actual error" (Capitani 1970: 78; Mazzini 1999: 315).

down "to posterity for the sake of their advantage, so that they not be deceived in the same way as anyone earlier was deceived" (*utilitatis causa posteris ... ne qui decipiantur eadem ratione qua quis ante deceptus est*). This brief digression – compelled, Celsus says, by his "regard for the great teacher" (*memoria magni professoris*) – demonstrates the fallibility of even the greatest physician in medical practice and illustrates how the proper attitude can turn *error ueri* to advantage.[165] A suture is deceptive (cf. 8.4.5 *potest ... sutura ... fallere*) because it is rough (*aspera*) like a fracture, so the physician may mistakenly believe a fracture is a suture where it is "probable" (*uerisimile*) that a suture should be found. The adjective *uerisimile* is carefully chosen, because (as Celsus reminds us), "there is no entirely fixed seat of the sutures, and the natural join of the bone and the fracture caused by a blow can be in the same place, or the fracture can be nearby" (8.4.5 *neque utique certa sedes ... suturarum est, et potest idem et naturaliter commissum et ictu fissum esse iuxtaue aliquid fissum habere*). To avoid being "deceived" (8.4.5 *decipi*), it is "safest" (*tutissimum*) to lay bare the skull for visual inspection. Not only does this remove the confusion, but it also allows the physician to discover fractures too fine to be felt by the *specillum* (8.4.6; cf. 8.4.2). Sometimes these are visible, but if the crack is not "clear" (*manifesta*), the physician can apply black writing-ink to aid detection (8.4.6).[166] Then, too, the physician must consider that a blow to one part may result in a fracture in another, so that if a crack is not discovered (*reperitur*) where the skin is broken, the physician should "examine in another area" (*parte altera considerare*) for evidence for injury.[167] Once the fracture is found, the doctor may proceed to offer treatment.

This third example reveals further practical challenges encountered by the physician which require a combination of knowledge of *natura* and careful empirical assessment. Diagnosing skull fractures reveals the potentially treacherous character of the phenomena: it is not just the manifestation of pathologies which varies according to the individual, but also the underlying anatomy (here, sutures). The diagnostic process itself is observationally intensive and multisensory, requiring both sight and feel. Knowledge and experience may inoculate the physician against some degree of error but cannot eradicate all confusion, as Hippocrates' anecdote shows. Even the knowledgeable physician may be deceived by natural differences in the positions of the sutures from individual to individual, or by the coincidence of injury site and suture. Methodical investigation,

[165] For Celsus and Hippocrates, see also n.48.
[166] Cf. Hippocr. *VC* 14 L 3.238–42 (with Camoletto 1990: 17).
[167] See further Camoletto 1990: 17–18.

including laying bare the bone for visual inspection, testing with ink to reveal hidden fractures, and careful examination of other possible sites of fracture, affords the "safest" (8.4.5 *tutissimum*) approach, both for the physician grappling with uncertain phenomena and for the patient.

Not all the medical precepts in *DM* are so involved as the cases discussed in this chapter, of course. Often Celsus moves swiftly through an algorithm, so to speak, of diagnosis, prognosis, and treatment, with scant mention of ambiguities or alternatives. But even in apparently straightforward cases, the physician must bring the same judgment to practice; it is just that his critical faculty becomes conspicuous when prompted by methodological controversies, variable and dubious phenomena, and the limits of the art. The examples we considered shed further light on Celsus' conceptualization of *medicina* within his epistemology. Cutting out the kind of explanatory knowledge of nature professed by the Rationalists – and well attested in Imperial *artes* before and after Celsus' time – by no means makes medicine simple or easy. To the contrary, Celsus' pragmatic empiricism demands as much of the practitioner as do theoretically weightier approaches. In *DM*, each patient presents the doctor with a unique challenge, as he strives to adapt general rules to the particular situation while navigating the ineluctable imprecision and uncertainty of the art. This dynamic physician–patient interface throws the burden for successful practice onto the judgment of the physician, who must carefully observe the circumstances of the illness or injury and combine this data with his experience; his knowledge of health, disease, and the body; and his grasp of the accounts and methods (*rationes*) that belong to medicine to effect successful treatment.

5.5 The *Artes* beyond Medicine: The Agricultural Fragments

To close, we will return to Celsus' *Artes* as a whole, with which we began, and ask what our study of *DM* suggests about his treatment of other disciplines and his conception of *ars* as such. There is much we would like to know. For example, what did Celsus have to say about *ratio* and *usus* for other disciplines? Did he bring the same attitude to evaluating and synthesizing earlier sources? Did he maintain a similarly moderate view of the efficacy of *ars* in other spheres? Was the practitioner's judgment in other domains also regulated by a strict empiricism? Even a guess at such questions is complicated by the peculiar goal of *DM*: to make a Roman *ars* of medicine out of an overwhelmingly Greek *technê*. Indeed, we suggested that the unique challenge of this tradition shaped Celsus' presentation of *medicina*, affording the basic terms of his medical epistemology. Yet, as we also argued, Celsus' medical epistemology tends toward an intellectual independence that serves his own purposes. His epistemology can indeed stand apart from the Greek inheritance as a cogent framework for the constitution of expertise.[168] Hence, whether or not Celsus provided methodological preambles for other *artes* as he did for *DM*, we can still extrapolate from *DM* to his treatment of other disciplines – especially where fragments are extant to fill out the picture.

Substantial fragments of other books of the *Artes* are preserved only for agriculture and rhetoric; here we focus on agriculture, which we studied in the Chapter 4. As it turns out, approaching the agricultural fragments with *DM* in mind supports the idea that a similar outlook on specialized knowledge informed his treatment of *res rustica*. Several common themes are tentatively reflected in the fragments: a moderate approach to *ars* that tends toward the mean in practice and balances the rule and the exception; an empirical and descriptive conception of nature, serviceable for *ars*, that avoids deeper explanatory claims; and emphasis on experience in the constitution of the discipline. The themes are not, of course, realized in an identical way, nor should we expect them to be. Moreover, the obligatory caveats are required when dealing with the fragments: we are not certain how much later authors altered or distorted Celsus' words and sense; we lack fuller context that might reveal subtleties in his use of significant concepts; and, above all, we are simply missing too much of the work to draw definite conclusions about what it contained.[169]

Hints of a moderate approach to agriculture not unlike that in *DM* can be gleaned from a handful of fragments. A preference for the mean appears in Celsus' recommendations for the planting location of vines. According to Columella, "as Celsus very rightly holds, the best soil is neither too dense nor crumbly . . . neither thin nor very rich . . . neither flat nor steep . . . neither dry nor wet" (*RR* 3.1.8 = Cels. *Agr.* Marx B.VIII *sicut censet uerissime Celsus, optimum est solum nec densum nimis nec resolutum . . . nec exile nec laetissimum . . . nec campestre nec praeceps . . . nec siccum nec uliginosum*). The mean (natural) condition is frequently normative in *DM*, too.[170] Columella's criticisms also occasionally suggest that Celsus favored a cautious view on agricultural investment. For plowing Celsus

[168] As we observed, none of the concepts Celsus uses to structure the *ars* would have required from his Roman audience awareness of the Greek background, although they would appear more significant against it.

[169] For the agricultural books, see the beginning of this chapter plus Ch. 4, n.24.

[170] See on *natura* as a norm, Ch. 5.3.1; cf. Mielke 2023: 137–8 on this reference to Celsus from a stylistic point of view.

recommended the use of a lighter plowshare and share beam that a smaller ox might pull, "fearing the expenditure" (*RR* 2.2.24 = Marx A.III *reformidans inpensam*), as Columella claims, of the larger ox for a heavier plow.[171] Celsus also thought that beans should not be hoed because, after they were later gathered by hand, the grass that remained between could be cut for hay (*RR* 2.11.6 = Marx A.V).[172] Finally, Celsus, along with Iulius Atticus, apparently recommended using a single quickset (*uiuiradix*) of vine to clothe two props successively, ostensibly at a "lesser cost" (*RR* 4.1.1 = Marx B.XIV *minore sumptu*) than Columella's use of one quickset per prop.[173] An analogy can perhaps be ventured between the conservative attitude in these passages, which Columella portrays as aiming to minimize expense rather than maximize returns,[174] and Celsus' cautious approach in medicine, which sees him avoiding risky or extravagant treatments in favor of conservative therapies.[175] A final suggestion of Celsus' moderate outlook is Columella's claim that Celsus "brought moderation" (*RR* 9.2.1 = Marx E. XXXVIII *adhibuit modum*) to the discussions of apiculture found in Hyginus and Virgil. No doubt stylistic moderation is meant, but the avoidance of Hyginus' erudite collation of older opinions (*placita secretis dispersa monumentis*) and Virgil's "poetic adornments" (*poeticis floribus*) implies a more sober treatment in the content, too.[176]

As in medicine, Celsus also seems to handle agriculture with healthy appreciation for exceptions to the rule. He gives generalized precepts (e.g. Col. *RR* 2.9.10 = Marx A.IV, 5.6.22 = Marx B.XVIII, 8.13.1 = Marx D. XXXVII) but is sensitive to a continuum of possible correct practices. Gargilius Martialis preserves a handful of examples. "Celsus," he reports, "is of the opinion that the peach tree may be planted in <u>any climate and even in meager soil</u>; <u>yet</u> the same author is confident that they become most fruitful if they receive relief from <u>a removed location</u>" (Garg. Mart. *Pom.* 2.4.1 = Marx B.XX <u>*quolibet caelo et quamuis macro solo*</u> *persicorum arborem poni Celsus existimat; idem* <u>*tamen*</u> *laetissimas fieri confidit si* <u>*remotione loci*</u> *blandimentum ceperunt*).[177] Of the chestnut, Gargilius says that "Celsus, keeping before his eyes what might benefit the fruit, believes that <u>warm, sunny lands</u> are most suited to the chestnut ... <u>yet</u> he most prudently added afterwards that it may be sown in <u>cool places, too</u>" (4.6.2–3 = B. XXIV *Celsus, ante oculos habens quid pomis expediret,* <u>*calidas apricas regiones*</u>

[171] Columella disagrees (2.2.24), though he allows that *illa Celsi ratio* may suit poorer soils (2.2.25).
[172] Also sharply criticized by Columella. [173] Their opinion is attacked by Columella in 4.2.
[174] Cf. Noè 2000: 434. [175] See Ch. 5.4.1 on 2.10.7 and n.127.
[176] Cf. the remarks below on Celsus and "bougonia."
[177] For *remotio loci* see Zainaldin 2020a: 163–4.

castaneae aptissimas credit . . . prudentissime tamen postea adiunxit quod et frigidis locis conseri posset).[178]

There is also evidence that Celsus entertained a descriptive view of *natura* akin to that in *DM* while excluding natural-philosophical inquiry. Columella says (*RR* 4.28.2 = Marx B.XVII):

> Celsus . . . et Atticus consentiunt tres esse motus in uite seu potius in omni surculo naturales: unum, quo germinet; alterum, quo confloreat; tertium, quo maturescat. hos ergo motus censent fossionibus concitari; non enim natura quod uult satis efficit, nisi eam labore cum studio iuueris.

> Celsus . . . and Atticus agree that there are three natural movements in the vine or rather in every shoot: the first, in which it sprouts; the second, in which it flowers; the third, in which it reaches maturity. Therefore, they believe that these (sc. natural movements) are stimulated by trenching; for nature does not sufficiently complete what it intends, unless you diligently aid it with effort.

The doctrine of "natural movements" ascribed to Celsus and Atticus is akin to the empirical accounts (*rationes*) of nature that inform medical practice. The account of shoot-growth simply describes observable changes and offers no further explanation, yet it is sufficiently robust to guide practice (trenching). It may be that the final sentence *non enim*, and so on, is Columella's addition (either his own belief or what he thinks Celsus' and Atticus' statements imply), though the personification of *natura* is paralleled in *DM*.[179] Celsus also seems to have purposefully omitted speculative topics concerning *natura*, such as "bougonia" (βουγονία), the theory of spontaneous generation of bees from the corpse of a cow that attracted much physical and biological interest.[180] Columella says: "I think it is superfluous to follow up this account more carefully, agreeing on this point with Celsus, who very prudently says that these creatures (sc. bees) are never afflicted by such losses as to stand in such need (sc. of generation by bougonia)" (*RR* 9.14.6 = Marx E.XLII *quam rationem diligentius prosequi superuacuum puto, consentiens Celso, qui prudentissime ait non tanto interitu pecus istud amitti ut sic requirendum sit*).[181] While the reason given for omitting bougonia is that it is unnecessary in practice,[182] defending the

[178] See the further remarks on chestnuts later in this section.

[179] For art aiding nature see Ch. 4, n.98; for Celsus' personification of *natura*, n.61.

[180] See Osorio 2020.

[181] The supposition that Celsus devoted no space to bougonia would explain why Democritus, Mago, and Virgil – but *not* Celsus – are said to have discussed the time of year for bougonia (9.14.6).

[182] Cf. Gummerus 1909–10: 49; Diederich 2007: 136.

method would presumably also have taken Celsus into questions touching on *contemplatio rerum naturae* that he viewed as falling outside the *ars*.

Finally, a few fragments suggest the significance of experience in the constitution of the art. When discussing the bewildering variety of vines, Columella approvingly quotes the precept of Celsus, who is said to follow Cato (*RR* 3.2.31 = Marx B.XI):

> quare prudentis magistri est eiusmodi nomenclationis aucupio, quo potiri nequeant, studiosos non demorari, sed illud in totum praecipere, quod et Celsus ait et ante eum M. Cato, nullum genus uitium conserendum esse nisi fama, nullum diutius conseruandum nisi experimento probatum.

> Therefore it is the part of a prudent teacher not to detain his students by questing after nomenclature (sc. of different vines) that they cannot master, but rather to teach the general advice, which Celsus and before him Cato pronounced, that no kind of vine should be sown except unless with the approval of tradition, and that none should be preserved any longer unless it is verified by experience.

The two principles for selecting and cultivating vine types (*genera*) are (1) *fama* – that is, "what is said" of a vine in the sense of its popular reputation based on cumulative experience of its successful growth; and (2) *experimentum* – that is, proof of the vine's productivity personally ascertained by the cultivator. Celsus' notion of medical experience in *DM* covers both senses, as we saw, since it can refer to either time-tested medical precepts or the practitioner's habituated skill. On another subject, Gargilius Martialis has Celsus appealing to experience to defend one of his precepts. When justifying the advice to plant the chestnut in warm, sunny regions – the opposite of the cold locations which "which nearly all have believed should be preferred" (*Pom.* 4.6.3 = Marx B.XXIV *quae fere omnes anteponenda senserunt*) – Celsus "proves it on the evidence that Naples (the city in Campania) especially prides itself on that fruit" (4.6.2 = Marx B.XXIV *id illo argumento probat, quod Neapolis, Campaniae ciuitas, eo fructu maxime glorietur*).[183] This is roughly what Columella might have called *fama*, "tradition" based on experience, and what Celsus himself could have called *usus* or *experimentum* in *DM*.

5.6 Conclusion

Only so much can be extrapolated from Celsus' agricultural fragments, but this brief look supports the idea that the epistemology of *DM* is in some respects representative of his epistemology of *ars* generally. We can reasonably expect

183 See Zainaldin 2020a: 303–4.

that other parts of his *Artes* evinced the same balanced and moderate approach to their subjects by, inter alia, privileging the data of observation and experience, eschewing natural-philosophical explanation, and foregrounding the practitioner's capacity to rationally assess and infer the appropriate methods from the evident facts. The attitude toward the *artes* propounded in Celsus' treatise therefore helps us to better appreciate the diversity of specialized knowledge in the early Imperial period. Celsus' epistemology and his presentation of his discipline are no less sophisticated than those of Vitruvius or Columella, but his notion of *ars* appears very different in several ways. The theoretical burden of explanation from natural first principles and, accordingly, the importance of interdisciplinarity at the level of explanatory accounts or principles is minimized in *DM* and, seemingly, in the *Artes* as a whole. Celsus' skepticism of the explanatory accounts which in Vitruvius' *DA* and Columella's *RR* underwrite best practice throws even greater weight in *DM* onto the judgment of the practitioner, who must wrestle with the contingent, variable, and often uncertain nature of the phenomena as well as of his own *rationes*. To be sure, Vitruvius and Columella also locate the seat of expertise in the judgment of the *artifex* (the *architectus* and *agricola* respectively), but this faculty is aided by the possibility of securing a definite account based on nature. In the end, while Celsus dealt with problems in composing his *ars* – or rather *Artes* – very like those faced by Vitruvius and Columella, his differing solutions highlight the considerable variety that the scientific culture of the *artes* could sustain. In the following chapters, on the art of war and land-surveying, we will see aspects of Celsus' reserved and pragmatic attitude to *ars* alongside the interdisciplinary and explanatory interest in *natura* traced in Vitruvius and Columella. Chapter 6 will show this to be true especially of the art of war, where intradisciplinary difficulties much like those thematized in *DM* were responsible for the growth of the *ars*.

PART IV

CHAPTER 6

The Character and Growth of the Latin Art of War
From ars *to* exempla

Of all the *technai* or *artes*, medicine possessed the largest and most fully developed methodological tradition. In composing his *ars*, Celsus sought to formulate an approach that paid deference to traditional problems while also establishing his independence and imparting a fundamentally moderate character to practice. Many of the epistemological issues debated fiercely in medicine were not unique to the discipline, of course. The dispute over the contributions of experience (ἐμπειρία) and theory or reason (λόγος) that left a particularly deep mark on medicine was formative for other branches of specialized knowledge in different ways. We have already seen, on numerous occasions, that the Roman *artes* strove to balance both sides of this problematic, cleaving to neither of the extremes manifested in medicine by the Hellenistic sects.[1] Nonetheless, the attempt to work out alternative methodologies under the epistemological tensions dramatized in Greek medicine sometimes did more profoundly shape the *artes*. In this chapter, we will examine another field in which controversy around the sources of expertise was responsible for a tectonic shift in the discipline: the art of war.

6.1 Reconstructing the Latin Art of War in the Early Empire

We will use "art of war" or "polemology" as broad terms of convenience for the specialized knowledge involved in the conduct of war, including strategy and generalship, tactics, troop type and formation, siege warfare, and so on.[2] In stark contrast with medicine, the art of war was regarded by

[1] See Ch. 1 on "balance," Vitruvius on *fabrica* (*DA* 1.1.1–2, Ch. 3.1.1), and Columella on *usus et experientia* (*RR* 1.pr.32, 1.1.16, Ch. 4.2.3).

[2] Relevant ancient lexica are varied. To describe their treatises Greek authors often append the relevant descriptors (e.g. τακτικός, στρατηγικός, βελοποιικός) to a standard term for literature of knowledge (e.g. ἐπιστήμη, θεωρία, μάθημα, τέχνη). Plato, Xenophon, and Aristotle all speak of a τέχνη or ἐπιστμήμη στρατηγική or πολεμική. Latin terminology is similarly diverse. The *ars duellica* is

the Romans as among the greatest forms of their native expertise – the proof, of course, being found in their world rule.[3] It is perhaps ironic, therefore, that no full-scale Latin *ars* is extant from before the time of Vegetius.[4] The unfortunate accident does not owe to an original dearth of material. Certainly, the tradition was as venerable as that of any Roman *ars*: no less an authority than Cato the Elder stood at the head of it, and other Republican authors are known.[5] Moreover, the early Empire saw two major further works: the military books of Celsus' *Artes* and, some decades later, Frontinus' treatise on the *scientia rei militaris* (his words).[6] As it is, however, the only Latin work of polemology before AD 100 that survives is the *Strategemata* (Domitianic) of Frontinus, an *exempla*-collection presented as a pendant to the earlier *ars*.

The Imperial *ars* of war is not entirely obscure, however. First, we have an unusually rich body of extant Greek military texts which illustrate the varied character of the *technê* and show different ways in which the discipline might be handled.[7] Five works before AD 100 survive. Xenophon (c. 430–354 BC) wrote a short work, *Cavalry Commander* (Ἱππαρχικός), with practical advice for the Athenian cavalry commander ranging from the formation and drill of troops to leadership qualities to tactical maneuvers.[8] Aeneas, called Tacticus, composed *Siege Matters*

mentioned by Plautus (*Epid.* 450), and expertise in warfare could be called *ars, disciplina, ratio, scientia*, or *studium* (*studia*). The branch of knowledge could also be named directly, for example by *militia* (*TLL* VIII 961.61–75 [Bulhart]) or *res militaris* (cf. *res rustica*). Cato's lost work was probably called *De re militari* (less likely *De disciplina militari*), for which n.5 (and Cincius mentioned therein); Frontinus speaks of his own treatise on warfare as being on the *scientia rei militaris* (n.6).

3 This common sentiment is memorably expressed by Veg. *Mil.* 1.1 *nulla enim alia re uidemus populum Romanum orbem subegisse terrarum nisi armorum exercitio, disciplina castrorum usuque militiae*.

4 Before Vegetius the only extant works even adjacent to the art besides Frontinus' *Strategemata* are pseudo-Hyginus' *De munitionibus castrorum* (probably Trajanic: Ch. 2, n.140) and the anonymous *De rebus bellicis* (after AD 337: Fleury 2017: xxviii–liii).

5 For Cato's work, probably called *De re militari*, see Nap 1927; Astin 1978: 184–5, 205; Wheeler 2004: 336–42. Cato was a key source for Vegetius (cf. *Mil.* 1.8.10, 2.3.6–7), even if not directly, and traces of his work may be recovered thence: Schenk 1930; Milner 1991: 253–75; 292–7; 1996: xvii–xxvii. The Republican author C. Cincius (probably first century BC) wrote at least six books *de re militari* (Gell. *NA* 16.4.1–6): S–H I 174–6.

6 For Celsus' *Artes*, see Ch. 2, n.62; for the military books, Schenk 1930: 26–39; Barwick 1960: 236–9 (arguing *contra* Krenkel 1959: 116–21, who acknowledges a military work but denies that it belonged to the *Artes*); Krenkel 1973: 24–5; Wheeler 1977: 366–7; Milner 1991: 255–8, 311; Schulze 2011. For Frontinus' earlier treatise, see *Strat.* 1.pr.1 (Ch. 1 and quoted again in Ch. 6.6). Aelian Tacticus knew this treatise περὶ τακτικῆς (1.2), and Vegetius mentions Frontinus' writings *de disciplina militari* as a source at *Mil.* 1.8.11, 2.3.7. The relation of the lost work to the *Strategemata* is discussed later in this section; for conjecture as to its contents, see n.22.

7 For overviews of Greco-Roman military writing, see Campbell 1987: 13–18; Gilliver 1999: 173–7; Paniagua 2007: 2–5; Whitehead 2008; Chiritoiu 2018: 11–47; Whately 2020.

8 Probably composed late in Xenophon's career (360s): Dillery 2017: 209.

(Πολιορκητικά) in the mid-fourth century BC.[9] This treatise, which taught how to ensure that a city not fall into enemy hands from action within or without and interlarded its precepts with historical illustrations, was a military work on an altogether different scale from Xenophon's. It was in fact only one of several military treatises written by Aeneas, a seminal figure.[10] Probably in the middle of the first century BC, Asclepiodotus composed his *Tactical Art* (Τέχνη τακτική) offering a spare, quasi-mathematical treatment of the Greek phalanx and likely hewing closely to a work by Posidonius.[11] Onasander wrote a work *On Generalship* (Στρατηγικός, *Stratêgikos* = *Str.*) in the mid-first century AD with comprehensive advice for the general before, during, and after battle.[12] Finally Aelian, called Tacticus, composed his own *Tactical Theory* (Τακτική θεωρία) in Trajan's time; the work shares much with that of Asclepiodotus and likely also goes back to Posidonius.[13] There are, furthermore, several extant treatises concerning military engineering and siege weaponry; the mathematical and mechanical nature of such works sets them apart from the other *technai* mentioned, but it would be wise not to posit too sharp a break.[14]

These Greek works together represent an accidental cross-section of the *technê*; some offer more plausible *comparanda* for the Latin *artes* than others. What is particularly significant here is that the Greek and Latin arts of war seem gradually to have come into closer contact with one another – certainly by the end of the first century AD, and probably much earlier. There is no

9 For the dating, see Bengtson 1962: 459–61; Whitehead 2001: 8–10; Lane Fox 2018: 35–45. For the possible identification of Aeneas Tacticus with the Arcadian general Aeneas of Stymphalos (a suggestion of I. Casaubon argued by Hug 1877), see Whitehead 2001: 10–13; Lane Fox 2018.

10 For Aeneas' writings and influence, see Whitehead 2001: 13–17; Vela Tejada 2004: 141–3. Aelian (1.2) says his many books were epitomized by Cineas of Thessaly, minister of Pyrrhus of Epiros (*RE* (1) XI 1.473.23–476.53 [Stähelin]).

11 Asclepiodotus' Τέχνη closely resembles those of Aelian (n.13), Arrian (n.19), and a later *glossarium militare*: the model for all seems to have been a tactical work by Posidonius, who might have been Asclepiodotus' teacher. See further *RE* (1) II 2.1637.59–1640.32 (K. K. Müller); Oldfather and Oldfather in Illinois Greek Club 1923: 230–8; Dain 1946: 26–40; Wheeler 1977: 342–50; Stadter 1978: 117–18; *DPA* no. 456 (Goulet); Wheeler 2016: 575–7; Netz 2017: 382n21.

12 The Suda refers to Onasander's military writings as τακτικὰ περὶ στρατηγημάτων; some punctuate after τακτικά, making two separate treatises (Wheeler 2010: 33n103), but the words taken together likely refer to the *Stratêgikos* (cf. Oldfather in Illinois Greek Club 1923: 343n1; Schellenberg 2007: 182–4).

13 Date: Devine 1989: 31; Wheeler 2016: 580–1; and Posidonius, n.11.

14 Biton wrote Κατασκευαὶ πολεμικῶν ὀργάνων καὶ καταπαλτικῶν in the third or second century BC (Marsden 1971: 61–103). The Βελοποιικά of Philo (c. 200 BC) survives (Marsden 1971: 105–84), along with substantial fragments of his siege writings (Whitehead 2016); these works belonged to his nine-volume Μηχανικὴ σύνταξις. Athenaeus, called Mechanicus, composed Περὶ μηχανημάτων in the first century BC (Whitehead and Blyth 2004).

reason to suppose that Celsus, so learned in the Greek medical tradition, would have been ignorant of the Greek *technê* of war. We can trace the connections more easily later. Onasander, who addressed the consular Quintus Veranius in his preface (thus allowing his work to be dated more narrowly to the period AD 49–58),[15] dedicated his *Stratêgikos* "to the Romans, and above all to those Romans who have obtained senatorial distinction and by the wisdom of Caesar Augustus are honored with the office of consul or praetor" (pr.1 Ῥωμαίοις καὶ μάλιστα Ῥωμαίων τοῖς τὴν συγκλητικὴν ἀριστοκρατίαν λελογχόσι καὶ κατὰ τὴν Σεβαστοῦ Καίσαρος ἐπιφροσύνην ταῖς τε ὑπάτοις καὶ στρατηγικαῖς ἐξουσίαις κοσμουμένοις).[16] Onasander claims that his work is based upon a study of Roman military history; true or not, he obviously hoped for a Roman readership too, and his precepts are framed with sufficient generality to be relevant to Roman warfare.[17] While the latter point cannot be made of Aelian's *Tactical Theory*, an apparently antiquarian treatment of the Hellenistic phalanx, this work too was issued into Roman circles: Aelian dedicated his treatise to Trajan and claimed that he was encouraged to compose it by none other than Frontinus, who Aelian says expressed interest in Greek tactical theory when he was visiting Frontinus' villa at Formiae.[18] Such cultural cross-fertilization in the art of war continued into the second century AD.[19] The relevance of this convergence will become clearer in a moment.

Other valuable sources for the Imperial *ars* of war come from Latin sources, including Frontinus' *Strategemata* and (much later) Vegetius' *Epitoma rei militaris* (between AD 383 and 450). The *Strategemata* (*Stratagems*: *Strat.*) itself may be searched for clues to the structure of Frontinus' earlier treatise on the plausible hypothesis that it was ordered on similar principles.[20] Corroborating evidence comes from Vegetius, who used Frontinus as a source at some remove but seems not to have relied

[15] The biographical details of Q. Veranius (cos. 49) and his family most relevant to Onasander are handled by Smith 1998: 152–8; see further Gordon 1952; Birley 1953: 1–19; Syme 1957; Birley 2005: 37–44. The range AD 49–58 for the *Stratêgikos* is the period from when Veranius was made consul until his death in Britain while governing; Bayer 1947 less plausibly believes that the elder Q. Veranius is the dedicatee and hence posits a wider timeframe: AD 18–59. Onasander may have entered Veranius' *clientela* while the latter governed in Lycia (Smith 1998: 154–5).

[16] The precise significance of the Roman administrative vocabulary is uncertain, especially στρατηγικαῖς ἐξουσίαις: generals? praetors? propraetors? See Eramo 2021.

[17] For Onasander's sources, see n.39. [18] Ael. *Tact.* pr. (esp. pr.3), with König 2020: 135–46.

[19] Apollodorus' Πολιορκητικά, addressing Trajan or Hadrian (Whitehead 2010: 17–24); the Ἔκταξις κατὰ Ἀλανῶν (after AD 135) and Τέχνη τακτική (AD 136; see Devine 1993: 315–16) of the Roman general and statesman Arrian (with König 2020: 147–56); and Polyaenus' Στρατηγικά, addressing Marcus Aurelius and Lucius Verus (probably before AD 166; see Schettino 1998: 33–5; Wheeler 2010: 8–15).

[20] So Wachsmuth 1860: 574–5; cf. Schenk 1930: 41; Grosse 1934: 64; Bendz 1938: 82; Lammert 1939: 238.

upon the *Strategemata*.[21] Parallels between *Strat.* and *Epitoma rei militaris* thus point rather toward the lost work on the *scientia rei militaris*.[22] The contemporary Greek situation further enriches this picture, particularly Onasander's *Stratêgikos*. There is a suggestive coincidence between Onasander's disposition and Frontinus'. By itself this may not mean much, but it has also been observed that Vegetius' *Epitoma* appears to owe something to the *Stratêgikos*: because Vegetius almost certainly did not read Onasander directly, and because these parallels occur in parts of the work evincing the influence of Frontinus, it is likely that Frontinus' work on the *scientia rei militaris* assimilated Onasander.[23] There are no historical reasons to deny this possibility, especially given the convergence of the Greek and Latin traditions mentioned earlier; it has furthermore been pointed out that Onasander's dedicatee Veranius was an older contemporary of Frontinus and indeed his predecessor in the governorship of Britain.[24]

Source criticism is a treacherous business, but we cannot afford to ignore hints at the margins of the Latin tradition, given the paucity of direct evidence. In this chapter, we will creatively scour both the Greek and Latin sources to paint as detailed a picture of the development of the *ars* of war in the early Empire as the situation allows. Our major source to fill the gap in the Latin *artes* will be Onasander's *Stratêgikos*. Stronger and weaker claims can be made for this approach. The stronger claim depends on the hypothesis that Onasander's work substantively informed Frontinus' *ars*, as seems plausible.[25] If this hypothesis is correct, Onasander's work can be used as a rough proxy for Frontinus' in its structure, content, and methods (although there is no reason to suppose Frontinus' work merely reproduced Onasander's). The weaker claim does not rely on any influence. Instead, we can simply treat Onasander's work as representative of a universalizing approach to the discipline germane in its purpose and scope to the Roman *ars*. While we could not in this case infer as much about Frontinus' work from Onasander's, the *Stratêgikos* would compellingly model an account of warfare that presents generalized rules of command. At any rate, Frontinus entertained similar ambitions of setting out generalized *praecepta* to construct a "theory" (*Strat.* 1.pr.1 *scientia*) of

[21] No direct use of the *Strategemata*: Schenk 1930: 43–59; Bendz 1938: 81–8; Lammert 1939: 238; Milner 1991: 267–8, 301; 1996: xxiii n.1.

[22] Schenk 1930: 39–81; Sander 1932: 375; Milner 1991: 255–8, 297–301, 311, and *passim*; 1996: xviii–xxiii.

[23] Sander 1929: 1231; Schenk 1930: 81–3; Lammert 1939: 236–7; Wheeler 1977: 370–1; Milner 1991: 298–300; 1996: xxii–xxiii; cf. Peters 1972: 10, 254, Ambaglio 1981: 366.

[24] Schenk 1930: 82. [25] See n.23.

warfare – analogous to Onasander's "theory of generalship" (*Str.* pr.1 στρατηγικῆς ... θεωρίας). Moreover, Onasander's *Stratêgikos* is wholly devoid of historical illustrations, and it seems from the fact that the *Strategemata* was intended to furnish *exempla* that Frontinus' lost treatise shared this abstraction.[26] Even on the weaker claim, we might treat the presuppositions of Onasander's method as representative, up to a point, of any similarly generalized account. Apart from Onasander's *Stratêgikos*, our other major source in this chapter will of course be Frontinus' *Strategemata*. This inclusion is easier to justify. Although not an *ars* proper, Frontinus presents this work as a vital part of his attempt to survey specialized knowledge of warfare and generalship. Accordingly, we stand to learn much about the Roman *ars* of war by appreciating not only why Frontinus thought this collection of military *exempla* belonged with his earlier work, but also how expertise is thereby constructed and communicated.

6.2 Two Arts of War

Onasander's *Stratêgikos* and Frontinus' *Strategemata* offer alternative, and in some sense complementary, paradigms of the art of war – in a word, different ways of structuring the knowledge that belonged to a domain of the greatest importance for the Romans. While ostensibly founded on the facts of Roman history, Onasander's *Str.* embodies an abstracting approach to the discipline that offers highly generalized precepts of command. These precepts are in turn based in psychological principles regarded as valid for all soldiers – that is, in an account of (human) nature. The actions of the general (στρατηγός) are related more or less explicitly to this explanatory psychological foundation. In this sense, Onasander's epistemology is roughly analogous to that of the Rationalist physicians as presented by Celsus: generalized rules for practice are to be derived from an understanding of nature (Chapter 5). Just as the Rationalists were attacked for this commitment, however, there are indications that an abstracting art of war such as Onasander's likewise met with challenges. One source of criticism came from outside the discipline, in Roman culture more broadly: from Republican times down to the late-first century AD, authors articulated doubts that written *praecepta* were useful for cultivating military expertise. Besides that, Onasander himself reveals a sophisticated awareness of the

[26] Isolated *exempla* have sometimes been conjectured for Frontinus' earlier work (cf. Schenk 1930: 48–9; Bendz 1938: 87–8), but these are doubtful exceptions.

limitations of his work, which he indeed believes apply to any theoretical account of military matters: Warfare is an indefinite and chancy matter, and there necessarily remains a gap between what generalized rules can provide for and what the intelligent leader must devise in the messy realities of combat. Concerns strikingly like those voiced by Onasander were echoed vis-à-vis the art of war in other quarters of the Roman *artes*, including by Vitruvius and Quintilian.

When Frontinus composed his own work on the *scientia rei militaris*, it would, in virtue of its abstracting approach, have been no less vulnerable to such objections than Onasander's *Str.* I argue that *Strat.* responded to these challenges by presenting a systematic but fully particularized body of historical *exempla* that would cultivate the general's deliberative faculty. These *exempla* enrich and indeed complete the *praecepta* of the earlier work by supplying a set of materials whereby the general's judgment can be trained. But in avoiding the abstraction of an *ars* proper (such as evinced in Onasander's *Str.*), *Strat.* adopts an alternative didactic paradigm that would have mitigated concerns about the applicability of the written art of war. If this interpretation is correct, Frontinus' *Strat.* represents a novel and important generic extension of the scientific culture of the Roman *artes*. In what follows, we will begin by characterizing the epistemology of Onasander's *Str.* so as to shed light on the nature of the *technê*/*ars* of war more generally. After addressing the limitations of any such *ars*, we will turn to Frontinus' *Strat.* and examine how his treatise serves to address such challenges and complement the military *praecepta*.

6.3 The Epistemic Basis of Onasander's Art of War

Very little is known of Onasander[27] – even his name has not been securely transmitted[28] – but the preface to *Str.* provides valuable clues to the

[27] Two Cyprian inscriptions (Pouilloux et al. 1987 no. 108 = *IGRRP* 3.995; Pouilloux et al. 1987 no. 109) bear dedications to one Servius Sulpicius Pancles Veranianus by an 'Ονήσανδρος son of Artabates and a Zeno son of 'Ονήσανδρος. Some (e.g. de Vogüé 1866: 443; Dain 1930: 140–3) try to identify this 'Ονήσανδρος with our author, not impossibly but without real proof: *RE* (1) XVIII 1.403.43–58 (Daly–Oldfather); *RE* (2) VIII A.1.956.21–959.4 (Gordon); Le Bohec 1998: 171–2. Regardless of the identification, Smith 1998: 151–8 offers a plausible reconstruction of the sociohistorical context of Onasander's activity; cf. Le Bohec 1998. The Suda (386 O, and n.12) says that Onasander also wrote Ὑπομνήματα εἰς τὰς Πλάτωνος Πολιτείας. Oldfather (Illinois Greek Club 1923: 343–5), who denies any unambiguous trace of Platonic philosophy in *Str.*, suggests that it predated the commentary (345); Peters 1972: 73–4, however, finds some traces of Plato's influence, and Ambaglio 1981 does not doubt that Plato was one of the "maestri di Onasandro" (375). The question requires further investigation.

[28] MSS also show 'Ονόσανδρος (cf. Suda 386 O) and 'Ονήσανδρος (cf. Lyd. *Mag.* 1.47, Leo VI, *Tactica* 14.98 = Dennis 2014: 344). For the spelling, see Oldfather in Illinois Greek Club 1923: 345–7; Dain

cultural context of his writing and establishes the epistemological framework for his treatment of generalship. Onasander begins thus (*Str.* pr.1):[29]

> ἱππικῶν μὲν λόγων ἢ κυνηγετικῶν ἢ ἁλιευτικῶν τε αὖ καὶ γεωργικῶν συνταγμάτων προσφώνησιν ἡγοῦμαι πρέπειν ἀνθρώποις, οἷς πόθος ἔχεσθαι τοιῶνδε ἔργων, στρατηγικῆς δὲ περὶ θεωρίας, ὦ Κόϊντε Οὐηράνιε, Ῥωμαίοις καὶ μάλιστα Ῥωμαίων τοῖς τὴν συγκλητικὴν ἀριστοκρατίαν λελογχόσι καὶ κατὰ τὴν Σεβαστοῦ Καίσαρος ἐπιφροσύνην ταῖς τε ὑπάτοις καὶ στρατηγικαῖς ἐξουσίαις κοσμουμένοις διά τε παιδείαν, ἧς οὐκ ἐπ' ὀλίγον ἔχουσιν ἐμπειρίαν, καὶ προγόνων ἀξίωσιν.

> I believe that it is fitting to address a work on horsemanship or hunting or a treatise on fishing or farming to men who desire to engage in such activity, but a work on the theory of generalship to the Romans, Quintus Veranius, and above all to those Romans who have obtained senatorial distinction and by the wisdom of Caesar Augustus are honored with the office of consul or praetor owing to their training, in which they have no little experience, and the dignity of their ancestors.

The opening priamel establishes the subject matter of the work as the "theory of generalship" (στρατηγικῆς . . . θεωρίας) and foregrounds the status of generalship as a *technê* through comparison with other disciplines.[30] Onasander asserts the appropriateness of dedicating his treatise to the Romans, whose exploits he will claim as the basis for his precepts; the dedication is enacted specifically in his address to Quintus Veranius, a distinguished general and official in the time of Tiberius, Caligula, Claudius, and Nero.[31]

The Roman connection becomes more significant as Onasander sets out his intentions and explains his methods of research and composition. He has dedicated (cf. pr.2 ἀνέθηκα) the work to the Romans because "the ignorant soul fails to understand even what is done correctly in another's case, but a knowledgeable one confirms that which is done well" (τὸ μὲν ἀμαθὲς τῆς ψυχῆς καὶ τὸ παρ' ἄλλῳ κατορθούμενον ἠγνόησεν, τὸ δὲ ἐν

1930: 137–44; Korzenszky and Vári 1935: xvii–xviii; *RE* (1) XVIII.1 403.21–67 (Daly–Oldfather). I follow Oldfather in writing Ὀνάσανδρος, which has become standard; Dain preferred Ὀνήσανδρος, maintained by Le Bohec 1998: 169–70.

[29] Text of *Str.* is that prepared by Oldfather for the Loeb edition (Illinois Greek Club 1923); other readings, including those of Korzenszky and Vári 1935, are mentioned as relevant.

[30] Onasander's selection of disciplines is considered (cf. Petrocelli 2008: 128–9): horsemanship, hunting, and agriculture have martial associations and place *Str.* into conversation with Xenophon, who wrote Περὶ ἱππικῆς, Ἱππαρχικός (sc. λόγος), Κυνηγετικός (sc. λόγος), and in the Οἰκονομικός famously praised the value of agriculture. Xenophon is an important source for *Str.* (n.39).

[31] See n.15.

ἐπιστήμῃ τῷ καλῶς ἔχοντι προσεμαρτύρησεν). The Romans are the readers par excellence because they are already knowledgeable (ἐν ἐπιστήμῃ) about military matters and will thus be capable of judging a *logos* or *theôria*. The basis for the Romans' competence is suggested by Onasander's equivocal statement of his intentions in the following sentence: on the one hand, he aims to "marshal precepts of generalship" (pr.3 στρατηγικὰς συνεταξάμην ὑφηγήσεις) and to "aim at the art of generalship and the practical wisdom in this sphere" (στρατηγικῆς ἐστοχασάμην καὶ τῆς ἐν αὐτοῖς φρονήσεως); on the other, he will simply "encompass in writing those things which have actually been accomplished by the Romans through their power and deeds" (ἃ δὴ Ῥωμαίοις δυνάμει καὶ δι' ἔργων πέπρακται ... λόγῳ περιλαβεῖν).[32] The effect of balancing these expressions is to put the abstract theory or principles of military art on a par with historical summary of Roman deeds: Onasander "aims at" (στοχάζεσθαι) the lessons figured implicitly in Roman history.

In the opening passages of *Str.*, Onasander introduces a cluster of programmatic concepts, as he uses the fact of Roman military dominance to justify an abstracting presentation of rules of military conduct. These concepts, which include *empeiria* (ἐμπειρία, "experience"; pr.1), *ergon* (ἔργον "deed," "action"; pr.2), and *praxis* (πρᾶξις "action," "technique"; cf. pr.2 πέπρακται), capture the historical facticity of the unfolding of "Roman prowess" (pr.4 τὴν Ῥωμαίων ἀρετήν) throughout the entire world (ἐπὶ πέρατα γῆς). These facts are in turn presented as the epistemic foundation for Onasander's work (pr.4–5):

> τὸ δὲ σύνταγμα θαρροῦντί μοι λοιπὸν εἰπεῖν ὡς στρατηγῶν τε ἀγαθῶν ἄσκησις ἔσται παλαιῶν τε ἡγεμόνων κατὰ τὴν σεβαστὴν εἰρήνην ἀνάθημα, εἰσόμεθά τε καὶ εἰ μηδὲν ἄλλο, παρ' ἣν αἰτίαν οἵ τε πταίσαντες ἐσφάλησαν τῶν στρατηγησάντων, οἵ τε εὐπραγήσαντες ἠγέρθησαν[33] εἰς δόξαν· μάλιστα δὲ τὴν Ῥωμαίων ἀρετὴν ἐννοήσομεν, ὡς οὔτε βασιλεὺς οὔτε πόλις οὔτε ἔθνος μεῖζον ἡγεμονίας ἐκρατύνατο μέγεθος, ἀλλ' οὐδ' εἰς ἶσον ἤλασεν, ὥστε τοσούτοις βεβαιώσασθαι χρόνοις ἀκίνητον δυναστείαν.

[32] The paradosis ἃ δὴ Ῥωμαίοις ... πρᾶξαι is corrupt. Köchly's conjecture πέπρακται for πρᾶξαι is preferable (pace Peters 1972: 260) to Schwebel's solution of retaining πρᾶξαι and emending δὴ Ῥωμαίοις to δεῖ Ῥωμαίους (followed by Korzenszky–Vári). Onasander is interested in Roman πράξεις and ἔργα as a source for military precepts; far from teaching the Romans how to wage battle, he claims to learn from them. The vulgate prints περιβαλεῖν ("comprehend") but περιλαβεῖν is preferable, exemplifying Onasander' penchant for using military terminology figuratively (cf. Petrocelli 2008: 132)

[33] Oldfather prints the unaugmented ἐγέρθησαν found in the MSS; I follow Korzenszky–Vári in writing ἠγέρθησαν. There are other conjectures, for example ἤρθησαν (Wyttenbach); see Lucarini 2010: 222, who proposes ἐπήρθησαν.

> [5] οὐ γὰρ τύχῃ μοι δοκοῦσιν ὑπεράραντες τοὺς τῆς Ἰταλίας ὅρους ἐπὶ πέρατα γῆς ἐκτεῖναι τὴν σφετέραν ἀρχήν, ἀλλὰ πράξεσι στρατηγικαῖς.
>
> As I confidently undertake this composition it remains to say that it will be a training for good generals and a source of delight to retired commanders in this holy peace; and we will know from it, if nothing else, for what reason those who were unsuccessful as generals erred and those who did well were raised to a position of high repute. But the chief object of our consideration will be the prowess of the Romans, since there has been neither king nor city nor people that has acquired a greater extent of power or even drawn equal to it so as to secure rule unshaken for so many generations. [5] For it does not seem to me to be by chance that they ventured beyond the borders of Italy to spread their reign to the ends of the Earth, but by the actions of their generals.

Onasander says that instruction (ἄσκησις) and entertainment (ἀνάθημα) are among the aims of his composition (punningly called σύνταγμα[34]) before underscoring that his purpose is fundamentally aetiological, as he aims to disclose the causes (αἰτίαι) for successful and unsuccessful military action.[35] Is this kind of account possible? The contrast between *tuchê* and *praxis* allows Onasander to address this question head on.

Tuchê had long been conceived as a potent force in determining the outcome of historical events, and it was traditionally invoked by history writers to explain events that appeared either outside human control or beyond rational explanation. Yet it would be wrong to see Onasander's dichotomy between *tuchê* and *praxis* (or *aretê*) as a mere commonplace: his avowal that it was "by the actions of their generals" (πράξεσι στρατηγικαῖς) and not merely "by chance" or "by fortune" (τύχῃ) that the Romans came to dominate the world substantiates his own methodology. First, it wards off a potential objection that it was not owing to military methods that the Romans achieved success, but to their good fortune (τύχη). Were this true, Onasander's use of Roman military history as a source for precepts of generalship would be misguided: if Roman victories were simply a matter of luck, there could be no criterion for determining which techniques or actions (πράξεις) were decisive in war and hence no criterion for extracting precepts. Second, it addresses an even more potent challenge: The argument might be applied to the art of war, as

[34] A military term for an array of troops; cf. Petrocelli 2008: 132–3; Formisano 2011: 47; and n.32.

[35] The dual purpose of pleasure and utility recalls (so Peters 1972: 87–8) Polybius' "constantly repeated antithesis between τὸ χρήσιμον and τὸ τέρπνον" (Walbank 1957–79: I 7), for example Polyb. 1.4.11, 7.7.8, 9.2.6, 11.19.a.1–3, 15.36.3, 31.30.1. The fact that Polybius is a key source for Onasander suggests influence here (n.39); thus, note also the stress on aetiology in historical inquiry.

it was to other branches of specialized knowledge in antiquity, that the essential chanciness of the subject matter precludes the possibility of a true *technê* or *theôria* at all.[36] If outcomes are ultimately governed by *tuchê*, what room is there for a scientific treatment of generalship that would teach successful *praxis*?

These concerns explain in part why Onasander is eager to place constraints on *tuchê* in military affairs (pr.5–6):

> συνεπιλαμβάνεσθαι μὲν γὰρ εὔχεσθαι δεῖ καὶ τὴν τύχην, οὐ μὴν τὸ παράπαν οἴεσθαι ταύτην κρατεῖν. [6] ἀλλ᾽ ἀνόητοι οἱ καὶ τὰ σφάλματα τῆς τύχης ἐγκλήματα μόνης ποιούμενοι, οὐ τῆς τῶν στρατηγούντων ἀμελείας, καὶ τὰ κατορθώματα ταύτης, οὐ τῆς ἐμπειρίας τῶν ἡγουμένων.
>
> For one must pray that fortune assists, but not believe that it has complete power: [6] it is rather those who lack intelligence who attribute defeat to chance alone, and not to the negligence of those in command, or likewise attribute success to it (sc. fortune), but not to the experience of those in charge.

Tuchê influences events on the battlefield but only the foolish person (ἀνόητος) believes that it holds sway alone; the competence of the commander – his negligence (ἀμέλεια) or experience (ἐμπειρία) – is also decisive. While Onasander thus eschews the extreme path of depriving *tuchê* of any role,[37] the limitations he places upon it are critical: these not only open up the possibility of a technical knowledge of generalship (στρατηγική) as such, but also reclaim historical Roman *praxis* as a foundation for it, for it is *empeiria* (against ἀμέλεια) by which the general succeeds – and *empeiria* is exactly what the Romans possess (cf. pr.2 ἔχουσιν ἐμπειρίαν). With *praxis* in the picture, *tuchê* becomes a difficulty that must be reckoned with or an unexpected boon, rather than an overwhelming force; in a sense, *tuchê* is even subordinated to *praxis*, for the right action or technique can mitigate, if not negate, its disadvantages. It is in this spirit that Onasander denies that chance is wholly responsible for any result in warfare: it is neither fitting (pr.6 ἐπιεικές) in defeat to blame *tuchê* completely (ὡς πάντων αἰτιᾶσθαι τὴν τύχην), nor is it just (δίκαιον) in victory to pay gratitude to *tuchê* for everything (ἁπάντων ἀνατιθέναι τῇ τύχῃ τὴν χάριν).

The fact that *tuchê* is not sovereign makes a posteriori aetiological investigation into military history legitimate. Onasander thus reiterates

36 For τέχνη and τύχη, see also Ch. 1, n.48.
37 Indeed, the chancy nature of warfare is thematic in *Str.*: see Ch. 6.5.2 and esp. nn.67, 68.

the epistemic value of *praxis* and *ergon* for the art of war, asserting the value of Roman history in particular as a repository of precepts (pr.7–8):

> ἀναγκαῖον ἡγοῦμαι περὶ τῶν ἐν τῷδε τῷ λόγῳ στρατηγημάτων ἠθροισμένων τοσοῦτο προειπεῖν, ὅτι πάντα διὰ πείρας ἔργων ἐλήλυθεν καὶ ὑπὸ ἀνδρῶν τοιούτων, ὧν ἀπόγονον ὑπάρχει Ῥωμαίων ἅπαν τὸ γένει καὶ ἀρετῇ μέχρι τοῦ δεῦρο πρωτεῦον. [8] οὐθὲν γὰρ ἐσχεδιασμένον ἀπολέμῳ καὶ νεωτέρᾳ γνώμῃ τόδε περιέχει τὸ σύνταγμα, ἀλλὰ πάντα διὰ πράξεων καὶ ἀληθινῶν ἀγώνων κεχωρηκότα μάλιστα μὲν Ῥωμαίοις.
>
> I think it is necessary to say concerning the principles of generalship[38] collected in this work that they all proceed from experience of actual deeds and from such sort of men as has descended to the present the Romans' whole primacy in race and prowess. [8] For this treatise contains nothing invented by a newfangled judgment untested in war, but all things derive from actions and genuine battles, chiefly those of the Romans.

Source criticism has complicated Onasander's claim to rely upon Roman history, since much material seems to have been drawn from the Greek tradition.[39] The problem is instructive. While some have thought Onasander adopts a subversive or ironic stance, it is better to understand him as simply presenting the best possible case to a Roman audience for his highly abstract and theoretical approach to the *res militaris*.[40] As we will see, bare precepts – especially those originating from a nonmilitary man, as Onasander might have been – faced a contentious reception at Rome. In arguing that his *theôria* is a formulation of the strategic principles embodied in Roman history, Onasander interprets the past in a way that could be easily accepted by the Romans at the same time as he underscores his own contribution, which is to formally express these principles.[41]

[38] For στρατηγήματα = "principles of generalship" (~ στρατηγικά), see Wheeler 1988: 3–7; the specialized use of the word for a "stratagem" (cf. *Str.* 32.4) is discussed in Ch. 6.5.2.

[39] For Onasander's sources (Xenophon and Polybius are most important), see Peters 1972: 84–253; Ambaglio 1981: 358–65. Roman influence cannot be ruled out (*RE* (1) XVIII 1.404.63–405.1 (Daly–Oldfather); Gugel 1970: 368–71; Wheeler 1977: 368n15; Galimberti 2002), but Onasander draws mainly on Greek texts.

[40] For subversive readings, see for example Ambaglio 1981: 375; Le Bohec 1998, esp. 178–9; Smith 1998: 165; cf. Peters 1972: 86. *Contra*, see Chiritoiu 2018: 69–85. On Greek–Roman interactions in Imperial military writings, see also Chiritoiu 2018; König 2020.

[41] For Onasander's conception of Roman history as the source of the military art, cf. Liu. 9.17.10 (with Oakley 1997–2005: II 445–6 on Liu. 8.7.16). For Onasander's contribution as a (Greek) intellectual one, cf. König 2020: 135 on Aelian.

6.4 From Psychology to Warfare: Onasander's Abstracting Approach to Generalship

Having striven to close the gap between theory (θεωρία, λόγος) and practice (πρᾶξις, ἔργον) and to place his work on an epistemically secure foundation, Onasander takes up the subject of war from the perspective of the general (ὁ στρατηγός). He begins by discussing the criteria germane to choosing a commander (1–2): these are qualities primarily moral and psychological, including moderation (1.2 σώφρων), self-control (1.3 ἐγκρατής), sobriety (1.4 νήπτης), frugality (1.5 λιτός), and so on; more important than wealth and high birth are character and personal capacity.[42] Once virtually conscripted in the opening chapters, this exemplary general is assumed as the subject for the remainder of the treatise (3–42). *Str.* thereafter moves briskly, covering a remarkably wide range of topics: The *stratêgos* declares war, moves the troops, sets up camp, prepares for battle, forms and drills the ranks, conducts maneuvers on the battlefield, lays siege to cities, manages victory or defeat, and does sundry other things.[43] If some of this material is treated rather conventionally, that is in large part because Onasander's principal interest is in the psychological aspects of warfare and command, which recur to anchor his discussions throughout the treatise.[44] He pays much attention to the commander's and soldiers' own experience of battlefield phenomena, and is interested to teach how the good general may control the emotions and passions of his own men as well as of those of the enemy. Onasander's psychological interests provide the treatise with its distinctive cast, as he sifts human perception, emotion, and motivation for insight into the conduct of war and brings out the latent psychological content even of military precepts and phenomena that might seem at first blush far removed from the mental life of the soldier.

We can gain insight into the epistemology of Onasander's work – and, more generally, of an abstracting approach to the art of war in terms of a systematic body of generalized precepts – by tracing certain psychological principles that animate *Str.* As we will see, Onasander is careful to ground his precepts in more abstract facts concerning human nature from which they may be deduced or inferred. At their root, these facts take the form of psychologically primitive cognitive states that Onasander assumes (but does not demonstrate) as

[42] For Onasander's criteria, see Momigliano 1944: 111–12; Plöger 1975: 280–4; Peters 1972: 99–117; Smith 1998: 159; Galimberti 2002: 147–8; Petrocelli 2008: 15–17, 138–61.

[43] For an outline of *Str.*, see Campbell 1987: 13–14; more fully, Peters 1972: 18–22; Ambaglio 1981: 355–7.

[44] Cf. Ambaglio 1981: 355–6, 360–1; Goldmann 1988: 50–75 (*passim*); Galimberti 2002: 152; Petrocelli 2008: 8; Formisano 2011: 44 ~ 2017: 141, Chlup 2014: 54–6; and especially Smith 1998: 158–65.

principles explanatory of human behavior in war. He presumably wishes us to understand that these principles are drawn from (Roman) military history, as advertised in the preface, but they often appeal to intuitions from ordinary human experience such as any reader might share. To the extent that Onasander aims to uncover the causes (pr.4 αἰτίαι) of generalship, it is these psychological principles which most basically account for the general's actions before, during, and after battle. To understand the epistemology of Onasander's approach, we will examine the function of two such principles that play especially important and recurring roles in his work: *phobos* (φόβος, fear) and *tharsos* (θάρσος, courage, confidence).

6.4.1 Fighting with Fear, Fighting with Confidence

Chapter 14 of *Str.* deals explicitly with *phobos* and *tharsos* and can thus serve as a key to their appearances elsewhere. In the previous chapter (13) Onasander discussed how the general through his bearing and words can encourage (13.2 ἀναθαρρεῖ) an army that is affected by despair (13.1 δυσθυμία) or fear (φόβος). Now he addresses the opposite situation (14.1):

> καθάπερ γε μὴν ἐν καιρῷ στρατεύματος ἀναθάρσησις ὤνησεν, οὕτως καὶ φόβος ὠφέλησεν.
>
> Just as encouraging the army at the right moment is profitable, so too fear (sc. at the right moment) is beneficial.

The assertion states the theme of the chapter: neither *tharsos* nor *phobos* is essentially good or bad; the usefulness of each depends rather on the proper moment (ἐν καιρῷ). Onasander will state the conditions (καιροί) under which each part of this statement is true, but since he has just finished discussing how the general can restore the army's confidence (ch. 13), he first puts emphasis on the salutary effects of *phobos* (14.1):

> ὅτ᾿ ἂν γὰρ ῥᾳθυμῇ στρατόπεδον καὶ ἀπειθέστερον ᾖ τοῖς ἡγουμένοις, τὸν ἀπὸ τῶν πολεμίων ὑποσημαίνειν δεῖ κίνδυνον, οὐχ ἥκιστα φοβεροποιοῦντα τὴν ἐκείνων ἐφεδρείαν· οὐ γὰρ δειλοὺς ἔσται ποιεῖν οὕτως, ἀλλὰ ἀσφαλεῖς· ἐν μὲν γὰρ ταῖς δυσθυμίαις θαρρεῖν ἀναγκαῖον, ἐν δὲ ταῖς ῥᾳθυμίαις φοβεῖσθαι· τοὺς μὲν γὰρ δειλοὺς ἀνδρείους ποιεῖ, τοὺς δὲ θρασεῖς προμηθεῖς.
>
> For whenever an army is careless (*or* idle)[45] and becomes less obedient to its leaders, one must intimate a danger from the enemies, above all by instilling

[45] ῥᾳθυμεῖν (ῥᾳθυμία) can refer to either idleness/inactivity or carelessness/negligence; the meanings are related and both are relevant. The army is "idle" (i.e. not engaged in or threatened by combat),

fear of an ambush: this will make them not cowardly, but steady. For it is necessary in despair rather to feel confident, and in carelessness (*or* when idle) rather to feel fear, since (sc. fear) makes the cowardly courageous and the bold prudent.

This passage exhibits several distinctive features of Onasander's method. First, we note the generalized nature of the treatment, which is reflected also in the style.[46] The precepts are apothegmatic, carefully presented in balancing antitheses.[47] Absent are detailed illustrations or historical examples (we will return to this point shortly). We are presented instead with a play of abstract forces: generic agents (τὸ στρατόπεδον, οἱ ἡγούμενοι) are dropped into static situations (ἐν ταῖς δυσθυμίαις, ἐν ταῖς ῥᾳθυμίαις); experience simple emotions or cognitive states (θαρρεῖν, φοβεῖσθαι); and undergo categorical changes in character, attitude, or disposition (δειλούς, ἀσφαλεῖς, ἀνδρείους, θρασεῖς, προμηθεῖς). The psychological drama, stripped of any particularizing features, proceeds as though an exercise in the mind of the general (or the reader); the *stratêgos* hides in plain sight as the unnamed figure who "intimates danger" and "instills fear."[48] Even the unnamed general drops out when the maxims are compressed into impersonal statements of necessity (ἀναγκαῖον) and the passions become the subject ([sc. φόβος] ποιεῖ). Onasander's discussion thus proceeds swiftly to a high level of psychological abstraction.

The second aspect of the passage deserving remark is Onasander's use of *phobos* and *tharsos*. Onasander does not explain here, or anywhere else in *Str.*, how these cognitive states govern behavior. Yet their efficacy for effecting changes in mental disposition – from cowardice to courage, for example – is never questioned, and their influence is taken for granted in considerations of how the general is to handle the troops. This is all to say that Onasander takes *phobos* and *tharsos* (see further in what follows) as indemonstrable principles. These primitive psychological states are the starting points of an investigation into the art of war: they are not

and hence grows "careless" or "negligent," so also disobedient (cf. ἀπειθέστερον). Onasander prioritizes the psychological dimension, so I lead with the translation "careless."

46 On Onasander's style, see Peters 1972: 25–83; also, Ercolani 1997: 43–7; Smith 1998: 151–2; 152n5. Von Wilamowitz-Moellendorff 1911: 220, although disparaging the contents (not uniquely: Peters 1972: 8n4), characterizes it as a "sorgsam stilisierte Schrift"; cf. Oldfather in Illinois Greek Club 1923: 351.

47 For antithesis in *Str.*, see Peters 1972: 67 (comparing Thucydides); for his predilection for gnomic statements, Peters 1972: 81–3.

48 For the ἅπαξ φοβεροποιοῦντα, see Ercolani 1997: 52. Onasander's continuous focalization through the *stratêgos* is unique among polemological treatises, but was likely paralleled in others not extant: Polybius, for example, sustains this perspective in the digression on generalship in the *Histories*, where he remarks that he has more to say in his work on tactics (9.20.4).

themselves the objects of inquiry or demonstration, but instead are used to demonstrate and explain other facts. One could of course quarrel with Onasander about which psychological principles he adopts and how he describes them, but the method is coherent, and he would presumably justify his principles by appealing to the record of historical military actions (πράξεις) and deeds (ἔργα), as discussed in the previous section.

A third noteworthy aspect of this passage is its epistemic structure. We are presented with a situation in which the army is idle or careless (14.1 ῥᾳθυμῇ) and growing disobedient (ἀπειθέστερον). The implicit question is how the general should regain control. Onasander says that *phobos* will make careless men not "cowardly" (δειλούς) but "steady" (ἀσφαλεῖς). This effect is taken as a principle; if granted, it follows (cf. ἀναγκαῖον) that the army must "in carelessness (*or* when idle) feel fear." The general may produce such an emotion by "intimating a danger" from the enemies, and above all by "instilling fear" (φοβεροποιοῦντα) of an ambush. In all, Onasander thus begins from a certain problem or situation, introduces the relevant psychological principles, and proceeds to increasingly specific actions that the general may take. The order of our presentation of this sequence is not identical to Onasander's, which is influenced also by didactic expediency. It is intended, rather, to demonstrate the epistemic structure of his advice, which is represented in Table 6.1.

Table 6.1 *The epistemic structure of Onasander's advice (*Str. *14.1): Instilling fear*

[Problem]	[The army is idle/careless (14.1 ῥᾳθυμῇ) and becoming disobedient.]
Psychological principle	*Phobos* has a positive effect on soldiers who are idle/careless by making them steady (ἀσφαλεῖς) and turning them either courageous (ἀνδρείους) or prudent (προμηθεῖς).
↓	↓
Precept	The soldiers must be made to feel *phobos* when idle/careless.
↓	↓
Command action (more general)	The general should intimate some danger from the enemy (sc. that will cause fear).
↓	↓
Command action (more specific)	The general should make the soldiers afraid of an ambush.

The psychological principle stands at the most abstract level, as an observation or assumption about human emotions and behavior in combat situations. It is not demonstrated but simply asserted. One level below, at the precept, we move from description to prescription, formulating a statement of obligation or necessity: the *is* of the principle becomes an *ought* in a given situation. Finally, at the levels of command action, there are given specific measures (actions) that allow the general to satisfy the imperative of the precept. These command actions can be divided into more or less general versions, but, in any event, they represent the most particular level of Onasander's instructions, which never descend to historical observations or *exempla*.

The complete omission of historical *exempla* in *Str.* has occasioned the surprise of commentators,[49] but in view of the schema presented it is neither mysterious nor troubling: Even if history stands behind Onasander's work as the raw material for his psychological principles, it plays no role in the deduction, so to speak, that establishes the precepts (which depend on the principles) or actions (which depend on the precepts) of generalship. To be sure, Onasander might have included historical illustration for didactic purposes, but he could not have done so without also doing some violence to the economy and universality of his presentation. As it is, his statements, rooted in (human) nature, range over all possible conflicts, past, present, and future, but are tethered to none – a crucial advantage of an abstracting treatment of the material. The epistemological paradigm captured in the diagram is representative of Onasander's approach in *Str.*, in which principles ground precepts and precepts ground actions in a movement from greater to lesser abstraction – but never all the way down to the historically concrete.

After arguing that the general's grasp of the relevant principles will cause him to know when to encourage his soldiers and when to induce healthy fear (14.2),[50] Onasander returns (14.3–4) to the alternative situation mentioned earlier, in which the army is "in despair" (ἐν ταῖς δυσθυμίαις) rather than "in carelessness (*or* idleness)" (ἐν ταῖς ῥᾳθυμίαις). He now explains how, with battle looming (14.3 μελλούσης . . . μάχης), the general can raise the spirits of his troops "when the army is disturbed by fear because it considers the outcome of the engagement uncertain" (ὅτε ἄδηλον ἔχοντα τὰ στρατεύματα τὴν κρίσιν τοῦ πολέμου διατετάρακται τῷ φόβῳ). *Phobos* no longer enforces discipline but rather "disturbs" or "confuses"

[49] Cf. Ambaglio 1981 (esp. 367, 375); Smith 1998: 162–4; Campbell 2004: 65; Whitehead 2008: 142–3; Formisano 2011: 44–5, 50 ~ 2017: 142; Chlup 2014: 44, 51; also Petrocelli 2008: 15.

[50] For knowledge "when" as characteristic of τέχνη, cf. Pl. *Phdr.* 268b7–8 (Ch.1, n.59).

the soldiers.[51] Importantly, the source of *phobos* is uncertainty about the outcome of battle. Onasander does not say whether the army is right to doubt a successful event, because that question is irrelevant: no matter the circumstances, an army that is frightened to confusion (as opposed to healthy caution) will be less effective than one that is confident. The general's concern is in the state of ignorance that unsettles the troops.

The gist of Onasander's advice is to stage a demonstration to convince the soldiers of the contemptible nature of their opponents (14.3–4). The general should assemble a group of choice men to abduct some of the enemy: if those captured are noble in spirit and body, they are to be slain immediately or kept from the army's sight; but if they are weak (ἀσθενεῖς), ignoble (ἀγεννεῖς), and small-souled (μικροψύχους), the general should privately "enslave their souls with fear" (προδουλώσας σφῶν τῷ φόβῳ τὰς ψυχάς) before parading them in front of his men and making much of their cowardice and worthlessness (14.4 ὡς ἀγεννεῖς καὶ ταπεινοὶ καὶ οὐδενὸς ἄξιοι). Onasander continues (14.4):

> ἐπαναθαρρεῖ γὰρ ἐπὶ τούτοις ὁ στρατὸς ἤδη προκατανενοηκὼς τῶν πολεμίων ὄψεις τε καὶ πάθη ψυχῆς· ἀεὶ γάρ, ὃ μηδέπω τις ἑώρακεν, ἐλπίζει μεῖζον γενήσεσθαι τῆς ἀληθείας, ἔτι καὶ τῷ τοῦ μέλλοντος φόβῳ τὴν ἐλπίδα μετρεῖ πρὸς τὸ χαλεπώτερον.
>
> For the soldiers thereby regain their confidence, since they have observed beforehand the sight of the enemy and the passions of their souls: for what one has not seen one always expects to be greater than in truth it is, and besides, for fear of the future, one takes the worse as the measure of one's expectation.

The clause ἀεὶ γάρ, κ.τ.λ., presents an observation about human psychology not restricted to a military context: Humans tend to exaggerate what they have not seen and imagine that what they will experience is worse than it is. This psychological principle explains Onasander's earlier statement (14.3) that an army may become unsettled when it considers the outcome of battle uncertain (ἄδηλον) and also indicates a remedy for the situation, as outlined in Table 6.2.

This diagram captures the basic epistemic structure of Onasander's treatment, again showing a movement from greater to lesser abstraction and from description to prescription. As compared to the earlier schema, there is only one level of command action, but there are two levels of

[51] Ταραχή (ταράσσειν), when among the general's own troops, is a major foe (cf. 10.3, 19.1, 26.1, 26.2), and hence the general must himself be ἀτάραχος (2.2; cf. 1.7); when ταραχή is turned against the enemy, it is a useful weapon (cf. 21.6, 28.1, 41.2).

Table 6.2 *The epistemic structure of Onasander's advice (*Str. *14.3–4): encouraging the army*

[Problem]	[The army is fearful to the point of confusion (14.3 διατετάρακται τῷ φόβῳ) because it is uncertain about the outcome of battle.]
Psychological principle (higher)	It is human nature to expect what one has not seen (14.4 ὃ μηδέπω τις ἑώρακεν) to be greater than it is and to anticipate the worse outcome of a future event (τὸ μελλόν).
↓	↓
Psychological principle (lower)	Soldiers who consider the outcome of battle uncertain may grow frightened to the point of disturbance, because they exaggerate the enemy they have not seen and anticipate the worse outcome of the coming battle (14.3 μελλούσης . . . μάχης).
↓	↓
Precept	The soldiers must be given a calculated sight (ὄψις) of the enemy to remove the uncertainty that stokes their fear to confusion (ταραχή), thus restoring their confidence (14.4 ἐπαναθαρρεῖ).[52]
↓	↓
Command action	The general should capture an enemy soldier and stage a demonstration for the troops that will replace their uncertainty with visual proof (cf. ὄψις) of the enemy's wretchedness.

psychological principles: of higher and lower order. The lower-order principle is the one most important in the military context, but the higher-order principle is valuable in showing that the lower is only a special case. The higher-order principle resides at the most general level to which Onasander appeals in *Str.*: that of the human mind, as such.

6.4.2 *The Explanatory Economy of the* Stratêgikos

Onasander's abstracting approach to the art of generalship (στρατηγική) that privileges generalized precepts based in psychological first principles is evinced in his discussion of individual points of strategy. Importantly, it is also instantiated in his reuse and extension of the same principles to various aspects of warfare throughout *Str.*: the continual application of *phobos* and *tharsos* in particular help to construct an explanatory economy in which a limited set of principles does a relatively large amount of work. One noteworthy effect is to unify apparently disparate wartime phenomena by reducing them to a shared premise or premises. Consider the following example. We have seen that

52 For ὄψις, significant in *Str.*, see Formisano 2011: 48–50 ~ 2017: 145–6; the usage at 32.10, discussed in Ch. 6.5.2, is somewhat different.

phobos is not unqualifiedly bad for the general's army but can also be beneficial (cf. 14.1 ὠφέλησεν) in the right situation (ἐν καιρῷ). Onasander returns to this idea in discussing the general's conduct after battle (36.6):

> εἶτ' αὖ νικῶν τὴν αὐτὴν ἐχέτω προμήθειαν ὑπὲρ τοῦ μὴ παθεῖν ἀμελῶν, ἣν ἂν εἰς τὸ δρᾶσαί τι τοὺς ἐχθροὺς ῥᾳθυμοῦντας εἰσενέγκαιτο. φόβος γὰρ εὔκαιρος ἀσφάλεια προμηθής, ὡς καὶ καταφρόνησις ἄκαιρος εὐεπιβούλευτος τόλμα.
>
> But if he is victorious, let him exercise the same prudence against suffering careless loss that he would himself exploit in conducting an action against an enemy that is careless (*or* idle). For fear at the right moment is a prudent source of safety, just as contempt at the wrong moment is recklessness that leaves one open to treacherous attack.

This advice relies on the same principle as the earlier precept for dealing with a "careless" army (14.1). As we saw, *phobos* makes the soldiers "steady" or safe (14.1 ἀσφαλεῖς) and turns those who are rash "prudent" (προμηθεῖς). Onasander now compares the vulnerability of a victorious army to an enemy that is "careless (*or* idle)" (36.6 ῥᾳθυμοῦντας) and, recalling his earlier words, says that *phobos* is a "prudent source of safety" (ἀσφάλεια προμηθής). The correlative phrase ὡς καὶ καταφρόνησις, κ.τ.λ., further shows that Onasander is thinking of the earlier dichotomy, where he warned that an army can "so despise (sc. the enemy) that it lets down its guard" (14.2 οὕτως καταφρονεῖν ὥστε μηδὲν φυλάττεσθαι). Characteristically, Onasander abstracts from any concrete situation and sets the psychological phenomena in the foreground, making φόβος and καταφρόνησις the subjects of the gnomic statement φόβος γάρ, κ.τ.λ. These cognitive states and dispositions are accordingly treated as having, under the relevant conditions (36.6 εὔκαιρος ... ἄκαιρος; cf. 14.1 ἐν καιρῷ), a regular causal relationship with human behavior, producing, respectively, a "prudent source of safety" (when ἐν εὐκαιρίᾳ) and "recklessness that leaves one open to treacherous attack" (when ἐν ἀκαιρίᾳ). The gnomic formulation reinforces the impression that it is a universally valid observation. Although specific contexts of military action may vary, the same facts about human psychology undergird the general's conduct.

A survey of everywhere in *Str.* where *phobos* and *tharsos* serve as explanatory principles would not significantly revise our findings, instructive though it would be.[53] We will conclude with a different exercise intended to illustrate the extent of Onasander's reuse of psychological principles,

[53] Cf. 9.3, 10.9, 10.14, 10.20, 10.26, 13, 21.9, 22.4, 23.2, 33.5, 34.3, 39.6, 42.1, 42.19–20 (not an exhaustive list).

tracing the ramifications of one principle that we have already encountered: Humans exaggerate what they have not seen and, for fear of the future, expect the worse outcome (14.4). When this principle, which we will call the "principle of future fear" (PFF) for convenience, is first introduced (*Str.* 14), it pertains chiefly to the morale of the troops before battle, but it is not limited to this application throughout *Str.*

Onasander's discussion of reserve troops demonstrates the broader significance of the PFF. He recommends that the general draw up a group of picked soldiers and hold them back when battle begins (22.1). Such reserve troops are advantageous for launching a fresh assault once the enemy has been worn down by the fighting. But "even more useful" (22.2 καὶ τούτου χρησιμώτερον) is to conceal this picked corps and send them out once battle is engaged. The enemy will thus sometimes flee even before arms are joined, "believing that the force attacking is greater than it actually is" (οὐ τοσοῦτον, ὅσον ἐστίν, ἀλλὰ πλεῖον ἐπιέναι πλῆθος νομίζοντες). The tactic is effective because, as Onasander adds in a climactic statement (22.3):

> ἐν αὐτοῖς τοῖς δεινοῖς ἐπιφάνειαι πολεμίων ἀπειράστων[54] ἐκπλήττουσι τὰς ψυχάς· προλαμβάνουσαι[55] γάρ τι χεῖρον οὗ πείσονται, φοβερώτερον ἐκδέχονται τὸ μέλλον.
>
> The arrival of unexpected enemies at the moment of greatest suspense strikes the soul with fear: The soldiers, anticipating something worse than what they will suffer, expect some future event even more terrible.

This explanation depends closely on the PFF. According to the first clause of the PFF, "what one has not seen one always expects to be greater than in truth it is" (14.4). Although the enemy soldiers here do apprehend the additional forces, the reserves are "unexpected" (22.3 ἀπειράστων), and, because the enemy lack a clear notion of them, they exaggerate their number, mistakenly believing them more than they are. According to the second clause of the PFF, "for fear of the future, one makes what is worse the measure of one's expectation" (14.4). Hence the enemy, facing the reserve troops, anticipates (22.3 προλαμβάνουσαι) something even worse to come and thus fears the future (φοβερώτερον ἐκδέχονται τὸ μέλλον).[56]

[54] It is uncertain whether ἀπείραστος can furnish the required sense; Köechly conjectures ἀπροοράτων (cf. 19.3, 22.2, 40.1), Vári ἀπρολήπτων (cf. 1.7, 8.1, 39.6). Oldfather compares [Gal.] *Int.* K 14.678.18–697.2, but inappositely. See also Petrocelli 2008: 234–5.

[55] Korzenzsky–Vári print ὑπολαμβάνουσαι with some MSS; this reading obfuscates the important relationship between expectation/anticipation (ἐλπίζειν, ἐκδέχεσθαι) and φόβος maintained throughout *Str.*

[56] With προλαμβάνουσαι, κ.τ.λ., Peters 1972: 209 compares Thuc. 5.9.8; apt also is Aen. Tact. 38.2.

Apart from tactics in pitched battle, the PFF is also important in siege warfare, perhaps unsurprisingly, since the siege was recognized as one of the most psychologically intensive moments of warfare in Greco-Roman antiquity.[57] Yet Onasander's treatment differs from other military and nonmilitary authors' remarks on the emotion of the siege insofar as his is founded on psychological principles developed already in other areas of war. These principles accordingly invite the possibility for richer explanations of certain phenomena. Consider, for example, his account of why only the besiegers, never the besieged, can carry out a strategy of continual assault (42.7–13). This strategy consists in establishing a system for rotating fresh troops into the ranks of the attackers so that the pitch of the assault remains constant.[58] Onasander explains (42.12):

> τοὺς μέντοι πολιορκουμένους, μηδ᾽ ἂν πάνυ πολλοὶ τυγχάνωσιν, οἰέσθω τις τὸ αὐτὸ στρατήγημα ἀντεισοίσεσθαι· τὸ γὰρ κινδυνεῦον, οὐδ᾽ ἂν ἐπιτρέπῃ τις, ὕπνῳ χαρίζεσθαι βούλεται· φόβῳ γὰρ τοῦ δεινοῦ, παρ᾽ ὃν ἀναπαύεται χρόνον, ὡς ἁλωσομένης τῆς πόλεως ἐγρήγορε· καὶ τὸ πολιορκούμενον, κἂν ὀλίγον ᾖ τὸ πολιορκοῦν αὐτό, πασσυδὶ προσαμύνει, καὶ πᾶν ὅσον ἐντειχίδιόν ἐστι κεκίνηται, ὅτι καὶ τὸ μέλλον φοβερώτερον, ὡς, εἰ παρὰ μικρὸν ἀμελήσαιεν, ἀπολούμενοι πάντες.
>
> But no one should think that the besieged, even if there are very many of them, could implement the same strategy (sc. of continual assault) in return. For those in danger are unwilling to enjoy sleep, even if permitted them, because they remain awake for fear of the terrible event that the city will be captured while they are resting. Furthermore, even if the besieging force itself is small, all hands among the besieged come to the defense and the entire city is disturbed throughout, because indeed the future is more frightening to them, since they believe that if they are remiss in any small respect, they will be ruined completely.

The result of the constant tension is that the besieged people, weary and fearful for the future (42.13 κάμνοντας . . . καὶ πρὸς τὰ μέλλοντα τεταλαιπωρηκότας), will surrender.

On the one hand, we can understand Onasander's explanation for this strategy without appealing to the PFF: The stakes really are in a sense higher, "the future indeed more frightening" (cf. 42.12 καὶ τὸ μέλλον φοβερώτερον), for the besieged, since their lives and property are threatened in a way not

[57] See Levithan 2013, esp. 47–79; also Chaniotis 2013: 441–3, 451–4.

[58] The exact rotation varies based on the number of troops and the size of the city besieged (cf. 42.7): supposing that there are five cohorts of troops, then while the first division attacks, the second will stand ready, and the third, fourth, and fifth rest; the first will be relieved after an interval by the second, and the third made ready. This rotation continues day and night (42.11).

true for the besieging.[59] Yet his account of the psychological state of the besieged is more satisfying if we suppose that the principle is operative: the defenders are unable to rest, constantly fearing the worst and imagining that the city will be taken before they awake (φόβῳ ... τοῦ δεινοῦ ... ὡς ἁλωσομένης τῆς πόλεως); the entire citizenry is in commotion, and, with a touch of hysteria, it is believed that even a small (παρὰ μικρόν) error will result in utter ruination (cf. ἀπολούμενοι πάντες) – all of this "even if the besieging force itself is small" (κἂν ὀλίγον ᾖ τὸ πολιορκοῦν αὐτό), suggesting that the defenders exaggerate the threat. Reading the PFF into Onasander's remarks here bolsters the credibility of his precepts and connects them with the psychological account developed across *Str.*

6.5 The Advantages and Limits of Abstraction

In sum, Onasander's *Str.* embodies an intensely abstracting approach to generalship and the art of war. He sets out his precepts in a highly generalized fashion; they are in turn grounded in a set of universal psychological principles ultimately accounting for certain phenomena in war. While Onasander says that his teachings are drawn from the material of history, he never descends to the level of the particular or historically real; rather, the principles are taken as the undemonstrated basis for a *theôria* of warfare, as he systematizes principles and precepts into a *technê*. There are noteworthy advantages to Onasander's abstracting approach. First, he can explain specific military precepts in terms of basic facts about human nature (psychology) that his readers could be reasonably presumed to accept. This presumption of consensus, strengthened by Onasander's claim to have drawn his principles from history, means that his explanations double as justifications, appealing to plausible intuitions and observations to persuade readers that the precepts founded upon them are sound. Second, the universalizing character of Onasander's psychological principles suggests that the general can acquire a broadly applicable understanding of military command if he masters a few certain facts about human behavior; he need not contemplate all possible battlefield situations as discrete events if he can instead reduce them to a smaller number of psychological fundaments. Third, the widespread reuse of principles throughout Onasander's work brings out the systematic structure of the knowledge of generalship, providing

59 For the treatment of an enemy town taken by siege, see Gilliver 1996: 222–6; 1999: 154–60; Kern 1999: 323–51.

a coherent and economical organization of material: here is a distinct benefit, both for the author and the audience, in a textualized art of war. Finally, the explanatory bent of Onasander's advice, which furnishes not just precepts but also ground-level reasons for those precepts, yields benefits akin to those enjoyed by Vitruvius' approach to architecture (Chapter 3): the careful derivation of precepts from secure first principles sets the *ars* on a sound epistemic foundation.

From *Str.* we can also plausibly frame some conclusions about the Latin *artes* of war in the first century AD, for Celsus' and Frontinus' writings are likely to have resembled Onasander's work in adopting an abstracting approach to the discipline. The highly systematic and generalizing character of Celsus' *DM* and fragments from other parts of the *Artes* implies a similar treatment for warfare, and we have already observed the reasons for supposing that Frontinus' treatise looked in some respects like Onasander's. It is, of course, impossible to say how much their works departed from Onasander's. Celsus or Frontinus – or both – might, for example, have included more historical illustration than Onasander or adopted a more culturally specific (Roman) perspective.[60] They might also have eschewed Onasander's interest in human psychology in favor of alternative organizing principles, or sought to ground or explain their precepts by other means. Be that as it may, any sufficiently general and systematic approach to the art of war was bound to be marked by features evident in *Str.* (and which were at stake in other *technai*/*artes*, too). As we saw in Chapter 5, it was possible for Celsus to reject the Rationalist view of medicine that put knowledge of nature at the root of expertise and still to produce a systematic *ars* in which reasoned explanation and universalized rules are indispensable.

Its advantages notwithstanding, an abstracting approach to the art of war would have also faced certain global limitations. In Chapter 5 we saw that the Rationalists, who attempted to derive definite rules for medicine from an explanatory account of nature, were attacked, inter alia, for their pretensions to certainty and universality. It might be guessed that Onasander's method would have faced analogous criticisms – and, in fact, there is a good deal of evidence that the Roman art of war was marked by tensions between the theoretical ideals embodied in generalized military precepts and the realities of actual warfare. These tensions were given voice in various rhetorical and historiographic contexts in the late Republic and

[60] Probably Frontinus' treatise did not have many *exempla* however (Ch. 6.1 and n.26), and if *DM* is anything to go by, Celsus' military books likely had few if any as well.

articulated in more sophisticated ways by the *scriptores rei militaris* themselves. In fact, basic questions about the character and efficacy of the written art of war endured at Rome through Frontinus' time and importantly influenced the form of his *Strategemata*.

We will begin by briefly considering the sentiments regarding military *praecepta* and experience voiced by Cicero and Sallust to gain a rough appreciation for some of the relevant criticisms. We will then return to Onasander, who acknowledges that the precepts in *Str.* are not always sufficient to guide military action owing to the ineliminable uncertainties of warfare. For Onasander, these uncertainties crystallize in the "stratagem," the brilliant piece of strategy that exploits the contingencies of circumstance to great effect. Because the stratagem is always particular and dependent on a more or less unique configuration of circumstances, it paradigmatically resists incorporation into a *technê* (*ars*). While this problem is developed at greatest length by Onasander, he is not alone in recognizing it: both Vitruvius (on siege warfare) and Quintilian (on education and the *artes*) recognize versions of this difficulty for the written *ars* of war. Importantly, both authors also suggest that it can be remedied by an alternative approach which privileges historical *exempla* over generalized *praecepta*. I argue that Frontinus' *Strat.* should be understood in the light of the tradition represented in Vitruvius, Onasander, and Quintilian, and that it represents precisely the kind of compilation of military *exempla* indicated by Vitruvius and Quintilian – indeed, the first such independent Latin collection of which we know. *Strat.* thus complements Frontinus' earlier work on the *scientia rei militaris*, redressing the shortcomings of the abstracting approach to the art of war.

6.5.1 Popular Roman Criticism of Military Writing

By the late Republic, Greek and Latin literature on military themes that could be read by those interested in the art of war – whether *artes* proper, historiographic writings, or something else – was popular enough at Rome to be controversial. As much is indicated by passing remarks in authors such as Cicero and Sallust, whose criticisms capture one important complaint about the written art of war: that the discipline was essentially hollow or incomplete absent practical experience.[61] To be sure, this

[61] Cf. n.1 (experience in other *artes*); on the following passages (and others), see also Campbell 1987: 21–3; Bettalli 2010: 216–19; Formisano 2011: 43–4; König 2017: 163–7; and esp. Formisano 2017: 129–39.

complaint could be leveled against any *ars*, but the political significance of military command for Romans brought special attention to the discipline. In *Pro Fonteio*, for example, Cicero laments the poor quality of military men of his day, contrasting them with generals of the previous generation, men "very experienced in waging war" (43 *belli gerendi peritissimos*) who were "educated in the knowledge of warfare not by books, but by their exploits and victories" (*non litteris ... ad rei militaris scientiam, sed rebus gestis ac uictoriis eruditos*).[62] As if in confirmation, Sallust's Marius remarks disapprovingly that he knows of *nobiles* who "after they were made consuls began to read about the deeds of their forefathers and the military precepts of the Greeks" (*Iug.* 85.12 *qui postquam consules facti sunt et acta maiorum et Graecorum militaria praecepta legere coeperint*). He dismisses such men as follows (85.13–14):

> quae illi audire aut legere solent, eorum partem uidi, alia egomet gessi; quae illi litteris, ea ego militando didici. nunc uos existumate facta an dicta pluris sint.

> What they are used to hearing or reading about, some of those things I actually witnessed, others I did myself; what they learned from books, I learned from military service in the field. Consider now whether deeds or words are worth more.

Cicero's Pompey might have said the same: he "waged more wars than others have read about" (*Leg. Man.* 28 *plura bella gessit quam ceteri legerunt*), Cicero says in praise of his military ability, and "as a young man was educated in the art of war not by the precepts of others but by his own commands" (*cuius adulescentia ad scientiam rei militaris non alienis praeceptis sed suis imperiis ... est erudita*).[63] Elsewhere, Cicero considers it a point of national pride that Roman accomplishments on the battlefield fell into the category of what was achieved *non litteris* but by prowess, discipline, and natural character or spirit (*natura*) (*Tusc.* 1.2):[64]

> quid loquar de re militari? in qua cum uirtute nostri multum ualuerunt, tum plus etiam disciplina. iam illa, quae natura, non litteris adsecuti sunt, neque cum Graecia neque ulla cum gente sunt conferenda.

62 Cf. Dyck 2012: 74–5, comparing *Verr.* 2.5.2, 2.5.25, *Leg. Man.* 28 (quoted in this section), *De or.* 3.136.

63 *Praecepta* need not refer exclusively to written rules, but it very likely does here: cf. Sallust's *Graecorum militaria praecepta ... legere* (quoted earlier), and for *praecepta* "de ipso opere vel libro," *TLL* X 2.456.69–457.3 (Baumgartner). Cf. Quint. *Inst* 2.5.15 *praecepta communia*, discussed in Ch. 6.5.3.

64 For *natura* here, see *OLD* 12a; *TLL* IX 1.157.56–67 (Hajdú); for *Tusc.* 1.pr, Gildenhard 2007: 90–156 (*Tusc.* 1.2 on pp. 109–30).

> What is there to say about war? In our prowess as well as our discipline we (sc. Romans) have proved our worth here. Indeed, we are superior not simply to Greece but to every nation in those things which are attained by natural character, not by writing.

This theme finds a different setting in a memorable anecdote in *De oratore*. After the Peripatetic Phormio was received by Hannibal, "he is said to have spoken for some hours on the duty of the general and on warfare" (2.75 *locutus esse dicitur . . . aliquot horas de imperatoris officio et de re militari*).[65] Hannibal thought Phormio deranged, and rightly so, Cicero has Catulus say (2.76):

> quid enim aut adrogantius aut loquacius fieri potuit quam Hannibali, qui tot annis de imperio cum populo Romano omnium gentium uictore certasset, Graecum hominem, qui numquam hostem, numquam castra uidisset, numquam denique minimam partem ullius publici muneris attigisset, praecepta de re militari dare?
>
> For what could be more insolent and prattling than for a Greek man, who had never seen an enemy or a military camp, had never in the slightest part engaged in public service, to issue precepts about the art of war to Hannibal, who had for many years vied for power with Rome, conqueror of all nations?

The statements gathered in this section (more could be added[66]) intersect with broader issues, such as Roman attitudes toward the Greeks and popular views on *dicta*/*uerba* and *facta*/*res*, but what is germane here is that written *praecepta* could be a flashpoint for controversy about the possibility of teaching and learning the *res militaris*. That these references are drawn mostly from rhetorical contexts does not vitiate the point, but rather shows that general agreement could be expected.

6.5.2 *Onasander's Optimistic Epistemology and the Limits of* technê

Needless to say, such opinions did not deter Greeks or Romans from attempting to write *artes* of war (nor, we presume, Romans from reading them). Yet the *scriptores rei militaris* did acknowledge concerns about the applicability of textualized *praecepta* and emphasize the limitations of their works. To some extent this response may have resulted from

[65] On the episode (Catulus' rejoinder to Antonius), see Leeman et al. 1981–2008: II 288–90; Li Causi et al. 2015: 474–5.
[66] See n.61.

extradisciplinary pressures, but it was also a conclusion that could be reached by the internal logic of the subject matter. No author illustrates this better than Onasander. The tension between generalized precepts and practice is present in specific ways throughout *Str.* It often takes the form of the ubiquitous "if" or "when" (εἰ, ἐάν/ἄν) with which Onasander prefaces his advice: *if* the general must march by night . . . (10.22); *if* there is uneven land . . . (18); *if* light troops are lacking . . . (20); *if* the enemy are superior in cavalry . . . (31); *if* the general is defeated . . . (36). These "if" clauses, along with other linguistic devices, signal the conditional nature of Onasander's advice: they are an admission that the *stratêgos* will not always encounter the situations that are described in the text – in short, that there is chance (τύχη) and uncertainty in any conflict.[67]

Most interesting, however, are moments when the tension comes to the fore in the shape of a self-conscious statement about the variable nature of war. Among numerous examples,[68] consider Onasander's preamble to the discussion of battle formations (τάξεις) (15):

> τάξις δ' οὐ μία πολέμου, πολλαὶ δὲ καὶ διάφοροι καὶ παρὰ τοὺς ὁπλισμοὺς καὶ παρὰ τοὺς στρατευομένους καὶ παρὰ τοὺς τόπους καὶ παρὰ τοὺς ἀντιπολέμους, ὧν τὰς διαφορὰς ὁ στρατηγὸς ἐπ' αὐτῶν εἴσεται τῶν καιρῶν· ἃ δ' ἂν οὐχ ἥκιστα πολλαῖς ἁρμόζοι παρατάξεσι δίχα τῶν ἐπ' αὐτῶν τῶν πραγμάτων ἀνάγκην ἐχουσῶν νοεῖσθαι, ταῦθ' ὡς ἐν κεφαλαίῳ δίειμι.
>
> There is not one formation for battle, but many kinds that vary according to armament, soldiers, place, and enemy; the general will in the actual moment apprehend the differences in these things. I will give a summary account of what is appropriate for most battle formations, setting aside those things which require consideration in the actual circumstances.

As we observed, it is typical of Onasander's approach to omit the particularities of a situation in framing his advice. In this passage – tantamount to an admission of this method – he promises to relate what is relevant to "most battle formations" (πολλαῖς . . . παρατάξεσι), ignoring differences (διαφοράς) in armament, soldiers, place, and enemy. Yet this is precisely where Onasander could be challenged: every battle is determined by actual configurations of armament, soldiers, place, and enemy, and his precepts insofar as they apply to "most" (cf. πολλαῖς) formations do not describe

[67] Other hedging expressions (not exhaustive): ἂν δυνατὸν ᾖ (6.2), ἔστιν ὅτε vel sim. (6.5, 6.7), ἐνίοτε (6.5, 42.15), μάλισθ' ὅτ' ἂν (6.7, 22.2), πολλάκις (6.7, 6.9, 6.11, 10.14, 10.20, 11.1, 12.2, 27), ἐπειδάν / ὅτ' ἂν (7.1, 9.3. 13.1, 14.1), κἂν vel sim. (8.1, 10.27), ποτε (21.9, 23.1, 29.1, 29.3).

[68] Besides 15, 21.4, 32.9–10 (discussed in this section), cf. 6.12, 11.2, 14.2, 16, 18, 30, 31.1, 42.3.

any specific situation. How is the general to adapt himself to the myriad circumstances that might appear before him? The abstraction that from one perspective might appear a virtue could from this angle appear a flaw.

Onasander is of course cognizant of the need for the general to recognize and react to the unique circumstances of a conflict. His remarks in this vein touch on the dichotomy between *tuchê* and *technê* handled in the preface. Consider his statements about terrain (21.4):

> οὐ μὴν ἡ στρατηγικὴ φρόνησις ἐνταῦθα συλλαμβάνεται μόνον, ἀλλὰ καὶ ἡ τύχη· δεῖ γὰρ ἐπιτυχεῖν τοιούτων χωρίων· οὐ γὰρ αὐτῷ γε κατασκευάσασθαι δυνατὸν τοὺς τόπους· τῶν ὄντων μέντοι τοὺς ἀμείνους ἐκλέξασθαι καὶ τοὺς συνοίσοντας ἐννοῆσαι φρονίμου.
>
> It is not the practical wisdom of generalship alone which is relevant here, but chance too, because one must chance upon such (sc. advantageous) country and one cannot prepare the terrain for oneself; yet it is the part of the wise (sc. general) to choose the better of what is present and to recognize which will be advantageous.

As acknowledged in the preface, *tuchê* has no inconsiderable significance for the general's efforts; yet it is far from determining the outcome, and the general's *phronesis* (practical wisdom) operates on what *tuchê* has furnished to maximize the chances of success. Onasander here applies a version of this statement to the selection of a topographical situation. Even so, the basic problem remains: How is the general actually to "recognize" (ἐννοῆσαι) and choose what is better?

In the face of this quintessential problem of theory versus practice, Onasander adopts a strikingly optimistic epistemology: he does not regard as fundamentally problematic the cognitive work of real-time decision-making, whereby the general correctly infers or deduces the appropriate application of a generalized rule to the specific situation. That is, for Onasander theory appears fundamentally *actionable*, in the sense that knowledge of the universals automatically equips one to choose prudently among particulars. This viewpoint is virtually given as a slogan when he is discussing the right way to encamp the army (10.19–21). After advising that the general should compress his army into a small camp so as to induce rash action from the enemy, he claims that "he (sc. the general) who has such knowledge will know how to manage his command" (10.21 ὁ δὲ εἰδὼς οὕτως στρατηγεῖν εἴσεται). This is not meant tautologically (that a general is a general in virtue of knowing what he knows), but as an assertion of the general's ability to encamp properly in any given situation in virtue of knowing this precept. Hence, Onasander continues by maintaining that, even if the enemy commander employs the same tactic, nonetheless the general will, thanks to his knowledge, be "wise in taking action and prudent in defending himself" (10.21 δρᾶσαί τι φρόνιμος … καὶ φυλάξασθαι προμηθής). Elsewhere Onasander similarly qualifies precepts with statements that elide the difference between theoretical knowledge and situational decision-making. In the choice of topography (21.4), the whole of Onasander's advice is to exploit terrain, either a river or mountains, to prevent the army's being encircled (21.3): this rule is sufficient to make the commander "shrewd" (21.3 ἀγχίνους) or "wise" (21.4 φρονίμου). As a final example, Onasander later compresses the complex factors that determine whether and how the general's soldiers should be allowed to plunder the defeated enemy into the brief admonition that "he will know what is best in the moment" (35.3 γνώῃ δ' ἂν αὐτὸς ἄριστα πρὸς τοὺς καιρούς).

But do Onasander's generalized precepts really confer such "practical wisdom of generalship" (21.4 ἡ στρατηγικὴ φρόνησις) enabling the right decision in any given situation? In his defense, Onasander does discuss alternative strategies tailored to different situations, as we have already pointed out: such variability is signaled by conditional clauses, among other linguistic devices (especially the innocuous ἤ, "or," listing alternatives), and it is clear that when he hedges his advice with a phrase such as ἐὰν ᾖ δυνατόν (31.1, "if it is possible"), he means to introduce a measure of flexibility. Even so, it is equally clear that the totality of such qualified statements in *Str.* can hardly cover the contingencies of war, since *Str.* does not and could not possibly embrace all the unique circumstances that the general might encounter. The general is bound to wage war in situations differing in ways small and large from those found in *Str.*, as well as those in which different precepts might clash. The epistemology of *Str.* is "optimistic" in that Onasander places much confidence in the general's capacity for spontaneous decision-making to resolve these situations. It seems a necessary postulate of *Str.* that the "wise" (φρόνιμος) or "clever" (ἀγχίνους) general will by and large deduce the correct course of action from the precepts stored in his memory.

An important chapter in *Str.* makes explicit the aforementioned postulate and sheds light on the limits of the commander's spontaneous judgment. The issue crystallizes in Onasander's discussion of the gambits that generals sometimes use to encourage their troops to fight more vehemently, such as destroying their own defenses, crossing rivers, or placing the army with its back to a cliff (32.1). These pieces of strategy have less to do with "thought" (32.2 γνώμη) and "judgment" (κρίσις) than with "recklessness" (τόλμη) and "chance" (τύχη). Accordingly, to venture such ploys is to "roll the dice with uncertain chance" (32.3 τὴν ἄδηλον ἐκκυβεύειν τύχην) – perhaps a legitimate risk when the ruin of the army is manifest (32.5), but otherwise an unconscionable gamble. Having thus emphasized the chancy and desperate nature of such gambits – henceforth we will call them simply *stratêgêmata* (στρατηγήματα, "stratagems")[69] – he explains in the following sections how they may be hazarded (32.5–8). Then, in an important concluding passage (32.9–10), he steps back to compare tactics planned in advance and those improvised in the midst of battle. The latter, evidently referring to the *stratêgêmata* that are the subject of the chapter, have a peculiar and sometimes brilliant nature quite unlike that of prepared strategies (32.9–10):

> τῶν δ' ἐκ προλήψεως καὶ πρὶν ἢ συμβαλεῖν ἐπινοουμένων στρατηγοῖς[70] αἱ παρ' αὐτὸν τὸν τῆς μάχης καιρὸν ἐπίνοιαι νίκης καὶ ἀντιστρατηγήσεις ἔστιν ὅτε καὶ πλείους καὶ θαυμασιώτεραι γίγνονται τοῖς τὴν στρατηγικὴν ἐμπειρίαν ἠσκηκόσιν, ἃς οὐκ ἔστιν ὑποσημῆναι λόγῳ ἢ προβουλεῦσαι. [10] ὥσπερ γὰρ οἱ κυβερνῆται πρὸς μὲν τὸν πλοῦν ἐκ λιμένων ἀνάγονται πάντα ἐξηρτυμένοι τὰ κατὰ τὴν ναῦν, ἐπειδὰν δ' ἐμπέσῃ χειμών, οὐχ ὃ βούλονται ποιοῦσιν, ἀλλ' ὃ ἀναγκάζονται, πολλὰ καὶ πρὸς τὸν ἀπὸ τῆς τύχης ἐπείγοντα κίνδυνον εὐτόλμως παραβαλλόμενοι, καὶ οὐ τὴν ἀπὸ τῆς μελέτης εἰσφερόμενοι μνήμην, ἀλλὰ τὴν ἐκ τῶν καιρῶν βοήθειαν· οὕτως οἱ στρατηγοὶ τὴν μὲν δύναμιν ἐκτάξουσιν,[71] ὅπως σφίσι νομίζουσι συνοίσειν, ἐπειδὰν δ' ὁ τοῦ πολέμου περιστῇ χειμὼν πολλὰ θραύων καὶ παραλλάττων καὶ ποικίλας ἐπάγων περιστάσεις, ἡ τῶν ἀποβαινόντων ἐν ὀφθαλμοῖς ὄψις ἐπιζητεῖ τὰς ἐκ τῶν καιρῶν ἐπινοίας, ἃς ἡ ἀνάγκη τῆς τύχης μᾶλλον ἢ ἡ μνήμη τῆς ἐμπειρίας ὑποβάλλει.
>
> For those who have acquired much experience in generalship, sometimes plans for victory and counter-stratagems[72] (sc. conceived) in the actual moment of battle are superior to and more marvelous than those conceived by generals in anticipation and before battle is engaged. It is impossible to indicate such maneuvers in writing (*or* by reason, λόγῳ), or to frame them beforehand. [10] So consider the ship's captain, who makes all the appropriate preparations for his vessel before leaving harbor for a voyage but when a tempest strikes must act as the situation demands rather than as he had

[69] Onasander uses the word thus at 32.4 (cf. 42.12); cf. n.38. Stratagems as such are discussed later in this chapter.

[70] For στρατηγοῖς Korzensky conjectures στρατηγιῶν, a change which would make τῶν easier to construe; I retain and treat it as a dative of agent with ἐπινοουμένων, admittedly rare with verbs not in the perfect or pluperfect (Smyth §1490; Schwyzer II 150).

[71] Lucarini 2010: 225 argues for ἐκτάσσουσιν, pointing to the tense of the nearby verbs.

[72] For ἀντιστρατήγησις (an ἅπαξ) = "counter-stratagem," see Wheeler 1988: 9–10; cf. Ercolani 1997: 48.

> planned; he courageously ventures whatever is required against the danger that has by chance fallen upon him, relying not on memory born of experience but whatever source of aid there is in the moment. In the same way, the general will array his forces however he thinks will be most advantageous, but when the tempest of war is upon him, dashing and altering his plans and introducing chaotic changes in circumstance, the sight of the events before his eyes seeks plans according to the moment, suggested more by the necessity of chance than by memory of accumulated experience.

The key difference between stratagems and ordinary pieces of generalship lies in the fact that stratagems cannot be accounted for by generalized rules: Onasander says that it is impossible to indicate (32.9 ὑποσημῆναι) *stratêgêmata* in or by *logos* or frame them in advance (προβουλεῦσαι); by λόγῳ he probably means not merely "in writing" or "speech," but "in argument" or even "by reason."[73] It follows that it is impossible for the general to retrieve and apply the right rule from the "memory of accumulated experience" (32.10 ἡ μνήμη τῆς ἐμπειρίας).[74] So how does the general decide on an appropriate course of action? Onasander proposes a sort of perceptual cognition (ὄψις) responsible for invention in the face of the uncertain particulars of the "tempest of war."

Despite the emphasized differences, *stratêgêmata* in fact share more with broader issues concerning the art of war discussed here than might first appear. This is because, from a different vantage point, these "plans for victory and counter-stratagems (sc. conceived) in the actual moment of battle" are simply a special case of the spontaneous decision-making that the general requires in other situations on the battlefield, too.[75] As we saw, even where generalized rules can be given (as apparently they cannot for στρατηγήματα), a gap remains between the universals and what the

[73] The unusual usage of ὑποσημαίνειν (some MSS have a form of ἐπισημαίνειν) is another example of Onasander's fondness for using military terminology figuratively (nn.32, 34): the verb regularly means "give the signal" in battle (LSJ 2).

[74] His initial proviso that experienced generals (32.9) are more capable of devising στρατηγήματα shows, however, that practice/experience is somehow related to improvisatory ability; the dismissal here of ἡ μνήμη τῆς ἐμπειρίας simply establishes that experience cannot provide any model for the situation, since there is no correlate in the general's memory. This claim will be contested implicitly by Frontinus: see Ch. 6.6.

[75] The language describing improvised strategies recalls earlier passages concerning the variable particulars in warfare that we have considered: (1) the "actual" or "opportune moment" of battle (15 τὰς διαφορὰς . . . ἐπ' αὐτῶν . . . τῶν καιρῶν + 35.3 πρὸς τοὺς καιρούς ~ 32.9 παρ' αὐτὸν τὸν τῆς μάχης καιρὸν); (2) chance or fortune (21.4 ἡ τύχη, ἐπιτυχεῖν ~ 32.10 *bis* τῆς τύχης); (3) necessity (15 τῶν ἐπ' αὐτῶν τῶν πραγμάτων ἀνάγκην ἐχουσῶν νοεῖσθαι ~ 32.10 ὃ ἀναγκάζονται, ἡ ἀνάγκη); (4) and multiplicity or variety (15 πολλαὶ δὲ καὶ διάφοροι ~ 32.10 πολλὰ θραύων καὶ παραλλάττων καὶ ποικίλας ἐπάγων περιστάστεις).

general will actually encounter and how he will act in the field. It is no more possible to capture in *logos* the infinity of particulars in warfare than to capture the infinity of situations in which *stratêgêmata* might be deployed – something Onasander often implicitly admits when declining to enumerate all the possible situations with which the general must deal.[76] Granting this similarity, we can see *opsis* (sight) as analogous to the intellectual capacity – we will call it *phronêsis*, for convenience – elsewhere assumed to allow the general to turn universal rules into action. To the extent that they are different, *opsis* is presented as chiefly inventive, devising plots or schemes in the moment, while *phronêsis* is chiefly deliberative, taking counsel on how to apply a general rule. But both faculties operate at the level of spontaneous decision-making and both suggest actions appropriate to a unique situation.

Onasander's discussion of *opsis* shows that he is aware of the limitations of a textual account of warfare vis-à-vis stratagems. The likeness between the problem of devising stratagems and that of deliberating over particulars or battlefield differentiae (διαφορά) suggests that he is also cognizant of the broader limitations of his abstracting approach. Just as he posits *opsis* as a faculty of creative invention on the battlefield, so *phronêsis* (still understood narrowly as a faculty of spontaneous decision-making) must be postulated as a means of translating general rules into action. But this implies that *phronêsis* thus lies, like *opsis*, somewhere outside *logos*, and hence outside the *theôria* of his work. This explains why Onasander only gestures at the variety in real-world warfare: The flexibility of his instructions, conveyed in disjunctives and conditionals, may in some measure address this complexity, but it can never wholly close the gap. We are thus left with the general's *phronêsis* as an optimistic postulate. This conclusion is perhaps less a flaw than a feature of *Str.* – a necessary tradeoff of the abstracting method in exchange for its virtues.

6.5.3 *From* praecepta *to* exempla*: Vitruvius and Quintilian on Military Instruction*

Onasander thus demonstrates sensitivity to the limits of an abstract *technê* or *ars* of war. But while the postulate of a faculty for situational, spontaneous decision-making defines these limits, it does not address head-on how

[76] See the remarks earlier in this section and n.68. Xenophon, perhaps influencing Onasander, makes a point about the impossibility of textualizing precepts that applies not merely to some actions but all matters regarding the cavalry commander's skill: *Eq. mag.* 9.1 γράψαι δὲ πάντα ὁπόσα δεῖ ποιεῖν οὐδὲν μᾶλλον οἷόν τέ ἐστιν ἢ τὰ μέλλοντα πάντα εἰδέναι.

the reader may cultivate this capacity. We can investigate this crucial articulation between theory and practice if we cast a wider net in the Latin polemological tradition. One promising source is Book X of Vitruvius' *DA*, which is devoted to the construction of machines (*machinatio*).[77] The latter part of this book (10.9.7–16.12) discusses engines of war, euphemistically described as those machines "which have been invented as a defense against danger and for securing safety" (10.10.1 *quae ad praesidia periculi et necessitatem salutis sunt inuenta*).[78] After discussing various siege weapons (10.10–15), Vitruvius turns from their construction to defense against them.[79] He adds this remarkable qualification (10.16.2):

> itaque his praescriptionibus si qui attendere uoluerit <et> ex uarietate eorum eligendo in unam comparationem conferre, non indigebit auxiliis, sed quascumque res ex rationibus aut locis opus fuerit, sine dubitatione poterit explicare. de repugnatoriis uero non est scriptis explicandum. non enim ad nostra scripta hostes comparant res oppugnatorias, sed machinationes eorum ex tempore sollerti consiliorum celeritate sine machinis saepius euertuntur.

> Therefore if anyone attends carefully to these instructions (sc. those given in 10.10–15) and in choosing among their variety undertakes to compare them all at once, he will not be at a loss, but will rather be able to explain without hesitation which will be required on the basis of the circumstances or localities. Defensive measures, however, cannot be explained in writing. For our enemies do not prepare their siege weaponry according to our writings, but their engines are more often thwarted by swift and clever planning in the moment, without the aid of engines of our own.

Vitruvius identifies a problem for *res repugnatoriae* very similar to that Onasander would describe for *stratêgêmata*. Although he believes that his precepts for siege engines afford the sort of explanatory knowledge that

[77] In *DA* 10.1 Vitruvius gives the principal types of machines; in 10.2–9 he details the workings of various machines (Callebat and Fleury 1986: xxii–xvi).

[78] The euphemism, not uncommon among mechanical writers, is a species of the "classic paradox" *si uis pacem, para bellum* (Milner 1996: 63n3; cf. Otto 1890: 54). Vitruvius' contemporary Athenaeus, with whom he may have been in competition (Callebat and Fleury 1986: xxvii–xxxi; Whitehead and Blyth 2004: 28–31; Cuomo 2011: 315–16), argues (39) that his belopoetic treatise aims not at the destruction (ἀναίρεσις) of other cities but the defense (ἀσφάλεια) of one's own. Hero takes the cake when he describes βελοποιία as a sure way of obtaining the "greatest and most necessary part of the study of philosophy, that concerning peace of mind (περὶ ἀταραξίας)" (*Bel.* 1); see Pleket 1967: 4; Marsden 1971: 44; Krafft 1973: 13.

[79] Vitruvius' account of siege weaponry owes, like that of Athenaeus, to the first-century BC mechanician Agesistratus (Thiel 1895, and Marsden 1971: 4–5; Callebat and Fleury 1986: xxvii–xxxi; Whitehead and Blyth 2004: 25–9, 172–4), though the defensive measures may be his own (cf. Ath. Mech. 39, with Whitehead and Blyth 2004: 168).

allow for an a priori account of the right weapon for a given situation (cf. *ex rationibus aut locis*), the same cannot be furnished for *res repugnatoriae*. Thus, just as Onasander maintains that it is impossible to "indicate (sc. stratagems) in writing (*or* by reason), or to frame them beforehand" (*Str.* 32.9 ὑποσημῆναι λόγῳ ἢ προβουλεῦσαι), Vitruvius avers that defensive matters "cannot be explained in writing" (10.16.2 *non est scriptis explicandum*). Moreover, Vitruvius, like Onasander, argues that the expert (here *architectus*, rather than *dux*/*stratêgos*) must fall back on improvisatory skill. For Onasander this capacity is *opsis* (or, in other situations, *phronêsis*), while for Vitruvius it is *sollertia* (10.16.8, 10.16.12), the innate faculty represented throughout *DA* as responsible for invention, discovery, and extemporaneous decision-making.[80] So far, Onasander and Vitruvius seem to be in lockstep – but they differ in one illuminating respect.

Whereas Onasander leaves his discussion at emphasizing that memory (μνήμη) and experience (ἐμπειρία) are insufficient (*Str.* 32.10), Vitruvius amplifies on *sollertia* in a striking way: by the provision of historical *exempla*. He goes through six anecdotes that exhibit the working of the architect's *sollertia*. The first and longest (*DA* 10.16.3–8) tells of Diognetus, an architect of Rhodes, who foils Demetrius Poliorcetes' attempt to conquer his city using an enormous siege tower.[81] At night, Diognetus directs the city secretly to pump water and liquid waste onto the earth before the city walls; when the attackers move the tower forward the next day, it sinks under its own weight in the morass and is rendered useless. In the second *exemplum* (10.16.9), Chian citizens facing a naval assault secretly create underwater barriers; the enemy's ships, once grounded, are easily destroyed.[82] In the third (10.16.9–10), the architect Trypho devises a cunning scheme to detect mines (subterranean passages) with bronze sounding vessels and is thereby able to kill enemy sappers before they breach the city of Apollonia.[83] The fourth, fifth, and sixth *exempla* (10.6.11–12) concern the Massilians, who flood enemy mines, set fire to a siege ramp, and

80 See Ch. 3, n.24.

81 For the siege (305–4 BC), see Diod. Sic. 20.82–100, Plut. *Vit. Demetr.* 21–2, with Marsden 1969: 105–8; for Demetrius' tower (ἑλέπολις), Marsden 1971: 84–5; Garlan 1974: 228–34, Callebat and Fleury 1986: 281; Whitehead and Blyth 2004: 134–5; for the episode in *DA*, König 2009: 48–9; Nichols 2017: 26–8; Oksanish 2019: 88–93.

82 This siege, mentioned probably also by Ath. Mech. 27 (cf. nn.78, 79), may be that of 412–411 BC by the Athenians (Thuc. 8.40, 8.56, 8.61) or 201 BC by Philipp V (Polyb. 2.16.1 with Walbank 1957–79: II 503–4); cf. Callebat and Fleury 1986: 286.

83 Possibly the siege by Philipp V in 214 BC; Trypho is otherwise unknown (Callebat and Fleury 1986: 287–8).

snare and upturn an approaching ram.[84] Vitruvius concludes by remarking of the *exempla* that "these victorious cities were delivered not by machines, but against the contrivance of machines, by the cleverness of architects" (10.16.12 *eae uictoriae ciuitatum*[85] *non machinis, sed contra machinarum rationem, architectorum sollertia sunt liberatae*). Although he has disavowed explanation (*explicare*) of these defensive stratagems (10.16.2) and accordingly does not isolate any general *rationes* on which they are based, he evidently expects that the *exempla* will impart to the reader a better sense for the architect's *sollertia*. That is, the *exempla* stand in for the rules which are impossible to communicate.

Vitruvius does not spend much time discussing the didactic value of his *exempla*, but precious evidence from the *Institutio oratoria* of Quintilian, a contemporary of Frontinus, sheds light on his approach. In Book II, when discussing the merits of a pedagogical approach that requires the pupil to answer questions put by the instructor, Quintilian compares the situation in oratory with that in the *res militaris* (2.5.14–15):[86]

> hoc diligentiae genus ausim dicere plus conlaturum discentibus quam omnes omnium artes, quae iuuant sine dubio multum, sed latiore quadam comprensione per omnes quidem species rerum cotidie paene nascentium ire qui possunt? [15] sicut de re militari quamquam sunt tradita quaedam praecepta communia, magis tamen proderit scire qua ducum quisque ratione in quali re tempore loco sit sapienter usus aut contra: nam in omnibus fere minus ualent praecepta quam experimenta.

> I would venture to say that this kind of effort (sc. the student answering questions) will be more beneficial to the students than would be all the technical treatises (*artes*): These no doubt help a great deal, but how can they be of broad enough scope to embrace all the sorts of things which arise on nearly a daily basis?[87] [15] Take the *res militaris*: Although there are certain general precepts which have been passed down, it will be more beneficial (sc. for a pupil) to know what method each general used to good effect or not in a given situation, time, and place. For in almost everything, precepts are less valuable than experience.

84 Massilia was besieged in 49 BC by Caesarian forces. Vitruvius' account is somewhat discrepant from Caesar's (*BCiu.* 1.34–6, 56–8, 2.1–16, 22, with Kraus 2007); cf. Lucan 3.358–508. For analysis of the siege, see Marsden 1969: 112–13.

85 Retaining the paradosis and construing *eae uictoriae ciuitatum* as an abstract-for-concrete expression (= *eae uictrices ciuitates*), with Callebat and Fleury 2003: 292–3; others emend to some such expression as *ea uictoria ciuitates* (Rose and Müller-Strübing).

86 For the pedagogical context, see Keeline 2018: 22–8, esp. 23–4.

87 For *latiore quadam comprensione* as modal ablative, see Reinhardt and Winterbottom 2006: 133; for *compre(he)nsio*, Reinhardt and Winterbottom 2006: 133.

Quintilian's statements show that the late-Republican controversy around *praecepta* in the *res militaris* persisted through the time of Onasander and Frontinus, since it is unlikely that he would have looked to the art of war as confirmation for the superiority of experience (2.5.15 *experimenta*) over precepts (*praecepta*) if it were a problematic or contentious example.[88] But this passage also bears more directly on the problem observed in both Onasander and Vitruvius. The context shows that the *praecepta communia* of warfare are analogous to oratorical *artes* (2.5.14 "technical treatises"[89]), suggesting that *praecepta* refer not merely to notional instructions, but to writings *de re militari* that would organize and transmit such precepts.[90] *Communia* (2.5.15) might mean "commonplace" or "ordinary," but a better interpretation given the comparison is "made or expressed in general terms, of general application" (we have written both "general" and "generalized" to capture this sense):[91] As is shown also by the contrast with *qua ducum*, and so forth, Quintilian has in mind military precepts that are universal – that is, framed so as to apply to most or all *res*, *tempora*, and *loci*. While we do not know exactly what form the *praecepta communia* to which he alludes took, they seem naturally to point toward systematic treatises (*artes*) on the art of war, for nothing could better describe a work such as Onasander's *Str.* – and, we might infer, other Latin polemological writings including Celsus' books in the *Artes* and Frontinus' *scientia rei militaris* – than as presenting these kinds of generalized *praecepta*.[92]

Granting that Quintilian's *praecepta communia*, whether written or oral, indicate an abstracting approach to the art of war like that espoused by Onasander, it is especially interesting to note the alternative: "it will be more beneficial (sc. for a pupil) to know what method each general used to good effect or not in a given matter, time, and place." Quintilian can hardly be referring to battlefield experience, for one would not say that through one's personal experience in war one learns how "each general" (*ducum quisque*) acted; his statements point rather to the historical study of warfare with a view to determining how generals conducted themselves in specific situations. This interpretation is confirmed by the sentence

88 For Quintilian's figurative/comparative use of the *res militaris* in general, see Aßfahl 1932: 83–98; Mastrorosa 2000.

89 See Ch. 1, n.11 for *ars* = treatise.

90 This question cannot be decided on purely lexical grounds: we have seen *praecepta* of the *res militaris* referring both to treatises (n.63) and oral teachings (*De or.* 2.76).

91 *OLD* 6c; *TLL* III 1970.84–7 (Bannier)

92 Mastrorosa 2000: 288–300 argues that Quintilian's analogy reflects awareness both of historical changes in the Roman army and literary foment around the art of war exemplified by Celsus, Pliny, and Frontinus.

following the quoted passage, which returns to oratory and illustrates the general claim *minus ualent praecepta quam experimenta*: "Suppose the teacher declaims to serve as a model for his students: would it not be more beneficial for them rather to read Cicero or Demosthenes?" (2.5.16 *an uero declamabit quidem praeceptor ut sit exemplo suis auditoribus: non plus contulerint lecti Cicero aut Demosthenes?*). Hence, by *experimenta* Quintilian intends something more like study of historical *exempla* than personal practice:[93] Just as the student of oratory will read a Demosthenes or Cicero, so the pupil of war might study the exploits of an Alexander or Caesar.

6.5.4 *Toward an Exemplary Approach to the Art of War: Frontinus'* Strategemata

Quintilian's testimony, which splits military instruction into general precepts and historical *exempla*, provides an excellent vantage from which to consider different approaches to the art of war. More specifically, it suggests that (at least for some) the limitations of an abstracting approach could be avoided with a treatment establishing instead "what method each general used to good effect or not in a given situation, time, and place" (Quint. *Inst.* 2.5.15). Polemological authors had of course long recognized the didactic value of historical example: Aeneas frequently furnished historical illustrations for his advice, calling them "proofs" (μαρτύρια) or "examples" (παραδείγματα – that is, *exempla* in Latin).[94] More striking, however, are those cases where *exempla* actually supplant *praecepta*, reflecting a deeper epistemological divide. I suggest that Vitruvius' sudden pivot to *exempla* in Book X after he has rejected general rules for siege defense instantiates just such a choice of *exempla* over *praecepta*. Vitruvius' and Quintilian's views differ in some respects, but the key similarity is the privileged place accorded to *exemplum* for exhibiting successful practice. Thus, Vitruvius aims at the *sollertia* contained in the historical anecdotes (10.16.2, 10.16.12), while Quintilian focuses on the general's prudent use (2.5.15 *sapienter usus*) of strategy.

In a larger sense, military *exempla* were scattered throughout much of Roman literature, especially historiography, which is densely studded with

93 I use "*exempla*" carefully, for *praecepta* are opposed to *exempla* elsewhere in *Inst.* (Reinhardt and Winterbottom 2006: 132–3): e.g. 12.2.30.

94 For example, 2.2–6, 4.1–4, 4.7–11 (ἐπὶ παραδείγματος καὶ μαρτυρίου καθαροῦ), 5.2, 10.21–2, 11.3–15, 12.3–5, 15.8–10, 16.14, 17.2–6 (παράδειγμα), 18.8–11, 18.13–19, 20.4–5, 22.20, 24.3–19, 27.7–10, 28.5–7, 29.2–12 (παραδείγματος ἕνεκεν), 31.24–9 (μαρτύριον), 31.33–5, 36.6–9, 37.6–7, 40.2–5.

narratives of war and descriptions of battle. Even when not presented as *exempla*, such episodes could lend themselves to exemplary reinterpretation.[95] They were no doubt read and studied by those who fancied themselves erudite in the *res militaris*, and historiographic accounts were thus probably among the targets of contempt in Sallust and Cicero (Ch. 6.5.1). It was a bold but perhaps natural leap to the notion that military *exempla* could serve a useful didactic function detached from precepts (as they are found in Aeneas Tacticus) or outside a narrative frame (as in historiography). Vitruvius seems to provide the earliest Latin experiment in this approach, but it was repeated not long after, again on a relatively small scale, in the *Facta et dicta memorabilia* of Valerius Maximus. Valerius prefaces a short chapter (7.4) on stratagems in the following way (7.4.pr):

> illa uero pars calliditatis egregia et ab omni reprehensione procul remota, cuius opera, quia appellatione <Latina> uix apte exprimi possunt, Graeca pronuntiatione strategemata dicantur.

> Here is a laudable part of cunning far removed from all censure. Its performances can hardly be properly expressed by a Latin name, so let them be called in Greek parlance "stratagems." (transl. Shackleton Bailey)

The chapter sets out five clever military actions of the Romans, one of the Greeks, and one of the Carthaginians; the previous chapter (7.3) on *uafre dicta aut facta* contains some military *exempla* too. Valerius' notion of *strategema* less interests us here than the fact that his collection represents a compilation of historical military exploits presented independently of theoretical precepts, as in Book X of Vitruvius' *DA*.[96]

In hindsight, Vitruvius' and Valerius Maximus' efforts were but the fits and starts of a project that would reach its logical conclusion only toward the end of the first century AD, with Frontinus' *Strategemata* – the first work in Latin polemology that rigorously instantiated the preference for *exempla* over *praecepta*. *Strat.* comprises four books containing 583 historical *exempla*, called by Frontinus "stratagems" (Gk. στρατηγήματα, Lat. *strategemata*).[97] It is organized neatly according to the stages of war: Book I deals with matters before battle (1.pr.5), II in and after battle (2.pr), and III siege warfare; Book IV contains material "of miscellaneous content"

[95] Cf. Langlands 2008: 162.

[96] See Wheeler 1988: 14–17 on Valerius and the history of the stratagem.

[97] Editors print στρατήγημα (etc.) in Greek letters, with MS support; the word is recognized by Frontinus (1.pr.2), like Valerius Maximus (7.4.pr), as an *appellatio Graeca*. See also König 2017: 158.

(4.pr *diuersae . . . substantiae*) that does not fit under any earlier headings.[98] Each book is structured in the same way: a brief preface (longer and more discursive before Book I), a table of contents with thematic section headings called *species* (e.g. 1.pr.5 *de occultandis consiliis*, 3.pr.3 *de simulatione regressus*, 4.pr *de effectu disciplinae*), and the main text of the chapter consisting of the *exempla* organized in sections according to these *species*.[99] The exempla vary in length, one of the shortest being a mere six words (4.1.42 *Q. Fabius Maximus transfugarum dextras praecidit*), the longest a substantial paragraph (2.5.31), but most two or three sentences; within a given section each *exemplum* is presented independently, and the *exempla* are not ordered strictly either by chronology or, as Valerius Maximus does, by nationality, although there is a tendency to have one or more Romans named first, and the order of the *exempla* is by no means random (see Ch. 6.6.2).[100] Whatever the prehistory of the stratagem collection, it is a remarkable fact that Frontinus' *Strat.* is the first work we possess, even know of, to present such a body of historical military *exempla* on their own terms – that is, not embedded within a larger work with separate aims.[101] As a work squarely in the polemological tradition, it

[98] Book IV is peculiar in several respects and escapes the tripartite schema announced in the first preface (1.pr.2 *in tres libros ea diduximus*). Wachsmuth 1860 and Wölfflin 1875 argued it is spurious; the early Teubner (Gundermann 1888: xiii; cf. 1888b, esp. 319–26) and Loeb (Bennett and McElwain 1925: xix–xxv) followed. The *communis opinio* turned after Bendz 1938, who judiciously defended Frontinus' authorship (for earlier attempts, see Del Chicca 2002: 209n28, adding Connor 1921: 93–104). Most subsequent commentators have been persuaded by Bendz (König 2017: 174n54, adding Kuchma 1984; Del Chicca 2002: 209–10), but Ireland (editor of the new Teubner) remains unconvinced. Sallmann, *NP* s.v. "Frontinus, S. Iulius," doubtless thinks of the latter's edition in saying that the authenticity of Book IV is not settled; cf. also Cusmano 2005: xix–xvi.

[99] For *species*, see Ch. 6.6.1.

[100] 56 percent of the *exempla* feature Roman generals (Gallia 2012: 197n56), but by my calculation 88 percent (44/50) of the *species* begin with a Roman *exemplum*, and Roman *exempla* constitute approximately 75 percent of the first three *exempla* in each *species*. For the variety of generals in *Strat.*, see König 2017: 169–73.

[101] For the stratagem collection before Frontinus, see Wheeler 1988: 1–24; Schettino 1998: 45–68; Wheeler 2010: 19–25. A standard view is that Frontinus was merely "updating" an extant "genre" dating to Hellenistic times (Wheeler 2010: 22), so that while *Strat.* might inaugurate the "golden age" of the stratagem collection (19), it is not fundamentally original. Yet to call the stratagem collection a "genre" is to overstate the stability and coherence of the tradition, which claims an astonishingly wide range of authors and text types (cf. Schettino 1998: 61–8). Frontinus' work, in its independent (or quasi-independent) and highly systematic presentation of *exempla*, may be more formally innovative than usually appreciated, though this is impossible to confirm given the dearth of evidence. Frontinus' own statements do not necessarily attest to earlier stratagem-collections, as sometimes supposed. He does mention "the books of others promising the same things" (1.pr.3 *aliorum libros eadem promittentium*), but these could be broader collections of *exempla*, such as Valerius': consider thus Frontinus' claim that earlier authors of *exempla*-collections *ipso uelut aceruo rerum confuderunt legentem* (1.pr.2), indicating perhaps that they did not sufficiently distinguish categories/themes. Frontinus' emphasis on the orderly, systematic, and appropriately judicious

is a paradigm case of the didactic use of *exempla* and *experimenta* to the exclusion of *communia praecepta* that we saw hinted in Quintilian's *Inst.*

In the remainder of this chapter, we will draw out the implications of the form and contents of Frontinus' *Strat.* for the development of the Roman *ars* of war. The various threads studied until this point can help to locate *Strat.* in this tradition. We first considered the abstracting approach embodied in Onasander's *Str.*, which constructs an explanatory framework for specialized knowledge of warfare based on psychological first principles. Although this approach has certain advantages, it is vulnerable to objection based on the contingency and spontaneity of war: The tension between generalized precepts and messy reality comes to the fore in Onasander's treatment of the innumerable particulars of combat as well as of *stratêgêmata*, the improvised battlefield decisions for which rules cannot be framed in word or by reason (λόγῳ). This gap between theory and practice plays into wider concerns about the utility of the art of war reflected in Latin authors. When Vitruvius identifies a similar problem for defensive tactics in siege warfare, he turns to the provision of *exempla*. On the later evidence of Quintilian, we can recognize Vitruvius' move as reflecting an epistemological divide in the art of war between *praecepta*- and *exempla*-based instruction.

This is precisely the context in which Frontinus' *Strat.* must be understood. While we do not know how closely his treatise on the *res militaris* resembled Onasander's, it must have given general rules. It would therefore have faced challenges similar to those encountered by Onasander and referenced by Quintilian. I suggest that Frontinus intended *Strat.* to counter this epistemological challenge: drawing upon the diffuse but well-established tradition of military *exempla* in historiographic writings and whatever earlier collections there may have been (if any), he composed a work to redress the shortcomings of theoretical treatises such as Onasander's and presumably his own earlier composition on the *scientia rei militaris*. Predicated on something like the belief that "it will be beneficial (sc. for a pupil) to know what method each general used to good effect or not in a given matter, time, and place" (Quint. *Inst.* 2.5.15), Frontinus' stratagem collection eschews generalized precepts in favor of setting before the general's (reader's) eyes battlefield exploits in all their historical particularity. *Strat.* can thus be understood as complementing the earlier work by providing a mode of cultivating the general's expertise independent of the contemplation of *praecepta*. Frontinus himself asserts at the outset that he thought this collection of *exempla* was

selection of *exempla* (see Ch. 6.6.1) implies that he saw his work as a formal advance over similar texts. See also Chiritoiu 2018: 42–5 for renewed argument for the originality of *Strat.*

"owed" (1.pr.1 *deberi*) to his undertaking; he probably held this belief because of the limitations of the abstracting approach highlighted earlier. In this way, *Strat.* represents a generic extension of the scientific culture of the *artes* in the early Empire: Although formally different from the *artes*, Frontinus' work stands as an organic outgrowth of the *artes* under epistemological pressures both within and without the discipline. Together, *Strat.* and the earlier work offer a complete treatment of generalship. We now turn to consider how Frontinus presents *Strat.* as solving the problem occasioned by abstracting, *praecepta*-centered treatments.

6.6 Cultivating the Faculty of Command: The "Open" Didactic System of the *Strategemata*

We return to the opening of Frontinus' *Strategemata*, which we considered briefly in Chapter 2 (1.pr.1):

> cum ad instruendam rei militaris scientiam unus ex numero studiosorum eius accesserim, eique destinato quantum cura nostra ualuit satisfecisse uisus sim, deberi adhuc institutae arbitror operae ut sollertia ducum facta, quae a Graecis una στρατηγμάτων appellatione conprehensa sunt, expeditis amplectar commentariis. ita enim consilii quoque et prouidentiae exemplis succincti duces erunt, unde illis excogitandi generandique similia facultas nutriatur; praeterea continget ne de euentu trepidet inuentionis suae, qui probatis eam experimentis comparabit.
>
> Since I, alone of so many students of this subject, have furnished a theoretical treatment (*scientiam*) of warfare and, as it seems to me, completed what I set out to do to the best of my ability, I believe that it is still owed to the task that I have undertaken that I survey in a convenient aide-mémoire the clever deeds of generals which the Greeks call "stratagems." For generals will thus be equipped with examples of judgment and foresight that will nourish their faculty for devising and producing similar (sc. deeds). Moreover, (sc. the general) will not worry about the outcome of his plan if he can test it against proven experience.

Frontinus' notion of a stratagem is broader than Onasander's, including all "clever deeds of generals" (*sollertia ducum facta*), not merely all-or-nothing gambits.[102] Nonetheless, for both authors the stratagem importantly shares the feature of situational particularity. This quality is underlined by Onasander when he says that stratagems are always "of the moment" (*Str.* 32.10 ἐκ τῶν καιρῶν) and governed by "the necessity of chance"

[102] For the stratagem in antiquity, see Wheeler 1988 (1–3, 17–21, on Frontinus).

(ἡ ἀνάγκη τῆς τύχης); it is realized more concretely in Frontinus' *Strat.* by the historical facticity of his *exempla*, which are accordingly always determined by specific and highly contingent configurations of circumstances. The preface connects in further ways to the tradition we have considered: the emphasis on *sollertia* recalls Vitruvius (*DA* 10.16.8, 10.16.12), and the connection between *exempla* and experience (*Strat.* 1. pr.1 *experimentis*) anticipates Quintilian (*Inst.* 2.5.15 *experimenta*).

Frontinus distinguishes two aims for *Strat.*: to nourish the general's inventive *facultas* for military actions and to provide a reservoir of "proven experience" (*probata experimenta*) to guide his future conduct. These objectives bespeak a didactic approach fundamentally different from that represented in Onasander's *Str.* and presumably Frontinus' earlier work. While Onasander maintains that his work is based on the experience of war (Roman πράξεις and ἔργα), he leaves these data in the background and only furnishes us with the precepts ostensibly abstracted from experience and organized on epistemically prior psychological principles. This is a didactically "closed" system: while its implicit explanatory economy has great power, it also perforce excludes indeterminate and nongeneralizable elements of warfare, including the particularities of actual battlefield situations as well as *stratêgêmata*. By contrast, Frontinus' remarks invite us to see the *exempla*-centered approach of *Strat.* as constructing a didactically "open" system. It is "open" in that Frontinus imposes no generalized rules that would limit or "close" the potential meanings of the historical data of warfare. Although the *exempla* of the work may furnish the raw materials required for the induction of precepts or principles, Frontinus never frames any; he allows the military *exempla* set before the reader to provoke what response they will. In this way, Frontinus implicitly acknowledges the problem identified by both Vitruvius and Onasander that stratagems are not capable of textualization. Yet he does not regard this as an insurmountable barrier to the systematic treatment of this department of warfare.

Whereas Onasander forbore to examine more closely how the general could cultivate his capacity for improvisation, Frontinus explicitly foregrounds this "faculty for devising and producing (sc. stratagems)" (1.pr.1), arguing that it can be "nourished" by "examples of judgment and foresight." Although the precise workings of this *facultas* remain obscure, Frontinus' key contribution is to assert that *exempla* can strengthen it. Onasander denied that memory (μνήμη) or experience (ἐμπειρία) could equip the general with stratagems; the point was that each situation is sui generis and therefore the general's response must be as well. Frontinus might agree: his *probata experimenta* are not the very stratagems that the general is to implement, but a yardstick by which to test new inventions (hence, *comparabit*). Yet he takes a more constructive and nuanced approach to memory insofar as he allows that the general if "equipped" (cf. *succincti*) with *exempla* will be able to innovate strategies "similar" (*similia*) to those that he knows. In offloading the cognitive work of translating experience into practical skill onto the general's memory of *exempla* and his "faculty for devising and producing," Frontinus avoids setting out generalized rules to govern the stratagems but still explicitly retains the possibility of cultivating the general's capacity for invention and decision-making.

6.6.1 Species *and the Structure of Exemplary Education*

How exactly are the *exempla* of *Strat.* to cultivate the general's practical wisdom? While Frontinus does not extract generalized rules, he is far from haphazardly piling up stratagems. Indeed, he asserts that authors who have previously presented collections of "noteworthy" (1.pr.2 *notabilia*) things (= *exempla*) culled from history "have confused the reader by the disordered heap as it were of material" (*ipso uelut aceruo rerum confuderunt legentem*).[103] His plan deserves consideration (1.pr.2):

> nostra sedulitas inpendet operam ut, quemadmodum res poscet, ipsum quod exigitur quasi ad interrogatum exhibeat. circumspectis enim generibus praeparaui opportuna exemplorum ueluti consilia. quo magis autem discreta ad rerum uarietatem apte conlocarentur, in tres libros ea diduximus. in primo erunt exempla quae competant proelio nondum commisso; in secundo quae ad proelium et confectam pacationem pertineant; tertius inferendae soluendaeque obsidioni habebit στρατηγήματα. Quibus deinceps generibus suas species adtribui.

> I have gone to painstaking efforts so that, as the circumstances demand, precisely that which is required is presented as if summoned for examination; for I have made ready to hand the counsels so to speak of *exempla*, surveying them by type (*genera*). Furthermore, so that they might be appropriately distinguished according to their different subject matters, I have divided them into three books: in the first will be those *exempla* which are relevant when battle has not yet commenced; in the second, those which pertain to battle and after peace is made; and the third will contain stratagems for laying or breaking a siege. I have in turn assigned to these types (*genera*) their own themes (*species*).

[103] Cf. n.101.

Frontinus' work is systematized by a didactic plan that organizes *exempla* according to both type (*genus*) and theme (*species*), which make the *exempla* effectual as "counsels ready to hand." *Strat.* instantiates this plan with tables of contents that precede each book; these provide the *genus* and then enumerate the *species*. The *species* subsequently reappear as section headings throughout each book. Considered macroscopically, *genera* and *species* are an important organizational device: keyed to the tables of contents, they segment the contents of the treatise, either guiding the reader's engagement with the *exempla* if read continuously through or else allowing perusal of select topics.[104]

More intriguing for present purposes, however, is the microscopic function of the *species*. Frontinus may not directly articulate rules that would embrace or explain the *exempla*, but he nonetheless exerts considerable control in their presentation. Most significantly, he can choose how to group particular stratagems into *genera* or *species*, what section headings (*species*) to write, how to order the *exempla* under a *species*, and which details of the *exempla* to emphasize.[105] Of these, the *species* represent the first and arguably most important guide to how Frontinus means the reader to understand the significance of the given *exempla* and can be thought of as analogous to *praecepta* within the "open" didactic system of the stratagem collection. The *species* is the only place where Frontinus offers something like a direct intervention at a level above the particular. Even so, *species* are merely orientational and remain ancillary to the *exempla* themselves, through which the didactic purpose of the work is realized. Owing to the absence of *praecepta*, there remains in the "open" system of *Strat.* a fundamental indeterminacy in the meaning of the *exempla* – that is, as regards what lessons or guidance we are intended to understand from them (see Ch. 6.6.3). Yet Frontinus' ordering and presentation of the stratagems establish a positive didactic context for the *exempla* that invite us to draw certain conclusions from them – something that remains true despite the impossibility of framing fixed rules. The best way to understand the epistemological workings of the "open" system of *Strat.* as against the "closed" system of a *praecepta*-based work is to examine Frontinus' work more closely.

104 For tables of contents in the earlier Imperial period, see Bodel 2015: 28–35; as an early information technology, Riggsby 2007 (89–90 on the uses alluded to here); 2019: 22–9. Wheeler 2010: 30 suggests Frontinus' tables of contents may be his own innovation (cf. n.101); see also König 2020: 143–4 on Asclepiodotus and Aelian (the latter may respond to Frontinus).

105 Frontinus' choices were presumably also guided by the structure of the work *scientia rei militaris* (n.20).

6.6.2 *Learning by Example: How to Let the Enemy Go*

Here we will examine a modestly sized *species* from the first half of Book II, whose *genus* contains "(sc. *exempla* for the instruction of the general) pertaining to matters usually operative during battle itself" (2.pr [sc. *exempla instructura ducem*] *pertinentia ad ea quae in ipso proelio agi solent*). The *species* we will study contains all the features typical of other *species* plus some unusual features marking the limits of Frontinus' method. The *species* is (2.6):

> De emittendo hoste, ne clausus proelium ex desperatione redintegret
>
> On letting the enemy host go, so that they do not desperately renew battle because they are trapped

The first point to be made is that the *species* does not, as a rule, set out a generalized principle or precept. The *ne*-clause here has some flavor of an injunction but is anomalous and attached only to an otherwise enigmatic heading "on letting the enemy go." *Species* more typically only state a bare theme or situation, such as, for example, 2.1 *de tempore ad pugnam eligendo* ("on choosing a time for battle") or 2.9 *si res prospere cesserit, de consummandis reliquiis belli* ("on concluding a war, if things have gone well"), although there are a handful that present an indirect question implicitly commending a course of action, such as 1.7 *quemadmodum ea quibus deficiemur uideantur non deesse, aut usus eorum expleatur* ("how we can conceal our lack of something, or manage something to fulfill its use") or 3.5 *quemadmodum persuadeatur obsidionem permansuram* ("how one can convince the enemy that the siege will continue"). The common thread among *species* is a notional response to a question – *quasi ad interrogatum* (1.pr.2) – about war. The slightly directive force of some *species* is far from rising to the level of a principle or precept. In 2.6, the *species* announces what situation the following *exempla* illustrate, "letting the enemy go," while the *ne*-clause glosses the reason, to avoid the desperate struggle of a trapped opponent.[106]

Let us now consider the stratagems under the *species*. The first two are given thus (2.6.1):

> [1a] Gallos, eo proelio quod Camilli ductu gestum est, desiderantes nauigia quibus Tiberim transirent, senatus censuit transuehendos et commeatibus quoque prosequendos.

106 For "battle from desperation" (ἐξ ἀπονοίας μάχη) in the later stratagem collection of Polyaenus, see Wheeler 2010: 39–44.

[1b] Eiusdem generis hominibus postea per Pomptinum agrum fugientibus uia data est, quae Gallica appellatur.[107]

[1a] When in the conflict waged under Camillus the Gauls were desirous of boats to cross the Tiber, the senate proposed to ferry them across and even furnish them with provisions.

[1b] The same ones when retreating through the *ager Pomptinus* were given passage, now called the "Gallic way."

If these *exempla* are read out of context, the senate's decision to assist the Gauls seems paradoxical: why would one help (as it seems) the enemy? Of course, no such worry will occur to Frontinus' audience, for he presents the *exempla* in a way that encourages us to understand that the senate acted in view of the danger that the Gauls, hemmed in by the river, would wage desperate combat if pressed. This encouragement manifests itself both in the provision of the gloss in the *species* (*ne clausus*, etc.) and, in a more general fashion, in Frontinus' choice to collocate these *exempla* with other stratagems in which the stakes are less ambivalent.

Consider the next three *exempla* (2.6.2–4):

[2] T. Marcius[108] eques Romanus, cui duobus Scipionibus occisis exercitus imperium detulit, cum circumuenti ab eo Poeni, ne inulti morerentur, acrius pugnarent, laxatis manipulis et concesso fugae spatio dissipatos sine periculo suorum trucidauit.

[3] C. Caesar Germanos inclusos, ex desperatione fortius pugnantis, emitti iussit fugientesque adgressus est.

[4] Hannibal, cum ad Trasumennum inclusi Romani acerrime pugnarent, diductis ordinibus fecit eis abeundi potestatem euntesque sine suorum sanguine strauit.[109]

[2] When the Roman knight T. Marcius, who had been granted command of the army after the two Scipios were killed, surrounded the Carthaginians, and they began to fight more fiercely so that they would not die unavenged, Marcius had the maniples open up and make room for them to flee; after they scattered, he slaughtered them with no danger to his troops.

[3] C. Caesar ordered that certain Germans who were surrounded and fighting more valiantly out of desperation be let go, and then attacked them as they fled.

[107] A supplementary apparatus in Ireland's Teubner offers dates and ancient testimonia; I reproduce this information in nn. *ad locc. Strat.* 2.6.1a–b date to 349 BC (Liu. 7.25–2). It is useful to compare other extant sources with Frontinus' *exempla* (Gallia 2012: 198–20; Malloch 2015) but beyond the scope of present interests; for one example, cf. n.113.

[108] Editors usually emend "*T.*" to *L.*" from Livy (following n.): *RE* (1) XIV 2.1591.27–47 (Münzer).

[109] *Strat.* 2.6.2 dates to 212 BC (Liu. 25.37–40), 2.6.3 to sometime 58–3 BC, 2.6.4 to 217 BC.

> [4] When the Romans were hemmed in at Lake Trasimene and began to fight very fiercely, Hannibal had his ranks separate so that the Romans could leave, and then slew them without any loss of life to his own troops.

These three *exempla* establish a close connection, absent from the first two (2.6.1a–b), between the enemy's confinement and a more intense pitch of battle (2.6.2 *circumuenti* ~ *acrius pugnarent*, 2.6.3 *inclusos* ~ *fortius pugnantis*, 2.6.4 *inclusi* ~ *acerrime pugnarent*). One adds the psychological detail *ex desperatione* (2.6.3). It should be noted that the *exempla* provide a minimum of circumstantial historical detail, offering only enough for the reader to identify the moments as belonging to famous persons, wars, or conflicts;[110] the second (2.6.3) is stinting even here, identifying C. Caesar but offering little to specify the actions against an anonymous group of *Germani*. The parsimonious character of the *exempla* is instructive if interpreted within the "open" didactic framework of the *species*: by reducing extraneous information, Frontinus avoids cluttering the account with details that could distract the reader's attention from the aspects of the *exempla* selected in advance (cf. 1.pr.2 *praeparaui*) for their salience to the *species*. Bearing in mind this economical approach, we may note also that all three stratagems involve open (if not necessarily pitched[111]) battle, arguably the least complicated situation *in proelio* to exemplify these maneuvers. Studying these relatively unproblematic cases, the general (reader) is brought by the shortest path to the essential lesson. We note also that the *exempla* are unanimous in reporting that once the enemy had fled, pursuit was given, with two *exempla* specifying that they were slaughtered advantageously (2.6.2 *sine periculo suorum*, 2.6.4 *sine suorum sanguine*). Repetition renders this detail makes thematic.

If the simplicity of 2.6.2–4 makes these *exempla* a good introduction to the strategic advantage of "letting the enemy go," the three that follow gradually introduce more variety and complexity. In key respects they continue the themes of 2.6.2–4, but under different conditions (2.6.5–7):

> [5] Antigonus rex Macedonum Aetolis, qui in obsidionem ab eo conpulsi fame urguebantur statuerantque eruptione facta commori, uiam fugae dedit, atque ita infracto impetu eorum insecutus auersos cecidit.
>
> [6] Agesilaus Lacedaemonius aduersus Thebanos, cum acie confligeret intellexissetque hostes locorum condicione clausos ob desperationem fortius dimicare, laxatis suorum ordinibus apertaque Thebanis ad euadendum uia, rursus in abeuntis contraxit aciem et sine iactura suorum cecidit auersos.

110 The function of the *cui*-clause in 2.6.2 is presumably to identify which Marcius is meant; for the minimization of historical detail, cf. Gallia 2012: 199.

111 Hannibal ambushed the Romans, trapping them against the lake; n.113 on the omission of this detail.

[7] Cn. Manlius consul, cum ex acie reuersus capta ab Etruscis Romanorum castra inuenisset, omnibus portis statione circumdatis inclusos hostes in eam rabiem efferauit ut ipse in proelio caderet. quod <ut> animaduerterunt legati eius, ab una porta remota statione exitum Etruscis dederunt et effusos persecuti, occurrente altero consule Fabio, ciciderunt.[112]

[5] Antigonus, king of the Macedonians, allowed the Aetolians passage for flight when, pressed by siege and burdened with famine, they had decided to die making a sally; having quashed their ardor, he slew them as they fled.

[6] When Agesilaus the Lacedaemonian was engaged with the Thebans in pitched battle and noticed that the enemies were blocked by the lay of the land and had because of their desperation begun to struggle more valiantly, he had his men open ranks and make way for the Thebans to flee; drawing his battle line back up he then slew them from behind with no losses of his own.

[7] After returning from battle the consul Cn. Manlius found that the Roman camp had been captured by the Etruscans; stationing a guard outside all of the gates, he roused the enemy thus enclosed to such a state of frenzy that he himself perished in the ensuing combat. When his generals learned this, they allowed the Etruscans an escape by removing a guard from one of the gates and, pursuing them once they had dispersed, slew them with the help of another consul who had come to their aid, Fabius.

The situation of these stratagems contrasts significantly with the assumption of open battle in the previous three: in Antigonus' case, the setting is a city under siege; in Agesilaus', it is emphasized that the lay of the land (2.6.6 *condicio locorum*), and not only his troops, hems in the opponent;[113] and in Manlius', he surrounds his own camp occupied by the enemy. Despite these differences, there are definite elements of continuity with previous *exempla*: above all, the relationship between confinement and the intensity of battle (2.6.5 *conpulsi ~ facta eruptione commori*, 2.6.6 *clausos ~ ob desperationem fortius dimicare*, 2.6.7 *inclusos ~ in eam rabiem efferauit ut*, etc.), but also emphatic reports that the enemy, once put to flight, was expeditiously slain (2.6.5 *insecutus auersos cecidit*, 2.6.6 *sine iactura suorum cecidit auersos*, 2.6.7 *effusos persecuti . . . ciciderunt*). This repetition ensures that the reader will follow the thread established in the simpler *exempla*, but the novel contexts of action extend the limits of applicability of the *species*.

[112] *Strat.* 2.6.5 dates to sometime 223–21 BC, 2.6.6 to 394 BC (Xen. *Hell.* 4.3, Plut. *Vit. Ages.* 18, Polyaen. 2.1.19), 2.6.7 to 480 BC (Dion. Hal. *Ant. Rom.* 9.12, Liu. 2.47).

[113] The same situation could have been illustrated by the battle at Lake Trasimene (*Strat.* 2.6.4), where Hannibal waited until the Roman army was *clausum lacu ac montibus* (Liu. 22.4.5) (see Goldsworthy 2000: 186–9; Fronda 2011: 244). Frontinus' omission of the topography at 2.6.4 only to emphasize it here indicates his authorial control over the *exempla.*

The general is trained on gradually more complex *exempla* and is led to appreciate the variety of situations in which it could be beneficial to allow the enemy to escape.

The final two *exempla* do much the same as 2.6.5–7, pushing the boundaries of the *species* still further to incorporate other settings; they are followed by a kind of theoretical coda (2.6.10) to the stratagem of 2.6.9 that is unique in *Strat.* (2.6.8–10):

> [8] Themistocles uicto Xerxe uolentes suos pontem rumpere prohibuit, cum docuisset aptius esse eum expelli Europa quam cogi ex desperatione pugnare. idem misit ad eum qui indicaret in quo periculo esset nisi fugam maturaret.
>
> [9] Pyrrhus Epirotarum rex, cum quandam ciuitatem cepisset clausisque portis ex ultima necessitate fortiter dimicantes eos qui inclusi erant animaduertisset, locum illis ad fugam dedit.
>
> [10] Idem inter praecepta imperatoria memoriae tradidit non esse pertinaciter instandum hosti fugienti, ne non solum ea re fortius ex necessitate resisteret, sed ut postea quoque facilius acie cederet, <cum sciret> non usque ad perniciem fugientibus instaturos uictores.[114]

> [8] After Xerxes was conquered, Themistocles prevented those who wanted to destroy the bridge from doing so, since he knew that it was better for Xerxes to be expelled from Europe than to be driven by desperation to fight. He also sent a messenger to Xerxes to tell him what danger threatened unless he hastened his retreat.
>
> [9] After Pyrrhus, king of the Epirotes, had taken a certain city and noticed that the men who had been trapped inside when the gates were closed were fighting valiantly in a last-ditch effort, he allowed them to flee.
>
> [10] He also has handed down among other precepts of generalship that one ought not stubbornly pursue a fleeing enemy, not only to avoid their struggling more valiantly in such extremity, but also so that they would flee more easily in the future, since they knew that the victors would not pursue them to their utter destruction.

Sections 2.6.8–9 not only illustrate new situations *in ipso proelio* in which the stratagem proves advantageous, but also reintroduce from the initial *exempla* (2.6.1a–b) the notion that the enemy be allowed to leave without pursuit. In 2.6.10, Frontinus reports a precept from the writings of Pyrrhus that does not so much explain the stratagem of 2.6.9 as complement it, as well as those of 2.6.1 and 2.6.8, indicating why it is sometimes desirable to allow a fleeing enemy to escape. This theoretical coda or excursus is

[114] *Strat.* 2.6.8 dates to 480 BC (Hdt. 8.110, Nep. *Them.* 5.1, Diod. Sic. 11.19, Pomp. Trog. *Iust.* 2.13.5, Plut. *Vit. Them.* 16, Polyaen. 1.30.4); 2.6.9 is unknown.

absolutely unique in *Strat.*:[115] The generalizing nature of Pyrrhus' advice, which Frontinus explicitly calls *praeceptum imperatorium* – that is, *not* a *stratêgêma* – is difficult to reconcile with the "open" didactic structure of *Strat.* However, this incongruity is somewhat ameliorated by the fact that the *praeceptum* is given in reported speech: Frontinus attributes both the substance of the precept (*tradidit non esse*, etc.) and the explanation for it (<*cum sciret*>) to Pyrrhus.[116] This is less a precept enjoined by Frontinus than an adjunct or element of the preceding stratagem.[117] In a broader sense, 2.10 is also instructive to the extent that it throws into sharper relief the concreteness and historical particularity of the preceding stratagems; just as Onasander's encounter with *stratêgêmata* (*Str.* 32) demarcates the limits of his project in *Str.*, so Frontinus' brush with *praecepta imperatoria* does for *Strat.*

6.6.3 Thinking with Stratagems

As this reading of one *species* (2.6) in *Strat.* shows, Frontinus' "open" collection of *exempla* does not entail any abdication of didactic control. Rather, he guides the reader's interpretation of the *exempla* contained under a *species* in several ways. The chief techniques that we examined are his choice of title, ordering of the *exempla*, selection and emphatic repetition of narrative and descriptive elements, and gradual introduction of complexity and variety in all aspects of the *exempla*. So much for a sequential reading. If we look at the stratagems contained under this *species* synoptically, what conclusions are we entitled to draw?[118] Most obviously, the *exempla* effectively illustrate the theme of the *species* itself: that it may be useful in battle to allow a trapped enemy to retreat. Besides that, Frontinus also seems to want us to notice the ease with which the enemy who thus flees may be destroyed, although as 2.6.1 and 2.6.8–10 show, pursuit is not always the best policy. He perhaps also encourages us to reflect on the many ways in which the enemy can be hemmed in to the point of dangerous ferocity (by troops or topography, under siege, in a captured city, even in a military camp). We learn these lessons inductively, by contemplating the *exempla* presented for this

[115] Cf. König 2017: 163.

[116] The supplement <*non sciret*> is found in late, interpolated MSS; other possibilities with MS support include *ratus*, *credens*, or the omission of the clause *non . . . uictores* (so also Oudendorp).

[117] It likely reflects the fact that Pyrrhus' military treatise joined exemplary accounts of stratagems (Wheeler 1988: 13–14; 2010: 37) with commentary and generalized precepts. In reproducing this *exemplum*, Frontinus evidently felt moved to add Pyrrhus' own commentary.

[118] For the cumulative effect of reading *exempla* within and across *species*, see König 2018 (esp. 150, 155).

species and considering which aspects of the situations and what actions taken by the generals seem to contribute to a successful outcome; but we do so under the supervision of Frontinus, whose didactic role obtains in the narration or editing, so to speak, of the accounts. This is not the raw experience of warfare but a curated collection of clever deeds of generalship that Frontinus has shaped for our edification.[119]

While it is thus possible to identify certain preferred conclusions, an important result of the "open" system of *Strat.* is the impossibility of framing for the *exempla* under a given *species* the kind of fully generalized principles or precepts that we identified in Onasander's *Str.* To be sure, we are encouraged, even invited, to make the effort to induce "precepts" of a more limited scope: we would not be violating the spirit of 2.6.2–7 if we inferred an imperative such as "One ought to allow an enemy host that is surrounded on all sides to flee so that it can be pursued and destroyed more safely." Yet this injunction would be true only for this subset of *exempla*: indeed, it is contradicted by 2.6.1 and 2.6.8–9 (and of course by Pyrrhus' precept in 2.6.10), which suggest that it is sometimes better to allow the enemy to leave unmolested. It is no solution to scrub from our directive the offending conclusion "so that the enemy can be pursued and destroyed more safely," yielding a precept of the form "One ought to allow an enemy who is surrounded on all sides to flee." This is not reliably prudent advice, for on a little reflection, many situations come to mind in which a cornered enemy would be better pressed without respite. Nor is the difficulty ameliorated even if one were to seek a more precise (and seemingly less objectional) formulation, such as "One ought to allow an enemy to flee who is surrounded on all sides *and therefore fights more valiantly and dangerously in their desperation.*" Consider a situation in which a general's troops have separated and enclosed one part of a larger enemy force: it may be better to endure the frenzied struggles of this contingent than to open ranks and allow it to withdraw, since that would risk its rejoining the opposing army with potentially worse consequences.

It is always possible in this way to present a counterexample that will frustrate the precept induced from some or all of the exempla of a *species* – to find, in other words, uniquely different battlefield circumstances that will provide an exception to the rule. By their very nature, the historical *exempla* do not rigorously establish the conditions under which the stratagems are valid; what they offer instead are particular situations that indicate the possible value of a given course of action. The content of a *species* is

[119] Cf. also Chiritoiu 2018: 90–4.

never fully determinate, even if it is suggestive. Thus, we are nowhere told in 2.6 that we ought at all events to allow an enemy who is surrounded to escape, but, rather, are shown some situations in which that was perspicaciously done. Whatever limitations this fuzziness may entail, it is also epistemologically fruitful insofar as it allows Frontinus to avoid the difficulties met by an abstracting approach, whose generalized *praecepta* always remain at a remove from the conditions of any actual situation of warfare.[120]

The "open," inductive experience of reading *Strat.* has other consequences owed not only to the conceptual indeterminacy of a given *species* but also to the implicit richness of its *exempla*. Perhaps the most radical of these is the fact that the field of possible inferences presented by a group of stratagems is (in theory) unlimited. To use only one example from 2.6, the general might find in the *exempla* the suggestion that it could be a useful gambit to station his troops where, *locorum condicione clausi* (2.6.6), they will be induced to fight "more valiantly because of their desperation" (2.6.6 *fortius ob desperationem*). There would be nothing unusual about drawing such a conclusion: the tactic belongs, after all, to the class of stratagems to which Onasander gives qualified approval (*Str.* 32.1–8). Whether or not this suggestion is among those prioritized, Frontinus does not and could not intend all such possible interpretations, yet he enables them by leaving open-ended the meaning of the *exempla*. The stratagems are meant to cultivate the general's creative *facultas*, which will invent "things similar" (1.pr.1 *similia*) – but similar *in respect of what*? Even if Frontinus forecloses certain possibilities by choosing which narrative and descriptive details to include, those same details remain subject to free reuse and extension by our faculty for invention. This implicit richness is one of the virtues of an *exempla*-collection and aids in developing the general's capability for spontaneous decision-making in the uncertain particulars of war.

This last feature is among the most important of the "open," inductive method of *Str.* and should be connected with the abstracting approach of the earlier work on the *scientia rei militaris*. We saw that Onasander does not address directly how to develop the faculty of situational judgment that the general needs to deduce the appropriate instance of a rule in a unique battlefield situation. Since Frontinus' treatise must also have furnished generalized *praecepta*, he would have faced the same problem of inference or deduction. But instead of taking the optimistic postulate that the general's intelligence suffices, and thus leaving his teachings in some

120 Cf. also König 2017: 174–7.

sense incomplete, Frontinus attaches a systematic stratagem collection to aid the general (reader) in learning how to think flexibly and creatively on the battlefield. While the *exempla* in *Strat.* can be read indefinitely on their own terms, they would have been even more effective as illustrating the manifold ways that the *praecepta* in the earlier work could be applied in different situations.[121] By contemplating the *exempla* in *Strat.* in tandem with the earlier precepts, the general could develop his own *facultas* for innovation faced with the variable conditions of battle. It is important to remark that, even on this assumption of complementarity, Frontinus' approach differs from others which interwove exemplary anecdotes as proofs for precepts (such as Aeneas Tacticus did).[122] This latter approach embeds *exempla* within a "closed" didactic system and thereby more tightly circumscribes their meaning. By allowing the *exempla* to stand on their own, apart from the *praecepta*, Frontinus encourages greater independent judgment from his reader. This remains true even if he did edit and shape these *exempla* to lead toward lessons consonant with the precepts in the earlier treatise. The desire to close the gap between general rules and the unique realities of war – between theory and practice – thus motivates the form of the *Strategemata*.

6.7 Conclusion

To conclude, we have traced in this chapter the development of the Latin *ars* of war in the first century AD. While a paucity of evidence makes it difficult to access the *ars* directly, enough evidence remains to paint a relatively nuanced picture of its growth. We can form some notion of the virtues and limitations of a *praecepta*-based method such as Celsus' or Frontinus' treatises would have adopted with the help of Onasander, who offers an abstracting approach to the *technê* that privileges general rules derived from psychological principles. As we saw, there is evidence internal and external to the polemological tradition that this abstracting approach was regarded as inadequate to the task of fully embracing the discipline. Specifically, the variety and contingency of battlefield phenomena frustrate attempts to articulate rules that would guide the general's action in every instance. We argued that Frontinus' *Strat.* meets this challenge by proposing an alternative and complementary didactic framework based on the

[121] Chiritoiu 2018: 63–4 rightly emphasizes that *Strat.* has "stand-alone value," but attacks a straw man in denying that the *exempla* have any relation to the earlier precepts.

[122] See n.94; contrast Chiritoiu 2018: 63.

open-ended study of *exempla*. These *exempla* provide the material whereby the general can cultivate precisely the faculty for spontaneous decision-making that will allow him to respond successfully to the contingencies of warfare.

The development of the art of war in the first century AD sheds new light on the growth of the scientific culture of the *artes* in the early Empire. As we saw in Chapter 5, not all authors responded to methodological and specifically epistemological challenges by foregrounding the explanatory role of natural first principles. Celsus rather sought a methodological middle ground, with an epistemologically reserved approach to nature and explanation. The art of war was marked, like medicine, by controversy regarding the possibility of establishing generalized rules for practice. There are to be sure significant differences between the traditions, but a central tension in both cases were the questions of whether precepts of universal applicability could be framed and, if so, whether they sufficed for practice. The Latin art of war offers a noteworthy solution to these generic problems of specialized knowledge: The attempt to present an *ars* proper with first principles and generalized rules was never abandoned, but the *exempla*-collection emerged as a literary form to complement and indeed complete such treatments. Frontinus' *Strat.* thus represents a novel generic extension of the *ars* of warfare specifically, and of the *artes* more generally. As we turn to our final *ars* – land-surveying – we will see elements of the epistemological issues examined in this and the previous chapters united with a fresh concern for the justificatory importance of explanatory knowledge of *rerum natura.*

CHAPTER 7

The Emergence of an Ars mensoria

Frontinus and Hyginus on the Historical Realities and Theoretical Ideals of Roman Land-Surveying

The Roman land-surveyors were latecomers to the *artes*: no evidence for systematic treatises akin to those we have studied exists before the end of the first century AD.[1] Various reasons could explain the earlier dearth of literature, but it is effect rather than cause that interests us, for when the first authors undertook to write *artes* on surveying – probably in the wake of Imperial policy under the Flavians[2] – they faced the challenge of creating a systematic literature more or less ex nihilo. This is not to say that they invented the substance of the *ars*: the essence of surveying always resided in the conjunction of instruments and expertise that allowed the surveyor to measure, mark out, and assess land. Hence, practitioners could be called either after their activity (*finitor*, *mensor*, *metator*) or tools (*decempedator*, *gromaticus*).[3] There were indeed centuries of surveying practice and countless square miles of divided land that attested to the robustness of the profession. Nonetheless, to present the atoms of practice as belonging to the composite whole of an *ars*, the early authors did ultimately have to establish a unified intellectual identity for the discipline. Their attempt to do so not only brought the *ars mensoria* within the scientific culture of the Imperial *artes*, but also inscribed into nature uniquely Roman ways of knowing and dividing the physical world. How their writings sought to accomplish these goals is the subject of this chapter.

[1] Certainly, there were various archival documents, maps, laws, registries, and so forth, that dated to Republican times (cf. Moatti 1993, esp. 31–104), but pace Nicolet 1991: 157 there is no firm evidence for "theoretical treatises" from the Augustan period: von Cranach 1996: 139–41; Chouquer and Favory 2001: 33; cf. Fuhrmann 1960: 156–7, 162, 181–2.

[2] Ch. 2, n.141.

[3] *Finitor* is chiefly Republican, occurring in Plautus (*Poen.* 49) and 5× in Cicero: *TLL* VI 1.803.75–83 (Bacherler); cf. Non. Lindsay p. 11.22 *finitores dicebantur, quos nunc agrimensores dicimus*. *Mensor*, first in Varro, *Rust.* 1.10.2, is the most common word for "surveyor" in the *CAR*: *OLD* b; *TLL* VIII 753.41–50 (Brandt). *Metator*, first at Cic. *Phil.* 11.12, often refers to military surveyors: *TLL* VIII 878.82–879.10 (Dittmann). *Decempedator* (from *decempeda*, surveyor's ten-foot rod), is probably a nonce word (only Cic. *Phil.* 13.37, contemptuously). *Gromaticus*, from *groma* (surveyor's staff, n.18), is popular among scholars but appears only once of surveyors: [Hyg.] *Mun. castr.* 12 *professores eius artis . . . gromatici sunt cognominati*. *Agrimensor* (a calque on Gk. γεωμέτρης) became current only in the late Empire: *TLL* I 1428.64–78 (Vollmer). See further Nadjo 2005.

7.1 Clearing the Ground: Some Caveats for the Study of the *corpus agrimensorum Romanorum*

First, however, a further round of caveats is in order. In Chapter 2, we stressed the challenges of studying the texts transmitted in the *corpus agrimensorum Romanorum* (*CAR*). Its composite nature entails grave difficulties for the attribution, dating, and constitution of surveying writings, which extend to the texts of two authors whom we introduced there and who are the focus of this chapter: Frontinus and Hyginus. Certain questions that we touched on are especially urgent: Did Frontinus really write one or more works on surveying now represented by the passages in the *CAR* under his name? Given the similarities among the writings of "Frontinus" and Hyginus, which author depended on the other? How extensively were the writings of "Frontinus" and Hyginus later redacted or interpolated? Our editorial and interpretative choices will differ greatly depending on how we answer such questions. To take one example, if the etymologies of surveying terminology found in both "Frontinus" and Hyginus that we will discuss later in this chapter are genuine in part or whole, they can elucidate early representations of the *ars mensoria*; if they are late-antique interpolations, however, they are less germane to the first-century *ars*.[4] In view of these complexities, I will spell out a few assumptions that will inform the remainder of this chapter.

First, I assume that the writings attributed to "Frontinus" bear some relation to an authentic work (or works) by Sex. Julius Frontinus.[5] Supposing that they do not go back to Frontinus, I nonetheless date them, along with Hyginus' *Constitutio limitum*, to the late first or early second century AD.[6] Second, I attach no particular significance to the order of dependence between Frontinus and Hyginus (although Frontinus' writings are probably earlier).[7] Rather, I treat similarities as reflecting the shared use of certain strategies in the construction of the *ars mensoria*: the salient point is that the texts demonstrate a collaborative effort to define and present an *ars* of surveying and can thus be read in a complementary fashion. Third, I adopt a generally conservative attitude to the transmitted texts of Frontinus and Hyginus. These texts have certainly been corrupted, and in producing a sensible text the editor must make many choices about the extent and nature of this corruption; as a rule, I prefer the fewest changes required for a cogent solution, even where there are reasonable grounds to suspect wider corruption or

[4] See n.41. [5] Ch. 2, n.146. [6] Ch. 2, n.149. [7] Ch. 2, n.155.

interpolation. This is because I judge the possibility of losing material that could shed light on the early surveying treatises worse than the alternative of retrojecting later steps in the process of the intellectualization of the *ars* to an earlier period.[8] Credulity does not always pay, of course, so greater caution is required in handling the *CAR* than the more soundly transmitted *artes* studied in earlier chapters; I note contested readings throughout.

7.2 Creating an *ars* of Surveying

Qualifications aside, there is abundant evidence to show that both Frontinus and Hyginus were concerned to present land-surveying in such a way as to demonstrate the essential coherence and rationality of the discipline and to set its methods on a secure theoretical foundation. While the *CAR* has no single treatise on the scale of other works we have considered, the assemblage of Frontinus' and Hyginus' writings reveal many of the basic concerns animating the Imperial *artes* studied to this point. We will focus chiefly on their discussion of centuriation, which is cast as the central operation of land-surveying, as paradigmatic of the ambitions of the early generation of surveying authors. Centuriation is the central topic of Hyginus' *Constitutio limitum* (*Const.*) and of one of the four works attributed to Frontinus, *De limitibus* (*Lim.*).[9] After briefly sketching centuriation, we will investigate two fronts on which Frontinus and Hyginus advance their project. The first front can be understood in terms of the historical justification of surveying. The messy realities of surveying language and practice inherited by the Imperial surveyors could pose a threat to a unified conception of the discipline; accordingly, both Frontinus and Hyginus develop sophisticated strategies for handling variable and seemingly arbitrary aspects of the art. These strategies are intended to show that surveying language and practice are both rationally founded and historically consistent with the origins of the discipline. Importantly, these origins are represented as a fundamentally naturalistic way of conceptualizing and dividing space: the result is an example of what we have called the Romanization of specialized knowledge (Chapter 2), whereby the principles of Roman surveying are reformulated as an expression of the cosmic order of nature. The second front can be understood as the extension, by interdisciplinary means, of this naturalization of surveying practice. Beginning from the assumption of the practice that is "best" and "rational" (Frontin. *Lim.* L 31.1 = T 14.11 *optima ... ac rationalis*),

[8] Long development of the *CAR*: Ch. 2, n.142. [9] Titles: Ch. 2, nn.145, 150

Frontinus and Hyginus ground the ideals of surveying in a set of principles furnished by other branches of specialized knowledge, chiefly astronomy and mathematics. This interdisciplinary engagement creates a theoretical foundation for a consistent, precise, and rigorous practice in accord with nature. From the uncertain situation in the early Empire, a well-defined "image of expertise" for the *ars mensoria* takes shape in the writings of Frontinus and Hyginus.

7.3 A Sketch of Centuriation (*centuriatio, limitatio*)

The fundamental practice of Roman land-surveying was that of dividing territory into plots for distribution to colonists and others, a process that could be called *limitatio* ("delimitation"[10]), after the balks or roadways (*limites*) that constituted the boundary lines for these plots, or *centuriatio* ("centuriation"), after the regular square or rectangular plots (*centuriae*) into which the land was delimited.[11] A brief sketch of centuriation will set the stage for the following discussion. This simplified account is based chiefly on Hyginus' *Constitutio limitum*, which contains the fullest exposition of the technique, and is corroborated by details in Frontinus' *De limitibus* and other surveying texts.[12] The surveyor first established two broad, central *limites* aligned at right angles to one another on the north–south and east–west axes; the main *limes* running from east to west was called the *decumanus* (or *decimanus*) *maximus* (henceforth DM), and that running from north to south the *kardo* (or *cardo*) *maximus* (henceforth KM).[13] Their intersection formed the center of a grid established by tracing out at regular intervals *limites* parallel to the DM and KM. The intersections of the *limites* would close off the large plots called *centuriae* ("centuries") that would receive boundary stones (*lapides*) on their angles inscribed with numbers indicating their positions relative to the DM and KM (see Figure 7.1).[14] *Centuriae* varied in size, but were often squares of twenty *actus* on a side (where one *actus* is 120 Roman feet), thus containing

10 The rendering of Dilke 1971: 87.

11 In modern scholarship *limitatio* and *centuriatio* are common, but the *CAR* has many other expressions: for example, *agere/constituere/ducere/facere/instituere/mittere/regere limites*, *delimitare* (*delimitatio*), *accipere limites* (of an *ager*), and so forth.

12 Modern overviews: Dilke 1971: 82–97; Campbell 1996: 80–90; Guillaumin 2005: 18–20; cf. von Cranach 1996: 153–83.

13 Cf. Frontin. *Lim.* L 28.5–10 = T 11.9–14, L 31.1–3 = T 14.11–13, Hyg. *Const.* L 167.3–9 = T 132.6–12, L 181.5–182.7 = T 145.10–146.8.

14 Cf. Hyg. *Const.* L 171.14–175.14 = T 136.11–140.11, L 194.17–196.14 = T 157.18–159.17, Hyg. *Lim.* L 111.16–112.21 = T 71.10–72.26. Cf. n.88.

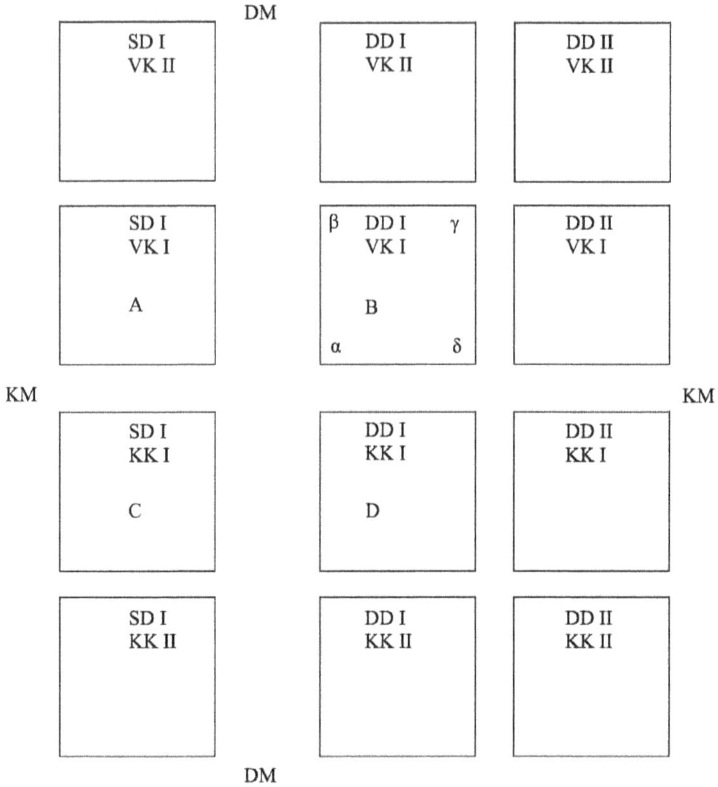

Figure 7.1 Centuriated land (after Campbell). ABCD are the four central *centuriae*. DM = *decumanus maximus*, KM = *kardo maximus*. In the abbreviations in the centuries, in the first place S = *sinistra* (to the left), D = *dextra* (to the right), V = *ultra* (beyond), K = *citra* (on this side); in the second place, D = DM and K = KM. The stones at the angles of the centuries would be inscribed thus (as an example): α = DM KM, β = DM KII, δ = KM DII, γ = DDI VKI.

altogether 200 *iugera* (where one *iugerum* is a rectangle of 120×240 feet).[15] Every fifth *limes* in the survey, called *quintarius* (because enclosing five *centuriae*) or *actuarius*, was broader; the narrower *limites* intermediate to the *quintarii* were called *subrunciui* or *linearii*.[16] Laws granted the use of *limites* as public roads and passages and provided for the widths of specific

[15] Cf. Frontin. *Lim.* L 30.5–22 = T 13.13–14.10, Hyg. *Const.* L 200.16–201.6 = T 163.16–164.5, Balb. L 96.14; Sic. Flacc. L 159.9–25 = T 123.19–124.8.

[16] Cf. Hyg. *Const.* L 168.8–169.13 = T 133.10–134.14, Hyg. *Lim.* L 111.14–15 = T 71.8–9, L 112.9–12 = T 72.13–17, Sic. Flacc. L 158.19–20 = T 123.2–3.

types.[17] In conducting these operations, the surveyor employed the all-important tool called *groma* or *ferramentum*, a staff topped by a perpendicular crosspiece hung with plumb bobs which aided in establishing perpendiculars and sighting the straight lines, called *rigores*, which defined the *limites*.[18] After the land had been completely centuriated, the surveyor would aid in the sortition process,[19] classify the different plots,[20] and transfer the results to maps for safekeeping both in town and in the *tabularium* at Rome.[21]

7.4 The Historical Inheritance of Surveying

This centuriation procedure represents an ideal for Frontinus and Hyginus, as for other surveyors, but it was not the only way land was designated and divided. As both authors appreciate, there was much variety in surveying practices across the Empire, which could be explained chronologically, as a function of the historical development of surveying; geographically, as local convention or adaptation; or functionally, as meeting different purposes.[22] For example, in place of *centuriae*, land could be marked off in strips called *scamna* (wider than long) or *strigae* (longer than wide); Frontinus and Hyginus suggest that this was an ancient system.[23] Sometimes no internal organization was imposed on land at all, only the area of its irregular

[17] Cf. Frontin. *Contr.* L 24.6–7 = T 10.7–8, Hyg. *Const.* L 168.14–169.9 = T 133.16–134.10, L 194.9–13 = T 157.9–13, Hyg. *Cond. agr.* L 120.18–121.6 = T 83.18–84.7, Sic. Flacc. L 153.9–11 = T 117.7–9, *Lib. col.* L 212.4–7.

[18] Classic studies of the *groma* are *RE* (1) VII 2.1881.54–1886.14 (Schulten); Della Corte 1922. Its basic principles are understood, but reconstruction remains challenging: Dilke 1974: 569–73; Adam 1982; Schiøler 1994; Lewis 2001: 120–33; 2012: 142–4; Guillaumin 2015b. For *rigor*, n.145.

[19] Cf. (with Campbell 1995) Hyg. *Const.* L 199.11–201.6 = T 162.11–164.5, Hyg. *Lim.* L 113.1–18 = T 73.6–24.

[20] Cf. Hyg. *Const.* L 196.15–198.23 = T 159.18–161.24, L 201.7–202.10 = T 164.6–165.9.

[21] To previous n. add Hyg. *Const.* L 202.11–203.6 = T 165.10–166.2, Hyg. *Cond. agr.* L 121.10–18 = T 84.12–20, Sic. Flacc. L 154.22–155.2 = T 118.26–119.6, L 155.23–5 = T 119.28–120.3.

[22] Our concern is with variation reported by surveying authors, but a separate question concerns the consistency of surveyors' reports with archaeological and topographical evidence; the material record is intensely studied: for example, Chouquer 1996–7 (with bibliography); cf. also Dilke 1974; Clavel-Lévêque 1983; Chouquer and Favory 1992: 101–67; Clavel-Lévêque and Favory 1992.

[23] Cf. Frontin. *Agr. qual.* L 3.6–4.1 = T 1.14–15 *more antiquo*, Hyg. *Const.* L 206.9 = T 169.11 *sicut antiqui*. Such systems are called by scholars *scamnatio* or *strigatio* (neither an ancient word). Some (e.g. Hinrichs 1974: 23–48; Chouquer et al. 1987: 12–13; Gabba 1985: 270–2) argue that *scamnatio/strigatio* predated centuriation; this view has been nuanced but continues to receive support (cf. Chouquer and Favory 2001: 115–24, esp. 122–4). Others (Campbell 1996: 85–6; 2000: lx–lxi; Guillaumin 2005: 12–13; cf. Behrends 1992: 237–42) argue that the systems coexisted from an early time and that they were employed under different circumstances.

shape calculated.[24] Even granting centuriation, *centuriae* frequently varied in size, as did the width of *limites*.[25] Sometimes the orientation of the DM and KM were altered, or even reversed so that the KM ran east to west and the DM north to south.[26] There were alternative systems for inscribing the boundary stones numbering *centuriae*.[27] The properties of the location to be surveyed also necessitated adjustments: "certain obstacles" (Frontin. *Art. mens.* L 33.14 = T 17.10 *quaedam deuitanda*) presented by the "manifold nature of places" (L 32.7 = T 16.5–6 *multiplex locorum natura*), such as mountains or rivers, demanded modified techniques. It also mattered whether one was delimiting "uncultivated provincial land" (Hyg. *Const.* L 203.7 = T 166.3 *agrum rudem prouincialem*) or if the place were a *municipium* granted colonial status. In the latter case, where there was already significant human development, the surveyor was enjoined to "inspect the condition of the place and distribute (sc. the land) according to its specific requirements" (L 203.9–10 = T 166.5–6 *condicionem regionis excutiemus et secundum suam postulationem adsignabimus*).

While surveying authors disapproved of some techniques that they believed were illogical and contrary to the norms of practice, they were far from attempting to stamp out all variation. To the contrary, there was a productive flexibility in practice that Frontinus and Hyginus viewed as a virtue of the discipline. Frontinus thought that the "resources of the art" (*Art. mens.* L 32.11 = T 16.10 *artis copiam*) would enable the *mensor* to work around natural obstacles, and he trusted to the surveyor' independent judgment in stating that he will survey difficult terrain "in whatever way it will be best able to be measured" (L 33.17–18 = T 17.13–14 *quacumque ratione optime poterint mensuram accipere*). Hyginus' remarks on different kinds of land are programmatic (*Const.* L 205.5–9 = T 168.5–9):[28]

> nam quem ad modum illis (sc. agris inmunibus et uectigalibus) condicio diuersa est, mensurarum actus dissimilis esse debet. nec tam anguste professio nostra concluditur, ut non etiam per singulas prouincias priuatas limitum obseruationes dirigere possit.

[24] Frontinus calls such land *ager per extremitatem conprehensus* (L 1.4 = T 1.4); it is discussed in Ch. 7.5.5 à propos of Frontin. *Art. mens.*

[25] Variation in *centuriae*: Frontin. *Lim.* L 13.18–20 = T 14.5–8, Hyg. *Const.* T 135.15–20, Sic. Flacc. L 159.9–25 = T 123.19–124.8; Campbell 2000: 378–9 for a synopsis of the archaeological evidence. Variation in *limites*: Hyg. *Lim.* L 12–15 = T 71.6–9.

[26] See n.106; these reversals are discussed in Ch 7.4.4.

[27] Cf. Hyg. *Const.* L 171.14–173.15 = T 136.11–138.13.

[28] I use Thulin 1913 as the base text for Hyg. *Const.* (Ch. 2, n.157).

> For just as they (sc. lands exempt and lands yielding revenue) have different conditions, so the conduct of the survey ought to be different. Our profession is not so narrowly defined that it cannot establish *limites* in accordance with the usage distinctive to particular provinces.

Hyginus embraces local variation (*singulas . . . priuatas*) in practice and makes it a characteristic excellence of the art that it can adapt to these differences.[29] The spirit of these remarks is borne out throughout the *CAR*, from the systematic works attributed to another Hyginus and Siculus Flaccus to tracts such as the *Liber coloniarum* that are essentially lists of local surveying arrangements.[30]

As suggested by Frontinus' and Hyginus' remarks and by other writings in the *CAR*, the *gromatici* did not so much condemn or ignore local variations in practice as acknowledge and incorporate them into their conception of the discipline. This inclusive attitude was practically compelled by the diverse realities of surveyed lands that formed a part of the disciplinary inheritance of centuries of practice. Yet this same variety also posed a threat to the coherence of the *ars mensoria*, since it might seem that Roman land-surveying lacked an appropriately unified and systematic disciplinary structure. Could surveying really be established as one of the "higher *artes*" (Chapter 2), or was the discipline only a catch-all for a loose collection of methods and terminology, strewn over space and time, for measuring and dividing land? This problem was at the heart of early efforts to construct an *ars*, and Frontinus and Hyginus employed strategies that allowed them to engage constructively with the history of land-surveying. Some of their tactics are straightforward, such as division and definition. These belong to the rubric of "systematicity" distinguished for the higher *artes* in Chapter 1 and are well illustrated in application to land-surveying by Frontinus' fragment *De controuersiis*, for example, a small text that identifies and distinguishes fifteen *genera* of land disputes that embrace the range of situations a surveyor might encounter.[31]

In what follows, we will explore two more-elaborate strategies employed by Frontinus and Hyginus in presenting the *ars*. The first such strategy is etymological argument and explanation. The treatises of Frontinus and

[29] For *priuatus* meaning "individual to a person or thing, peculiar, special," see *OLD* 4; *TLL* X 2.1397.63–1398.15 (Kruse).

[30] Cf. Hyg. *Gen. contr.* L 129.7–11 = T 92.13–16, Sic. Flacc. L 139.9–10 = T 103.9–10, L 139.20–2 = T 103.19–104.2, L 140.5–6 = T 104.8–9. The *Liber coloniarum* (Brunet et al. 2008) records variation in *limites*, size and type of plots, boundary markers, and so forth.

[31] L 9.2–26.2 = T 4.2–10.18 (see also Conso 2005: 21). Note that Balbus frames his treatise as an effort to classify the *species qualitates condiciones modos et numeros* of shapes (L 93.14–15), which it is "shameful" (L 93.11 *foedum*) for the surveyor not to know.

Hyginus both offer etymologies for core technical terms of surveying which purport to demonstrate agreement between the original meaning and its contemporary usage in practice. These etymologies are not merely antiquarian but can rather be seen as arguments demonstrating how surveying terminology encodes a primitive conceptual framework for the discipline. The second strategy is rationalizing historical investigation. Historical inquiry into the practices of land-surveying on the one hand reveals a deep and natural origin for the discipline, and on the other serves as a tool for investigating and critiquing variation in surveying norms, which can be understood either as errors or conscious departures from established principle. By revealing the origins of surveying language and practice, highlighting their historical continuity, and demonstrating that the variety of the discipline is capable of explanation on rational grounds, these two strategies shape a coherent identity for the "higher *ars*" of surveying.

7.4.1 The Etymological Horizon of Surveying: Decumanus, kardo, limes

An expression of the particolored reality of surveying practice inherited by the *agrimensores* was the bewildering diversity in technical terminology.[32] The basic units of the survey could, for example, vary according to local and historical *consuetudo* and took different names. The other Hyginus (c. AD 100) remarked that "I have found in a great many places that they call the measure of the field not a *iugerum* but by some (sc. other) name" (*Cond. agr.* L 121.5–122.1 = T 84.27–85.2 *plerisque locis inueni, ut modum agri non iugerum sed aliquo nomine appellarent*). He mentions the names for measures in Dalmatia, Gallia Narbonensis, and Spain besides those of "various other" peoples (L 122.7 = T 85.8–9 *alii ... alii*). The surveyor had to adapt himself to local practice, but to avoid gromatic Babel, Hyginus suggests reporting two sets of measurements, one in the local *modus* and the other with the *iugerum* as a universal standard: "hence, once the *iugera* have been noted in addition, one can make some use of the foreign terms in the operation, but the *iugerum* will remain undisturbed as our standard (*or* principle)" (L 122.11–14 = T 85.13–15 *ita cum iugera adscripta fuerint, licet peregrinis uocabulis possit <agi> aliquid, iugerum nobis ratio sui fidem seruabit*).

[32] For technical terminology in the *CAR*, see Conso 2005 (generally Conso et al. 2005); Fögen 2014: 219–28; Guillaumin 2015a: 10–17; for a glossary, Chouquer and Favory 2001: 406–69. (The claim of Campbell 2000: xxxviii; 2005: 173–4 that the language of the *CAR* is "straightforward and non-technical" is untenable.)

The *CAR* is full of such reflections on the variety of surveying terminology, conveying an impressive linguistic self-awareness.[33]

Besides *peregrina uocabula*, another area of linguistic interest for the *mensores* was the reputedly old and native core of surveying vocabulary. The surveying authors scrutinize certain key terms to explain their contemporary usage, often proposing etymological accounts that clarify potentially obscure senses from the putatively original meanings or formations of the words. Thus, Frontinus, both Hygini, and Siculus Flaccus offer etymological accounts for, inter alia, the names of types of land (e.g. *ager arcifinius*, *territorium*), designations of surveyed areas (e.g. *centuria*, *subseciuum*), and boundaries and roads (e.g. *decumanus*, *kardo*, *limes*).[34] The impulse to etymologize was of course widespread and ancient in Greece and Rome, and served diverse purposes. Beyond their intrinsic interest, etymologies could embody artistic play or literary flourish; they might invite amusement or serious reflection, and could elevate the stylistic or intellectual level of a text through a display of erudition.[35] In the hands of philosophers (especially the Stoics), etymologies could be wielded as tools to investigate the essence of things named, thus uncovering insights forgotten with time. This last usage is closely related to the function of etymology that we will investigate here.[36] We have already seen similar interests manifested in Vitruvius' treatment of the word *architectôn* and Columella's analysis of terms for vine parts.[37]

[33] Cf. Conso 2006: 18–22; Fögen 2014: 223–5. Other good examples: Sic. Flacc. L 154.13–19 = T 118.16–22, Agenn. L 44.17–20 = T 34.13–17.

[34] *Arcifinius/arcifinalis*: Frontin. *Agr. qual.* L 6.1–3 = T 2.12–14, Hyg. *Cond. agr.* L 284.8–9 = T 78.7–8, Sic. Flacc. L 138.8–10 = T 102.6–8; for *territorium*, Frontin. *Contr.* L 19.5–20.2 = T 8.4–6 (possibly interpolated), Sic. Flacc. L 137.17–18 = T 101.9–10; *centuria*: Frontin. *Lim.* L 30.17–18 = T 13.3–5, Sic. Flacc. L 153.26–30 = T 117.26–118.3; *subseciuum*: Frontin. *Agr. qual.* L 6.5–6 = T 2.16–17, Hyg. *Cond. agr.* L 284.1–4 = T 77.22–78.3, Sic. Flacc. L 155.27–8 = T 120.3–4. *Limes*, *decumanus*, and *kardo* are discussed in this section. Other etymologies in the older treatises include *iugum* (Hyg. *Gen. contr.* L 128.11–12 = T 91.12–13), *municipium* (Sic. Flacc. L 135.18–19 = T 99.7–8), and various adjectives for *limes* (e.g. *quintarius*, *maritimus*, *montanus*). The *mensores* share many etymologies, but differ sharply for some words such as *centuria* and *decumanus* (see the subsequent discussion and n.52).

[35] For ancient etymological theories, see Amsler 1989: 15–56; O'Hara 2017: 7–56. Varro's *De lingua latina*, although fragmentary, remains the most important Latin etymological work: Schröter 1960; Dahlmann 1964; Cavazza 1981; now De Melo 2019. See also the following n.

[36] Since Plato's *Cratylus* (Sedley 2003), philosophical etymologizing had aimed to recover knowledge possessed by earlier generations, but Stoic etymology offers perhaps the best *comparandum* for the *CAR*. For the Stoics, "etymological explanation affords a glimpse of what the early makers of names thought about the items on which they imposed the names" (Allen 2005: 34); such knowledge is valuable because it reflects "pristine cosmological beliefs" (Long 1992: 56) or "primitive wisdom" (Allen 2005: 15). Frontinus and Hyginus may have been influenced directly or indirectly by Stoic practices, perhaps mediated by Varro.

[37] Ch. 3.1.1 (n.19), Ch. 4.4.6 (n.163).

Modern reactions to etymologies in the *CAR* have diverged, often dramatically. Some scholars have focused on the question of their historical and linguistic veracity and stressed that they are historically unreliable.[38] Others set such issues aside and emphasize the sense in which the etymologies are an index for the intellectual ambitions of their authors.[39] Along these lines, it has been suggested that certain etymologies carry political or ideological content meant to advance the surveyors' view of their *ars* in Roman history.[40] As suggested earlier, others are skeptical of etymologizing as such in the older treatises of the *CAR*, maintaining that almost all of the etymologies are later interpolations.[41] In line with our assumptions, we will handle the etymologies as though they were authentic; if they are partly or wholly interpolated, this would not indicate that they lack intelligence, only that they are the product of a more gradual and collaborative process in the intellectualization of surveying.[42] In examining the etymological accounts offered by Frontinus and Hyginus, we will focus on the core vocabulary for the balks or roads that served as borders in centuriation: *decumanus*, *kardo*, and *limes*. Both authors offer similar etymologies for these words and present them in similar contexts; this is of a piece with the broader structural parallelism between Frontinus' *Lim.* and especially the opening passages of Hyginus' *Const.* that we noted. These etymologies are clearly not a pragmatic strategy for dealing with the complications of surveying terminology, as in the inscriptional technique of the other Hyginus mentioned earlier; yet it is too simplistic to see them only as the products of antiquarianism. Rather, the etymologies function in the context of a larger intellectual strategy as arguments for a rational and historically coherent view of the discipline. In particular, Frontinus and Hyginus treat the key surveying terms as linguistic artifacts that preserve traces of ancient concepts upon which the practice of the discipline is based.

Frontinus' *Lim.* and Hyginus' *Const.* both begin with a historical account of the origins and development of centuriation. We will return to this account later, after considering the etymologies embedded within it. Frontinus and Hyginus remark that "the ancients . . . first established two *limites*" (Hyg. *Const.* L 157.3–5 = T 132.6–8 *antiqui . . . primum duos limites constituerunt*; cf. Frontin. *Lim.* L 28.5–7 = T 11.9–11 *maiores nostri . . . primum duo limites duxerunt*). Mention of the Roman "ancestors" or "ancients"

[38] Gaide 2006, esp. 39. [39] Pikulska 2004: 213; Guillaumin 2010: 161; Fögen 2014: 221–2.
[40] Cuomo 2001: 207; Hermon 2003: 148–50, 155; Guillaumin 2004, esp. 45; 2009c: 134–5; 2010: 162–3.
[41] Lindermann 2013; cf. 2018: 20–1, 39. Accordingly, Lindermann et al. athetize most etymologies in their edition (see *passim*).
[42] See Ch. 7.1 and n.8.

recalls us to a foundational moment in land-surveying; the etymologies that follow provide access to the language and thought of the early progenitors of the art. Stating that the main *limites* established by the ancients were the *decumanus* and *kardo*, Frontinus and Hyginus give etymologies for these terms that are meant to square with contemporary usage.

We begin from *decumanus*, the main east–west *limes*. Here is Frontinus' version (*Lim.* L 28.11–15 = T 11.15–12.2):[43]

> quare decumanus a decem potius quam a duobus, cum omnis ager eo fine in duas diuidatur partes? ut duopondium et duouiginti, quod dicebant antiqui, nunc dicitur dipondium et uiginti, sic etiam duodecimanus (*v. l.* duocimanus) decumanus est factus.

> Why is *decumanus* from ten (*decem*) rather than from two (*duo*), although the whole field is divided into two (*duo*) parts by that boundary? Just as the words pronounced *duopondium* and *duouiginti* by the ancients are now pronounced *dipondium* and *uiginti*, so too has *duodecimanus* become *decumanus*.

There are difficulties with the text which we will discuss momentarily, but the terminological worry at least is obvious: the spelling *dec-* suggests a derivation from the number *decem*, but this connection seems irrational if the *decumanus* (the DM) splits the field not into tenths but halves.[44] The solution to this problem depends on the constitution of the text. One possibility is to write *duocimanus* for *duodecimanus*.[45] In this case, Frontinus would be postulating an original word somehow incorporating the meaning "half" or "two" – a sense in agreement with the fact that the *decumanus* divides the field into two (*duas*) parts – and showing how *duo-* had changed to *de-* over time.[46] The analogy *duopondium* > *dipondium* appears more relevant here than *duouiginti* > *uiginti*, so the latter might be deleted (but does not have to be).[47] An alternative is to retain *duodecimanus*. This choice is prima facie difficult because, though it

[43] I use Thulin 1913 as the base text for Frontinus' surveying writings; see Ch. 2, n.145, for other editions consulted.

[44] This reasoning is circular, however, because it is only on the assumption that *decumanus* must be related to *duo* that the apparent trace of *decem* is problematic (cf. n.52).

[45] So Goesius 1674: 215; Lachmann et al. 1848–52: II 342; Mommsen 1892: 91 = 1905–13: V 96; *duocimanus* is printed by Thulin but not Lachmann. *Duocimanus* is found in **P** in the corresponding sentences in Hyg. *Const.* (L 167.6, 10, 13 = T 132.9, 13, 16). I prioritized this reading in Zainaldin 2020b: 252–6.

[46] Rudorff in Lachmann et al. 1848–52: II 342 believes this etymology presupposes *duo* + *caedere* or *decidere*. The illustrations in the MSS (Lachmann fig. 27 = Thulin fig. 25 = Campbell ill. 25) are based on this etymology, which is not to say that they are authoritative.

[47] For *duopondium* > *dipondium*, cf. Varro, *Ling.* 5.169, 9.83, and see Maltby 167; for spellings of *dupondius* (or *-um*), *TLL* V 1.2285.9–32 (Bannier). Thulin deletes *et duouiginti* and *et uiginti* at Frontin. *Lim.* 11.17–12.1 (but not at Hyg. *Const.* 132.16) as glosses explaining *duodecimanus* (which he regards as the false reading).

restores *duo-*, it replaces the original problem ("why from ten?") with another ("why from twelve?").[48] In support, we might say that the analogy *duouiginti* > *uiginti* presupposes that *duodecimanus* rather than *duocimanus* was written.[49] A stronger reason to retain the paradosis, however, is the possibility of a cogent interpretation of *duodecimanus* based on a feature of the DM emphasized by both Frontinus and Hyginus: in the sense "twelfth," the word would refer not to a fraction of land but to the hour of the day (*duodecimana hora*) when the sun set, from which could be derived the east–west bearing that this *limes* traced (see also later in this section).[50] If Frontinus believed this and wrote *duodecimanus*, he was not intending to show thereby that *decumanus* originally meant that the line divided the field in "half," only that the apparent relation of *decumanus* to *decem* is specious. But of course, if Frontinus did write and understand *duodecimanus*, why would he highlight the notion of division (*in duas diuidatur partes*) and not bother to explain its true meaning?

Hyginus' text shows similar problems.[51] Multiple explanations for corruption can be entertained (and different answers might be given for each author), but whatever one prefers, the basic function of the etymological argument remains intact. *Decumanus* will either derive from *duocimanus*, so called because it cuts the field into "two" (*duas*) parts, or from *duodecimanus*, because of its patterning after the east–west course of the sun. Either explanation would defuse concerns about the obscure relation with "ten" (*decem*) and shed light on some fundamental property of *decumanus* which remains true in surveying practice. Finally, it bears remarking that other etymologies for *decumanus* are found in the *CAR*, including one given by Siculus Flaccus that preserves the meaning "tenth" but suggests that, as practice changed, the word evolved to acquire a new meaning distant from its root (*decem*).[52] It is perhaps noteworthy that

48 *Duodecimanus* is not otherwise attested but looks like it would mean "related to that which is twelfth," as *decumanus* can mean "related to that which is tenth" (though note the peculiarities in this word's meaning: Otto 1903: 18–19; Leumann 1977: 325; *OLD decimanus*[1] and *decimanus*[2]).

49 Cf. n.47; of course, in this case *duopondium* > *dipondium* loses its force, so this datum may seem equivocal.

50 Lindermann 2013: 129–32 argues that the whole passage etymologizing *decumanus* and *kardo* in Frontinus (*Lim.* L 28.11–17 = T 11.15–12.4) is an interpolation; he maintains that the derivation *duodecimanus* > *decimanus* in Hyginus (*Const.* 132.9–17) is authentic, adducing, inter alia, the association between *kardo* and *hora sexta* (L 170.8 = T 135.6), but would delete material relating to *duocimanus* (Lindermann et al. 2018: 113). For *duodecimanus* as pointing to the twelfth hour, see Salmasius 1629: 675–6; Forcellini 1771: I 390; Josephson 1950: 55; Lindermann 2013: 131–2; 2018: 20, 39, 214.

51 Lachmann again prints *duodecimanus* everywhere and Thulin again *duocimanus*; see also previous n.

52 According to Siculus, captured lands were sold through quaestors in fifty-*iugera* squares, and because the *limites* marked ten *actus* each, all *limites* were originally called *decumani* (L 152.25–7 = T 116.20–4; cf. L 153.12–13 = T 117.9–10). Siculus makes clear (T 117.9–21 = L 153.11–22) that his account implies a change in the meaning of *decumanus*. Isidore gives a third, completely different etymology: *Orig.* 15.14.4 = L 367.2–3 *qui* (sc. *decumanus*) *pro eo quod forma X faciat, decumanus est appellatus*.

Frontinus and Hyginus suppose a greater degree of consistency in the meaning of the word over time. In any case, their etymologies represent the present usage of the word as both rational and grounded in practice.

The other main *limes* established by the *antiqui* was the *kardo*. The etymological explanation for this word is more straightforward but equally revealing. Frontinus' version is as follows (*Lim.* L 28.8–9, 15–17 = T 11.12–13, 12.2–3):

> alterum (sc. limitem duxerunt) a meridiano ad septentrionem, quem kardinem appellauerunt . . . kardo nominatur quod directus a kardine caeli est.[53]

> (sc. They drew the other *limes*) from south to north, which they called the *kardo* . . . The *kardo* is so called because it was oriented from the "hinge" (*kardo*) of the heavens.

Hyginus' account is virtually identical.[54] The information preserved in *kardo* is that of orientation or directionality.[55] *Kardo* means "pivot," "axis," or "hinge," as for example of a door; it was extended on the basis of the geocentric model of the universe to the figurative "hinge" on which the universe revolves around the Earth.[56] Because this revolution occurs from east to west, the *kardo* was conceived (in the northern hemisphere) as the north pole. The extension of the name of the celestial *kardo* to the *limes* is at the same time a conceptual transfer, so that the *kardo-limes* comes to share the orientation of the *kardo mundi*; a corollary is that the *decumanus*, which runs perpendicular to the *kardo*, must be oriented from east to west. The etymological explanation, which consists in the shift of a metaphor from one sphere to another,[57] may be simple, but it establishes a practical imperative by stipulating the proper orientation of the *limes*. Thus, both Frontinus and Hyginus appeal to the homology of the celestial *kardo* and the *kardo-limes* to censure those who adopt other orientations in centuriation. Frontinus, for example, says that it was because "some were later ignorant of this fact" (*Lim.* L 29.1 = T 12.5 *postea hoc ignorantes non nulli*) – that is, the origin of

[53] Thulin deletes as a gloss the short sentence that follows: L 28.16–17 = T 12.3–4 *nam sine dubio caelum uertitur in septentrionali orbe.*

[54] Hyg. *Const.* L 167.8–9 = T 132.11–12 *alterum* (sc. *limitem*) *a meridiano ad sepentrionem; quem kardinem nominauerunt a mundi kardine.* Siculus Flaccus (L 153.20–2 = T 117.19–21) and Isidore (*Orig.* 15.14.4 = L 366.23–367.1) have the same account; see Maltby 108.

[55] A virtue of retaining *duodecimanus* is to connect the etymology of *decimanus*, too, with the orientation of the *limes* (n.50).

[56] *OLD* 1, 3; *TLL* III 442.79–83, 443.80–3 (Bannier).

[57] Compare the second of four categories of Stoic etymological derivation given by Augustine at *Dialect.* 6, *ipsarum inter se rerum similitudo*, with Jackson 1975: 128; Allen 2005: 16–17. Of course, the transfer of *kardo* entails no phonological change (unlike Augustine's example *crux* > *crura* [*crus*]).

kardo – that they reversed the direction of the KM and DM or measured out the DM in whichever direction the field is longest.[58] Moreover, when noting variations in the direction of the central *limites*, Hyginus complains that "it is irrational (*or* it lacks a principle) to orient the *kardo* from east to west, because the name itself of the *kardines* is called after the *kardo* of the universe" (L 182.1–3 = T 146.3–5 *cum ipsa kardinum appellatio a kardine mundi nominetur, quare ab oriente in occidentem dirigatur, nulla est ratio*).

The final example we will consider is *limes*, etymologized by Frontinus and Hyginus following *decumanus* and *kardo*. The word is ubiquitous in surveying of the uncultivated strips of land, typically used as public roads or passageways, within or between fields. As we noted, the establishment of *limites* was virtually synonymous with centuriation, so that it was usually called *limitatio* or *constitutio limitum* among similar expressions.[59] It is not surprising, therefore, that Frontinus and Hyginus would think *limes* worthy of linguistic scrutiny, especially because it does not transparently bear in its parts the origin of its meaning. *Limites* had many specific appellations,[60] but two ancient ones that Frontinus and Hyginus distinguish are *limites prorsi* and *limites transuersi*. The distinction relates to the orientation, those facing east called *prorsi* ("straight") and those south *transuersi* ("across").[61] The etymology given for *limes* piggybacks on the class of *limites* called *transuersi*. Frontinus' version runs as follows (*Lim.* L 29.9–10, 13–16 = T 12.13–15, 13.2–5):

> qui (sc. limites) spectabant in orientem, dicebant (sc. maiores nostri)[62] prorsos: qui dirigebant in meridianum, dicebant [et] transuersos . . . limites autem appellati transuersi s<unt> a limo, [id est] antiquo uerbo [transgressa]; a quo dicunt poetae 'limis oculis'; item limum cinctum, quod purpuram transuersam habeat, et limina ostiorum.

> (sc. *Limites*) which looked to the east (sc. our ancestors) called "straight" (*prorsos*); those which faced toward the south, they called "crosswise" (*transuersos*) . . . As for *limites*, being "across" (*transuersi*) they were named after (sc. the word) "oblique" (*limus*), an ancient word that is used by the poets in the phrase "sidelong eyes" (*limis oculis*); (sc. consider) also the *limus cinctus*, which has a purple stripe across it, and the thresholds (*limina*) of doors.

[58] Cf. Frontin. *Lim.* L 29.1–6 = T 12.5–10, and see Ch. 7.4.4. [59] See n.11.

[60] Chouquer and Favory 2001: 438–41; cf. the *nomina limitum* at L 247.21–249.31.

[61] The use of *prorsus* for east–west and *transuersus* for north–south assumes a particular orientation: for the (ritual) significance of the surveyor's perspective, see Ch. 7.4.3 on *dexter, sinister, antica*, and *postica*.

[62] Although this subject far precedes (L 28.5 = T 11.9), it is the only real possibility.

The passage is textually desperate, but the thrust of Frontinus' remarks is clear; Hyginus' explanation is neater but omits the poets.[63] The argument relies on two premises: (1) *transuersus* is a proper description of *limites*, and (2) *limus* is a synonym of *transuersus*. The first premise occasions some difficulty, since it is only some *limites* that are called *transuersi* (others are *prorsi*): do *limites* in general acquire their name by a sort of synecdoche with this subclass, or is *transuersus* taken differently here – that is, oblique not with respect to the east–west axis but to other *limites* in the grid system? However one chooses to resolve this issue (Hyginus' solution is given in a moment), admitting that *limites* are in some sense "crosswise" or "across" (*transuersi*) opens the door for a connection with *limus*, an *antiquum uerbum* whose meaning is preserved in the poetic expression *limis oculis*, in the formulaic *limus cinctus*, and indirectly in the noun *limen* (the threshold, which runs "across" a door).[64] The etymology elucidates *limes* by showing how the stem *lim-* linguistically encodes one of the pertinent features of *limites* – that is, that they run "across" (*transuersus* ~ *limus*) the surveyed territory.

Frontinus and Hyginus also offer a second etymology for *limes* from *limen* in slightly different presentations. For Hyginus, the etymology from *limen* seems to supplant that from *limus*, updating the story to explain the meaning not only of *limites transuersi* but also *prorsi* (*Const.* L 168.1–3 = T 133.2–4):

> postea et prorsos et transuersos limites appellauerunt a liminibus, quod per eos agrorum itinera seruentur.

> Later, they called both *limites prorsi* and *transuersi* after *limina*, because they provide ways into the fields.

Frontinus' version has *alii* ("others") in place of *postea*.[65] The difference is significant, for in Frontinus' account the etymology is thus split into alternative explanations, whereas in Hyginus' the narrative is continuous. What both accounts share, however, is to diminish the contribution of the adjective *limus*

[63] Cf. Hyg. *Const.* L 167.16–168.1 = T 132.18–133.2 *qui* (sc. *limites*) *spectabant in orientem prorsos, qui ad meridianum, transuersos appellauerunt* (sc. *antiqui*). *limites autem appellati a limo, id est antiquo uerbo transuersi: nam et limum cinctum ideo quod purpuram transuersam habeat, item limina ostiorum.* Lachmann replaces *transgressa* in Frontinus (omitted in **F**) with *transuersi* from Hyginus, but still secludes.

[64] *Limis oculis* is poetic in Republican Latin; it appears in prose from the first century AD: *OLD limus*[2]; *TLL* VII 2.1427.34–64 (Balzert); for the *limus cinctus*, *OLD limus*[3]; Timpanaro 1986: 58–63 and Tarrant 2012: 121–2 on Verg. *A.* 12.120; Zollschan 2011: 57–62. For the etymology relating *limus* and *limen*, cf. Paul. Fest. p. 116 *limis obliquus, id est transuersus; unde et limina*; see Maltby 341.

[65] *Lim.* L 29.16–18 = T 13.6–7 *alii et prorsos et transuersos dicunt limites a liminibus, quod per eos in agro intro et foras eatur.*

and make the pertinent aspect of *limen* not the fact that it runs "across" (as of doors, so *limina ostiorum*), but that it marks a place of coming and going (cf. *per eos agrorum itinera*).[66] (This association alludes to the fact that *limites* typically served as roads or passageways.) Even in Frontinus' explanation, which more clearly separates the second etymology from the first, there is a natural economy to the presentation of the two etymologies thanks to the reuse of *limen*. The implicit phonological relationship through *lim-* underpins the conceptual transfer regardless of whether one chooses to emphasize that *limen* means "across" (*limus/transuersus*) or that it marks a passageway. Hyginus supposes that both meanings influenced the development of the word, and we should not assume that because Frontinus distinguishes the etymologies he thinks them contradictory.[67] However that may be, the second explanation for *limites* is germane insofar as it reveals the coincidence between the original meaning of the word (a passageway) and a fundamental property of the balks.

Altogether, the etymologies for *decumanus*, *kardo*, and *limes* in the texts of Frontinus and Hyginus demonstrate a concern to account for the linguistic inheritance of land-surveying. By laying bare the origin of key surveying terms, the etymological arguments recover essential features of the objects they describe: The words are thus treated as historical vehicles for concepts and practices that can be traced to the early history of the discipline, when the *maiores* or *antiqui* bequeathed the elements of the *ars mensoria* with their proper names. Put another way, etymological analysis allows for insight into the ancient principles of the discipline – principles which originally defined Roman land-surveying and still inform practice. The successful prosecution of these etymologies thus not only reinforces the rational character of the core surveying vocabulary but also promotes a historically coherent view of the art.

We noted that the authenticity of the etymologies we studied in Frontinus and Hyginus is uncertain, and they are complicated by severe textual problems. Even so, these are only a handful of the numerous etymologies scattered throughout the early surveying treatises and capture the spirit of linguistic self-reflection in the *CAR*.[68] Of course, the historical

[66] The etymology is given also by Siculus Flaccus (L 153.7–9 = T 117.5–7) and Isidore (*Orig.* 15.14.2 = L 366.14–19); the latter's narrative is similar to Hyginus'. A different derivation of *limes* from *limen* is given at *Inst. Iust.* 1.12.6.

[67] Varro does not avoid double etymologies, and it was believed that "one and the same word could have two separate origins" (De Melo 2019: I 43): cf., for example, *Ling.* 5.131, 6.31, 7.16, 7.57, 7.94.

[68] Cf. n.34. Further etymologies could be added from later glossary material, for example Isid. *Orig.* 15.13 = L 368.19–370.30.

and geographic variety of surveying terminology was such that it would have been a Herculean task to furnish etymologies for every word that the surveyor might encounter. But that did not compel Frontinus, Hyginus, or others to compromise on their rationalizing view of the discipline. To the contrary, they leave us with the suggestion of a larger phenomenon governing the local evolution of surveying language. Frontinus thus says of *limites* that "they have received different names from residents of different areas based on their location and the nature of the place" (*Lim.* L 29.18–30.2 = T 13.7–9 *hi ab incolis uariis ac dissimilibus uocabulis a caeli regione aut a loci natura sunt cognominati*).[69] The *regio caeli* and *natura loci* are thus sources – not necessarily the only ones – for regional investigation of surveying *uocabula.* Having shown the way, Frontinus and Hyginus leave it to the surveyor (reader) to work out the details of their open-ended project.

7.4.2 *Order from Chaos: Accounting for Surveying Practice*

Surveying practice in the first century AD exhibited the same if not a greater degree of variety that characterized surveying terminology. Frontinus indeed remarks that "throughout the whole world there are individual systems for *limites*" (*Lim.* L 31.9–10 = T 14.19–20 *per totum orbem terrarum est unaquaeque limitum constitutio*). Although the thought is truncated by textual disturbance,[70] the point of this statement, which follows a discussion of local differences in orienting *limites,* is apparently to connect the varied systems of *limites* exemplified in many surveyed regions with the local factors that would explain them. The surveyors were acutely conscious of such variation in practice, and much of the *CAR* is indeed devoted to capturing different historical and local configurations of land measurement. While Frontinus and Hyginus, like other authors in the *CAR,* did not reject this variety – it was simply a reality to be accepted – they strove to mitigate concerns that the principles of surveying were arbitrary or, worse, contradictory. To that end, both authors developed strategies for handling obscure and variable elements in the key practices of centuriation that are comparable to the strategies used to explain surveying

[69] So Hyg. *Const.* L 168.3–4 = T 133.4–5, Sic. Flacc. T 117.14–16; the examples given are *limites montani* and *maritimi.* Frontinus says the *limites* of Fanum Fortunae in particular were so called: cf. *Lib. col.* L 256.13–15, with Alfieri 1976–7; Campbell 2000: 326–7.

[70] The following clause begins *ubi proxima* (L 31.10 = T 14.20–1) before trailing off; a copyist supplemented *sibi plagarum caeli conuersio fuit* in the twelfth-century London MS Add. 47679 (Guillaumin 2005: 229).

terminology. Etymology allowed the surveyors to reconstruct a core vocabulary that was consistent both with the putatively proper significations of the words given by their forebears (*maiores*/*antiqui*) and with the facts of contemporary practice. By similarly inquiring into surveying history and practice (and not merely language), Frontinus and Hyginus could establish a standard for the discipline that would attribute meaning to its principles and methods congruent with the origins of the *ars* and provide a way of accounting for deviations from this standard.

As we observed in the sketch of *limitatio* and, indirectly, regarding the etymologies for *decumanus* and *kardo*, the ideal of centuriation was based on an orthogonal grid system centered on the intersection of the east–west *decumanus maximus* (DM) and the north–south *kardo maximus* (KM). Frontinus and Hyginus offer broadly similar accounts of this ideal, which can be analyzed into two parts. The first part is positive: the orthogonal system derives, ultimately, from the Roman adaptation of an Etruscan augural practice rooted in the natural order (*ratio*) of the universe. The origin of the orthogonal system thus establishes a principle for surveying practice whose normative implications are continued to the present. Furthermore, because this principle derives its authority from the cosmic order, it naturalizes the Roman model of centuriation – or, we might equally say, Romanizes a natural partitioning of space, fusing an objectively cognized pattern of natural phenomena with a subjectively (that is, culturally) preferred means of organizing space.

The second part is negative: Frontinus and Hyginus maintain that Roman surveyors who have deviated from the ideal of centuriation did so either because they erred in applying the method or because they were guided by exigencies peculiar to a specific survey. The upshot is that differences in local surveying configurations do not reflect a chaotic multiplicity of principles all with equal claims to be a genuine basis for centuriation; they merely represent departures in the practical application of theory, whether unintentional or intentional. These departures are not all equally culpable but are rather censured only to the degree that they present a genuine challenge to the theoretical principles of the ideal. The two parts of the historical account provided by Frontinus and Hyginus, positive and negative, together reconstruct the descent of disciplinary practice to the present: by deriving a principle for centuriation from an ancient augural procedure grounded in nature and showing how differences in practice can, with the aid of rationalizing explanation, be understood in relation to this principle, they construct a theoretical foundation for an *ars* of surveying.

7.4.3 The Celestial Origin of Centuriation

I begin with the positive part of the account. Frontinus and Hyginus proffer similar though not identical narratives of the relationship between Etruscan haruspicy and the orthogonal axes used in centuriation. Both accord the narratives a prominent place in framing and motivating the discussion of *limitatio*. This is especially easy to see in the case of Hyginus, who weaves the Etruscan story into the rhetorical opening of the work (*Const.* L 156.3–167.2 = T 131.1–132.5):[71]

> inter omnes mensurarum ritus [siue actus][72] eminentissima traditur limitum constitutio. est enim illi origo caelestis et perpetua continuatio, cum quadam latitudine recturae diuidentibus ratio tractabilis, formarum pulcher habitus, ipsorum etiam agrorum speciosa designatio. constituti enim limites non sine mundi ratione, quoniam decumani secundum solis decursum diriguntur, kardines a poli axe. unde primum haec ratio mensurae constituta ab Etruscorum haruspicum [uel auctorum habet, quorum artificium] disciplina; quod illi orbem terrarum in duas partes secundum solis cursum diuiserunt, dextram appellauerunt quae septentrioni subiacebat, sinistram quae ad meridianum terrae esset, <ab oriente ad>[73] occasum quod eo sol et luna spectaret; alteram lineam duxerunt a meridiano in septentrionem, et a media ultra antica, citra postica nominauerunt. ex quo haec constitutio liminibus templorum adscribitur.

> Among all of the rites of surveying, the loftiest that has been handed down to us is that of the establishment of *limites*: it possesses an origin in the heavens and an unbroken tradition, as well as a method for those dividing (sc. land) that is easily implemented in conjunction with straight lines of a certain breadth;[74] it also yields a beautiful appearance for maps and an attractive[75] disposition for the fields themselves. For *limites* were established in concert with the order (*or* principle) of the universe, since *decumani* are oriented according to the course of the sun and *kardines* from the axis of the

71 Lindermann et al. 2018: 39 (cf. 14–15) delete certain statements regarding Etruscan haruspicy (in particular *quod . . . nominauerunt*) as interpolations drawn from Frontinus.

72 **A** transmits *ritus siue actus*, **B** *siue actus* in superscript after *eminentissima*; Thulin retains, but Lachmann's deletion (as a gloss) is preferable.

73 Frontinus' text has a lacuna in the same place (L 27.16–17 = T 11.3, n.76); the supplement *<ab oriente ad>* in both instances is Lachmann's. Hübner 1992: 156–70 argues that uncertainties in text and interpretation (especially prepositions) may have sprung from difficulties interpreting an original Etruscan document.

74 Clavel-Lévêque et al. 1996: 3n1 (followed by Guillaumin 2005: 167–8) implausibly suggest that *cum quadam latitudine* refers to "*copia artis*" (cf. Frontin. *Art. mens.* L 32.11 = T 16.10), translating "avec une certaine latitude." The phrase is awkward to render but makes sense (compare the translations of Campbell and Lindermann et al.).

75 Lindermann et al. translate *speciosus* as "deutlich," but the word does not usually mean that; an aesthetic appraisal is appropriate after *formarum pulcher habitus.*

> heavens. From here (sc. the celestial phenomena), this principle (*or* method) of surveying was established first in the discipline of the Etruscan *haruspices*. Inasmuch as they (sc. the Etruscans) divided the Earth into two parts according to the course of the sun, they called that (sc. part) "right" which fell under the north, and that "left" which was toward the south of the Earth, (sc. proceeding) from east to west because the sun and moon face there (sc. to the west). They (sc. the Etruscans) drew another line from south to north, and from the middle called the further (sc. part) "in front" (*antica*) and the nearer (sc. part) "behind" (*postica*). Hence the same establishment (sc. of parts) is ascribed to the thresholds of temples.

Hyginus praises the received *constitutio limitum*, by which he means the orthogonal system of centuriation based on *decumani* (east–west *limites*) and *kardines* (north–south). It is commended for aesthetic and practical reasons and, of particular moment, for its "origin in the heavens and an unbroken tradition." The rest of the passage explains this system by reference to the augural practice of Etruscan *haruspices*, who are said to have been the first to spatialize the world in this natural way. Frontinus' version is close to Hyginus' except that it lacks an encomium of the traditional *constitutio*, does not so clearly stress a natural (celestial) principle for *limites* independent of the *disciplina Etrusca*, and has a slightly different order of material.[76] Frontinus also adds a parenthetical statement crediting Varro with drawing the connection between the *prima origo limitum* and Etruscan augury; Varro should therefore stand somewhere behind Hyginus' account as well.[77]

The "Etruscan account," as I will call Frontinus' and Hyginus' narrative relating the orientation of the *limites* and Etruscan haruspicy, functions in multiple ways to lend significance to the orthogonal system on which Roman centuriation was based. (This remains true even if one doubts the historical veracity of direct Etruscan influence on Roman surveying practices.[78]) First, it grounds the orthogonal system in nature by locating

[76] Frontinus' version: *Lim.* L 27.13–28.4 = T 10.20–11.8 *limitum prima origo, sicut Varro descripsit, a disciplina Etrusca; quod aruspices orbem terrarum in duas partes diuiserunt, dextram appellauerunt <quae> septentrioni subiaceret, sinistram quae a meridiano terrae esset <ab oriente ad> occasum, quod eo sol et luna spectaret, sicut quidam †carpiunt† architecti delubra in occidentem recte spectare scripserunt. aruspices altera linea a septentrione ad meridianum diuiserunt terram, <et> a me[ri]dia[no] ultra antica, citra postica nominauerunt.* The words *sicut . . . scripserunt* have no equivalent in Hyginus' prefatory statements but are paralleled later in *Const.*, L 169.14–170.2 = T 134.15–21.

[77] Previous n.

[78] The question of influence is difficult: cf. Thulin 1905: 26–30; Hinrichs 1974: 80–1; Le Gall 1975, esp. 301–8; Dilke 1976; Gabba 1985: 267–8; Dilke 1987: 202–4; 1988: 158–9; Behrends 1992: 218–19; Hübner 1992 (survey of earlier opinions on 144); Gargola 1995: 42–3; Clavel-Lévêque et al. 1996: 3–4n2; Gargola 2004: 127–9; Guillaumin 2005: 16–17. The firmest evidence remains the probability

its principle in the celestial phenomena. This is clearest in Hyginus' version: even before introducing the *Etrusca disciplina*, Hyginus says that *limites* are established "in concert with the *ratio mundi*," following the path of the sun and the axis of the heavens. The *ratio mensurae* is modeled after the *ratio mundi*. Hyginus casts the Etruscan haruspices as mediating the transfer of this *ratio* from the heavens to the Earth, because in their divinatory practice, they used the course of the sun to divide the world into ritually significant spaces. It bears emphasizing that Hyginus does not say that the Etruscans invented the *ratio mundi*, which belongs by definition to the natural structure of the universe; they were simply the first (cf. *primum*) to draw it down from this source (cf. *unde*) and show how it could ground a human *ratio*.[79] Having established northern and southern *partes* of the world from the sun, the Etruscans next drew a north–south line parallel to the *axis poli*, partitioning the world into four spaces. These spaces were designated according to the perspective of the auspicant, who is imagined as standing in the middle (*a media*) of the intersection and facing west: the northern and southern halves are thus *dextra* and *sinistra*, respectively, and the western and eastern *antica* and *postica*. In this way, the divisions patterned after the *cursus solis* and *axis poli* allow the auspicant to assign a local identity to all points in his field of vision. The chief virtue of the Etruscan procedure is thus to offer a fundamentally natural way of spatializing the world which the *agrimensores* could turn to their own purposes. More on this shortly.

A second reason that the Etruscan origin story is significant for Frontinus and Hyginus is that it affiliates the practice of surveying with other spheres of official activity at Rome, both religious and political, that employed similar spatializing procedures based on orthogonal axes.[80] Roman rituals involved in dividing and designating spaces formally sanctioned them for their intended use and imposed order and law upon them. A striking comparison with surveying is found in the rites for the creation of the *templum*, the multipurpose ritual space which might take the form of a temporary field of augural vision projected onto land or sky or a clearly demarcated space on the ground in which official actions would occur, such as the establishment of a temple structure or the meeting of the

that the word *groma* enters Latin from Greek γνώμων (or γνῶμα?) via an Etruscan intermediary: *TLL* VI 2.2335.26–7 (Krohn); De Simone 1970: II 271–3; Leumann 1977: 55.

[79] Cf. Lindermann 2018: 14–15.

[80] See Gargola 1995: 25–50, esp. 39–50; 2004; 2005; 2017: 154–86, esp. 175–8; cf. Behrends 1992: 213–44, esp. 227–37.

Senate.[81] In creating *templa*, Roman augurs divided the space into areas that were also designated *sinistra* or *dextra* and *antica* or *postica*.[82] Temple structures were constructed on *templa* spaces and could accordingly bear similar divisions. Hyginus invites comparison of this practice with centuriation when he notes that the thresholds of temples are based on the same principle of spatial division described in the Etruscan account (*Const.* L 167.1–2 = T 132.4–5).

Both Frontinus and Hyginus also explicitly connect the physical perspective adopted in the Etruscan account with the orientation of temples. When drawing a line to divide the Earth, the Etruscan *haruspices* are said to trace it "from east to west because the sun and moon face (*spectaret*) there."[83] Some ancient architects, Frontinus and Hyginus add, have thus also prescribed that temples "correctly face (*spectare*) west."[84] In fact, Vitruvius does state that temples ought to face (*spectare*) west so that the worshipper may be turned toward the east when at the altar.[85] The recurrence of *spectare* calls attention to the shared ritual importance of directionality, highlighting an important similarity between Roman augury and religion and land-surveying via the Etruscan account.[86] It is easy to see why orientation is so significant when one considers that the designations *sinistra*, *dextra*, *antica*, and *postica* are relative, not absolute, positions, and thus depend on the direction where the haruspex or augur looked (*spectare*). The details of visual orientation in augury are complicated, but it was of fundamental importance in demarcating *templa* and temple spaces.[87]

While the Etruscan account effects an important transfer of meaning to centuriation, it does not make centuriation itself a religious or augural practice. Indeed, neither Frontinus nor Hyginus says anything of the sort when deriving the *constitutio limitum* from Etruscan haruspicy; the Etruscan practice is emphatically only a foundation (*fundamentum*) or

[81] For *templa*, see Catalano 1960: 248–72; Humm 2014, esp. 319–21; Gargola 2017: 124–30, 154–67, and following n.

[82] Cf. Varro, *Ling.* 7.7, Liu. 1.18.7; on these passages and *inauguratio*, see Norden 1939: 3–106; Catalano 1960: 211–319; Magdelain 1969; Catalano 1978: 467–79; Linderski 1986: 2256–96; Gargola 2004: 130–1; Short 2008: 122–6; Guittard 2009; De Melo 2019: II 904–9. Cf. Paul. Fest. pp. 244–5 and L 303.22–304.1 (from the extract *ex libris Dolabellae* in the *CAR*).

[83] For <*ab oriente ad*>, see n.73.

[84] Frontin. *Lim.* L 28.1 = T 11.5; cf. Hyg. *Const.* L 169.16–17 = T 134.18–19 *in occidentem recte spectare* (n.76).

[85] Cf. Vitr. *DA* 4.5.1 (*spectare* 3×), with Gros 1992: 152–6.

[86] See also Gargola 2017: 176, and cf. n.88.

[87] Hübner 1992; Hänger 2001: 73–6; Gargola 2004: 134–5; 2005: 136–9; 2017: 158–160, 176–7.

model (*exemplum*) from which the *antiqui/maiores* borrowed the *ratio* of orthogonal axes in surveying:

> ab hoc fundamento maiores nostri in agrorum mensura uidentur constituisse rationem. (Frontin. *Lim.* L 28.5–6 = T 11.9–10)
>
> From this foundation our ancestors seem to have established their method (*or* principle) in the survey of fields.
>
> ab hoc exemplo antiqui mensuras agrorum normalibus longitudinibus incluserunt (Hyg. *Const.* L 167.3–4 = T 132.6–7).
>
> From this model the ancients encompassed the measures of fields with perpendicular lengths.

The *disciplina Etruscorum* does not bequeath to the surveyors a practice whose meaning and function conform completely to religious observances; its purposes are different. It rather locates the origins of centuriation in nature by connecting the DM with the course of the sun and the KM with the celestial axis. This association with *natura* in turn points toward a spatialization of the world in terms of uniquely identifiable areas designated in accordance with the celestial divisions. The value of this spatializing system may be seen most clearly in the surveyors' method for designating the location of *centuriae*, which were closed off with boundary markers labeled *sinistra* or *dextra* with respect to the DM and *ultra* and *citra* with respect to the KM. Just as each space in the augural field could be said to be "to the right" (*dextra*) or "to the left" (*sinistra*) and "in front" (*antica*) or "behind" (*postica*), so each century could be identified as "to the right" (*dextra*) or "to the left" (*sinistra*) of the DM and "beyond" (*ultra*) or "on this side of" (*citra*) the KM (see Figure 7.1).[88] Along with numbers marking the relative distance from the central *limites*,[89] these designations created a spatial field in which all of the plots obtained in relation to the DM and KM a place not only in the civic but also the cosmic order, because the DM and KM were aligned along natural divisions of the *orbis terrarum*. In this sense, centuriation Romanized the natural order.

[88] Significantly, Hyginus continually uses *spectare* in his account of the system for inscribing boundary markers (*Lim.* L 111.16–112.8 = T 71.10–72.13); the verb, whose importance was noted earlier, occurs 12× in the short passage (4× by plausible conjecture; cf. *inspectare*, 1×).

[89] DD II VK II (*dextra decumanum* II, *ultra kardinem* II), for example, would mark the century second to the right of the *decumanus* and second beyond the *kardo*; assuming a westward orientation, the century so marked would lie to the northwest of the intersection of the DM and KM. Cf. n.14, see Figure 7.1, and illustrations at Dilke 1971: 90–2; Campbell 2000: 490–3.

7.4.4 *Departing from the Ideal: Errors and Alternate Realities*

The early history of *limitatio* establishes a natural principle for surveying practice, but, as suggested earlier, there was no shortage of surveying configurations which in practice departed from this ideal. In what we call the negative part of their account, both Frontinus and Hyginus aim to explain such deviations. In general, they attribute variation to one of two causes: either methodological error or deliberate abandonment of the natural orientation in response to the local exigencies of a survey. These causes are intended only as heuristics, and, as we will see, have some overlap. Indeed, both share a crucial feature: it is possible in either case to analyze the resulting variation in the survey as the result not of the destruction of the established principle of centuriation but as its alteration or abandonment. This distinction is subtle but important. To *destroy* the Etruscan principle would be to falsify it, whether by undermining its historical claim or supplanting it with another principle; to *alter* it (in error) or *abandon* it (deliberately) is to depart from theory but not necessarily to establish an opposing principle. While such departures are not desirable per se, they do not automatically threaten the theoretical basis for *limitatio*: it is thus noteworthy that while Frontinus and Hyginus report discrepancies in the practice of centuriation, they are careful not to authorize them by acknowledging a legitimate alternative *ratio*. More broadly, bringing surveying differences within the fold of rational explanation can also be viewed as a means of controlling the local and historical variety of disciplinary practice: *mensores* have accomplished surveys in many ways, but it remains possible to understand why or how they centuriated the land as they did; and, while they did not follow the established principle everywhere, neither did their departure from theory destroy it.

The first factor that leads surveyors to depart from the established principle is methodological error. Frontinus mentions "many" (cf. *multi*) who fail to orient the DM and KM according to the celestial divisions recognized by the Etruscans (*Lim.* L 31.4–7 = T 14.14–17):

> multi mobilem solis ortum et occasum secuti uariarunt hanc rationem. sic uti<que> effectum est, ut decumani spectarent ex qua parte sol eo tempore, quo mensura acta est, oriebatur.

> Many, following the mobile rising and setting of the sun, have introduced variety into this principle (*or* method). The inevitable result is that the *decumani* are oriented from that quarter in which the sun was rising at the time when the survey was accomplished.

Ratio here refers to the orientation of the DM and KM on the true east–west and north–south axes, as stipulated by the Etruscan account. The problem that Frontinus identifies is the misapprehension of the cardinal directions. The sun rises and sets at different points on the horizon throughout the year: on the equinoxes it will rise due east and set due west, but for the remainder of the year it rises and sets further north or south depending on its place between the summer and winter solstices. Hence, Frontinus calls the rising and setting of the sun "mobile" or "changing" (*mobilis*): bare observation on a single day cannot serve as a reliable guide for determining the cardinal directions (except by dumb luck).[90] Importantly, Frontinus says that *multi* have not undermined the *ratio*, only "introduced variety" (*uariarunt*); while their methods for orienting are faulty, they preserve the internal consistency of the orthogonal system. The system remains intact, but the orientation is approximate rather than exact. Hyginus tells a similar story, attributing such a mistake to those who, "ignorant of the order of the universe, have followed the sun, that is, its rising and setting, a thing which cannot be ascertained once and for all with the *groma*" (*Const.* L 170.3–5 = T 135.1–3 *multi ignorantes mundi rationem solem sunt secuti, hoc est ortum et occasum, quod is semel conprehendi ferramento non potest*). Accordingly, it is nature itself which gives the lie to their method when the noon sun reveals that the *kardo* does not face north (cf. L 170.7–8 = T 135.5–6 *limites emiserunt, quibus kardo in horam sextam non conuenerit*).[91]

Later, Hyginus returns to those whose limited knowledge of the *ratio mundi* leads them astray and expounds upon a slightly different set of problems. Recalling those surveyors who faithfully (but erroneously) follow the risings and settings of the sun,[92] he continues (L 182.8–183.4 = T 146.9–147.3):

> multi ita ut supra diximus solis ortum et occasum conprehenderunt, qui est omni tempore mobilis nec potest secundum cursum suum conprehendi, quoniam ortus et occasus signa a locorum natura uarie ostenduntur. sic et limitum ordinatio hac ratione conprehensa semper altera alteri disconuenit. hos qui ad limites constituendos hac ratione sunt usi, fefellit mundi

[90] See Ch. 7.5.3 for the procedures for ascertaining the cardinal directions on any date.

[91] For sundials as exposing incorrect methods or construction, cf. Gal. *Aff. pecc. dig.* K 5.68.12–69.5, 80.6–82.10, 88.8–10.

[92] A large interval separates the following passage from the one just discussed (*Const.* L 170.3–8 = T 135.1–6), but it the same group is evidently meant (so *ita ut supra diximus*): cf. L 170.3 = T 135.1 *multi* ~ L 182.8 = T 146.10 *multi*; L 170.3–4 = T 135.1–2 *solem sunt secuti, hoc est ortum et occasum* ~ L 182.9–10 = T 146.10–11 *solis ortum et occasum conprehenderunt*; L 170.4–5 = T 135.2–3 *is semel conprehendi . . . non potest* ~ L 182.10–11 = T 146.11–12 *nec potest secundum cursum suum conprehendi.*

magnitudo, qui [se] ortum et occasum peruidere crediderunt; aut forte scierunt errorem et neglexerunt, ei contenti tantum regioni ortum et occasum demetiri. immo contendisse feruntur ortum eum esse singulis regionibus, unde primum sol appareat, occasum, ubi nouissime desinat:[93] hactenus dirigere mensuram laborauerunt.

> Many, as we said above, have (sc. in centuriating) ascertained (*conprehenderunt*) the rising and the setting of the sun, which is in motion all the time and cannot be ascertained (*conprehendi*) merely according to its course, since the points of its rising and setting are exhibited in different ways by the nature of (sc. different) places. The result is that a given arrangement of *limites* ascertained (*conprehensa*) with this method (*hac ratione*) is discrepant with others (sc. so ascertained). Those who have employed this method (*hac ratione*) for the establishment of *limites* were either deceived by the extent of the universe, insofar as they believed that they had a direct view of the rising and the setting (sc. of the sun), or perhaps recognized their error but paid it no heed, contenting themselves with measuring out the rising and setting only for the (sc. particular) region. Indeed, it is said that they maintained that for individual regions the rising is where the sun is first visible and the setting where it finally ceases (sc. to be visible): this was the limit of their efforts to orient the survey.

Despite Hyginus' allusion to the mobility of the sun's course throughout the year (*omni tempore mobilis*), the central difficulty here is different: the local effect of topography on the perceived rising and setting of the sun.[94] That is what Hyginus means in saying that the "points" (*signa*[95]) of the rising and setting of the sun are exhibited or expressed "in different ways" (*uarie*) by the "nature of (sc. different) places" (*locorum natura*). Such topographic variation could take many forms. One obvious example, which Hyginus goes on to discuss (L 183.8–16 = T 147.8–16), is that of lofty mountains obscuring the sight of the sun; after the sun has apparently set at the foot of the mountain on one side, it continues to shine on the further territory plus on the other side as of the mountain (Figure 7.2). In this case, the course of the sun observed by a surveyor under the mountain could not, without correction, accurately orient the *limites*, even if the proper methods were employed (L 183.4–8 = T 147.3–7). Differences in the *natura locorum* thus furnish another reason why local observation will not always exactly orient the DM and KM. As Hyginus warns, the uncorrected effect of local topographic variation is that the *ordinatio limitum* of any

[93] Lindermann et al. conjecture *decidat* for *desinat*, but the latter can be understood with an infinitive *apparere* after *appareat*; cf. *Const.* L 183.10–12 = T 147.10–12 *sol . . . parere . . . desinat.*

[94] Cf. Lindermann 2018: 16–17; Knobloch 2018: 64.

[95] = Gk. σημεῖα: Guillaumin 1997: 90–4; Knobloch 2014: 85; 2018: 64; cf. n.112.

Figure 7.2 Illustrations in the manuscripts of Hyginus' *Constitutio* representing the difficulties of attaining an accurate orientation at the foot of a mountain based on the sun's course. (a) Herzog August Bibliothek Wolfenbüttel, Cod. Guelf. 36.23 Aug. 2°, folio 51r (Lachmann fig. 159 = Thulin fig. 98). (b) Biblioteca Apostolica Vaticana, ms. Pal. lat. 1564, folio 92r (= Thulin fig. 98a).

individual region will align neither with the cardinal directions nor with the *ordinatio* of any other region. The putative "method" (cf. *haec ratio*) of the *multi* is thus reframed as a mere "error" or "mistake" (*error*).

Hyginus' speculation on the source of these surveyors' errors is instructive. The size of the universe "deceived" (*fefellit*) some, he says, into believing that their sightings of the rising and setting of the sun were veridical indicators of its course.[96] This group seems to lack an appreciation of astronomy or natural philosophy that would allow them to grasp the *ratio mundi*. More intriguing, however, is the second group: those who recognize their error but choose to ignore it. These surveyors are not only satisfied to observe the rising and

[96] Cf. Ch. 5, n.124 on deceptive phenomena in Cels. *DM*.

setting of the sun peculiar to their region (cf. *tantum regioni ortum et occasum*), but even assert that *ortus* and *occasus* as such are defined by local observations of "where the sun is first visible" and "where it finally ceases (sc. to be visible)." This claim does not necessarily undermine the established principle, since it preserves the intent to orient the KM and DM along the appropriate axes, but it enshrines a carelessness in the methods used to ascertain the cardinal directions which, if followed to their conclusion, would yield a chaotic multiplication of *ordinationes limitum* (cf. *altera alteri disconuenit*). Hyginus thus dismisses their claims as a species of *error* – indeed, a professed one (cf. *scierunt*). Their neglect (cf. *neglexerunt*) of the mistake and their satisfaction (cf. *contenti*) with local measurement are culpable. As Hyginus suggests, the assertions of this second group are a theoretical dead end, a kind of counsel of despair: Their efforts to orient the survey extend only "so far" (*hactenus*), and no further, because they lack the knowledge to implement a more precise *constitutio*.[97] We will return later (Ch. 7.5.2–3) to what Hyginus considers the requisite knowledge for enforcing the standards of surveying.

We turn now to the second factor that causes discrepancies from the established principle of centuriation: deliberate modification of the *limites* in response to local circumstances. In some respects, this rubric overlaps with the first (especially in the second example; compare the surveyors who are *contenti* to accept the effects of local topography), but it differs typologically in that Frontinus and Hyginus represent the ensuing departures not as failure to achieve the desired orientation but as the intentional abandonment of it. An example from Hyginus illustrates this well (*Const.*L 170.9–12 = T 135.7–10):

> et quidam, ne proximarum coloniarum limitibus ordinatos limites mitterent, relicta caeli ratione mensuram constituerunt, qua tantum modus centuriarum et limitum longitudo constaret.
>
> And some, in order that they not lay *limites* aligned with those of neighboring colonies, have abandoned the order of the heavens in their survey, so that only the measure of the *centuriae* and the length of the *limites* are fixed.

In this situation (recorded also by Frontinus[98]), the *mensores* have not erred in their application of the celestial principle (*ratio caeli*) from the Etruscan account but have simply "abandoned" it to avoid a coincidence in the

[97] *Hactenus* implies a limitation or failing; for this "restrictive" connotation, see *OLD* 3; *TLL* VI 3.2750.76–2751.13 (H. H. Schmid-Bulhart-Ehlers).

[98] *Lim.* L 31.7–9 = T 14.17–19 *et multi, ne proximae coloniae limitibus ordinatos limites mitterent, exacta conuersione discreuerunt.*

alignment of the *limites* with those of adjacent colonies. Frontinus and Hyginus offer several other cases where surveyors have deliberately departed from the established orientation. Some have drawn the DM in the direction in which the field is longest.[99] Other modify the *constitutio* because obstacles such as mountains or the sea would impede the roadways if traced on the cardinal directions.[100] In still other cases the theoretical departure is left implicit, as when Hyginus mentions that in some colonies the DM is made to overlap with the *uia consularis.*[101]

Examples of such departures could be multiplied, but of greater moment is the question how Frontinus and Hyginus generally regard the abandonment of the principle evinced in these cases. It may be obvious that they disapprove of mistakes (*errores*), but what of these deviations in practice? On the one hand, many of their reports have a descriptive rather than prescriptive character, an observation that reflects a more neutral but nonetheless crucial dimension of their project: the collection and explanation of varieties of practice. On the other hand, we should not imagine that Frontinus and Hyginus are thereby endorsing fragmentation in practice: unavoidable departures from the principle are to be comprehended, not commended. Nor should it escape us that throughout their discussions the *ratio mundi/caeli* remains the standard by which differing systems are evaluated; other alignments are thus introduced as permutations of or deviations from the *ratio mundi*, never the other way around. It is for this last reason that Frontinus and Hyginus call those who adopt alternative systems "ignorant" (*ignorantes*) of the heavenly origin first embodied in Etruscan practice.[102] And they explicitly maintain on several occasions their preference for the ideal of centuriation that establishes the DM and KM in parallel with the north–south and east–west celestial divisions: Frontinus, for example, calls this orientation "the best, rational establishment of land" (*Lim.* L 31.1 = T 14.11 *optima … ac rationalis agrorum constitutio*), and Hyginus agrees that it is "best" (*optimum*) to direct the *limites* to achieve an orientation in alignment with the celestial divisions.[103] Hyginus furthermore calls a version of this *constitutio* "the most beautiful method for establishing *limites*" (*Const.* L 180.4–5 = T 144.13 *constituendorum*

[99] Frontin. *Lim.* L 29.1–3 = T 12.6–7, Hyg. *Const.* L 170.12–13 = T 135.10–12.

[100] Hyg. *Const.* L 180.12–182.7 = T 144.20–146.8, quoted in part later in this section.

[101] *Const.* L 179.11–13 = T 144.1–3. The example given is Anxur (Terracina) and the *uia Appia* (Chouquer et al. 1987: 105–9; Chouquer and Favory 2001: 52–3).

[102] Frontin. *Lim.* L 29.1 = T 12.5, Hyg. *Const. lim.* L 170.3 = T 135.1.

[103] Discussing the use of the sundial to orient the *limites* (see Ch. 7.5.2–3): *Const.* L 188.14–17 = T 152.3–7 *optimum est ergo umbram hora sexta deprehendere et ab ea limites inchoare, ut sint semper meridiano ordinati: sequitur deinde ut et orientis occidentisque linea huic normaliter conueniat.* See also Lindermann 2018: 20.

limitum ratio pulcherrima).[104] Nonetheless, the efforts of both authors to account for the diversity of surveying practice stands as a cautionary lesson that, in spite of their commitment to the celestial principle reflected in the Etruscan account, they are not tempted by sheer idealism. The orientation of *limites* is in fact only one (albeit paradigmatic) example in their treatises of an attitude finely calibrated to the balance between theory and practice, something we have met frequently in the *artes*.[105]

The careful equilibrium between the realities of surveyed land and the ideal principle for centuriation can be well illustrated by an example of a theoretical departure that may be placed under the second rubric (deliberate abandonment of the *ratio mundi*): some *mensores* have completely reversed the orientation of the *limites* such that the DM runs north–south and the KM east–west.[106] Frontinus' remarks suggest that the reversal results from the direction that the surveyor faces when tracing out the first *limites*.[107] Hyginus' fuller discussion is instructive (*Const.* L 181.5–182.1 = T 145.10–146.2):

> itaque si loci natura permittit, rationem seruare debemus: sin autem, proximum rationi; non quo minus aliquid de finibus fiat aut amissionis periculum habeat: si aliter egerimus, mensura sua uni cuique constabit, decumani suo nomine appellabuntur, tantundem kardines, fines terminis obligabuntur, nihil operi deerit nisi ratio, habebit tamen inter professores existimationem. nam nec illis coloniis hoc nomine quicquam iniuriae factum est, quod kardines loco decumanorum obseruantur, decumani loco kardinum: omnis limitum connexio rectis angulis continetur, extremitas mensuraliter obligata est, nihil res publica, nihil possessor de finibus queritur; constat illis ratio mensurae, limitum ratio non constat, et potest dici mensura orbis alterius aut certe sinistra, hoc est inuersa.

104 The remarks apply to instances of the *constitutio* in which the *decumanus maximus et kardo a ciuitate oriuntur et per quattuor portas in morem castrorum et uiae amplissimae limitibus diriguntur* (*Const.* L 180.2–4 = T 144.10–12).

105 Cf. Hyg. *Const.* L 205.5–9 = T 168.5–9, quoted in Ch. 7.4. Hyginus records many differences in practice, for example of counting *limites* (L 173.16–174.14 = T 138.14–140.11) or marking boundary stones (n.14). See also on Frontin. *Art. mens.*, Ch. 7.5.5. For "balance" and the *artes*, see Chapter 1.

106 Cf. Frontin. *Lim.* L 29.3–6 = T 12.7–10, Hyg. *Const.* L 170.14–16 = T 135.12–14, where Capua and the *ager Campanus* are mentioned (Chouquer et al. 1987: 199–231, 294–314; Campbell 2000: 326). Guillaumin 2005: 175 notes other examples of reversal in the *Liber coloniarum*: L 209.16–22 (Consentia, Vibo Valentia, Clampetia), 210.1–2 (Beneventum). Cf. also Seru. *ad* Verg. *G.* 1.126 *cum agri colonis diuiderentur, fossa ducebatur ab oriente ad occidentem, quae kardo nuncupabatur, et alia de septentrione ad meridiem, qui decimanus limes uocabatur*. As Servius' account hints, other surveyors might well reject the description of their own orientation of the DM and KM as a "reversal" of the norm.

107 L 29.3–4 = T 12.7–8 *non ortum spectant, sed ita aduersi sunt* (previous n.).

> So, if the nature of the location allows it, we should preserve the principle, but if that is not possible, what is closest to the principle. It is not that there is anything at stake for the boundaries or danger of loss: if we do otherwise (sc. by reversing the orientation of the *limites*), the appropriate measure will remain for each person, the *decumani* as well as *kardines* will have their own names, and the boundaries will be marked by limits; the work will lack nothing except reason (*or* a principle), though it will still be well regarded by professionals. For no harm comes to colonies just because *kardines* are observed in place of the *decumani* and the *decumani* in place of the *kardines*: every intersection of *limites* is still contained by right angles, the edge is still bound in measurable fashion, and there arises no dispute on behalf of the state or private owner concerning boundaries. They have a principle (*or* method) for surveying, but lack a principle for the *limites* – such a survey can said to belong to another world, or certainly reversed, that is, turned around.

The *ratio* that Hyginus urges the surveyor to preserve is the *ratio mundi* or *caeli*. He acknowledges that reversing the *limites* will have no practical consequences: such territory will still garner the "respect" or "esteem" (*existimationem*) of professional surveyors, divided according to a "principle" or "method of surveying" (*ratio mensurae*) as it is.[108] Yet he maintains that it will lack *ratio* both in a general way (*nihil operi deerit nisi ratio*) and specifically in the alignment of the DM and KM (*limitum ratio non constat*). Plainly *ratio* here does not refer just to the systematic division of land, but to conformity with the natural principle on which surveying practice is founded in the Etruscan account – that is, to the *ratio mundi/caeli*. The "best and rational" (Frontin. *Lim.* L 31.1 = T 14.11) mode of surveying thus agrees with nature: reversing the orientation of the DM and KM fails to inscribe Roman space properly into the cosmos.

The theoretical inadequacy of the reversed *constitutio* is illustrated by the fact that the *kardo-limes* so drawn will not be in alignment with the *kardo* of the universe's rotation. As Hyginus puts it, the alignment of the reversed *limites* will belong to "another world" (*orbis alterius*), apparently meaning a region whose *kardo* would align not with our north and south but with our east and west.[109] Thus, while the locality may sometimes oblige the surveyor to depart from the established principle – here is Hyginus'

[108] For the weight of *inter professores existimationem*, cf. Ch. 2, n.168.

[109] *orbis alterius* is Thulin's conjecture for the paradosis *urbis alterius* (printed by Lachmann, Lindermann et al.). It is hard to see what the point of *urbis alterius* would be, however (cf. Campbell 2000: 391), and *orbs* and *urbs* are not infrequently confused (*TLL* IX 2.907.23 [Bohnenkamp]). Guillaumin 2005: 191 argues that *orbis alterius* refers to the southern hemisphere (so Thulin), but that does not explain the reversal of the KM and DM.

concession to practical exigency against the rational principle of nature (cf. *proximum rationi*) – it is irrational (L 182.3 = T 146.4–5 *nulla est ratio*) to reverse the DM and KM or, we may extrapolate, to falsify the principle by needlessly altering their orientation in other ways.[110] Surveyors may get away with much in practice, but there remains only one *ratio mensurae* that accords with the *ratio mundi*.

A few remarks to conclude this discussion. The great variety in surveying practice that the early *gromatici* inherited could obstruct a coherent view of the *ars*. Like etymologizing, the historical account of the development of *limitatio* offered by Frontinus and Hyginus is properly read as a sophisticated intellectual response to this problem. Both the positive and negative components of their rationalizing investigations cooperate to this end. The Etruscan account enables them to recover a secure natural principle (*ratio*) for the central rite of centuriation that "possesses an origin in the heavens and an unbroken tradition" (Hyginus). Because this principle authorizes correct practice, it becomes a criterion for identifying and explaining instances in which surveyors have strayed from it. Some *mensores* have erred accidentally, while others have deliberately abandoned the *ratio mundi*: Although Frontinus and Hyginus acknowledge a gap between theory and practice in local *ordinationes limitum*, neither author sees these departures as posing a legitimate challenge to the ancient and natural principle for centuriation. Of course, we must remember that while the *constitutio limitum* has a claim to being the central procedure of surveying – the "loftiest" *ritus*, in Hyginus' words – it is only one of many parts of the discipline. The problems and solutions studied here are representative of those found in other areas of land-surveying, both in the treatises of Hyginus and Frontinus and in other early authors in the *CAR*.

7.5 The Interdisciplinary *ars* of Surveying: Nature, Mathematics, and Mensuration

Earlier in this chapter we described the picture of centuriation reconstructed from Frontinus' and Hyginus' writings as an ideal, against which the diversity of alternative surveying practices reported in the *CAR* is thrown into sharper relief. As we have now seen, this ideal indeed structures Frontinus' and Hyginus' engagement with the historical realities

[110] Cf. L 182.1–3 = T 146.3–5 (quoted in Ch. 7.4.1).

of surveying practice, both terminological and methodological. From the history of surveying emerges the "best, rational establishment of land" (Frontin. *Lim.* L 31.1 = T 14.11 *optima . . . ac rationalis agrorum constitutio*) – namely, that which parcels out land in conformity with divisions of earth and sky proposed by nature itself. The attempt to ground method in knowledge of nature has numerous parallels with other *artes* that we have studied, of course. For Vitruvius, Columella, and Celsus no less than the surveyors, *rerum natura* has a normative besides epistemological function: The principles of nature not only aid in determining appropriate techniques, but also set the goals for those techniques. The most obvious *comparandum* for the surveyors' theorization of *limitatio* is found in Vitruvius' method for orienting the town (Chapter 3); like Vitruvius, the surveyors construct the Roman city as what we called the "natural" or "universal city." There are further important similarities which deserve consideration, however. Especially Vitruvius (Chapter 3) and Columella (Chapter 4) underscore the interdisciplinary character of their *artes*, appealing to natural first principles furnished by other fields to explain their own methods. In the remainder of this chapter, we will see how the surveying authors similarly relate the *ars mensoria* to other branches of specialized knowledge to justify correct technique. While neither Frontinus nor Hyginus constructs an epistemology such as Vitruvius' (and which Columella and Celsus handle more allusively),[111] their search for first principles with the aid of other disciplines reveals a coincidence of intent: to create a theoretical basis for their *ars*, founded in nature, that will support what they present as best practices of surveying.

In the following sections we will consider two ways in which Frontinus and Hyginus connect the knowledge and practice of their *ars* with other branches of specialized knowledge, chiefly astronomy and mathematics. First, we will study the astronomical and natural-philosophical portion of Hyginus' *Const.*, where he invokes an "element of the most lofty and divine art of the sundial" (L 184.1–2 = T 147.19–20 *gnomonices summae ac diuinae artis elementum*) to investigate "the magnitude of the universe, the principle of the rising and setting (sc. of the sun), and the size of the Earth relative to the universe" (L 183.17–18 = T 147.17–18 *quae sit mundi magnitudo, quae ratio oriundi aut occidendi, quanta sit mundo terra*). These inquiries are propaedeutic for two methods of calculating true north with a sundial, knowledge of which is required to implement correctly the *constitutio* descending from the Etruscan account. Astronomy and

[111] See Ch. 3.1 on *DA* 1.1; cf. Ch. 4.2.3 on Col. *RR* 1.pr.21–33, Ch. 5.1 on Cels. *DM* 1.pr.47.

gnomonics thus provide the surveyor with an infallible basis for orienting the DM and KM. Next, we will examine Frontinus' fragmentary *De arte mensoria*. In this short work, which is focused on the elementary techniques required to establish precise and measurable lines and calculate the areas of irregular spaces, Frontinus theorizes surveying practice in terms of the concepts *natura*, *ratio*, and *ueritas*. The surveyor's knowledge, expressed in the equivocal use of *ratio* (both "method" and "calculation"), is realized in a set of geometrical procedures for handling the uncertainties and irregularities of land. Measurement and calculation mathematize both *ager* and *ars*, putting the basic surveying techniques on a precise and rigorous foundation.

7.5.1 Tracking Shadows: The Astronomical "Digression" of Constitutio limitum

While Frontinus and Hyginus recognize diversity in surveying practice, they remain committed to the *constitutio limitum* developed from the Etruscan account. We saw that, even if the surveyor had the best intentions, methodological errors and local difficulties could yield *limites* not in precise agreement with the cardinal directions. Hyginus is acutely aware of these challenges, which result from not just from ignorance of the celestial order but also the fact that "surveys are conducted in dissimilar parts of the world" (L 173.7–8 = T 147.7 *in diuersis orbis terrarum partibus mensurae aguntur*). The surveyor must therefore acquire the principles and methods that will allow him to determine the cardinal directions at any time and in any place. This need prompts what could be called an astronomical "digression" in *Const.*, the aims of which, as Hyginus says (L 183.17–18 = T 147.17–18), are to discover (1) the magnitude of the universe, (2) the principles of the sun's rising and setting, and (3) the size of the Earth relative to the universe. The digression comprises an invocation of the art of gnomonics and praise of Archimedes (L 184.1–13 = T 147.19–148.11); discussion of the size of the Earth relative to the universe and of the intervals between the heavenly bodies (L 184.14–185.9 = T 148.12–149.4); remarks on the relation of the sun to the Earth and description of the zones/circles of the heavens and Earth, with a quotation from Virgil's *Georgics* (L 185.10–186.13 = T 149.6–150.7); an account of the sun's motion through the sky and the perception of light and shadow on Earth (L 186.13–187.6 = T 150.7–16); and remarks on the effects of location on light and shadow, with a quotation from Lucan (L 187.6–188.13 = T 150.16–152.2). This lengthy exposition culminates in two different methods for using a sundial to calculate true north, one of which uses plane geometry (L 188.14–15 = T 152.3–152.20) and the other solid (L 189.16–191.11 = T 153.1–154.10). The ambition of the digression is obvious, as Hyginus undertakes to show that surveying no less than other disciplines is embedded in the interdisciplinary matrix of the higher *artes*.[112]

To appreciate Hyginus' account, it is useful to take a wider view. Implementing the *constitutio limitum* requires determining the cardinal directions. Strictly speaking, this orientation can be achieved through a plane-geometric procedure with a gnomon (sundial) that requires neither sophisticated mathematics nor a theoretical grasp of astronomy. Recall that when Vitruvius teaches how to orient the city in a healthful way, he offers no astronomical prolegomena, but merely narrates the procedure using a gnomon.[113] Hyginus' first method for employing a gnomon (L 188.14–15 = T 152.3–152.20) is very similar to Vitruvius'.[114] Why, then, does he not simply offer this procedure to his readers at the outset? Although the astronomical content that precedes may not be practically necessary, it plays an important role in Hyginus' theoretical presentation of the *constitutio limitum* and the art of surveying. The digression draws a close connection between the mechanism of the sundial – that is, the movement of shadows – and the astronomical phenomena which explain it. Hyginus exploits this connection to show how the sundial provides a window onto the physical structure of the universe: The sundial becomes a point of articulation between a deeper body of natural principles and the simple phenomenon of the shadow which instantiates them. Hyginus' line of argumentation is ultimately calculated to suggest the theoretical background for the use of orientational methods in the art of surveying and, thus, to set the *ars mensoria* into broader natural-philosophical context.

[112] For the poetic quotations as indicative of Hyginus' lofty conception of the *artes*, cf. Dilke 1971: 61–2; Schindel 1992: 383; Clavel-Lévêque et al. 1996: xiii; Pikulska 2004: 213; Fögen 2014: 234; Lindermann 2018: 11–12. Numerous Grecisms (e.g. *diastema* ~ διάστημα, "interval"; *punctum* = κέντρον, "center"; *tetartemorion* = τεταρτημόριον, "quarter") and other aspects of his cosmology reveal familiarity with Greek scientific traditions: Guillaumin 2005 *passim* and esp. 191–8; Knobloch 2014; 2015; 2018.

[113] Different, however, is the construction of the analemma in Book IX.

[114] For differences, see Knobloch 2014: 103; 2018: 73–4. The two most significant are: (1) whereas Vitruvius' construction depends on marking the furthest points of the shadow at specific times (early morning, evening), Hyginus' method obviates this step; and (2) whereas Vitruvius finds the meridian (north–south line) first, Hyginus finds the east–west line first and then constructs a meridian perpendicular to it.

7.5.2 *Taking the Measure of the Cosmos: Astronomical Principles*

From the outset, Hyginus is at pains to connect the sundial with more general astronomical questions. The subjects are linked already in the first two sentences (*Const.* L 183.17–184.3 = T 147.17–21):

> quaerendum est primum quae sit mundi magnitudo, quae ratio oriundi aut occidendi, quanta sit mundo terra. aduocandum est nobis gnomonices summae ac diuinae artis elementum: explicari enim desiderium nostrum ad uerum nisi per umbrae momenta non potest.
>
> First, we must investigate the magnitude of the universe, the principle of the rising and setting (sc. of the sun), and the size of the Earth relative to the universe. We must call down an element of the most lofty and divine art of the sundial, for it is impossible to accomplish our desire to the standard of truth except by the movements of shadow.

Only the "art of the sundial" can teach us about the course of the sun and the size of the universe and Earth "to the standard of truth" (*ad uerum*).[115] The art's grandiloquent epithets *summa ac diuina* seem justified in view of the explanatory power promised for it. We should also note how Hyginus is careful to gloss, as it were, his invocation of the sundial by drawing attention to its basic mechanism: *umbrae momenta.* The gnomon is, of course, essentially nothing more than a device for casting a shadow (*umbra*); observation of the motions (cf. *momenta*) of this shadow throughout the hours of the day and the days of the year furnishes the information for many useful calculations.[116] The *umbra* is everything in gnomonics, and the recurrence of *umbra* – the word as well as the concept – throughout the astronomical digression unifies its contents and puts the sundial at the center of it.

The passage that follows further emphasizes the importance of *umbra* and its central place in astronomy and cosmology (L 184.3–13 = T 147.21–148.11):

> ortum enim aut occasum ne ab extrema quidem parte orbis terrarum peruidere quisquam potest, cum a sapientibus tradatur terram punctum esse caeli et infra solem amplo diastemate spiritum sumere.[117] nam

[115] *ad uerum* is obelized by Lindermann et al.; the phrase is taken adnominally with *desiderium nostrum* by Clavel-Lévêque and Guillaumin ("l'élan qui nous porte vers le vrai") and Campbell ("our desire to find truth"). More probable is an adverbial interpretation with *explicari* (as translated): cf. Hyg. *Const.* L 208.2–3 = T 171.2–3 *cuius* (sc. *agri*) *totam positionem ad uerum formatam inspiciemus*; also, for example, Sen. *Ep.* 98.17, Iuu. 6.325, SHA *Heliogab.* 25.5.

[116] Proclus' report (*in Euc.* 38–42.8) of the division of μαθηματική given by Geminus makes γνωμονική one of the parts (μέρη) of ἀστρολογία.

[117] *spiritum sumere* is a curious expression; it is interpreted by Clavel-Lévêque et al. 1996: 71n63 (followed by Guillaumin 2005: 192) as a Stoic detail alluding to the life-giving force of the sun (cf. Cic. *Nat. D.* 2.41 = *SVF* 1.113.21–4). The expression hardly makes that obvious, however, and

> et Archimeden, uirum praeclari ingenii et magnarum rerum inuentorem, ferunt scripsisse, quantum arenarum capere posset mundus, si repleretur. credamus ergo illum diuinarum rerum magnitudinem ante oculos habuisse. qua ratione, dicamus, tot saeculis unus mortalium hoc scire potuerit? unus propter hoc laborauit et per incrementa umbrarum deprehendit.[118]
>
> It is not possible for anyone to have a direct view of the rising or setting (sc. of the sun), not even from the most extreme part of the Earth, because, as the philosophers teach, the Earth is the center of the heavens and draws in breath at a great interval below the sun. So they say that Archimedes, a man of extraordinary intelligence and the author of many great achievements, wrote on the question of how many grains of sand the universe could hold if it were filled up with them. Let us grant that this man had before his eyes the magnitude of the heavens: How, you ask, could this mortal alone in so many generations come to know this? He alone labored after it and discovered it by the growth (sc. movements) of shadows.

Hyginus introduces a cosmological frame of reference in remarking that the *sapientes* have handed down that the Earth is the "center of the heavens."[119] We also learn that the Earth lies at a "great interval" (*amplo diastemate*) below the sun.[120] The size of this interval explains why it is impossible "to have a direct view" (*peruidere*) of the sun's rising and setting.[121] Hyginus is gesturing here to the second and third questions in his agenda (*quae ratio oriundi aut occidendi* and *quanta sit mundo terra*); in a moment he will dilate on the *amplum diastema* that separates the Earth and sun, but he first interjects an appeal to Archimedes, whose treatise *Sand Reckoner* (Ψαμμίτης, *Arenarius*) calculates the number of grains of

more often in Stoic cosmology it is the sun which receives effluences from the Earth as sustenance. Campbell prints *spiritum sumere* but does not translate; Lindermann et al. print a tacit conjecture *spatium* for *spiritum*.

118 Lindermann et al. excise the allusion to Archimedes (*nam ... deprehendit*) as a gloss: see Lindermann 2018: 40–4; cf. Knobloch 2018: 64–5. The arguments for deletion do not fully convince. Lindermann's most significant objections are that the eulogy of Archimedes (1) interrupts Hyginus' agenda; (2) misunderstands the contents and significance of Archimedes' treatise, which are inconsistent with the doctrines that Hyginus himself expounds; and (3) accordingly makes no substantial contribution in its context. (2) is inconclusive (Hyginus also admits he does not know Archimedes' work directly: *ferunt scripsisse*). Against (3), I argue here that the appeal to Archimedes has a definite function, even if it is only the idea of the *Arenarius* and not its contents that serves Hyginus' purpose. (1) is too weak to demand alteration by itself.

119 For *punctum* = Gk. κέντρον, see *TLL* X 2.2648.4–19 (Harbsmeier-Beikircher); cf. Cic. *Tusc.* 1.40, Guillaumin 1997: 90–4; 2005: 192; Knobloch 2014: 86.

120 For *diastema* = Gk. διάστημα, anticipating the musical intervals used of the distances between planets (Hyg. *Const.* L 185.8–9 = T 149.4–5), see also Lindermann 2018: 41–2.

121 Failure to appreciate this fact led some surveyors to err: cf. *Const.* L 182.15–16 = T 146.16–17 *fefellit mundi magnitudo* (sc. *eos*), *qui* [*se*] *ortum et occasum peruidere crediderunt*.

sand that would fill the universe.[122] Hyginus seems not to know this work directly (hence *ferunt scripsisse*), but he treats Archimedes generally as one of the *sapientes* who have investigated the size and structure of the universe. This is not the main purpose of Archimedes' treatise, but he does address the dimensions of the universe and the distances of the Earth, sun, and moon.[123]

More to the point, Hyginus' allusion to the great mathematician should be seen as impressionistic rather than rigorous, rhetorical rather than philosophical: Archimedes held a special place in Roman intellectual culture, and to invoke him here not only dignifies Hyginus' discussion but also signals his intention to draw his first principles from the most authoritative sources.[124] The full value of the appeal to Archimedes is demonstrated only in the rhetorical culmination to the passage, however: After asking in awe how "this mortal alone in so many generations" could achieve such a spectacular result, Hyginus states that the discovery was made *per incrementa umbrarum*. Although the relationship between gnomonics and Archimedes' work is left vague, the purpose of the claim is clear: to cause the sundial to play a decisive role in his calculation of the dimensions of the universe.[125] Hence, Archimedes' discoveries *per incrementa umbrarum* illustrate Hyginus' claim at the outset of the digression that it is only *per umbrae momenta* – that is, by the *ars gnomonice* – that questions of great natural-philosophical import can be investigated. Hyginus' early appeal to Archimedes may thus be read as a first step in his case for the broader astronomical significance of the sundial.

7.5.3 Between Light and Shadow: Astronomy, Horology, Surveying

Hyginus next turns to the first and third points in his outline: *quae sit mundi magnitudo* and *quanta sit mundo terra*. In particular, he seeks to

[122] On the *Arenarius*, see Heath 1921: II 81–5; for its goal, Netz 2003.

[123] Cf. *Arenarius* H 2.264.21–2; consider also the concluding remarks to Gelon, where he says that his results will seem trustworthy (πιστά) to those περὶ τῶν ἀποστημάτων καὶ τῶν μεγεθέων τᾶς τε γᾶς καὶ τοῦ ἁλίου καὶ τᾶς σελήνας καὶ τοῦ ὅλου κόσμου πεφροντικότεσσιν (2.290.20–2). Archimedes' discussion of the dimensions of the universe is ancillary to his attempt to calculate the number of grains of sand that would fill it (Heath 1921: II 81–2; Netz 2003: 254); the main achievement is strictly mathematical (a system to express very large numbers).

[124] On Archimedes' memory at Rome, see Jaeger 2008. With Hyginus' laudatory words, Guillaumin 2005: 192 compares Cic. *Verr.* 2.4.131, Liu. 24.34.2. Hyginus' sources are more generally the unnamed *sapientes*; in his discussion, he does not in fact follow Archimedes closely (cf. Lindermann 2018: 41–3; Knobloch 2018: 65), but the effect of invoking him remains.

[125] The sundial plays no significant role in Archimedes' *Arenarius*; Hyginus' belief that it does may reflect lack of familiarity with its contents (see the remarks earlier in this section and n.123).

provide a picture of the *mundus* that will corroborate his earlier reports that the Earth is the "center of the heavens" and that it is located at a "great interval" (*amplo diastemate*) below the sun. Drawing on terminology that belongs also to the "art of music" (L 185.8 = T 149.4 *ars musica*), he reports from the *sapientes* the intervals (*diastemata* or *interualla*) between the five planets, sun, moon, and Earth (L 184.14–185.9 = T 148.12–149.4).[126] These remarks give a notion of the magnitude of the universe itself as well as of the Earth's position and relative size. Having dispatched the first and third questions on his agenda, Hyginus can devote himself entirely to the second question, *quae ratio oriundi aut occidendi.*[127] In the remainder of the digression (L 185.10–188.13 = T 149.6–152.2), he devotes himself to explaining how the sun's motion is manifested in the phenomena of light and shadow on Earth. Some of the particulars are obscure, but the basic thrust of the discussion is clear enough: Hyginus first sets up an astronomical model that describes the motion of the sun around the Earth; next, he relates this model to the phenomena of shadow and light, thus laying the groundwork for a deeper understanding of the mechanism of the sundial (*umbrae momenta/incrementa umbrarum*).[128]

By way of preliminaries, Hyginus reports that the sun is said by the *sapientes* to be several times larger than the Earth.[129] Its illumination produces day (L 185.11 = T 149.6 *ab eo* [sc. *sole*] *inluminari diem*) and its shadowing over (L 185.12 = T 149.7 *obumbratio*) creates night. These remarks underscore the significance of light and shadow, making them the visible representation of the relationship between the sun and the Earth. Hyginus next says that experts "divide the heavenly sphere itself by five circles into six parts" (L 185.13 = T 149.8 *polum ipsum quinque circulis diuidunt in sex partes*). He quotes from Virgil's *Georgics* to illustrate this schema (*G.* 1.233–9 = L 185.15–186.2 = T 149.10–16). The citation is not entirely apposite, but as for the appeal to Archimedes, the allusion has

[126] See Guillaumin 2005: 193–4 and esp. Knobloch 2014: 86–91; 2018: 65–8. Guillaumin 2005: 97 believes that we are dealing with a version of the theory of the "musique des sphères," but Knobloch 2014: 88–91; 2018: 67–8 shows that the use of harmonic ratios and terminology does not necessarily imply that. Even so, the interdisciplinary appeal to harmonic theory is significant: consider the place of *harmonice* in Vitruvius' schema (*DA* 1.1.8–9) and the comparison between doctor and musician and musician and astronomer (1.1.15–16, Ch. 3.1.3).

[127] He has already partially answered this question by suggesting that the great size of the universe and the distance between the Earth and the sun make it impossible "to have a direct view" (*peruidere*) of the sun's rising and setting; some may therefore wish to see Hyginus as making a fresh start here.

[128] For the difficulties, see Knobloch 2014: 91–102; 2015: 51–62; 2018: 68–73; cf. Clavel-Lévêque et al. 1996: 79; Campbell 2000: 392.

[129] *Const.* L 185.10–11 = T 149.6–7; Guillaumin 2005: 194 notes the similarity between the vague estimate and Cic. *Luc.* 116.

rhetorical value.[130] The division of the heavens and/or Earth into zones with peculiar qualities long predated Virgil and was indeed commonplace by his time, but the authoritative poet is made an exemplar of the tradition, an artful expositor of the model that underpins Hyginus' account.[131] Hyginus names the five *circuli* that segment the heavenly sphere into six *partes*: From north to south, they are *septentrionalis*, *solistitialis*, *aequinoctialis*, *brumalis*, and *austrinalis*.

Hyginus next uses these *circuli* to describe the course of the sun and relate its motion to the hours (*horae*) of the day, which are in turn pegged to the motions of shadows (*umbrae*) on Earth. He continues (L 186.10–187.6 = T 150.4–16):

> circulus autem zodiacus, cuius fines sol negatur excedere, ex circulo solistitiali[132] ad brumalem per diagonum ostenditur[133] ita, ut meridianum circulum ex utraque parte medium secet. per hunc sol, hoc est infra, ire fertur et orbem terrarum uiginti et quattuor horis circumire. harum ferunt XXIIII horarum iunctarum semper unum esse interuallum: nam increscendi aut decrescendi inter ipsas horas alternam esse mutationem. hoc ipsum per umbrarum motus ostenditur. nam cum sol orbem medium conscendit, umbras omnium rerum in hoc nostro tetartemorio meridiano axi facit ordinatas. ab hoc enim exemplo sescontrariae partis, quae uidetur eisdem horis inluminari, umbra describitur.
>
> The zodiac circle, the bounds of which the sun is said not to go beyond, appears diagonally from the circle of the summer solstice to that of the winter solstice in such a way that it cuts the meridian circle in the middle from both sides. The sun is said to go through the (sc. meridian) circle, that is, below it, and to travel around the Earth in twenty-four hours. They say that the period made up of these twenty-four connected hours always remains one, because as it increases or decreases, there is a corresponding change in (sc. the length of) the hours themselves. This fact is illustrated by the motions of shadows. For when the sun has mounted to the middle of the sky, it makes the shadows of all things in this our quarter of the world fall parallel to the meridian (sc. north–south) axis.

130 For difficulties mapping Virgil's scheme onto Hyginus', see Knobloch 2014: 92–3; 2015: 53; 2018: 70–2.

131 Cf. Mynors 1990: 54 (and n.112).

132 Reading *solistitiali* from the codex Nansianus in place of *aequinoctiali* (cf. Guillaumin 2005: 194; Knobloch 2014: 96). Campbell 2000: 392 defends *aequinoctiali* by arguing that Hyginus means that the zodiac extends "right to the edge of the *aequinoctialis* zone, where it adjoins the *solistitialis* zone." Some confusion between celestial and terrestrial zones notwithstanding (n.128), *aequinoctialis* and *solistitialis* should refer to the celestial *circuli* ("circles") bounding the zodiac, not broader *partes* or *zonae*.

133 Thulin prints Goesius' conjecture *extenditur* for the paradosis *ostenditur*, but *ostendere* makes sense: the zodiac circle "appears" or "is visible" with respect to the reference points and along the orientation that Hyginus stipulates.

> From this standard the shadow is described for the opposite part, which it is apparent is illuminated for the same number of hours.

The sun's motion is described in terms of the zodiac circle. We learn (1) that the sun does not "go beyond" (*excedere*) the zodiac, an obvious result of the fact that the zodiac contains the ecliptic, the mean plane of the sun's annual motion; (2) that the zodiac is oriented "diagonally," as follows from the obliquity of the ecliptic; and (3) that the limits of the zodiac are the *solistitialis* and *brumalis* circles – that is, the summer and winter solstitial points – which mark the extremes of the sun's annual motion. By *circulus meridianus* Hyginus means a great circle encompassing the globe through its north and south poles; this line of longitude that is perpendicular to the course of the sun, and hence "cut" by the ecliptic (cf. *secet*), marks the point at which the sun has completed half of its daily journey through the sky – that is, noon or "midday" (so *meridianus*). The perpendicular alignment of the circle explains why the sun must always go "through" (*per*) or, in more careful physical terms, "under" (*infra*) the *circulus meridianus*.[134]

Hyginus remarks that the sun requires twenty-four hours to make its journey around the Earth and under the *circulus meridianus*. This period (*interuallum*) is "always one" (*semper unum*) in that, by definition, the day is the period in which the sun completes one revolution. The "connected" or "consecutive hours" (*horarum iunctarum*[135]) that divide this period (one day) into twenty-four units are accordingly not fixed but undergo a "corresponding change" (*alternam ... mutationem*) as the length of daylight increases or decreases seasonally.[136] Then comes an important moment in the exposition: *hoc ipsum per umbrarum motus ostenditur*. When the sun has completed half of its course (*cum sol orbem medium conscendit*), shadows (*umbrae*) cast by objects fall parallel to the north–south axis (*meridiano axi ... ordinatas*). As Hyginus suggests, this visible indication of the sun's position halfway through its diurnal motion allows for the determination of the relative length of the twelve daylight hours "in this our quarter of the world" (*in hoc nostro tetartemorio*).[137] For as he points out, the shadow cast on the "opposite part" (*sescontrariae partis*) of the

[134] For *infra*, cf. Guillaumin 2005: 195: "le soleil circule au-dessous de la voûte celeste." There is no reason to emend to *intra* (with Thulin): see Knobloch 2015: 56.

[135] Lindermann et al. adopt *iuncturam* (**E**), which will mean the "association (*iuncturam*) of twenty-four hours always makes one period (*semper unum ... interuallum*)."

[136] For the variable hour-lengths, see Marquardt 1886: I 256–8.

[137] *Tetartemorion* implies a model of the world divided into quarters, a popular theory originating with Crates of Mallus: Mette 1936: 76–9; Bilić 2012: 296 and n7; Knobloch 2014: 97–8; Geus 2014 (on Agennius Urbicus); Knobloch 2015: 57–8; 2018: 72–3.

sundial is symmetrical, so the sun's daily motion can be divided into equal parts.[138]

Hyginus' account of the motion of the heavenly bodies, admittedly highly schematic, thus frames *umbrae* as an index for the motion of the sun. As we learned from the beginning of the passage, the diurnal revolutions of the sun when projected onto a longer (annual) time scale are limited by the zodiac, the extent of which is defined by the celestial *circuli* (*solistitialis* and *brumalis*). Although the length of daylight changes seasonally, the sun's circuit can always be divided into twenty-four hours whose internal proportions vary accordingly. The *umbrae*, in turn, track the physical journey of the sun through the heavens and mark its passage under the noon circle (*meridianus circulus*) which divides the day into equal halves. The chronometric function of the sun articulates more general astronomical principles (the celestial *circuli*) and more specific ones (the annual and diurnal motions of the sun), which are in turn manifested in the phenomena of light and shadow. By making *umbrae* the end of this chain of astronomical principles, Hyginus invests them with deeper astronomical significance, causing their motion to reflect the organization of the celestial sphere.

A final point worth remarking is Hyginus' indication of the geographical value of *umbrae*, which is discussed in the final part of the digression (L 187.6–188.13 = T 150.16–152.2) relating variations in the apparent course of the sun to the regions of the Earth. There are interpretative challenges, but the central importance of *umbrae* is obvious. Hyginus observes that "in all the lands in this part" – that is, in his (our) quarter of the world – "the sun casts its shadows to the right for those looking westward" (L 188.4–5 = T 151.11–13 *omnibus terris in hac parte in occidentem spectantibus umbras in dextrum emittit*). In the quarter of the world which extends "from the border of Egypt all the way to the ocean" (L 188.5 = T 151.13–14 *ab Aegypti fine usque ad oceanum*), however, "shadows are cast to the left for those looking westward" (L 188.7–8 = T 151.16–17 *in occidentem spectantibus umbrae in sinistrum emittuntur*). To illustrate, Hyginus quotes Lucan saying that the Arabs who arrived in the Mediterranean to aid Pompey "marveled that the shadows of trees did not fall to the left" (3.248 = L 188.11 = T 151.19 *umbras mirati nemorum non ire sinistras*). The observation regards the reversal of the direction in which shadows fall in the northern and southern hemispheres. He concludes by remarking that "at midday in Egypt the shadow vanishes:

[138] For *sescontraria pars*, see Clavel-Lévêque et al. 1996: 81n71 (followed by Guillaumin 2005: 95); Knobloch 2014: 96–7; 2015: 56–7.

from this we know that the middle part of the Earth lies precisely at that place" (T 152.1–2 *Aegypto medio die umbra consumitur. ex hoc ibidem mediam terrae partem esse conprehendimus*). This statement is inaccurate,[139] but it is evident that Hyginus is attempting to associate *umbrae* in still another way with the larger astronomical picture he has sketched. The direction and movement of shadows are a function of space as well as time, so that we can deduce from them the position of the observer and draw conclusions about the location of fixed territories (Egypt, he argues albeit wrongly, is the *media pars terrae*). *Vmbrae* provide knowledge about the terrestrial as well as celestial spheres.

After this discussion, Hyginus finally turns to orienting the DM and KM with the use of a sundial. By this point, he has offered several reasons why the reader ought to be impressed by the power of the gnomon: The apparently commonplace phenomena of light and shadow (*umbra*) exploited by the device reflect more profound astronomical principles concerning the movements and relations among the heavenly bodies. Besides allaying worries about the soundness of the orientational technique, this exposition deepens the significance of the *constitutio limitum* by elaborating the *ratio mundi* of which Etruscan practice is an *exemplum*. We will not review in detail Hygnius' two methods for determining true north with a sundial. As mentioned, the first method (L 188.14–15 = T 152.3–152.20) is substantially the same as Vitruvius'.[140] The second method (L 189.16–191.11 = T 153.1–154.10) is more complicated and seems ill adapted to practical use.[141] That is not to gainsay its importance in Hyginus' treatise, however: Not an empty display of erudition, it functions also as an advertisement for the mathematical significance of the sundial, further extending the theoretical implications of the use of this device.

The remarks with which Hyginus frames the orientational methods are also revealing. We conclude by considering the final sentence of the second method (*Const.* L 191.6–11 = T 154.6–11):

139 The Tropic of Cancer runs through Egypt, around Syene (Aswan); Hyginus seems to generalize improperly the fact that on the summer solstice objects there have no shadows. But anyway, it is the equator which is more properly the "middle part of the earth" (where the sun is directly overhead on equinoxes). Lucan 2.587 has the same "généralisation hâtive" (Guillaumin 2005: 196) as Hyginus.

140 See n.114.

141 It was reconstructed by Mollweide 1813: 397–8, reproduced by Cantor 1875: 68–9 and (in paraphrase) Dilke 1967: 17–18. See Knobloch 2014: 105–8; 2015: 63–7; 2018: 76–9. This method apparently goes back to the first-century BC mathematician Diodorus' *Analemma*, as is known from an attribution of the same technique to him by the tenth/eleventh-century Arab polymath Al-Birūni: Neugebauer 1975: 840–2.

> ex hac (sc. linea recta quae erit ortus et occasus) in rectum rectam lineam eiciemus, hoc est normaliter: haec erit meridiano ordinata. eisdem signis et ipsum circulum quadrifidum constituimus et intuebimur quattuor caeli partes, quibus limitum ordinatio hac ratione constituta erit et omni tempore conuenit.[142]
>
> From this (sc. straight line which will be the rising and setting of the sun) we will project a straight line at a right angle, that is, perpendicularly, which will be aligned with the meridian. With the same marks, we establish the circle divided into four parts and (sc. thereby) will attain a view of the four parts of the heavens, by which the alignment of the *limites* will be established using this method (*or* on this principle) and with which it will agree at every time.

These statements bring the discussion of gnomonics full circle: The sundial offers a foundation, so to speak, for the *constitutio limitum* on astronomical and mathematical principles. The second method mathematizes the ritual division of the heavens, providing a rigorous means of accessing the *ratio mundi/caeli* on which the Etruscan account is based. Notably, the notions of vision and perspective so important in the *constitutio limitum* are reprised here as a privileged "sighting" or "view" (cf. *intuebimur*) of the *quattuor caeli partes* revealed by the mathematically constructed *circulus quadrifidus*. This *circulus* divides the heavens into four regions, which become a celestial paradigm for the terrestrial quadrants created by the intersection of the DM and KM. The upshot is to guarantee *omni tempore* a precise homology between centuriated land and the heavenly quarters, so long as the *limites* are correctly installed.[143] In the final analysis, the "divine art of the gnomon" crucially underpins the establishment of *limites* championed by both Frontinus and Hyginus.

7.5.4 Natura, ueritas, ratio*: Mathematizing the Land*

Once he has provided these methods for orienting the DM and KM, Hyginus enjoins the surveyor to undertake the physical establishment of the *limites* carefully so that no *error* will creep in (*Const.* L 191.14–18 = T 154.14–17). Even if the central *limites* are properly aligned, the whole

[142] With Lindermann et al. I print the text of the London manuscript **S**. The text of the first and second classes of MSS is *eisdem signis et ipsum constituemus et intuebimur quattuor caeli partes, quibus limitum ordinatio hac ratione constituta omni tempore conuenit*. In these MSS *circulum quadrifidum* and *erit et* were omitted. For *et* Lachmann conjectured *tetrantem* (printed by Thulin), which saves the transmitted text but is unnecessary in light of **S**; see Lindermann 2018: 25–6.

[143] *omni tempore* contrasts with the alignments based on the *mobilis* course of the sun, censured by Frontinus and Hyginus (Ch. 7.4.4).

system could be spoiled if the *quintarii* (fifth *limites*) and *linearii* (or *subrunciui*, intermediate *limites*) are crooked or lack the proper dimensions. Hyginus' injunction reminds us that the *ars mensoria* must always return, true to its name, to the basic operation of measuring (*metiri*) the Earth. Of course, the importance of that elementary practice of measurement itself – tracing out the lines that would divide the surface of the Earth and render its dimensions calculable – extends far beyond *limitatio*. While this may have been the exemplary "rite" (*ritus*) of the *mensores* and a focal point for the *ars*, it was far from the whole discipline. Surveyors were continually involved in tasks that did not require them to establish new *limites*, but to verify the area of territories already delimited, to find and confirm effaced or forgotten boundaries, and so on. What was shared by these procedures as well as centuriation was the technical apparatus of the *mensor* – that is, his literal and figurative toolkits. If centuriation provided a center of gravity for surveying, it was the surveyor's skills of measurement which unified the many parts of the discipline as a whole.

The contents of the surveyors' expertise were relatively few but extremely versatile and powerful. We touched upon them in our sketch of centuriation and will rehearse them briefly here. The indispensable instrument was the *groma* or *ferramentum*, a staff which was topped with a perpendicular crosspiece hung with plumb bobs at its four corners.[144] The *groma* allowed the surveyor to establish right angles and to sight out straight lines called *rigores*. The *rigor* is a sort of geometrical notion, a breadthless length traced over the land in order to make its surface calculable;[145] in practice, it would be materially represented by a rod or other measuring device.[146] Other surveying paraphernalia included portable sundials (for orientation), folding rules, temporary markers or milestones, and writing materials; the *mensor* could also have employed more complex instruments.[147] Finally, there was the cognitive equipment: this included, besides practical experience and know-how, skills in mapping and geometry. Mathematical techniques were crucial to

[144] See n.18.

[145] Cf. Balb. L 98.12–13; two *rigores* at the appropriate width define a *limes*. See further Lindermann 2014: 208–9; Roby 2014: 47–9.

[146] Lewis 2001: 21; the rod was called either a *decempeda* ("ten-footer") or *pertica* ("pole") (Balb. L 95.6–7). Other measuring devices, not apparently often used by the Romans, were the measuring cord (σχοινίον) and chain (ἄλυσις). See esp. Lewis 2001: 19–22; also Dilke 1971: 73; Chouquer and Favory 2001: 73; Guillaumin 2005: 10–11.

[147] For example, dioptra, *chorobates*, and *libra*. For surveying equipment, see esp. Lewis 2001; also Dilke 1971: 66–81; Chouquer and Favory 1992: 65–76; 2001: 65–71; Guillaumin 2005: 5–11; for portable sundials, Talbert 2017.

calculating the area of quadrilaterals (usually squares and rectangles) and triangles, and creative applications of similar triangles allowed the surveyors to accomplish many tasks.[148] With these tools, physical and intellectual, the surveyor performed his *professio*, from the humble duty of determining the straight-line borders and area of a small plot to the grand business of establishing an entire network of *limites* on a large, uneven territory, and many other operations besides.

In the remainder of this chapter, we will turn from *limitatio* to the skills of measuring as such to understand how surveying authors conceptualized the elements of their expertise. We will examine in particular how Frontinus theorizes the elementary operations of measurement in the short extract called *De arte mensoria* (*Art. mens.*).[149] In this work, he discusses the kind of land called *ager mensura per extremitatem comprehensus* ("land surveyed to its edge"); since such an area was not centuriated, it lacked the regularly sized plots enabling easy calculation of its dimensions.[150] The surveyor instead had to follow the irregular edge (*extremitas*) of the land, which was often settled or cultivated and thus presented challenges to his work, and ascertain the area of the irregular shape within. While Frontinus' text focuses on only this one kind of territory, it establishes a conceptual framework pertinent to other departments of the discipline as well.[151] *Art. mens.* is not the only surveying text to shed light on the geometrical elements of mensuration – the Trajanic "Exposition and Account of All Figures" (*Expositio et ratio omnium formarum*) by Balbus is another rich work in this vein – but it strikingly attempts to articulate the relationship between the mathematical principles and practical experience of surveying.

As I will argue, Frontinus' *Art. mens.* sets up a nexus of conceptual abstractions that structure the surveyor's experience: *natura*, *ueritas*, and *ratio*. We have seen that *natura* plays important normative and epistemological roles in the surveyor's art in that the best *constitutio limitum* is grounded in principles furnished by the order of the cosmos; on the other

148 For synoptic accounts of the surveyor's measuring activities, see Chouquer and Favory 1992: 77–90; 2001: 79–94 (cf. 264–6); for mathematics in the *CAR*, Folkerts 1992; Guillaumin 1995; Roth Congès 1996; 1998; 2000; Lewis 2001 *passim*; Guillaumin 2002; 2003; 2009a; 2009b; Folkerts 2014.

149 A modern title: Ch. 2, n.145 and n.151.

150 This is one of three kinds of land distinguished at Frontin. *Agr. qual.* L 1.2 = T 1.4, L 4.3–5.5 = T 1.18–2.7; see Hinrichs 1974: 114–15; Chouquer and Favory 2001: 109; Guillaumin 2005: 27–8. For *extremitas*, cf. Balb. L 98.3–4; see also Lindermann 2014: 207–8.

151 Cuomo 2000: 191 describes the treatise as dealing with "the land-surveyor's task in general," and Thulin's title <*De arte mensoria*>, though lacking ancient authority (Campbell 2000: 327; Lindermann 2016: 20n12), reflects a similar idea. In fact the scope of the text is very limited, though this does not prevent us from using it to reflect on the discipline generally.

hand, we have also seen that *natura* can manifest itself as an obstructive force, misleading the surveyor in the attempt to orient and trace out the *limites*. It is the latter aspect that is most significant in *Art. mens.*, for the surveyor's goal is to produce a geometric expression of the land that constitutes its *ueritas*, the "truth" of its calculable area, but *natura* presents many obstacles in this work. The crucial mediating link between *natura* and *ueritas* proves to be *ratio*. *Ratio* refers to the mathematical principles or methods that enable the surveyor to confront the difficulties posed by *natura* and to render uncertain land certain – in other words, to express its *ueritas*. The conceptual triad of *natura*, *ueritas*, and *ratio* offers a theoretically rich expression of the surveyor's task: by making *ratio* the fulcrum, it also links the most basic operations of surveying with the methods of geometry. In this way, Frontinus' account is analogous to the astronomical excursus of Hyginus' *Const.*, which we saw uses natural-philosophical and astronomical principles to establish a rigorous basis for practice.

7.5.5 *The Surveyor's Task: Truth and Method*

Art. mens. begins with a short sentence that frames Frontinus' discussion of the surveyor's art (L 31.12 = T 15.5–6):

> principium artis mensoriae in agendi[152] positum est experimento.
>
> The beginning of the surveyor's art is located in the practice of surveying.

It is tempting to read this statement as a general claim about the value of experience or practice (*experimentum*) against theory, but there are good reasons not to. First, Frontinus does not foreground any notion of theory with which *experimentum agendi* would contrast. Second, theory and practice are inseparable in Frontinus' mode of explanation, as we will see in the sentence that follows, and it is hard to see why he would thematize such a dichotomy only to abandon it. The opening rather has a twofold function more closely related to the purpose of his work. The first is merely to announce the subject of the discussion: what the surveyors *actually does* when he surveys a field, the elementary techniques of measurement that constitute his "experience" or "practice of surveying" (*agendi … experimento*).[153] The second is to foreground the priority of physically

[152] *agendi* is Lachmann's conjecture; MSS have *agendis*, for which Hinrichs 1992: 350–1 would prefer *agentis* (*in agentis … experimento*, "in the experience of the surveyor").

[153] With Hinrichs 1992: 351; Behrends et al. 1998: 53, I take *agere* in the technical sense "survey" (lit. "lead (sc. a survey)" or "trace (sc. a measurement)"), rather than the more general notion of "doing"

venturing into the land which is to be surveyed and tracing the "calculable lines" (*rationales lineae*) that will enable its dimensions to be measured. This emphasis is appropriate, as we will see, because the land is not a priori delimited with straight lines and right angles such that the surveyor's whole task amounts only to geometric calculation. It is therefore imperative that that the surveyor visit the *ager* to oversee the initial step (hence *principium*) of marking out its borders and recording their lengths.

If the first sentence exhorts us to take *groma* in hand, as it were, the second develops *in nuce* a view of the task that awaits, concisely articulating the challenges, aims, and methods of mensuration (L 31.13–18 = T 15.6–13):

> exprimi enim locorum aut modi ueritas sine rationalibus[154] lineis non potest, quoniam omnium agrorum extremitas flexuosa et inaequali[155] cluditur finitione, quae propter angulorum dissimilium multitudinem numeris suis manentibus et cohiberi potest et extendi: nam sola[156] mobile habent spatium et incertam iugerum enuntiationem.[157]

> For the truth of a place or rather of its measure cannot be expressed without the use of calculable lines, since the edge of all the land is contained by a curving and uneven boundary. This boundary, because of the multitude of its different angles, can be reduced or extended even as the number (sc. of the angles) remains constant; for land has a mutable area and a determination of *iugera* that is not fixed.

The objective of the survey is to express the *ueritas locorum aut modi.* The phrase *ueritas locorum* is peculiar, and I read *aut modi* as correcting or specifying the expression: What is meant by the "truth of a place" is the "truth of its measure" – that is, a mathematical statement of its area.[158] The mathematical connotation of *ueritas* is corroborated by the claim that this "truth" can only be obtained with *rationales lineae.* With the adjective *rationalis,* Frontinus appeals to what one might conceive of as

(cf. Campbell's "practical experience"). For this meaning, cf. *Art. mens.* L 33.7 = T 17.13 *mensura agenda*, and see Conso 2006: 16–17.

[154] **A** reads *rationabiles*, **E** *rationabilibus* (so printed by Lachmann); *rationalibus* is Thulin's correction. *Rationalis* has a better pedigree than *rationabilis* in the *CAR*: *TLL* XI 2.224.65–225.2 (Grossardt).

[155] Hinrichs 1992: 351, 370 retains the paradosis *inaequalis* and construes *flexuosa et inaequalis* with *extremitas*.

[156] For text and interpretation, see n.160.

[157] Guillaumin changes *enuntiatio* and *enuntietur* (L 31.20 = T 15.15) to *renuntiatio* and *renuntietur*, comparing (2005: 231) Hyg. *Cond. agr.* L 122.3–4 = T 85.4–5 (cf. also L 122.9–10 = T 85.11). But while *renuntiare* can be used of an "official report" in various political and juridical contexts (*OLD renuntiatio* 1–2, *renuntio* 2–4), it does not appear elsewhere in the *CAR* besides the two occurrences noted. *Enunt-* moreover yields appropriate sense (*OLD enuntiatio* 2a, *enuntio* 2a).

[158] Corrective *aut*: *OLD* 6b; H–S §269.c.β.

a constructive kind of equivocation, between *ratio* in the sense of "principle" or "method" on the one hand and "calculation" or "reckoning" on the other.[159] Thus, a *rationalis linea* is at the same time a line drawn "in accord with a principle" or "method" and a line that is "measurable" or "subject to calculation." These two interpretations indeed reflect not so much conceptual equivocation as coalescence of meanings: the surveyor's *ratio* is in this case identical with the geometrization of space.

Frontinus explains that the surveyor must employ *ratio* in his construction of the lines because the natural state of land is unequal, irregular, uncertain. The edge (*extremitas*) of the land is contained by a boundary (*finitione*) that is "curving" (*flexuosa*) and "uneven" (*inaequali*); without straight-line borders, it is impossible to determine its true area. Furthermore, land (*sola*) has a "mutable area" (*mobile ... spatium*) and a "specification of *iugera* that is not fixed" (*incertam iugerum enuntiationem*). This is a more general point independent of the unevenness of the *extremitas*: Before it is surveyed, land (*sola*) is ontologically and epistemologically amorphous.[160] That is, a physically and geometrically discrete *spatium* only comes into existence and is mathematically "specified" or "determined" (so *habent ... enuntiationem*) when the surveyor imposes *lineae* to delimit it.[161] Until that moment, the land could be delimited in many possible ways – hence its space is "mutable" (*mobile*) and its area "not fixed" (*incertam*). An angle made *here* would give such and such an area; an angle *there* would give another. Frontinus' remarks on the undefined state of *sola* prior to the survey should explain (hence *nam*) the preceding claim that the boundary of the land can be "reduced" (*cohiberi*) or "extended" (*extendi*): the precise details of the explanation *propter angulorum*, and so forth, are obscure, but Frontinus evidently wishes to emphasize that the shape of the space can be modified without breaking or interrupting the continuous border and angles that render it calculable.[162] The opening passage of *Art. mens.* thus presents the

[159] Cf. *TLL* XI 2.224.63–6 (Grossardt).

[160] **A** has *sol mobilem* (corrected by a later hand to *soli immobile*) and **E** *soli mobilem*; the reading *sola* is Scriverius'. If the plural does not simply stand for the singular, it should be interpreted distributively (H–S §31). Campbell 2000: 328 accepts *soli*, which will be genitive with *spatium*; he makes *anguli* the subject: "Moreover, they (the angles) have a variable area of land (sc. *spatium soli*)." Hinrichs 1974: 351n3 similarly makes "angles" the subject, but reads *soluti* for *soli*: "Denn wenn sie (sc. *anguli*) losgelöst sind, etc."

[161] For *spatium*, see Lindermann 2014: 201–3 (Frontinus' passage quoted on 201); cf. Behrends et al. 1998: 53n72.

[162] My translation follows that of Hinrichs 1992: 351 (cf. Cuomo 2000: 192) and Behrends et al. Campbell translates differently: "which (sc. boundary), because of its large number of angles of different sizes, can be reduced or extended while the complement of the angles remains the same." Guillaumin has another interpretation: "que (sc. limite) l'on peut resserrer et tendre en vue d'y

core elements that make up the conceptual nexus of the work: *ueritas* is the aim of the surveyor's art; *ratio* is the (mathematical) means of acquiring it; and *natura*, not yet named but present implicitly in the irregularities of the land, poses the challenge which demands the surveyor's skill.

Next, Frontinus elaborates the *ratio* that will express the *ueritas loci.* Although the constitution and interpretation of the text are very difficult,[163] the general tenor is clear (L 31.18–32.6 = T 15.13–16.4):

> sed ut omnibus extremitatibus species sua constet et intra clusi modus enuntietur, agrum quo usque loci positio permittet rectis lineis dimetiemur: ex quibus proximam quamque extremitatium obliquitatem per omnes angulos facta normatione conplectimur, et cohercitam[164] mensuralibus <lineis> statutis[165] †certo procentemate spatio simili†[166] futurae tradimus formae. modum autem intra lineas clusum rectorum angulorum ratione subducimus; subiectas deinde extremitatium partes, areas tangentium[167] nostrarum postulationum, podismis suis adaeramus, et adscriptis spatio suo finibus ipsam loci reddimus ueritatem.

placer un grand nombre d'angles d'espèces différentes, sans que la valeur de ses segments successifs en soit modifiée."

[163] The text I print is largely Thulin's, but I have daggered *certo procentemate spatio simili*; within the daggers I tentatively read *procentemate* (with Guillaumin *apud* Behrends et al.) for Thulin's *praecenturiatio*, which he adopted from Lachmann (n.166). For the passage, see Hinrichs 1992: 348–53 (updating 1974: 100–6); Behrends et al. 1998: 83–4; Campbell 2000: 327–9; Guillaumin 2005: 231–3; Peterson 2006: 158–60.

[164] MSS have *coh<a>erentem* (adopted by Hinrichs, Campbell, and Guillaumin), *cohercitam* being Lachmann's conjecture (followed by Thulin, Behrends et al.).

[165] Hinrichs and Campbell print the paradosis *mensuralibus statutis*, the former interpreting it to mean "Meßvorschriften" (1992: 351; cf. 352, 362), the latter "fixed elements of measurement" (cf. 2000: 328–9). Lachmann writes *moetis* for *statutis* (for *meta* in surveying language, see *OLD* 1d; Chouquer and Favory 2001: 444). I adopt Thulin's *<lineis>*, which yields an expression congruent with the earlier *rationalibus lineis*.

[166] **A** has *certo procentemato spatio simile*, **E** *certo proextimato spatio simile*. Lachmann and Thulin conjecture *certo praecenturiato spatio simili*, presumably to be translated as "we transfer it to what will become the map, which is similar to a fixed rectangular area" (*praecenturiatus*, otherwise unattested, supposedly means "centuriated beforehand"). Campbell follows but prints *similem*: "we transfer it, similar to a fixed rectangular area, to what will become the map." Hinrichs 1992: 351–2 hews more closely to **A** and reads *pro centesimato* or *procentesimato*, a word or expression otherwise unknown to Latin but supposed to mean "reduced (in size) a hundredfold"; see Hinrichs 1992 for details, and cf. Campbell 2000: 329. I have taken over but obelized the text of Guillaumin *apud* Behrends et al., who proposes *procentemate* from *procentemato* of **A**. This word is supposed to be a Greek dative from προκέντημα ("sketch" or "plan"); *spatio simili* is then construed in the ablative as meaning "to scale" ("à la échelle"): Behrends et al. 1998: 83–4; Guillaumin 2005: 231–2. These changes yield the translation: "et, chaque segment étant ainsi relié aux lignes de mesure établies, nous faisons, grâce à un plan précis, le report à l'échelle sur ce qui sera la *forma*." It is an ingenious interpretation, but the pileup of three distinct ablative phrases (*mensuralibus <lineis> statutis* + *certo procentemate* + *spatio simili*) seems too hard.

[167] **A** has *adrectamgentium*, the same hand writing *n* above the first *m*; **E** *are tangentibus*. Scriverius' *areas tangentium* seems to yield good sense, pace Campbell 2000: 329, who finds it obscure. (He

> But in order that appropriate form be fixed for all of the edges and the internal area of the enclosed (sc. space) be specified, we will measure out the field with straight lines so far as the disposition of the place permits. From these (sc. straight lines) we will embrace each bend in the edge by dropping perpendiculars at every angle; then we transfer (sc. the bend) bounded with fixed, measurable lines to what will become the map. Using the method of right angles, we calculate the area enclosed within the lines; next we measure in feet the attached parts on the edges, that is, the areas that belong to the contiguous demands (sc. imposed on us by the survey),[168] and by adding these to the space produce the very truth of the place.

The gist of Frontinus' advice is to use straight lines and right angles to render the space geometrically tractable, though it is uncertain how exactly we are to reconstruct the details of procedure and what precisely the *ratio rectorum angulorum* is. One plausible interpretation is as follows.[169] The *ratio rectorum angulorum* should be understood as the technique of tracing a grid of regular squares in such a way as to fill the interior area of the territory as far as possible. The surveyor next carefully traces a contiguous series of straight lines over the irregular borders of the territory; once these are established, he drops a line from each bend or angle perpendicular to the survey grid established in the interior, creating quadrilaterals and triangles to fill the space between the grid and the *extremitas*. Now all he must do to obtain an approximation of the area is (1) calculate the area contained within the grid of squares (~ *modum ... subducimus*); and (2) add to this the area of the figures created by the perpendiculars that were dropped from the edge to the grid (~ *subiectas ... adaeramus*). The sum of all these spaces makes the "very truth of the place" (*ipsam loci ... ueritatem*).

reads *ad recta tangentium*, construing *extremitatium* with *tangentium* and having this participle govern *ad recta nostrarum postulationum*: besides the improbable word order, the verb *tangere* does not govern *ad*.) On Scriverius' reading, let *areas* be in apposition to *partes*; construe the genitive as indicating possession or relation; and have *tangere* mean "border on," "be contiguous" (L&S I.2.b; *OLD* 2a); for *postulatio*, see the following n. Other solutions are less convincing. Hinrichs 1992: 351–2, following the earlier emendation to *procentesimato* (previous n.), prints the dubious *ad rectum centesimum* ("nach der richtigen hundertfachen Verkleinerung"). Behrends et al. (1998: 57n78) print *areas tangentibus nostris postulationibus* and take the ablative absolute, uncertainly as they admit, to mean "à l'aide de nos formules de calcul traitant des aires."

168 I take *postulationes* as the "demands" or "requirements" that the territory imposes upon the *mensor*: cf. *OLD* 1c; *TLL* X 2.258.63–8 (Korteweg). *Postulatio* occurs again thus at *Art. mens.* L 32.14 = T 16.12 (quoted later in this section); cf. also Hyg. *Const.* L 203.9–10 = T 144.18–19 (discussed in Ch. 7.4).

169 The following is based on Peterson 2006: 157–60; cf. Roth Congès 2000: 118–19, 124–8; Guillaumin 2003: 130–1; 2009b: 124–5; Bertoni 2017: 548, and n.163.

On any interpretation, the passage demonstrates the close relationship between the mathematical principle or method (*ratio*) of surveying and the expression of the *ueritas* of the land. Indeed, we see now that *ueritas* is an essentially mathematical notion, a quantity that follows from the calculations on the "measurable" (*rationales*/*mensurales lineae*) lines and angles imposed by the surveyor. As noted earlier, land (*sola*) before it is surveyed is ontologically and epistemologically indefinite; Frontinus' language in the second passage emphasizes how the surveyor's *ratio* resolves this state of potentiality into one of certainty and permanence. The surveyor prescribes the shapes or forms that will "fix" (cf. *constet*) the uneven edge and allow for the "specification" or "determination" (cf. *enuntietur*) of its area; its curves are "embraced" (cf. *conplectimur*) and turned into right angles; the edge is "bounded with fixed, measurable lines" (*cohercitam mensuralibus <lineis> statutis*). Within the obelized stretch of text notice also the adjective *certo*, "determined" or "sure," which should apply somehow to the results of the procedure. The territory, now physically defined by the lines set by the *mensores* and made mathematically explicit by the specification of its area, is transferred to the map (*forma*) so that, as Hyginus puts it in his own discussion, "it may be made permanent in bronze" (*Const.* L 202.15 = T 165.14 *in aere permaneat*).[170] The final triumph of *ratio* – or is it?

Immediately after this précis, Frontinus complicates the picture (L 32.7–14 = T 16.5–12):

> haec ubique una ratione fieri multiplex locorum natura non patitur, oppositis ex alia parte montibus, alia flumine aut ripis aut quadam iacentis soli uoragine, cum pluribus confragosorum locorum iniquitatibus, saepe et culturis, propter quae maxime ad artis copiam est recurrendum. debet enim minima quaeque pars agri in potestate esse mensoris et habita rectorum angulorum ratione sua postulatione constringi.

> The manifold nature of places does not allow these things (sc. the survey procedures) to be done in the same way everywhere, since from one side mountains might oppose, from another a river or embankments or a hollow of low-lying soil,[171] along with many uneven areas of broken ground, and often also cultivated lands. In view of all these things, there is a special need for us to have recourse to the resources of our art. For every piece of land, even the smallest, must lie in the surveyor's power and, contained by the method of right angles, be bounded according to its specific demand.

[170] See n.21. [171] For *iacere* ("be low-lying"), see *OLD* 14a.

Natura is introduced as an antagonist: The manifold obstacles unique to each place make a demand upon the surveyor (cf. *sua postulatione*),[172] compelling him to have recourse to the *copia artis*. *Natura* is not the contrary of *ueritas* but its substrate, the raw state of the land that must be wrestled within the bounds of the surveyor's straight lines. Yet, qua undelimited territory, *natura* is conceptualized chiefly in terms of its difficulties: areas that are impassable or must be excluded from the survey, that are treacherous and uneven, and that impose other stumbling blocks. The last item in Frontinus' list, "cultivated lands" (*culturae*), shows that *natura* is first and foremost a topographical notion here. While it may in a more limited way activate an antithesis between the "natural" features of the landscape and those which are the result of human artifice and skill, it chiefly marks a distinction between land in a *presurvey* state of potentiality and a *postsurvey* state of mathematical determination.

Only shortly after, Frontinus reprises this topographical description of the land in terms of its natural as well as artificial obstacles, confirming the suggestion just made about *natura*. Discussing the surveyor's use of the *groma* to draw out the *lineae*, he says (L 33.14–18 = T 17.10–15):

> sed si in rigore dictando quaedam deuitanda incurrunt, ualles, loca confragosa, arbores quas propter moram aut fructum succidere non oportet, item aedificia, maceriae, petrae aut montes et his similia, haec quacumque ratione optime poterint mensuram accipere debebunt.

> But if in fixing[173] the *rigor* certain impediments occur – valleys, uneven places, trees which are not to be cut down either because of the delay or their produce, likewise buildings, walls, crags or mountains, and things similar to these – these will have to be surveyed in whatever will be the best way possible.

Although the *deuitanda* are not explicitly identified with *natura*, there is sufficient overlap that they may be regarded as recapitulating the first description of *natura*.[174] In both passages, Frontinus focuses on the interface between the surveyor and the a priori topographic features of the land. The variety of obstructions specifies the causes for the curving *extremitas* which the surveyor approaches: the *mensor* must walk the twisting and irregular edge with his *groma*, plotting straight-line boundaries *above*, *below*, *along*, and *through* whatever complications the *multiplex natura* of the place

[172] See n.168.

[173] For *dictare* in surveying language see *OLD* 4b; cf. Hyg. *Const.* L 193.7 = T 156.7 *dictabimus metas*.

[174] L 32.10 = T 16.8 *confragosorum locorum* ~ L 33.15 = T 17.11 *loca confragosa*; L 32.11 = T 16.9 *culturis* ~ L 33.15–17 = T 17.11–13 *arbores . . . item aedificia, maceriae;* L 32.8 = T 16.6 *montibus* ~ L 33.17 = T 17.13 *petrae aut montes*.

harbors. These straight lines, established with the aid of the *copia artis*, geometrize the land. The indefinite *quacumque ratione* with which Frontinus qualifies his injunction does not imply any deeper uncertainty in the surveyor's principles or methods. As in the previous passage, where Frontinus remarked that the survey of the land could not be accomplished "in the same way" (*una ratione . . . non*) everywhere, *quacumque* only stresses that the *ratio rectorum angulorum* must be applied "according to the specific demand" (*sua postulatione*) of the territory. To approach the question from a different perspective: If we hear in *ratio* the meaning "calculation" that is so conspicuous in *Art. mens.*, then we could say that, while there is no identically worked calculation (*una ratione*) that will suffice in every case, the surveyor always has the resources to make "whichever calculation" (*quacumque ratione*) will express the *ueritas* of the particular field – for, ultimately, all these individually worked calculations are merely applications of the calculation of the right angles (*ratio rectorum angulorum*).[175] Here again we see the productive coalescence of different meanings of the word *ratio* highlighted earlier, linking the principle or method (*ratio*) of the *ars mensoria* with the mathematical treatment of the land.

The remainder of *Art. mens.* offers concise but detailed instructions for using the *groma* to sight out straight lines and establish right angles, as well as certain allusive instructions for handling the *deuitanda* presented by nature. This continuation supports Frontinus' conceptualization of practice by showing that the surveyor's physical equipment and skill are sufficient to meet the demands of the art: With the application of "diligence in measuring" (L 32.16 = T 16.14–15 *metiundi diligentiam*), the *groma* will allow the surveyor to overcome whatever obstacles he meets (cf. L 32.14–15 = T 16.13–14 *maxime prouidere debemus, quo usu ferramenta quidquid occurrerit transeamus*). One technique that Frontinus offers is meant to aid in tracing straight lines over sloping or uneven ground. This procedure, called *cultellatio*, helps equip the *mensor* to handle challenges such as mountains, valleys, and broken ground (*montes, ualles, loca confragosa*).[176] Although it is impossible to know, Frontinus may have continued by providing other methods enabling the *mensor* to trace out *lineae* in "whatever will be the best way possible" (*haec quacumque ratione optime poterint*) for any particular topographic situation (*natura*) of land under survey.

[175] Compare the various applications of *ratiocinatio* in Vitr. *DA* (Ch. 3.1.3).

[176] For *cultellatio*, see esp. Roth Congès 1996: 315–28; also Chouquer and Favory 1992: 86–8; Hinrichs 1992: 356–8; Lewis 2001: 124–5.

Art. mens. in sum theorizes some of the basic practices of surveying in terms of *ratio*, *natura*, and *ueritas*. This conceptual armature is developed in only brief compass – perhaps it was originally set out at greater length – but even so it is expansively suggestive. Together, *ratio*, *natura*, and *ueritas* express the goal, methods, and challenges of the *ars mensoria* as such. The mathematical principles that undergird *ratio* – indeed, that are in a sense identical with it – establish a rigorous basis for practice against the uncertainties projected by *natura*. Here *natura* is not identical with "nature" in the sense of the cosmic order that the surveyor emulates in the *constitutio limitum*, but rather a term for the ad hoc topographical difficulties of land in its raw, presurvey form. The process of physically and mathematically "determining" the land (*enuntiare*), and thus fixing its dimensions to make them permanent, culminates in the expression of *ueritas*, the accurate geometrical construction of the land that produces the "truth" (*ueritas*) of its measure. Because the fundamental measuring and calculation techniques that Frontinus discusses are relevant for almost all surveying business, including *limitatio*, his mathematization of practice helps to legitimate the discipline in toto by eliminating the possibility for error and doubt at its root. The *ueritas* that the surveyor expresses becomes in turn the principle for all other uses of the land, from its division and allocation to the adjudication of any disputes that might arise in the future.

7.6 Conclusion

At the beginning of this chapter, we proposed that the surveying writings of Frontinus and Hyginus provide us with a glimpse into the nascent intellectual formation of their *ars*. Their treatises thus offer a fresh take on the questions that we have studied in earlier chapters. The complex and various realities of surveying practice had been literally etched into the face of the Earth across Roman space over hundreds of years; these presented unique difficulties to the authors who came lately to fit the profession into the mold of the (higher) *artes*. Yet the problems posed, questions asked, and strategies framed by the early generation of surveying writers including Hyginus and Frontinus drew from a playbook common to the scientific culture of the *artes* in the Imperial period. From the pragmatic acknowledgment of differences in surveying practice and experience to the lofty explication of correct surveying methodology by natural-philosophical and interdisciplinary means, the *CAR* bears witness to the emergence of an *ars* of surveying proper.

Conclusion

rerum natura, hoc est uita, narratur

Nature, that is, life, is our subject. Plin. *HN* pr.13

Pliny's announcement of his theme in the *Historia naturalis* is a fitting epigraph. The *artes* are an operation of human *ratio*, which in its various senses – both instrumental and final – is perhaps the most conspicuous thread running through this book. As such, the *artes* belong to the domain of (human) "life" (*uita*).[1] Yet the *artes* of the early Empire were not merely the expression of Roman know-how: They were also a way of making explicit the equation between "life" and "nature" (*rerum natura*) – that is, of appropriating and understanding the world in which the Romans dwelt. The scientific culture that grew up along with the Imperial *artes* reflects the larger significance of this movement, which was at once a literary phenomenon and the realization of a Roman intellectual ideal.

C.1 The Meaning of *ars*

In the Introduction, we asked what constituted an *ars*, along several epistemological axes. As have seen, there are many interests and assumptions that distinguished the *artes* – as disciplines and as systematic treatises – from other kinds or expressions of specialized knowledge. The *artes* were the intellectual descendants of the notional "higher *artes*" which took shape in the late Republic and were marked by several premises that formed the basis for the scientific character of the *artes* in the early Empire: they were systematic, interdisciplinary, explanatory, and balanced, in the senses defined in Chapter 1. Nothing in the *artes* that we have studied controverts these premises, though they were developed along different lines in different

[1] For the ambiguity of *uita* – "life" in general but "human life" in particular – see Ch. 2.3 on Plin. *HN* pr.13.

disciplines. Indeed, early Imperial authors were not passive recipients of the idea of the higher *artes*: To the extent that they elaborated theories of knowledge as parts of the "images of expertise" that they aimed to communicate, they also elaborated the meaning of *ars*.

Systematicity remained a basic element of the Imperial *artes*, from Vitruvius' subdivisions of *architectura* to Celsus' partitioning of *medicina* into therapeutic modalities to the classification of land-types in the *corpus agrimensorum Romanorum*. Systematicity did not manifest itself only in the logical, didactic, or rhetorical organization of the material, however: it was also felt in the ordering of the scientific data upon which the methods of the *artes* depended. Authors draw distinctions among epistemic principles of varying orders, different cognitive processes or ways of thinking and learning, and different kinds of precepts or methods. Vitruvius, for example, makes *ratiocinatio* mediate between knowledge of *opera* and of *rationes*, while Celsus draws attention to subtle differences among *causae* and scrutinizes the physician's capacity for decision-making. Columella pits *ratio* and *consuetudo* against one another and examines the advantages and pitfalls of empiricism. Responding to the shortcomings of military *praecepta*, Frontinus proposes an alternative means of cultivating the general's faculty of command through contemplation of *exempla*. And Frontinus and Hyginus analyze multifarious surveying practices at different removes from their theoretical ideal for the *ars*. In the Imperial *artes*, systematicity did not therefore entail only an eloquent and logical exposition of the discipline. It also endowed expertise with an epistemological structure that gave it depth and breadth, grounding the methods of the *ars* in other types of knowledge and setting the *artes* into the sphere of Roman understanding of the world more widely.

The interdisciplinary character of the *artes* was worked out in various ways. Vitruvius, who stood closest to the Republican authors who fashioned the ideal of the "higher *artes*," represents perhaps the most faithful attempt to establish a canon of *artes* to which his own discipline belonged. This attempt is evinced in his novel epistemology of *ars* as well as his use of many other branches of specialized knowledge to explain the aims and techniques of architecture. No other Imperial *ars* shows quite the same degree of elaboration in this respect, but they all engage with the supposition of interdisciplinarity.[2] Columella is close to Vitruvius in imagining agriculture as one among many epistemologically worthy *artes*, and he not only devotes

[2] Absence of evidence restrains the claim only for the art of war, but Onasander does begin by comparing the discipline with other τέχναι (Ch. 6, n.30).

considerable attention to assembling a panoply of specialized knowledge that makes up the farmer's expertise, but also repeatedly negotiates the limits of agricultural knowledge in relation to other disciplines. Celsus' take on interdisciplinarity is both in theory and practice more conservative than Vitruvius' or Columella's, but he nonetheless emphasizes the importance of interdisciplinary study for the practitioner, and his own *Artes* textually embodies a unified vision of specialized knowledge. Finally, the interdisciplinary character of *ars* is on full display in the early surveying writings, which draw extensively on mathematics, astronomy, and gnomonics, among other disciplines, to justify surveying techniques.

While Cicero and Varro had both hinted at the idea of an explanatory program in the *artes*, neither had worked it out in any kind of systematic fashion.[3] This situation changed in the early Empire, as authors aimed to account for the methods of their disciplines through knowledge based in nature. We will return to nature and explanation just below. Finally, the *artes* remained balanced in that they strove to temper their theoretical claims with the demands of practice. Hence, in every *ars* that we considered, authors are careful to insist upon the incompleteness of theoretical knowledge without the know-how acquired through experience: "experience and practice are supreme in the *artes*" (Col. *RR* 1.1.16 *usus et experientia dominantur in artibus*). While much of this know-how stood necessarily outside the text, the striving for balance emerges in the *artes* in numerous ways: the nuanced attempts to account for variation in surveyed lands by the gromatical authors, the development of the stratagem collection by Frontinus, the *ratio-consuetudo* dialectic elaborated by Columella, Vitruvius' distinction between *opera* and *rationes*, and Celsus' use of the historical record of medicine and his attitude toward observation and experience in the medical sects. Besides all this, it was also realized in the Romanization of specialized knowledge, which embedded the scientific aspects of the *artes* in the matrix of Roman cultural practices. We will come back to Romanization momentarily.

So, what is the meaning of *ars* in the early Empire? There is no one answer, since the *artes* were the multifarious expressions of a living and growing intellectual ideal and staked out different positions on specialized knowledge (both within their discipline and in general) that in turn shaped the scientific culture in which they participated. Even so, it is plain that to be an *ars* in the sense studied in this book was not merely to be a set of techniques, a manual for practice, an exposition of method, or the like. It

[3] For Varro's different explanatory interests, see Ch. 4, n.27.

was, rather, to be a presentation of an image of expertise that brought with it a view of nature, knowledge, and society in which the *ars* acquired its full significance as a Roman way of engaging with the world.

C.2 The Science in the *artes*

The explanatory interest mentioned just now was largely responsible for the scientific character of the *artes*, because it was the desire to explain and justify method that impelled authors to engage with the fruits of systematic inquiry into the world – that is, to bring *science*, as defined in the Introduction, into the *artes*. Now that we have studied the *artes* more closely, we can describe their scientific character in greater detail vis-à-vis the theories of nature they articulate. In particular, the scientific culture of the *artes* is represented in the specific conceptualizations of nature that inform the *artes*; the use of the results of scientific inquiry drawn from other branches of specialized knowledge; and direct engagement with modes of scientific inquiry including observation, experiment, analogy, and more.

As we have seen, nature (*natura* or *rerum natura*) plays a vital role in all the *artes* in accounting for the methods conducive to best practice. Nature is conceived differently by different authors, however, and so too the kind of knowledge of nature constitutive of expertise. More ambitious approaches are found in Vitruvius and Columella. Drawing upon different philosophical doctrines, Vitruvius expounds an elemental theory of matter to explain the qualitative differences among building materials, among much else: the architect's knowledge of the elemental makeup of things allows for appropriate choices during construction. Columella makes knowledge of nature no less significant but puts the emphasis in a different place. He argues for a specific cosmological (or theological) picture of Nature, Earth, and a creator god. The properties of these entities as well as Nature's intelligent design of animals and plants inform the farmer's understanding of the phenomena of plant life and, accordingly, the correct methods for cultivation. Celsus rejects the sort of explanatory accounts of *ars* based in nature that Columella and Vitruvius propose. Even so, he does not lack an idea of nature as the descriptive essence of things, and knowledge of nature in this sense is crucial in the physician's decision-making. While we do not have a Roman military *ars* from the period, we have seen that Onasander grounded the precepts of warfare in psychology – that is, human nature – and we gave some reasons for supposing Celsus or Frontinus might have done the same. Finally, nature is prominent in the surveying writings, where it structures the central practice of centuriation and, in a different guise, embodies the topographic challenges prompting the *ars mensoria*. While knowledge of nature has a chiefly instrumental function in the *artes* – that is, it supplies the means used to accomplish the aims of the *ars* – it is frequently normative, too. Nature sets the physician's objectives for Celsus, teaches the farmer to manure his land in Columella, and establishes the pattern for Roman organization of space in Vitruvius and the surveyors.

Besides articulating theories of nature of varying complexity, the authors of the *artes* also draw upon different branches of scientific inquiry. This usage is warranted epistemologically by the interdisciplinary character of the *artes*. Vitruvius offers the most conspicuous example – unsurprisingly, perhaps, in view of his architectonic conception of architecture. The cases we studied show him appealing to various physical theories as well as to medicine, meteorology, geometry, and biology (although not necessarily under those names): His analysis of physical elements, his definition of health, his account of the winds, the geometry of his city plan, and his biological explanation of pregnancy in plants and humans all underwrite architectural method. Celsus has fewer examples to provide, although the anatomical knowledge with which he interlards his therapies are the fruit of a kind of medico-biological inquiry that falls outside the scope of his own work. Columella and the surveying authors offer richer material again. Columella owes the details of his physical theories and the teleological picture of life to natural-philosophical and biological inquiry; he also makes use of mechanics to explain plant physiology, as we saw, and appeals to mathematics (the geometry of surveying) and astronomy in other parts of his work that we did not examine as closely. Frontinus and Hyginus borrow from etymological research, augural and architectural practice, and astronomy in explaining the development of land-surveying. Frontinus also takes care to underscore the value of mathematics in the surveyors' work, but it is Hyginus' cosmological, astronomical, geographic, and horological appeals in expounding orienting techniques that are most notable. What he takes from these disciplines allows him to sketch a picture of the Earth and heavens that endows centuriation with deeper natural significance. Finally, while we are again hindered in the art of war by paucity of evidence, Onasander notably seems to have based himself on psychological principles that were sourced in part from outside military experience; it is possible that a version of this situation was true of Celsus' or Frontinus' work.

A third respect in which science came into the *artes* was through the use of methods characteristic of ancient scientific inquiry into the world. We find in the *artes* emphasis on the close observation of phenomena as a criterion for theoretical claims. Columella adopts an empirical perspective in refuting theories of the Earth's decline and in defending his approach to propagating the Aminnean vine, while Hyginus privileges observation when he maintains that incorrectly oriented *limites* will be revealed by their discordance with the shadows cast by the sun. A pragmatic empiricism saturates Celsus' medical books, as he looks both to the historical record and to the physician's observation as guides for therapy. Experiment, even merely probative, is rare in the *artes*, as it was in much of Greco-Roman science, but appears in Vitruvius' use of the aeolipile to prove his theory of winds. Analogy and modeling, both tools ubiquitous in ancient inquiry into nature, are much more common. From Vitruvius' use of human pregnancy to explain seasonal changes in trees to Columella's comparison between the vine and siphon or his analogizing of vine parts in terms of human limbs, analogies formed a part of the *disputatio communis* (as Vitruvius called it) shared among the *artes*. Models, such as the relatively detailed image of the heavens and Earth provided by Hyginus or Vitruvius' extrapolation from Andronicus' tower of the winds, also establish background for technique. Teleological reasoning, a hallmark of much ancient thought, is evinced in Vitruvius' physical theory and more thoroughly in Columella's understanding of Nature and Earth, where it allows for the proper grasp of ecological and biological processes. Finally, we should not miss in the *artes* the ubiquitous application of *ratio* to the natural and human worlds with a view to uncovering the mechanisms that underlie the phenomena. *Ratio* was hardly the province of scientific or technical thought alone, but one of the key aims to which it was put in the *artes* – the ordering, interpretation, and explanation of the evidence of the senses – belonged to philosophical and scientific modes of thought in antiquity.

The *artes* were not directed chiefly toward inquiry into the world and hence were not sciences in the sense of the word distinguished in the Introduction. Even so, they engage earnestly with the interests, results, and methods of scientific inquiry – so much so as to become vehicles for larger theories of nature, which are integral parts of the images of expertise that belonged to the *artes*. The science in the *artes* is not synonymous with modern science, but it is more than ordered knowledge of any character whatsoever: It is an ancient understanding of the world in which the Romans dwelt.

C.3 The Romanization of Specialized Knowledge, Once More

The balance between theory and practice in the *artes* allowed the Romans to accommodate their understanding of the world to culturally specific ways of organizing their society. In Chapter 2, we described the coalescence of the objective cognition of the world characteristic of scientific and technical thought with the subjective position of the Roman knower as the "Romanization" of specialized knowledge. The Romanization of specialized knowledge through the *artes* has the result that scientific knowledge is constituted as Roman knowledge. Also in Chapter 2, we explored such Romanization by way of the authors' explicit framing of their disciplines vis-à-vis Roman politics, society, and ideology. Now that we have sifted the scientific contents of the *artes* more closely, we are in a better position to grasp the influence of such Romanization upon the theories of knowledge and nature themselves presented in the *artes*. Particularly noteworthy is influence along the temporal and spatial axes of Roman civilization.

Temporally, the *artes* tended to locate their images of expertise within the unfolding history of Roman civilization. Thus, Celsus' appropriation of Greek medicine symbolizes a passage from an earlier period of Greek development to a Roman period of synthesis and maturation: So it is that from the historical archive of therapies handed down by the Greeks he develops a Roman *medicina*. Columella involves the discipline of agriculture in claims about successive epochs of Earth and the fate of the soil. As he believes, moral decay since the time of Rome's founding explains poor yields; his argument for the everlasting fertility of Earth underwrites his treatise on *res rustica* and hands the discipline back to the Romans for safekeeping. Frontinus, as we have seen, writes for a future audience of Roman generals and gives special place to Roman deeds among his *exempla*, and Onasander joins the Romans in seeing their military history as a repository for the precepts of the art of war – an opinion that was doubtless reflected in the *artes* of Celsus and Frontinus. Finally, the surveyors make the historical dimension of their *ars* central to its identity: Etymologies of surveying terms reveal the early origins of the discipline, and the coincidence between the spatializing practices of Etruscan and Roman augury, architecture, and surveying point to the deep religio-cultural wellspring from which the *ars mensoria* has flowed.

Spatially, the *artes* again and again imagine Roman ways of organizing space as natural ways of organizing space. We described Vitruvius' peculiarly Roman city plan as the "natural" or "universal city" because it purports to be founded on maximally salubrious natural principles. The surveyors' use of similar orientational procedures to establish a grid plan of *limites* makes the Roman organization of space in centuriation an image of the "celestial order" (*ratio caeli*). In this way, both Vitruvius and the surveyors pattern the organization of Roman space after nature, which has a normative function.

Besides this, we sometimes find in the *artes* apparently chauvinistic views of Italy and Roman land. Vitruvius, as we saw, makes Italy the climatic ideal, attributing Roman hegemony to God's providential situation of its people. Columella argues that Italy is a rich and fertile land capable of producing plants, animals, and humans as productively as any other. Indeed, the Aminnean vine – a native Italian cultivar – is the noblest of all grapevines, and with proper attention can bear high-quality fruit in sufficient quantities to exceed the profitability of foreign varieties. While the focus of this book is not how Roman *artes* mirror the values of Roman society, culture, and politics (an area that has already received considerable attention), even the few hints that we have observed allow us to confirm and add something new to this vein of study. Apart from the explicit situation of the *artes* in support of Roman *imperium* acknowledged in Chapter 2, the Romanization of specialized knowledge in the *artes* reveals the extent to which the Romans' scientific understanding of the world responded to the extrascientific imperatives of Roman culture. In other words, to understand the world through the *artes* was always to understand the world through Roman eyes.

C.4 The Idea of the *artes*

The *artes* have come down to us, like so much else from Greek and Roman antiquity, as the scattered fragments of what was once a brilliant whole. The proliferation of knowledge in the modern age, and the consequent drive toward specialization, have not been kind to this Roman ideal of polymathy. Distant as we are from ancient life, we find it strange that so many and such varied facets of knowledge could have once been joined together, crystal-like, in polyhedral unity – the many aspects of an ancient prism through which the Roman understanding of the world was reflected. It is my hope that this book has fitted those fragments together well enough that we can once again hold that prism to the light and, chipped and roughly joined as it may now be, catch a glimpse again of the world as the Romans saw it.

Appendix: Some Connections between Republican and Early Imperial artes

The following diagram reconstructs relationships of influence between three key Republican authors discussed in Chapter 1 (Cato, Cicero, Varro) and the authors of the Imperial *artes* considered in this study, plus Pliny the Elder. The authors adduced by Quintilian as exemplary of Roman polymathic learning (see the Introduction) are contained within the gray rhombus.

Solid lines indicate confirmed influence (citation by name in the text); **dashed lines** indicate probable influence strongly suggested by circumstantial evidence. I refrain from positing influence based only on comparisons of language or contents,[1] or where specific effects of influence germane to the *artes* cannot be easily adduced.[2] This diagram does not include all of the writings on or adjacent to the *artes* of the Republican or early Imperial period that can be shown or conjectured to have influenced the authors we have studied,[3] nor all those writings which we know to have belonged to the disciplines studied in this book.[4]

Additional Remarks: Although neither Lucretius nor Virgil is the author of an *ars* proper, their works influenced early Imperial *artes*. The following lines of influence can be posited:

Lucretius ➔ Vitruvius, ?Columella,[5] Pliny
Virgil ➔ Columella, Hyginus, Pliny

We cannot locate Celsus with respect to these poets because the books of the *Artes* in which Lucretius and Virgil are most likely to have appeared (agriculture and philosophy) are not extant.

1 Vitruvius, for example, possibly influenced Hyginus (cf. Knobloch 2015: 62; 2018: 64) and Columella Frontinus (cf. Rodgers 2004: 123, 125, 126, 208). It is also possible, for example, that the attitude toward interdisciplinary study evinced in Columella, Celsus, and Hyginus was mediated by Vitruvius' reception of Cicero, although this cannot be easily demonstrated, and influence, affinity, and mutual participation in an intellectual culture quickly become knotted together.

2 As an obvious example, it is impossible that Frontinus did not know Cicero, but it is harder to show how Cicero's writings influenced Frontinus' conception of *ars* (as can be done more securely for Vitruvius or Columella).

3 Consider, for example, the following lines of influence indicated *passim* throughout the text or notes of this book: Mago ➔?Cato, Varro, ?Celsus, Columella, Pliny; Fuficius ➔ Vitruvius; P. Septimius ➔ Vitruvius; Sasernae ➔ Varro, Columella, Pliny; Tremelius Scrofa ➔ Varro, Columella, Pliny; C. Iulius Hyginus ➔?Celsus, Columella, Pliny; Scribonius Largus ➔?Celsus, ?Pliny; Iulius Graecinus ➔ Columella, Pliny; Iulius Atticus ➔ Columella, Pliny.

4 For example, Cincius, who wrote six books *de re militari* but has left no trace among the extant Latin polemological writings (Ch. 6, n.5). For other Latin authors of scientific or technical works, see Keyser and Irby-Massie 2008: 1022–4.

5 Ch. 4, esp. n.39.

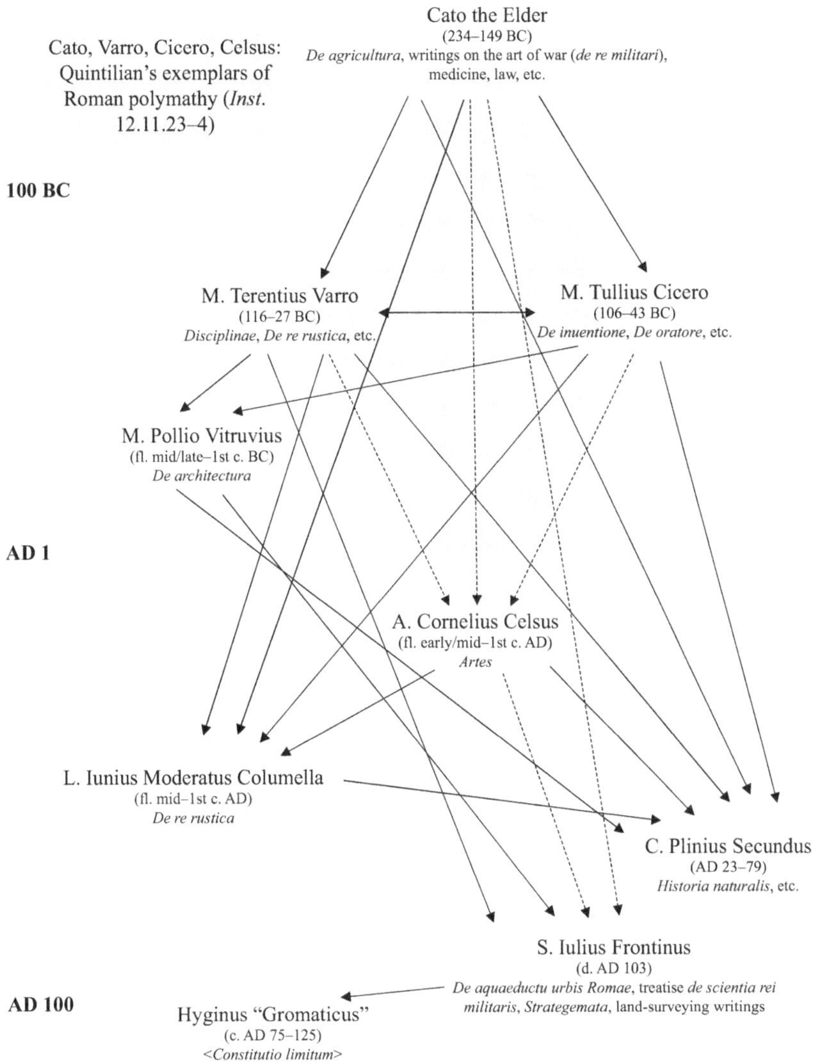
Cato, Varro, Cicero, Celsus: Quintilian's exemplars of Roman polymathy (*Inst.* 12.11.23–4)
Cato the Elder
(234–149 BC)
De agricultura, writings on the art of war (*de re militari*), medicine, law, etc.
100 BC
M. Terentius Varro
(116–27 BC)
Disciplinae, *De re rustica*, etc.
M. Tullius Cicero
(106–43 BC)
De inuentione, *De oratore*, etc.
M. Pollio Vitruvius
(fl. mid/late–1st c. BC)
De architectura
AD 1
A. Cornelius Celsus
(fl. early/mid–1st c. AD)
Artes
L. Iunius Moderatus Columella
(fl. mid–1st c. AD)
De re rustica
C. Plinius Secundus
(AD 23–79)
Historia naturalis, etc.
S. Iulius Frontinus
(d. AD 103)
De aquaeductu urbis Romae, treatise *de scientia rei militaris*, *Strategemata*, land-surveying writings
AD 100
Hyginus "Gromaticus"
(c. AD 75–125)
<*Constitutio limitum*>

References

Adam, J.-P. (1982) "Groma et chorobate: Exercices de topographie antique," *MEFRA* 94: 1003–29.

Adams, J. N. (2016) *An Anthology of Informal Latin, 200 BC–AD 900*, Cambridge.

Agache, S. (1998) "Construction dramatique et humour dans le Traité d'agriculture de Varron," in M. Trédé and P. Hoffmann, eds., *Le rire des anciens*, Paris: 201–30.

Ahrens, K. (1972) *Columella: Über Landwirtschaft. Ein Lehr- und Handbuch der gesamten Acker- und Viehwirtschaft aus dem 1. Jh. u. Z.*, Berlin.

Alfieri, N. (1976–7) "Per la topografia storica di *Fanum Fortunae* (Fano)," *RSA* 6–7: 147–71.

Alfonsi, L. and A. Ronconi, eds. (1982) *Plinio il Vecchio sotto il profilo storico e letterario*, Como.

Allen, J. (2005) "The Stoics on the Origin of Language and the Foundations of Etymology," in D. Frede and B. Inwood, eds., *Language and Learning: Philosophy of Language in the Hellenistic Age*, Cambridge: 14–35.

(2020) "Reason, Experience, and Art: The *Gorgias* and *On Ancient Medicine*," in L. Taub, ed., *The Cambridge Companion to Ancient Greek and Roman Science*, Cambridge: 39–57.

Ambaglio, D. (1981) "Il trattato *Sul comandante* di Onasandro," *Athenaeum* 59: 353–77.

Amsler, M. (1989) *Etymology and Grammatical Discourse in Late Antiquity and the Early Middle Ages*, Amsterdam.

Anderson, J. C. (1997) *Roman Architecture and Society*, Baltimore.

André, J. (1952) "Contribution au vocabulaire de la viticulture: Les noms des cépages," *REL* 30: 126–56.

(1958) *Pline l'Ancien: Histoire naturelle. Livre XIV*, Paris.

(2010) *Les noms des plantes dans la Rome antique*, 2nd ed., Paris.

André, J.-M. (1985) "Le prologue scientifique et la rhétorique: Les préfaces de Vitruve," *BAGB* 44: 375–84.

(1987a) "La rhétorique dans les prefaces de Vitruve: Le statut culturel de la science," in S. Boldrini, ed., *Filologia e forme letterarie*, Urbino: III 265–89.

(1987b) "Les écoles philosophiques aux deux premiers siècles de l'Empire," *ANRW* II.36.1: 5–77.

(1994) "Littérature technique et héritage de la rhétorique cicéronienne chez Columelle," in J. Dangel, ed., *Grammaire et rhétorique: Notion de romanité*, Strasbourg: 179–96 [= "Littérature technique et héritage de la théorie cicéronienne chez Columelle," in *Ktèma* 14 (1989): 255–72].

Angier, T. (2010) *Technē in Aristotle's Ethics: Crafting the Moral Life*, London.

Argoud, G. and. J.-Y. Guillaumin (1997) *Les Pneumatiques d'Héron d'Alexandrie*, Saint-Étienne.

Aßfahl, G. (1932) *Vergleich und Metapher bei Quintilian*, Stuttgart.

Ash, H. B. (1930) *L. Iuni Moderati Columellae rei rusticae liber decimus: De cultu hortorum*, unpublished PhD dissertation, University of Pennsylvania.

Ash, R. (2011) "Pliny the Elder's Attitude to Warfare," in Gibson and Morello 2011: 1–19.

Asper, M. (2007) *Griechische Wissenschaftstexte: Formen, Funktionen, Differenzierungsgeschichten*, Stuttgart.

ed. (2013) *Writing Science: Medical and Mathematical Authorship in Ancient Greece*, Berlin.

Astin, A. E. (1978) *Cato the Censor*, Oxford.

Baader, G. (1960) "Überlieferungsprobleme des A. Cornelius Celsus," *Forschungen und Fortschritte* 34: 215–18.

Baldwin, B. (1963) "Columella's Sources and How He Used Them," *Latomus* 22: 785–91.

(1990) "The Date, Identity, and Career of Vitruvius," *Latomus* 49: 425–34.

(1992) "The Career and Work of Scribonius Largus," *RhM* new series 135: 74–82.

(1994) "Notes on the *De aquis* of Frontinus," in C. Deroux, ed., *Studies in Latin Literature and Roman History* 7, Brussels: 484–506.

(2005) "Stylistic Notes on the Elder Pliny's Preface," *Latomus* 64: 91–5.

Bardon, H. (1952–6) *La littérature latine inconnue*, 2 vols., Paris.

Barwick, K. (1948) "Zu den Schriften des Cornelius Celsus und des alten Cato," *WJA* 3: 117–32.

(1960) "Die Enzyklopädie des Cornelius Celsus," *Philologus* 104: 236–49.

(1963) *Das rednerische Bildungsideal Ciceros*, Berlin.

Bayer, E. (1947) "Onasandros: Die Entstehungszeit des *Strategikos*," *WJA* 2: 86–90.

Beagon, M. (1992) *Roman Nature: The Thought of Pliny the Elder*, Oxford.

(2005) *The Elder Pliny on the Human Animal: Natural History, Book 7*, Oxford.

(2007) "Situating Nature's Wonders in Pliny's Natural History," in Bispham et al. 2007: 19–40.

(2011) "The Curious Eye of the Elder Pliny," in Gibson and Morello 2011: 71–88.

(2013) "*Labores pro bono publico*: The Burdensome Mission of Pliny's Natural History," in König and Woolf 2013a: 84–107.

(2018) "Science and Medicine in the Roman Encyclopedists: Patronage for Praxis," in Keyser and Scarborough 2018: 655–76.

Beare, R. (1978) "Were Bailiffs Ever Free Born?" *CQ* 28: 398–401.

Beaujeu, J. (1950) *Pline l'Ancien: Histoire naturelle*. Livre II, Paris.
Becher, G. (1897) *De Lucii Junii Moderati Columellae vita et scriptis*, Leipzig.
Beekes, R. (2010) *Etymological Dictionary of Greek*. 2 vols., Leiden.
Bees, R. (2010) "Die Einheit von Redekunst und Philosophie: Poseidonios bei Cicero, *De oratore* 3.19–24?" *Hermes* 138: 196–215.
Behrends, O. (1992) "Bodenhoheit und privates Bodeneigentum im Grenzwesen Roms," in Behrends and Capogrossi Colognesi 1992: 192–280.
Behrends, O. and L. Capogrossi Colognesi, eds. (1992) *Die römische Feldmesskunst: Interdisziplinäre Beiträge zu ihrer Bedeutung für die Zivilisationsgeschichte Roms*, Göttingen.
Behrends, O., M. Clavel-Lévêque, D. Conso, P. von Cranach, A. Gonzalès, J.-Y. Guillaumin, M. J. Pena, and S. Ratti (1998) *Frontin: L'œuvre gromatique*, Luxembourg.
Behrends, O., M. Clavel-Lévêque, D. Conso, A. Gonzalès, J.-Y. Guillaumin, and S. Ratti (2000) *Hygin: L'œuvre gromatique*, Luxembourg.
Bendz, G. (1938) *Die Echtheitsfrage des vierten Buches der frontinschen* Strategemata, Lund.
Bengtson, H. (1962) "Die griechische Polis bei Aeneas Tacticus," *Historia* 11: 458–68.
Bennett, C. E. and M. B. McElwain (1925) *Frontinus: Stratagems. Aqueducts of Rome*, Cambridge, MA.
Béranger, J. (1953) *Recherches sur l'aspect idéologique du principat*, Basel.
Berger, A. (1953) *Encyclopedic Dictionary of Roman Law*, Philadelphia, PA.
Berrey, M. (2017) "Ancient Medical Friendship," in A. H. Groton, ed., Ab omni parte beatus*: Classical Essays in Honor of James M. May*, Mundelein, IL: 75–96.
Berryman, S. (2009) *The Mechanical Hypothesis in Ancient Greek Natural Philosophy*, Cambridge.
Berthault, F. (2009) "*Biturica*, le vin des Bituriges," in J.-P. Brun, M. Poux, and A. Tchernia, eds., *Le vin: Nectar des dieux, génie des hommes*, 2nd ed., Gollion: 236–8.
Bertoni, D. (2014) *The Cultivation and Conceptualization of Exotic Plants in the Greek and Roman Worlds*, unpublished PhD dissertation, Harvard University.
(2017) "Geometry and Genre in Columella," *AJPh* 138: 527–54.
Bettalli, M. (2010) "Il militare," in G. Zecchini, ed., *Lo storico antico: Mestieri e figure sociali*, Bari: 215–29.
Bianchi, O., O. Thévenaz, and P. Mudry, eds. (2004) *Conceptions et représentations de l'extraordinaire dans le monde antique*, Bern.
Bilić, T. (2012) "Crates of Mallos and Pytheas of Massalia: Examples of Homeric Exegesis in Terms of Mathematical Geography," *TAPhA* 142: 295–328.
Birley, A. R. (1953) *Roman Britain and the Roman Army: Collected Papers*, Kendal.
(2005) *The Roman Government of Britain*, Oxford.
Bispham, E. (2007) "Pliny the Elder's Italy," in Bispham et al. 2007: 41–67.

Bispham, E., G. Rowe, and E., Matthews, eds. (2007) Vita vigilia est*: Essays in Honour of Barbara Levick*, London.

Blandenet, M. (2009) "Le savoir agronomique et sa transmission à Rome à la fin de la république," *Camenulae* 3: 1–19.

Blank, D. (1995) "Philodemus on the Technicity of Rhetoric," in D. Obbink, ed., *Philodemus and Poetry: Poetic Theory and Practice in Lucretius, Philodemus, and Horace*, Oxford: 177–88.

(2003) "Atomist Rhetoric in Philodemus," *Cronache ercolanesi* 33: 69–88.

(2009) "*Philosophia* and *technē*: Epicureans on the Arts," in J. Warren, ed., *The Cambridge Companion to Epicureanism*, Cambridge: 216–33.

Bliquez, L. J. (2014) *The Tools of Asclepius: Surgical Instruments in Greek and Roman Times*, Leiden.

Bloch, R. (2010) *Moses und der Mythos: Die Auseinandersetzung mit der griechischen Mythologie bei jüdisch-hellenistischen Autoren*, Leiden.

Bodel, J. (2015) "The Publication of Pliny's Letters," in I. Marchesi, ed., *Pliny the Book-Maker: Betting on Posterity in the Epistles*, Oxford: 13–108.

Boëthius, A. (1939) "Vitruvius and the Roman Architecture of his Age," in *Dragma, Martino P. Nilsson, A. D. IV id. Iul. MCMXXXIX dedicatum*, Lund: 114–43.

Bohlin, E. (2009) *Geometry in Varro, Cicero, and Vitruvius: A Philological Study*, unpublished PhD dissertation, University of Gothenburg.

Boldrer, F. (1996) *L. Iuni Moderati Columellae rei rusticae liber decimus (carmen de cultu hortorum)*, Pisa.

Bolton, R. (2018) "Two Conceptions of Practical Skill (τέχνη) in Aristotle," in D. Sfendoni-Mentzou, ed., *Aristotle: Contemporary Perspectives on His Thought*, Berlin: 279–96.

(2021) "*Technê* and *Empeiria*: Aristotle on Practical Knowledge," in Johansen 2021: 131–65.

Bömer, F. (1957–8) *P. Ovidius Naso: Die Fasten*, 2 vols., Heidelberg.

(1969–86) *P. Ovidius Naso: Metamorphosen*, 7 vols., Heidelberg.

Bonner, S. F. (1977) *Education in Ancient Rome: From the Elder Cato to the Younger Pliny*, London.

Borca, F. (2003) *Luoghi, corpi, costumi: Determinismo ambientale ed etnografia antica*, Rome.

Boscherini, S. (1970) *Lingua e scienza greca nel* De agri cultura *di Catone*, Rome.

(1986) "Considerazioni sulla *laus Italiae* di Varrone (*De re rust.* 1, 2, 3–8)," in *Studi in onore di Adelmo Barigazzi*, Rome: I 101–9.

(1993) "La medicina in Catone e Varrone," *ANRW* II.37.1: 729–55.

Bradley, K. (1994) *Slavery and Society at Rome*, Cambridge.

Brain, P. (1986) *Galen on Bloodletting: A Study of the Origins, Development and Validity of His Opinions, with a Translation of the Three Works*, Cambridge.

Bréguet, E. (1969) "*Urbi et orbi*: Un cliché et un thème," in J. Dumortier-Bibauw, ed., *Hommages à Marcel Renard*, Brussels: I 140–52.

Brehaut, E. (1933) *Cato the Censor on Farming*, New York.

Bretin-Chabrol, M. (2010) "Le Virgile de Columelle: Une lecture optimiste des *Géorgiques*," *Mosaïque* 3: 29–42.

(2012) *L'arbre et la lignée: Métaphores végétales de la filiation et de l'alliance en latin classique*, Grenoble.

Brisson, L. (2005) "*Epinomis*: Authenticity and Authorship," in K. Döring, M. Erler, and S. Schorn, eds., *Pseudoplatonica*, Stuttgart: 9–24.

Brown, F. E. (1963) "Vitruvius and the Liberal Art of Architecture," *Bucknell Review* 11.4: 99–107.

Brown, N. G. (2019) "A *res rustica* for All the People? Varro, *uillatica pastio*, and the Villa Publica," *TAPhA* 149: 317–51.

Bruère, R. T. (1956) "Pliny the Elder and Virgil," *CPh* 51: 228–46.

Brunet, C., D. Conso, A. Gonzalès, T. Guard, J.-Y. Guillaumin, and C. Sensal (2008) *Libri coloniarum (Livres des colonies)*, Besançon.

Bruno, M. G. T. (1969) *Il lessico agricolo latino*, 2nd ed., Amsterdam.

Brunt, P. A. (1972) "Cn. Tremellius Scrofa the Agronomist," *CR* 22: 304–8.

(1973) "Aspects of the Social Thought of Dio Chrysostom and of the Stoics," *PCPhS* 19: 9–34.

Bruun, C. F. M. (1991) *The Water Supply of Ancient Rome: A Study of Roman Imperial Administration*, Helsinki.

(2013) "Warum verfasste Frontin sein Werk über die Wasserversorgung Roms?" in Letzner 2013: 151–8.

Bubnov, N. (1899) *Gerberti postea Silvestri II papae Opera mathematica (972–1003)*, Berlin.

Bürge, A. (1990) "Der *mercennarius* und die Lohnarbeit," *ZRG* 107: 80–136.

Burkert, W. (1972) *Lore and Science in Ancient Pythagoreanism* (transl. E. L. Minar, Jr.), Cambridge, MA.

Butler, T. (2003) "*Empeiria* in Aristotle," *SJPh* 41: 329–50.

Cairns, F. (1975) "Horace, *Epode* 2, Tibullus I, 1 and Rhetorical Praise of the Countryside," *Museum Philologum Londiniense* 1: 79–91.

Callipo, M. (2011) *Dionisio Trace e la tradizione grammaticale*, Acireale.

Callebat, L. (1973) *Vitruve: De l'architecture. Livre VIII*, Paris.

(1982) "La prose du *De architectura* de Vitruve." *ANRW* II.30.1: 696–722 [cited in pagination of Callebat 2017a: 311–47].

(1989) "Organisation et structures du *De architectura* de Vitruve," in Geertman and de Jong 1989: 34–38 [cited in pagination of Callebat 2017a: 75–85].

(1994) "Rhétorique et architecture dans le *De architectura* de Vitruve," in *Le projet de Vitruve* (q. v.): 31–46 [cited in pagination of Callebat 2017a: 87–104].

(1997) "Encyclopédie et architecture," in J. Bouffartigue and F. Mélonio, eds., *L'entreprise encyclopédique: Études*, Nanterre: 169–181 [cited in pagination of Callebat 2017a: 105–16].

(1998) "'Architecte': Histoire d'un mot," in L. Callebat, ed., *Histoire de l'architecte*, Paris: 11–18 [cited in pagination of Callebat 2017a: 13–30].

(2001) "*Fabrica* et *ratiocinatio* dans le *De architectura* de Vitruve," in Courrént and Thomas 2001: 145–54.

(2003) "Les matériaux de l'architecte," in G. Ciotta, ed., *Vitruvio nella cultura architettonica antica, medievale e moderna*, Genoa: I 88–95 [cited in pagination of Callebat 2017a: 145–56].

(2005) "Le traité d'architecture vitruvien: Problème d'identité," in X. Lafon and G. Sauron, eds., *Théorie et pratique de l'architecture romaine*, Aix-en-Provence : 17–22 [cited in pagination of Callebat 2017a: 61–73].

(2011) "Le savant et le livre," in Jouanna et al. 2011: 473–80.

(2017a) *Le* De architectura *de Vitruve*, Paris.

(2017b) "*Fabrica* et *ratiocinatio*," in Callebat 2017a: 129–44.

(2017c) "Biologie, géographie et architecture: La théorie des tempéraments dans le *De architectura* de Vitruve," in Callebat 2017a: 157–66.

Callebat, L. and P. Fleury (1986) *Vitruve: De l'architecture. Livre X*, Paris.

Callebat, L., P. Gros, and C. Jacquemard. (2003) *Vitruve: De l'architecture. Livre II*, 2nd ed., Paris.

Cambiano, G. (2012) "The Desire to Know: *Metaphysics* A1," in C. Steel, ed., *Aristotle's* Metaphysics *Alpha*, Oxford: 1–42.

Camoletto, S. (1990) "Sul linguaggio traumatologico di Celso," *BStudLat* 20: 14–24.

Campbell, B. (1987) "Teach Yourself How to Be a General," *JRS* 77: 13–29.

(1995) "Sharing out Land: Two Passages in the *Corpus agrimensorum Romanorum*," *CQ* 45: 540–6.

(1996) "Shaping the Rural Environment: Surveyors in Ancient Rome," *JRS* 86: 74–99.

(2000) *The Writings of the Roman Land Surveyors*, London.

(2004) *Greek and Roman Military Writers: Selected Readings*, New York.

(2005) "Surveyors, Topography, and Definitions of Landholding in Ancient Rome," in Conso et al. 2005: 173–81.

Campbell, G. L. (2003) *Lucretius on Creation and Evolution: A Commentary on* De rerum natura, *Book 5, Lines 772–1104*, Oxford.

Canter, H. V. (1938) "Praise of Italy in Classical Authors, I." *CJ* 33: 457–70.

(1939) "Praise of Italy in Classical Authors, II." *CJ* 34: 396–409.

Cantor, M. (1875) *Die römischen Agrimensoren und ihre Stellung in der Geschichte der Feldmesskunst: Eine historisch-mathematische Untersuchung*, Leipzig.

Capitani, U. (1966) "La produzione letteraria di Aulo Cornelio Celso alla luce di un discusso passo dell'*Institutio oratoria*," *Maia* 18: 138–55.

(1967) "Note critiche al testo del *De medicina* di Celso," *SILTA* 39: 112–64.

(1970) "Note critiche al testo del *De medicina* di Celso (*continuazione*)," *SILTA* 42: 5–93.

(1974) "Il recupero di un passo di Celso in un codice del *De medicina* conservato a Toledo," *Maia* 26: 161–212.

(1980) "*Cornelius Celsus, mediocri uir ingenio* . . . : In margine a recenti interpretazioni di un giudizio di Quintiliano," *Prometheus* 6: 67–79.

(1991) "I Sesti e la medicina," in P. Mudry and J. Pigeaud, eds., *Les écoles médicales à Rome*, Geneva: 95–123.

Cardauns, B. (2001) *Marcus Terentius Varro: Einführung in sein Werk*, Heidelberg.

Carder, J. N. (1978) *Art Historical Problems of a Roman Land Surveying Manuscript: The Codex Arcerianus A, Wolfenbüttel*, New York.
Carey, S. (2003) *Pliny's Catalogue of Culture: Art and Empire in the* Natural History, Oxford.
Carra de Vaux, B. (1903) "Le livre des appareils pneumatiques et des machines hydrauliques, par Philon de Byzance, édité d'après les versions arabes d'Oxford et de Constantinople." *Notices et extraits des manuscrits de la Bibliothèque nationale et autres bibliothèques* 38: 27–235.
Carretero Gálvez, J. M. (1997) "La filosofía en Columela," in Maestre Maestre et al. 1997: 129–35.
Carroll, P. D. (1976) "Columella the Reformer," *Latomus* 35: 783–90.
Castiglioni, A. (1940) "Aulus Cornelius Celsus as a Historian of Medicine," *Bulletin of the History of Medicine* 8: 857–73.
Castillo Pascual, M.-J. (1996a) "El nacimiento de una nueva familia de textos técnicos: La literatura gromática," *Gerión* 14: 233–49.
(1996b) *Espacio en orden: El modelo gromático-romano de ordenación del territorio*, Logroño.
(2013) "Los honores de Sextius Iulius Frontinus y su literatura técnica: Una propuesta cronológica para su obra gromática," *DHA* 39: 35–52.
(2016) "Echt oder unecht? Frontin als gromatischer Autor aus historischer Sicht," *eTopoi Journal for Ancient Studies* 5: 46–62.
Castro Gasalla, M. P. (1993) "Aforismos, dichos, citas y sentencias en la obra de Columela," *EClás* 35: 65–80.
Catalano, P. (1960) *Contributi allo studio del diritto augurale*, Torino.
(1978) "Aspetti spaziali del sistema giuridico-religioso romano: *Mundus, templum, urbs, ager, Latium, Italia*," *ANRW* II.16.1: 440–553.
Cataudella, M. R. (2002) "Agronomia," in Mastrorosa et al. 2002: 31–62.
Cavazza, F. (1981) *Studio su Varrone etimologo e grammatico: La lingua latina come modello di struttura linguistica*, Florence.
Chandler, C. (2006) *Philodemus* On Rhetoric *Books 1 and 2*, New York.
Chaniotis, A. (2013) "Greeks under Siege: Challenges, Experiences, and Emotions," in B. Campbell and L. A. Tritle, eds., *The Oxford Handbook of Warfare in the Classical World*, Oxford: 438–56.
Chiasson, C. (2001) "Scythian Androgyny and Environmental Determinism in Herodotus and the Hippocratic περὶ ἀέρων ὑδάτων τόπων," *SyllClass* 12: 33–73.
Chiritoiu, D. A. (2018) *Commanding Texts: Knowledge-ordering, Identity Construction and Ethics in 'Military Manuals' of the Roman Empire*, unpublished PhD dissertation, Cambridge.
Chiron, P. (2007) "The *Rhetoric to Alexander*," in Worthington 2007: 90–106.
Chlup, J. T. (2014) "Just War in Onasander's ΣΤΡΑΤΗΓΙΚΟΣ," *JAH* 2: 37–63.
Choisy, A. (1909) *Vitruve*, 4 vols., Paris.
Chouquer, G., ed. (1996–7) *Les formes du paysage*, 3 vols., Paris.
(2004) "Une nouvelle interprétation du *corpus* des *gromatici veteres*," *Agri centuriati: An International Journal of Landscape Archaeology* 1: 43–56.

(2008) "Les transformations récentes de la centuriation: Une autre lecture de l'arpentage romain," *Annales (HSS)* 63: 847–74.

Chouquer, G., M. Clavel-Lévêque, F. Favory, and J.-P. Vallat. (1987) *Structures agraires en Italie centro-méridionale: Cadastres et paysages ruraux*, Rome.

Chouquer, G. and F. Favory (1992) *Les arpenteurs romains: Théorie et pratique*, Paris.

(1993) "*De arte mensoria*, 'Du métier d'arpenteur': Arpentage et arpenteurs au service de Rome," *Histoire & Mesure* 8: 249–84.

(2001) *L'arpentage romain: Histoire des textes, droit, techniques*, Paris.

Christmann, C. (1982) "Zur antiken *Georgica*-Rezeption," *WJA* 8: 57–67.

(2003) "Zum Verhältnis von Autor und Leser in der römischen Agrarliteratur: Bücher und Schriften für Herren und Sklaven," in Horster and Reitz 2003: 121–52.

Cichorius, C. (1922) *Römische Studien: Historisches, Epigraphisches, Literargeschichtliches aus vier Jahrhunderten Roms*, Leipzig.

Cid López, R. M., and E. Garcia Fernández, eds. (2013) Debita verba*: Estudios en homenaje al profesor Julio Mangas Manjarrés*, 2 vols., Oviedo.

Citroni Marchetti, S. (1982) "*Iuvare mortalem*: L'ideale programmatico della *Naturalis historia* di Plinio nei rapporti con il moralismo stoico-diatribico," *A&R* 27: 124–48.

(1991) *Plinio il Vecchio e la tradizione del moralismo romano*, Pisa.

(2011) *La scienza della natura per un intellettuale romano: Studi su Plinio il Vecchio*, Pisa.

(2017) "*Contingat aliqua gratia operae curaeque nostra*: An Ethic of Care in the *Naturalis historia*," in F. Bessone and M. Fucecchi, eds., *The Literary Genres in the Flavian Age: Canons, Transformations, Reception*, Berlin: 65–82.

Classen, C. J. (1994) "On the Training of the *agrimensores* in Republican Rome and Related Problems: Some Preliminary Observations," *ICC* 19: 161–70.

Clavel-Lévêque, M., ed. (1983) *Cadastres et espace rural: Approches et réalités antiques*, Paris.

Clavel-Lévêque, M. and F. Favory. (1992) "Les *gromatici veteres* et les réalités paysagères: Présentation de quelques cas," in Behrends and Capogrosso Colognesi 1992: 88–137.

Clavel-Lévêque, M., D. Conso, A. Gonzalès, J.-Y. Guillaumin, and P. Robin. (1996) *Hygin l'arpenteur: L'établissement des limites*, Naples.

Cole, T. (1967) *Democritus and the Sources of Greek Anthropology*, Cleveland, OH.

Connor, H. M. (1921) *A Study of the Syntax of the* Strategemata *of Frontinus*, unpublished PhD dissertation, Cornell University.

Conso, D. (2005) "Comment définir et délimiter le vocabulaire technique des arpenteurs latins?" in Conso et al. 2005: 11–23.

Conso, D., A. Gonzalès, and J.-Y. Guillaumin, eds. (2005) *Les vocabulaires techniques des arpenteurs romains*, Besançon.

Contino, S. (1988) *Auli Cornelii Celsi De medicina liber VIII*, Bologna.

Cooley, A. E. (2009) *Res gestae divi Augusti*, Cambridge.

Corbeill, A. (2002) "Rhetorical Education in Cicero's Youth," in J. M. May, ed., *Brill's Companion to Cicero: Oratory and Rhetoric*, Leiden, 23–48.

Corso, A. (1997) "Vitruvius and Attic Monuments" *ABSA* 92: 373–400.
(2016) *Drawings in Greek and Roman Architecture*, Oxford.
Cossarini, A. (1976–7) "Unità e coerenza del *De re rustica* di Varrone," *RAIB* 55: 177–97.
(1977) "Aspetti di Virgilio in Columella," *Prometheus* 3: 225–40.
(1980) "Columella interprete del suo tempo: Alcune considerazioni," *Giornale filologico ferrarese* 3: 97–108.
Courrént, M. and J. Thomas, eds. (2001) *Imaginaire et modes de construction du savoir antique dans les textes scientifiques et techniques*, Perpignan.
Courrént, M. (1999a) "À propos d'une polysémie problématique: Étude sémantique et lexicologique de *natura* dans le *De architectura* de Vitruve," *LEC* 67: 3–20.
(1999b) "Équilibre et changement dans le *De architectura* de Vitruve: Mécanique et chimie du monde appliquées à l'architecture," *Latomus* 58: 534–63.
(2001) "La construction du savoir par la confrontation interdisciplinaire: L'imaginaire vitruvien," in Courrént and Thomas 2001: 123–43.
(2004) "L'au-delà de la *ratio* dans le *De architectura* de Vitruve: Le discours de la méthode," in V. Naas, ed., *En deçà et au-delà de la* ratio, Villeneuve-d'Ascq.
(2005) "*Demonstrare atque explicare*: La fonction de l'écrit dans le *De architectura* de Vitruve," *Pallas* 69: 299–306.
(2011) De architecti scientia*: Idée de nature et théorie de l'art dans le* De architectura *de Vitruve*, Caen.
(2016) "Vitruve et les *naturales quaestiones*: La théorie de la matière dans le *De architectura*," *Arethusa* 49: 251–62.
(2019) *VITRVVIVS AVCTOR: L'œuvre littéraire de Vitruve et sa réception dans la littérature antique (Ier–Ve siècles)*, Bordeaux.
Cowan, R. W. (2009) "Virgil's Cucumber Again: Columella 10.378–92," *CQ* 59: 286–9.
Cuomo, S. (2000) "Divide and Rule: Frontinus and Roman Land-Surveying," *SHPS* 31: 189–202.
(2001) *Ancient Mathematics*, London.
(2007) *Technology and Culture in Greek and Roman Antiquity*, Cambridge.
(2011) "Skills and Virtues in Vitruvius' Book 10," in M. Formisano and H. Böhme, eds., *War in Words: Transformations of War from Antiquity to Clausewitz*, Berlin, 309–32.
(2016a) "A Case-Study in Roman Mathematics: The Description of the Analemma in Vitruvius' *De architectura*, Book IX," in A. Imhausen and T. Pommerening, eds., *Translating Writings of Early Scholars in the Ancient Near East, Egypt, Greece and Rome: Methodological Aspects with Examples*, Berlin, 439–64.
(2016b) "Tacit Knowledge in Vitruvius." *Arethusa* 49: 125–43.
Cusmano, G. (2005) *Sesto Giulio Frontino: Il IV libro degli* Strategemata, Penta.
Dahlmann, H. (1964) *Varro und die hellenistische Sprachtheorie*, 2nd ed., Berlin.
Dain, A. (1930) *Les manuscrits d'Onésandros*, Paris.

(1946) *Histoire du texte d'Élien le tacticien des origines à la fin du Moyen Âge*, Paris.
Dalby, A. (1998) *Cato: On Farming. De agri cultura*, Totnes.
Darab, A. (2012) "Lector in epistula – *studiorum otiosis* (Plin. *Nat. hist.* praef. 6)," *Antik tanulmányok* 56: 287–94.
Daube, D. (1974) "The Mediocrity of Celsus," *CJ* 70: 41–2.
De Melo, W. D. C. (2019) *Varro: De lingua latina*, 2 vols., Oxford.
De Nardis, M. (1994) *The Writings of the Roman Land Surveyors: Technical and Legal Aspects*, unpublished PhD dissertation, London.
De Rijk, L. M. (1965) "Ἐγκύκλιος παιδεία: A Study of Its Original Meaning." *Vivarium* 3: 24–93.
De Saint-Denis, E. (1969) *Columelle: De l'agriculture. Livre X (De l'horticulture)*, Paris.
(1971) "Columelle, miroir de Virgile," in H. Bardon and R. Verdière, eds., *Vergiliana: Recherches sur Virgile*, Leiden: 328–43.
De Simone, C. (1968–70) *Die griechischen Entlehnungen im Etruskischen*, 2 vols., Wiesbaden.
De Vaan, M. (2008) *Etymological Dictionary of Latin and the Other Italic Languages*, Leiden.
Debru, A. (2001) "*In respiritu* (*Nat. deor.* II, 136): L'expression de la respiration en latin," in A. Debru, N. Palmieri, and B. Jacquinod, eds., Docente natura*: Mélanges de médecine ancienne et médiévale offerts à Guy Sabbah*, Saint-Étienne, 43–67.
Deichgräber, K. (1965) *Die griechische Empirikerschule: Sammlung der Fragmente und Darstellung der Lehre*, reprint of the 1930 ed., with additions, Berlin.
Deinlein, B. (1975) *Das römische Sachbuch*, unpublished PhD dissertation, Erlangen-Nürnberg.
Del Chicca, F. (2002) "Personalità e interessi culturali di Frontino," in A. Isola, E. Menestò, and A. Di Pilla, eds., Curiositas*: Studi di cultura classica e medievale in onore di Ubaldo Pizzani*, Naples: 205–17.
(2004) *Frontino: De aquae ductu urbis Romae*, Rome.
DeLaine, J. (1996) "*De aquis suis*?: The *commentarius* of Frontinus," with discussion, in Nicolet 1996: 117–45.
Della Corte, F. (1982) "Tecnica espositiva e struttura della *Naturalis historia*," in Alfonsi and Ronconi 1982: 19–39.
Della Corte, M. (1922) "Groma," *Monumenti Antichi pubblicati per cura della R. Accademia dei Lincei* 28.
Dench, E. (2005) *Romulus' Asylum: Roman Identities from the Age of Alexander to the Age of Hadrian*, Oxford.
Dennis, G. (2014) *The* Taktika *of Leo VI*, rev. ed., Washington, D.C.
Deuse, W. (1993) "Celsus im *prooemium* von *De medicina*: Römische Aneignung griechischer Wissenschaft," *ANRW* II.37.1: 819–41.
Devine, A. M. (1989) "Aelian's *Manual of Hellenistic Military Tactics*: A New Translation from the Greek with an Introduction," *AncW* 19: 31–64.
(1993) "Arrians' *Tactica*." *ANRW* II.34.1: 312–37.
Di Benedetto, V. (1958) "Dionisio il Trace e la *Techne* a lui attribuita," *ASNP* 27: 169–210.

Di Lorenzo, E. (1986) "Il catalogo degli autori greci in Columella," in I. Gallo, ed., *Miscellanea filologica*, Salerno: 161–75.

Diederich, S. (2007) *Römische Agrarhandbücher zwischen Fachwissenschaft, Literatur und Ideologie*, Berlin.

(2013) "Humor, Witz und Ironie in Varros Dialog *De re rustica*," in S. Föllinger and G. M. Müller, *Der Dialog in der Antike: Formen und Funktionen einer literarischen Gattung zwischen Philosophie, Wissensvermittlung und dramatischer Inszenierung*, Berlin: 275–94.

Dilke, O. A. W. (1967) "Illustrations from Roman Surveyors' Manuals," *Imago Mundi* 21: 9–30.

(1971) *The Roman Land Surveyors: An Introduction to the* Agrimensores, Newton Abbot.

(1974) "Archaeological and Epigraphic Evidence of Roman Land Surveys," *ANRW* II.1: 564–92.

(1976) "Varro and the Origins of Centuriation," in *Atti del congresso internazionale di studi varroniani*, Rieti: II 353–8.

(1983) "De Frontini et Flaviorum communibus curis," in *Atti del congresso internazionale di studi flaviani*, Rieti, 125–32.

(1987) "Maps in the Service of the State: Roman Cartography to the End of the Augustan Era," in J. B. Harley and D. Woodward, eds., *Cartography in Prehistoric, Ancient, and Medieval Europe and the Mediterranean*, Chicago, IL: 201–11.

(1988) "Religious Mystique in the Training of *Agrimensores*," in Porte and Néraudau 1988: 158–62.

Dillery, J. (2017) "Xenophon: The Small Works," in M. A. Flower, ed., *The Cambridge Companion to Xenophon*, Cambridge, 195–219.

Dion, R. (1959) *Histoire de la vigne et du vin en France, des origines au XIXe siècle*, Paris.

Dodds, E. R. (1959) *Gorgias: A Revised Text*, Oxford.

Domínguez Petit, R. (2004) "Fuentes literarias para la agricultura cartaginesa: El tratado de Magón," *Habis* 35: 179–92.

Doody, A. (2001) "Finding Facts in Pliny's *Encyclopaedia*: The *summarium* of the *Natural History*," *Ramus* 30: 1–22.

(2007) "Virgil the Farmer? Critiques of the *Georgics* in Columella and Pliny," *CPh* 102: 180–97.

(2009) "Pliny's *Natural History*: The *enkuklios paideia* and the Ancient Encyclopaedia," *JHI* 70: 1–21.

(2010) *Pliny's Encyclopedia: The Reception of the* Natural History, Cambridge.

(2011) "The Science and Aesthetics of Names in the *Natural History*," in Gibson and Morello 2011: 113–29.

(2017) "The Authority of Writing in Varro's *De re rustica*," in König and Woolf 2017: 182–202.

Douglas, A. E. (1966) *M. Tulli Ciceronis Brutus*, Oxford.

Drachmann, A. G. (1948) *Ktesibios, Philon and Heron: A Study in Ancient Pneumatics*, Copenhagen.

Dueck, D. (2000) *Strabo of Amasia: A Greek Man of Letters in Augustan Rome*, New York.

Dumont, J.-C. (1993) *Columelle: De l'agriculture. Livre III*, Paris.

(2008) "Columella and Vergil," *Vergilius* 54: 49–58.

Dumont, J.-P. (1986) "L'idée de Dieu chez Pline (*HN* 2, 1–5, 1–27)," *Helmantica* 37: 219–37.

Duncan-Jones, R. P. (1982) *The Economy of the Roman Empire: Quantitative Studies*, 2nd ed., Cambridge.

Dyck, A. R. (1996) *A Commentary on Cicero,* De officiis, Ann Arbor, MI.

(2012) *Marcus Tullius Cicero: Speeches on Behalf of Marcus Fonteius and Marcus Aemilius*, Oxford.

Earl, D. C. (1961) *The Political Thought of Sallust*, Cambridge.

Eck, W. (2002) "An Emperor is Made: Senatorial Politics and Trajan's Adoption by Nerva in 97," in G. Clark and T. Rajak, eds., *Philosophy and Power in the Graeco-Roman World: Essays in Honour of Miriam Griffin*, Oxford: 211–26.

(2013) "Die Gestalt Frontins in ihrer politischen und sozialen Umwelt," in Letzner 2013: 111–34.

Eck, W. and A. Pangerl (2003) "Sex. Iulius Frontinus als Legat des niedergermanischen Heeres: Zu neuen Militärdiplomen in den germanischen Provinzen," *ZPE* 143: 205–19.

Edelstein, L. (1931) "The Hippocratic Physician," in Temkin and Temkin 1967: 87–110.

(1932) "The History of Anatomy in Antiquity," in Temkin and Temkin 1967: 247–301.

Edelstein, L. and I. G. Kidd (1972–99) *Posidonius*, 4 vols., Cambridge.

Elsner, J. (1995) *Art and the Roman Viewer: The Transformation of Art from the Pagan World to Christianity*, Cambridge.

Englund, J. (1935) *Kassusyntaxen hos A. Cornelius Celsus*, Gothenburg.

Eramo, I. (2021) "Precetti per gestire l'*imperium*: Nota a Onasandro, *Strategikos*, proemio 1," *DHA* 47: 239–50.

Ercolani, L. (1997) "La lingua di Onasandro: Ricerche sugli ἅπαξ λεγόμενα," *AFLS* 18: 43–53.

Espinosa Espinosa, D. (2013) "La *Historia natural* de Plinio el Viejo: Un proyecto 'augusteo' de época flavia," in Cid López and Garcia Fernández 2013: I 671–84.

Evans, H. B. (1994) *Water Distribution in Ancient Rome: The Evidence of Frontinus*, Ann Arbor, MI.

Evans, R. (2005) "Geography without People: Mapping in Pliny *Historia naturalis* Books 3–6," *Ramus* 34: 47–74.

Evans, J. and J. L. Berggren (2006) *Geminos'* Introduction to the Phenomena*: A Translation and Study of a Hellenistic Survey of Astronomy*, Princeton, NJ.

Fagan, G. G. (2006) "Bathing for Health with Celsus and Pliny the Elder," *CQ* 56: 190–207.

Favro, D. (1996) *The Urban Image of Augustan Rome*, Cambridge.

Fear, A. T. (2011) "The Roman's Burden," in Gibson and Morello 2011: 21–34.

Fensterbusch, C. (1976) *Vitruvii De architectura libri decem: Zehn Bücher über Architektur*, 2nd ed., Darmstadt.

Ferri, S. (1951) "Binomi obbligati nella critica d'arte degli antichi," *SCO* 11: 206–14.

(1960) *Vitruvio: Architettura (Dai Libri I–VII)*, Rome.

Festa, N. (1932–5) *I frammenti degli stoici antichi*, 2 vols., Bari.

Finkmann, S., A. Behrendt, and A. Walter, eds. (2018) *Antike Erzähl- und Deutungsmuster: Zwischen Exemplarität und Transformation*, Berlin.

Flach, D. (1990) *Römische Agrargeschichte*, Munich.

(1996–2002) *Marcus Terentius Varro: Gespräche über die Landwirtschaft*, 3 vols., Darmstadt.

Flemming, R. (2000) *Medicine and the Making of Roman Women: Gender, Nature, and Authority from Celsus to Galen*, Oxford.

Fleury, P. (1991) "Les roses des vents latines," *Euphrosyne* 19: 61–88.

(1994) "Le *De architectura* et les traités de mécanique ancienne," in *Le projet de Vitruve* (q. v.): 187–212.

(2003) *Vitruve: De l'architecture. Livre I.* 2nd ed., Paris.

(2017) *De rebus bellicis: Sur les affaires militaires*, Paris.

Fögen, T., ed. (2005) *Antike Fachtexte / Ancient Technical Texts*, Berlin.

(2009) *Wissen, Kommunikation und Selbstdarstellung: Zur Struktur und Charakteristik römischer Fachtexte der frühen Kaiserzeit*, Munich.

(2014) "Die Traktate römischer Agrimensoren im Kontext antiker Fachliteratur: Form, Funktion, Autorbewußtsein," in Knobloch and Möller 2014: 215–38.

(2018) "*L'architecte engagé*: Education, Morality and Politics in Vitruvius' *De architectura*," *GLO* 39–40: 17–46.

Folkerts, M. (1992) "Mathematische Probleme im *Corpus agrimensorum*," in Behrends and Capogrossi Colognesi 1992: 311–34.

(2014) "Die Mathematik der Agrimensoren – Quellen und Nachwirkung," in Knobloch and Möller 2014: 131–48.

Fontán, A. (1997) "Escritores hispanos en el siglo de Columela," in Maestre Maestre et al. 1997: 15–34.

Forcellini, E. (1771) *Totius Latinitatis Lexicon consilio et cura Jacobi Facciolati, opera et studio Aegidii Forcellini, alumni seminarii Patavini, lucubratum*, 4 vols, Padua.

Formisano, M. 2001. *Tecnica e scrittura: Le letterature tecnico-scientifiche nello spazio letterario tardolatino*, Rome.

(2006) Review of R. H. Rodgers, *Frontinus: De aquaeductu urbis Romae* (2004), *CR* 56: 132–5.

(2011) "The *Strategikós* of Onasander: Taking Military Texts Seriously." *Technai* 2: 39–52.

(2017) "Fragile Expertise and the Authority of the Past: The 'Roman Art of War,'" in König and Woolf 2017: 129–52.

(2018) "Beyond the Fence: Columella's Garden," in Finkmann et al. 2018: 501–14.

Formisano, M. and P. van der Eijk, eds. (2017) *Knowledge, Text and Practice in Ancient Technical Writing*, Cambridge.

Fortenbaugh, W. W. (2007) "Aristotle's *Art of Rhetoric*," in Worthington 2007: 107–23.

Franklin, L. (2005) "*Technē* and Teleology in Plato's *Gorgias*," *Apeiron* 38: 229–55.

Frede, M. (1980) "The Original Notion of Cause," in M. Schofield, M. Burnyeat, and J. Barnes, eds., *Doubt and Dogmatism: Studies in Hellenistic Epistemology*, Oxford: 217–49.

(1982) "The Method of the So-Called Methodical School of Medicine," in J. Barnes, J. Brunschwig, M. Burnyeat, and M. Schofield, eds., *Science and Speculation: Studies in Hellenistic Theory and Practice*, Cambridge: 1–23.

(1987) "The Ancient Empiricists," in M. Frede, *Essays in Ancient Philosophy*, Minneapolis, MN: 243–60.

(1988) "The Empiricist Attitude Towards Reason and Theory." *Apeiron* 21: 79–97.

(1990) "An Empiricist View of Knowledge: Memorism," in E. Stephen, ed., *Epistemology*, Cambridge: 225–50.

(1996) "Aristotle's Rationalism," in M. Frede and G. Striker, eds., *Rationality in Greek Thought*, Oxford: 157–73.

French, R. K. (1994) *Ancient Natural History: Histories of Nature*, London.

French, R. K. and F. Greenaway, eds. (1986) *Science in the Early Roman Empire: Pliny the Elder, His Sources and Influence*, London.

Frézouls, E. (1985) "Vitruve et le dessin d'architecture," *Le dessin d'architecture dans les sociétés antiques*, Leiden: 213–29.

(1989) "Fondements scientifiques, armature conceptuelle et *praxis* dans le *De architectura*," in Geertman and de Jong 1989: 39–48.

(1994) "Grammaire et rhétorique chez Vitruve," in J. Dangel, ed., *Grammaire et rhétorique: Notion de romanité*, Strasbourg: 151–69.

Fronda, M. P. (2011) "Hannibal: Tactics, Strategy, and Geostrategy," in D. Hoyos, ed., *A Companion to the Punic Wars*, Malden, MA: 242–59.

Fruyt, M. (2014) "Territoire, terrain, terre dans le vocabulaire des agronomes latines," in C. Brunet, ed., *Territoires et dépendances: Approches linguistiques*, Besançon, 17–50.

Fuhrmann, M. (1960) *Das systematische Lehrbuch: Ein Beitrag zur Geschichte der Wissenschaften in der Antike*, Göttingen.

Gabba, E. (1980) "La praefatio de Vitruvio e la Roma Augustea," *ACD* 16: 49–52.

(1985) "Per un'interpretazione storica della centuriazione romana," *Athenaeum* 63: 265–84.

Gaide, F. (2006) "À propos du vocabulaire des arpenteurs latins: Étymologies antiques et modernes; Analyses lexicologiques," in Conso et al. 2006: 33–9.

Gaines, R. N. (2007) "Roman Rhetorical Handbooks," in W. Dominik and J. Hall, eds., *A Companion to Roman Rhetoric*, Malden, MA: 163–80.

Gale, M. R. (2000) *Virgil on the Nature of Things: The* Georgics, *Lucretius, and the Didactic Tradition*, Cambridge.

(2013) "Piety, Labour, and Justice in Lucretus and Hesiod," in D. Lehoux, A. D. Morrison, and A. Sharrock, eds., *Lucretius: Poetry, Philosophy, Science*, Oxford: 25–50.

Galimberti, A. (2002) "Lo *Strategikòs* di Onasandro," in M. Sordi, ed., *Guerra e diritto nel mondo greco e romano*, Milan: 141–53.

Galinsky, K. (1996) *Augustan Culture: An Interpretive Introduction*, Princeton, NJ.

Gallia, A. B. (2012) *Remembering the Roman Republic: Culture, Politics, and History under the Principate*, Cambridge.

Galzerano, M. (2015) "Lucrezio, *De rerum natura* 2, 1173–1174: In difesa di *ire ad capulum*," *Paideia* 70: 243–53.

(2018) "Four Notes on Lucretius and a Note on Manilius," *Maia* 70: 280–90.

Garani, M. (2007) Empedocles redivivus*: Poetry and Analogy in Lucretius*, London.

García Armendáriz, J. I. (2004) *Columela: Libro de los árboles. La labranza: Libros I–V*, Madrid.

Gargola, D. J. (1995) *Lands, Laws and Gods: Magistrates and Ceremony in the Regulation of Public Lands in Republican Rome*, Chapel Hill, NC.

(2004) "The Ritual of Centuriation," in C. F. Konrad, ed., *Augusto augurio: Rerum humanarum et divinarum commentationes in honorem Jerzy Linderski*, Stuttgart: 124–49.

(2005) "Hyginus Gromaticus and Frontinus on the Installation of *limites*: Ritual, Law, and Legitimacy," *Caesarodunum* 39: 125–52.

(2017) *The Shape of the Roman Order: The Republic and Its Spaces*. Chapel Hill, NC.

Garlan, Y. (1974) *Recherches de poliorcétique grecque*, Paris.

Garrido-Hory, M. and A. Gonzalès, eds. (2003) *Histoire, espaces et marges de l'Antiquité: Hommages à Monique Clavel-Lévêque*, 4 vols., Besançon.

Gärtner, U. (2002) "Zum Lob Italiens in der griechischen Literatur," *Abhandlungen der Akademie der Wissenschaften und der Literatur Mainz. Geistes- und sozialwissenschaftliche Klasse* 2002 IV, Stuttgart.

Gauly, M. G. (2018) "Plinius' Zoologie und die römische Naturgeschichte," in G. M. Müller and F. Mariani-Zini, eds., *Philosophie in Rom – römische Philosophie? Kultur-, literatur- und philosophiegeschichtliche Perspektiven*, Berlin: 469–87.

Gautherie, A. (2017) *Rhétorique et thérapeutique dans le* De medicina *de Celse*, Turnhout.

Geertman, H. (1994) "Teoria e attualità della progettistica architettonica di Vitruvio," in *Le projet de Vitruve* (q. v.): 7–30.

Geertman, H. and J. J. de Jong, eds. (1989) Munus non ingratum: *Proceedings of the International Symposium on Vitruvius'* De architectura *and the Hellenistic and Republican Architecture*, Leiden.

Gemoll, A. (1876) "Über das Fragment *De munitionibus castrorum*." *Hermes* 11: 164–78.

Geus, K. (2014) "Agennius Urbicus und die Antichthonen: Ein stoisches Weltbild im *Corpus agrimensorum Romanorum*," in Knobloch and Möller 2014: 113–30.

Gibson, R. Y. and R. Morello, eds. (2011) *Pliny the Elder: Themes and Contexts*, Leiden.

Gildenhard, I. (2007) *Paideia Romana: Cicero's* Tusculan Disputations, Cambridge.

Gilliver, C. M. (1996) "The Roman Army and Morality in War," in A. B. Lloyd, ed., *Battle in Antiquity*, London: 219–38.

(1999) *The Roman Art of War*, Charleston, SC.

Gindhart, M. (2017) "Crossing Genres in römischen Fachtexten: Zu den Lehrdichtungen in den Agrarhandbüchern von Columella und Palladius," in A. Dresen and F. Freitag, eds., *Crossing: Über Inszenierungen kultureller Differenzen und Identitäten*, Bielefeld: 67–88.

Goesius, W. (Willem van der Goes) (1674) *Rei agrariae auctores legesque variae*, Amsterdam.

Goguey, D. (1978) "La formation de l'architecte: Culture et technique," in J.-M. André, ed., *Recherches sur les* artes *à Rome*, Paris: 100–15.

Goldmann, B. (1988) *Einheitlichkeit und Eigenständigkeit der* Historia Romana *des Appian*, Hildesheim.

Goldsworthy, A. (2000) *The Fall of Carthage: The Punic Wars 265–146 BC*, London.

Gordon, A. E. (1952) *Quintus Veranius Consul A.D. 49: A Study Based upon His Recently Identified Sepulchral Inscription*, Berkeley, CA.

Gourinat, J.-B. (2011) "Les définitions de l'*epistêmê* et de la *technê* dans l'ancien stoïcisme," in Jouanna et al. 2011: 243–56.

Gowers, E. (2000) "Vegetable Love: Virgil, Columella, and Garden Poetry." *Ramus* 29: 127–48.

Gozzoli, S. (2010) "Vitruvio e la basilica di Fano," *SCO* 56: 111–30.

Granger, F. (1931–4) *Vitruvius: On Architecture*, 2 vols., Cambridge, MA.

Green, C. M. C. (1997) "Free as a Bird: Varro *De re rustica* 3," *AJPh* 118: 427–48.

Greene, J. A. and D. P. Kehoe (1995) "Mago the Carthaginian," in M. H. Fantar and M. Ghaki, eds., *Actes du IIIe congrès international des études phéniciennes et puniques*, Tunis: 110–17.

Gregorić, P. and F. Grgić (2006) "Aristotle's Notion of Experience," *AGPh* 88: 1–30.

Griffin, M. T. (1972) "The Elder Seneca and Spain." *JRS* 62: 1–19.

(1976) *Seneca: A Philosopher in Politics*, Oxford.

(2000) "The Flavians," in A. K. Bowman, P. Garnsey, and D. Rathbone, eds., *The Cambridge Ancient History, Vol. IX, The High Empire, A. D. 70–192*, 2nd ed., Cambridge: 1–83.

Grillone, A. (2012) *Gromatica militare: Lo ps. Igino*, Brussels.

Grimal, P. (1986) "Pline et les philosophes," *Helmantica* 37: 239–49.

Gros, P. (1975) "Structures et limites de la compilation vitruvienne dans les livres III et IV du *De architectura*," *Latomus* 34: 986–1009.

(1982) "Vitruve: L'architecture et sa théorie, à la lumière des études récentes," *ANRW* II.36.1: 659–95.

(1989) "L'*auctoritas* chez Vitruve: Contribution à l'étude de la sémantique des ordres dans le *De architectura*," in Geertman and de Jong 1989: 126–33.

(1992) *Vitruve: De l'architecture. Livre IV*, Paris.

(1994) "*Munus non ingratum*: Le traité vitruvien et la notion de service," in *Le projet de Vitruve* (q. v.): 75–90.

(1996) "Les illustrations du *De architectura* de Vitruve: Histoire d'un malentendu," with discussion, in Nicolet 1996: 19–44.

(1997) "Vitruvio e il suo tempo," in Gros et al. 1997: ix–lxxvii.

Gros, P., C. Corso and E. Romano (1997) *Vitruvio: De architectura*, 2 vols., Torino.

Grosse, R. (1934) Review of D. Schenk, *Flavius Vegetius Renatus: Die Quellen der* Epitoma rei militaris (1930), *Deutsche Literaturzeitung* 55: 61–5.

Gruen, E. S. (1992) *Culture and National Identity in Republican Rome*, London.

Gugel, H. (1970) "Bemerkungen zur Darstellung von Catilinas Ende bei Sallust (Ein Beitrag zur antiken Feldherrntopik)," in D. Ableitinger and H. Gugel, eds., *Festschrift Karl Vretska zum 70. Geburtstag überreicht von seinem Freunden und Schülern*, Heidelberg: 361–81.

Guillaumin, J.-Y. (1994) "Géométrie grecque et agrimensorique romaine: La science comme justification d'une idéologie," *DHA* 20: 279–95.

(1995) "Les *agrimensores* romains et la géométrie grecque." In B. Jacquinod, ed., *Actes du XXVIIIème congrès international*, Saint-Étienne: 39–64.

(1996) *Balbus: Présentation systématique de toutes les figures. Podismus et textes connexes. Extraits d'Épaphrodite et de Vitruvius Rufus. La mesure des jugères*, Naples.

(1997) "Les noms latins du point géométrique," in P. R. Colace, ed., *Atti del II seminario internazionale di studi sui lessici tecnici greci e latini*, Naples: 85–106.

(2002) "L'éloge de la *geometria* chez Agennius Urbicus," *REA* 104: 433–43.

(2003) "Géométrie et arpentage: Le 'geometres' à l'époque néronienne d'après Sénèque, *ad Luc.* 88, §10–13," in Garrido-Hory and Gonzalès 2003: I 109–32.

(2004) "L'origine du terme *occupatorius* d'après Hygin," in M. Clavel-Lévêque and E. Hermon, eds., *Espaces intégrés et ressources naturelles dans l'empire romain*, Besançon: 39–47.

(2005) *Les arpenteurs romains. Tome I: Hygin le gromatique. Frontin*, Paris.

(2009a) "La classification des figures géométriques dans le traité gromatique d'Hygin," *Euphrosyne* 37: 177–91.

(2009b) "Quels rapports entre les *agrimensores* romains et la science grecque?" in F. Le Blay, ed., *Transmettre les savoirs dans les mondes hellénistique et romain*, Rennes: 119–32.

(2009c) "Regard sur le passé et identité romaine chez les auteurs du corpus gromatique latin," *Pallas* 31: 133–9.

(2010) "Le livre gromatique du Haut-Empire," in Y. Perrin and M. de Souza, eds., *Neronia VIII: Bibliothèques, livres et culture écrite dans l'empire romain de César à Hadrien*, Brussels: 156–64.

(2015a) "Le discours des *agrimensores* latins: Caractéristiques et sources, transmission et adaptation," *SPhV* 17: 9–34.

(2015b) "Quelques incertitudes et problèmes concernant la groma," in P. Fleury, C. Jacquemard, and S. Madeleine, eds., *La technologie gréco-romaine: Transmission, restitution, et médiation*, Caen: 177–89.

Guittard, C. (2009) "La délimitation du *templum* augural: Les formules d'*augur-atio* de Varron (*L.*, VII, 8) et de Tite-Live (I, 18, 6–10)," in O. Devillers and J. Meyers, eds., *Pouvoirs des hommes, pouvoirs des mots, des Gracques à Trajan*, Louvain: 77–89.

Gummerus, H. (1906) *Der römische Gutsbetrieb als wirtschaftlicher Organismus nach den Werken des Cato, Varro und Columella*, Leipzig.

(1909–10) *De Columella philosopho*, Helsinki.

Gundermann, G. (1888a) *Iuli Frontini Strategematon libri quattuor*, Leipzig.

(1888b) *Quaestiones de Iuli Frontini Strategamaton libris*, Leipzig.

Hadot, I. (2005) *Arts libéraux et philosophie dans la pensée antique: Contribution à l'histoire de l'éducation et de la culture dans l'antiquité*, 2nd ed., Paris.

Hahn, J. (1991) "Plinius und die griechischen Ärtze in Rom: Naturkonzeption und Medizinkritik in der *Naturalis historia*," *Sudhoffs Archiv* 75: 209–39.

Hamberg, P. G. (1965) "Vitruvius, Fra Giocondo and the City Plan of Naples: A Commentary on Some Principles of Ancient Urbanism and their Rediscovery in the Renaissance," *AArch* 36: 105–25.

Hänger, C. (2001) *Die Welt im Kopf: Raumbilder und Strategie im römischen Kaiserreich*, Göttingen.

Hankinson, J. (1987*a*) "Causes and Empiricism: A Problem in the Interpretation of Later Greek Medical Method," *Phronesis* 32: 329–48.

(1987b) "Evidence, Externality and Antecedence: Inquiries into Later Greek Causal Concepts," *Phronesis* 32: 80–100.

(1998) *Galen: On Antecedent Causes*, Cambridge.

(2003) "Causation in Galen," in J. Barnes, ed., *Galien et la philosophie*, Geneva: 31–66.

(2004) "Art and Experience: Greek Philosophy and the Status of Medicine," *Quaestio* 4: 3–24

Harig, G. (1971) "Zum Problem 'Krankenhaus' in der Antike," *Klio* 53: 179–95.

(1976) "Die Diätetik der römischen Enzyklopädisten," *NTM* 13: 1–15.

Harris, W. V. (1980) "Towards a Study of the Roman Slave Trade," *MAAR* 36: 117–40.

Harris-McCoy, D. (2017) "Making and Defending Claims to Authority in Vitruvius' *De architectura*," in König and Woolf 2017: 107–28.

Haselberger, L. (1989) "Die Zeichnungen in Vitruvs *De architectura*: Zur Illustration antiker Schriften über das Konstruktionswesen," in Geertman and de Jong 1989: 69–70.

(1999) "Geometrie der Winde, windige Geometrie: Städtebau nach Vitruv und Aristophanes," in E.-L. Schwandner and K. Rheidt, eds., *Stadt und Umland: Neue Ergebnisse der archäologischen Bau- und Siedlungsforschung*, Mainz: 90–100.

(2007) Urbem adornare: *Rome's Urban Metamorphosis under Augustus* (transl. A. Thein), Portsmouth, RI.

Hasper, P. S. and J. Yurdin (2014) "Between Perception and Scientific Knowledge: Aristotle's Account of Experience," *OSAPh* 47: 119–50.

Haß, C. D. (2018) "'Geerdetes Denken' (Teil 1): (Agri)Kultur als Formverfahren und Episteme bei Varro, *De re rustica*," in R. M. Erdbeer, F. Kläger, and K. Stierstorfer, eds., *Literarische Form: Theorien, Dynamiken, Kulturen. Beiträge zur literarischen Modellforschung*, Heidelberg: 305–46.

Hawkins, C. (2016) *Roman Artisans and the Urban Economy*, Cambridge.

Healy, J. F. (1999) *Pliny the Elder on Science and Technology*, Oxford.

Heath, T. (1921) *A History of Greek Mathematics*, 2 vols., Oxford.

Heinimann, F. (1961) "Eine vorplatonische Theorie der τέχνη," *MH* 18: 105–30.

Heisterhagen, R. (1952) *Die literarische Form der* Rerum rusticarum libri *Varros*, unpublished PhD dissertation, Marburg.

Heitland, W. E. (1921) Agricola*: A Study of Agriculture and Rustic Life in the Greco-Roman World from the Point of View of Labour*, Cambridge.

Henderson, J. (2002) "Columella's Living Hedge: The Roman Gardening Book," *JRS* 92: 110–33.

(2004) *The Roman Book of Gardening*, London.

Hentz, G. (1978) *Columelle: De re rustica. Livre I*, 3 vols., thesis, Strasbourg.

(1979) "Les sources grecques dans les écrits des agronomes latins," *Ktèma* 4: 151–60.

Hermon, E. (2003) "Les *gromatici* entre fictions et faits," in Garrido-Hory and Gonzalès 2003: I 133–60.

Heurgon, J. (1950) "L'effort de style de Varron dans les *Res rusticae*," *RPh* 24: 57–78.

(1976) "L'agronome carthaginois Magon et ses traducteurs en latin et en grec," *CRAI* 120: 441–56.

(1978–97) *Varron: Économie rurale*, 3 vols., Paris.

Highet, G. (1975) "The Mediocrity of Celsus," *CJ* 70: 57.

Hinrichs, F. T. (1974) *Die Geschichte der gromatischen Institutionen: Untersuchungen zu Landverteilung, Landvermessung, Bodenverwaltung und Bodenrecht im römischen Reich*, Wiesbaden.

(1992) "Die 'agri per extremitatem mensura comprehensi': Diskussion eines Frontintextes und der Geschichte seines Verständnisses," in Behrends and Capogrossi Colognesi 1992: 348–72.

Hofter, M. (1988) *Kaiser Augustus und die verlorene Republik*, Mainz.

Holmes, B. (2017) "Pure Life: The Limits of the Vegetal Analogy in the Hippocratics and Galen," in J. Z. Wee, ed., *Comparable Body: Analogy and Metaphor in Ancient Mesopotamian, Egyptian, and Greco-Roman Medicine*, Leiden: 358–86.

Holmes, B. and K.-D. Fischer, eds. (2015) *The Frontiers of Ancient Science: Essays in Honor of Heinrich von Staden*, Berlin.

Horster, M. and C. Reitz, eds. (2003) *Antike Fachschriftsteller: Literarischer Diskurs und sozialer Kontext*, Wiesbaden.

Howe, N. P. (1985) "In Defense of the Encyclopedic Mode: On Pliny's 'Preface' to the *Natural History*," *Latomus* 44: 561–76.

Hübner, W. (1992) "Himmel und Erdvermessung," in Behrends and Capogrossi Colognesi 1992: 140–70.
Hug, A. (1877) *Aeneas von Stymphalos: Ein arkadischer Schriftsteller aus classischer Zeit*, Zurich.
Humm, M. (2014) "Espaces comitiaux et contraintes augurales à Rome pendant la période républicaine," *Ktèma* 39: 315–45.
Ierodiakonou, K. (2019) "Dialectic as a Subpart of Stoic Philosophy," in T. Bénatouïl and K. Ierodiakonou, eds., *Dialectic after Plato and Aristotle*, Cambridge: 114–33.
Ilberg, J. (1907) *A. Cornelius Celsus und die Medizin in Rom*, Leipzig.
Illinois Greek Club. (1923) *Aeneas Tacticus. Asclepiodotus. Onasander*, Cambridge, MA.
Ireland, R. I. (1990) *Iuli Frontini Strategemata*, Leipzig.
Irwin, T. (1979) *Plato: Gorgias*, Oxford.
Isager, J. (1991) *Pliny on Art and Society: The Elder Pliny's Chapters on the History of Art*, London.
Jackson, B. D. (1975) *Augustine: De dialectica*, Dordrecht.
Jaeger, M. (2008) *Archimedes and the Roman Imagination*, Ann Arbor, MI.
Jahn, O. (1850) "Über römische Encyclopädien," *Berichte über die Verhandlungen der Königlich Sächsischen Gesellschaft der Wissenschaften zu Leipzig, Philologisch-historische Classe* 4: 263–87.
Janson, T. (1964) *Latin Prose Prefaces: Studies in Literary Conventions*, Stockholm.
Jocelyn, H. D. (1985) "The New Chapters of the Ninth Book of Celsus' *Artes*," in F. Cairns, ed., *Papers of the Liverpool Latin Seminar, Vol. 5*, Liverpool: 299–336.
Johansen, T. K. (2017) "Aristotle on the *Logos* of the Craftsman," *Phronesis* 62: 97–135.
ed. (2021) *Productive Knowledge in Ancient Philosophy: The Concept of* Technê, Cambridge.
Johnston, I. (2006) *Galen: On Diseases and Symptoms*, Cambridge.
Johnston, P. A. (1999) "*Huc pater O Lenaee veni*: The Cultivation of Wine in Vergil's *Georgics*," *Journal of Wine Research* 10: 207–21.
Jordan, H. (1860) *M. Catonis praeter librum de re rustica quae extant*, Leipzig.
Jori, A. (1996) *Medicina e medici nell'antica Grecia: Saggio sul* Perì téchnes *Ippocratico*, Naples.
Josephson, A. (1950) Casae litterarum*: Studien zum corpus agrimensorum Romanorum*, Uppsala.
Jouanna, J. (1999) *Hippocrates* (transl. M. B. DeBevoise), Baltimore, MD.
Jouanna, J., M. Fartzoff, and B. Bakhouche, eds. (2011) *L'homme et la science*, Paris.
Kaibel, G. (1885) "Antike Windrosen," *Hermes* 20: 579–624.
Kappelmacher, A. (1918) *Untersuchungen zur Enzyklopädie des A. Cornelius Celsus*, Vienna.
Kassel, R. (1971) *Der Text der aristotelischen* Rhetorik: *Prolegomena zu einer kritischen Ausgabe*, Berlin.
Keeline, T. J. (2018) *The Reception of Cicero in the Early Roman Empire: The Rhetorical Schoolroom and the Creation of a Cultural Legend*, Cambridge.

Kennedy, G. A. (1963) *The Art of Persuasion in Greece*, Princeton, NJ.

(1972) *The Art of Rhetoric in the Roman World, 300 B.C.–A.D. 300*, Princeton, NJ.

(1994) *A New History of Classical Rhetoric*, Princeton, NJ.

(2007) *Aristotle: On Rhetoric. A Theory of Civic Discourse*, 2nd ed., Oxford.

Kennedy, R. F. (2016) "Airs, Waters, Metals, Earth: People and Environment in Archaic and Classical Greek Thought," In R. F. Kennedy and M. Jones-Lewis, eds., *The Routledge Handbook of Identity and the Envrionment in the Classical and Medieval Worlds*, London: 9–28.

Keppie, L. J. F. (1983) *Colonisation and Veteran Settlement in Italy, 47–14 B.C*, London.

Kern, P. B. (1999) *Ancient Siege Warfare*. Indianapolis, IN.

Keßler, L. and K. Ott. (2017) "*Nec provident future tempori, sed quasi plane in diem vivant*—Sustainable Business in Columella's *De re rustica*?" in C. Schliephake, ed., *Ecocriticism, Ecology, and the Cultures of Antiquity*, Lanham, MD: 197–216.

Keyser, P. T. and G. Irby-Massie, eds. (2008) *Encyclopedia of Ancient Natural Scientists: The Greek Tradition and Its Many Heirs*, London.

Keyser, P. T. and J. Scarborough, eds. (2018) *The Oxford Handbook of Science and Medicine in the Classical World*, Oxford.

Kidd, I. G. (1978) "Philosophy and Science in Posidonius," *A&A* 24: 7–15.

Kienast, H. J. (2014) *Der Turm der Winde in Athen*, with contributions by P. Karanastasi and K. Schaldach, Wiesbaden.

Kier, H. (1933) *De laudibus vitae rusticae*, unpublished PhD dissertation, Marburg.

Knell, H. (1985) *Vitruvs Architekturtheorie: Versuch einer Interpretation*, Darmstadt.

Knobloch, E. (2014) "Wissenschaftshistorische Aspekte im Thulin'schen *Corpus agrimensorum Romanorum*," in Knobloch and Möller 2014: 83–112.

(2015) "Les écrits des arpenteurs romains et l'influence de la cosmologie," in P. Caye, R. Nanni, and P. D. Napolitani, eds., *Scienze e rappresentazioni: Saggi in onore di Pierre Souffrin*, Florence, 49–68.

(2018) "Wissenschaftsgeschichtliche Erläuterungen," in Lindermann et al. 2018: 64–84.

Knobloch, E. and C. Möller, eds. (2014) *In den Gefilden der römischen Feldmesser: Juristische, wissenschaftsgeschichtliche, historische und sprachliche Aspekte*, Berlin.

Kolendo, J. (1973) *Le traité d'agronomie des Saserna* (transl. J. Kasińska), Warsaw.

König, A. (2007) "Knowledge and Power in Frontinus' *On aqueducts*," in König and Whitmarsh 2007: 177–205.

(2009) "From Architect to *Imperator*: Vitruvius and his Addressee in the *De architectura*," in Taub and Doody 2009, 31–52.

(2013) "Frontinus' Cameo Role in Tacitus' *Agricola*," *CQ* 63: 361–76.

(2016) "Tracing the Ebb and Flow of *De architectura* 8," *Arethusa* 49: 161–81.

(2017) "Conflicting Models of Authority and Expertise in Frontinus' *Strategemata*," in König and Woolf 2017: 153–81.

(2018) "Reading Civil War in Frontinus's *Strategemata*: A Case-Study for Flavian Literary Studies," in L. D. Ginsberg and D. A. Krasne, eds., *After 69 CE – Writing Civil War in Flavian Rome*, Berlin: 144–77.

(2020) "Tactical Interactions: Dialogues between Greece and Rome in the Military Manuals of Aelian and Arrian," in A. König, R. Langlands, and J. Uden, eds., *Literature and Culture in the Roman Empire, 96–235: Cross-Cultural Interactions*, Cambridge: 134–56.

König, J. (2017) "Introduction: Self-Assertion and Its Alternatives in Ancient Scientific and Technical Writing," in König and Woolf 2017: 1–26.

König, J. and T. Whitmarsh, eds. (2007) *Ordering Knowledge in the Roman Empire*, Cambridge.

König, J. and G. Woolf, eds. (2013a.) *Encyclopaedism from Antiquity to the Renaissance*, Cambridge.

(2013b) "Introduction," in König and Woolf 2013a: 1–20.

(2013c) "Encyclopaedism in the Roman Empire," in König and Woolf 2013a: 23–63.

eds. (2017) *Authority and Expertise in Ancient Scientific Culture*, Cambridge.

Korzenszky, E. and R. Vári (1935) *Onasandri Strategicus*, Budapest.

Köves-Zulauf, T. (1973) "Die Vorrede der plinianischen Naturgeschichte," *WS* new series 7: 134–84.

Krafft, F. 1973. "Kunst und Natur: Die Heronische Frage und die Technik in der klassischen Antike." *A&A* 19.1: 1–19.

Kraus, C. S. (2007) "Caesar's Account of the Battle of Massilia (*BC* 1.34–2.22): Some Historiographical and Narratological Approaches," in J. Marincola, ed., *Greek and Roman Historiography*, Malden, MA: II 371–8.

Krenkel, W. (1959) "Zu den *Artes* des Celsus," *Philologus* 103: 114–29.

(1973) "A. Cornelius Celsus," in *Argentea aetas in memoriam Entii V. Marmorale*, Genoa: 17–28.

Krings, V. (2008) "Rereading Punic Agriculture: Representation, Analogy and Ideology in the Classical Sources," in P. van Dommelen and C. Gómez Bellard, eds., *Rural Landscapes of the Punic World*, London: 22–43.

Krings, V. and O. Devillers (1996) "Autour de l'agronome Magon," in M. Khanoussi, P. Ruggeri, and C. Vismara, eds., *L'Africa romana*, Ozieri: I 489–516.

Krohn, F. (1912) *Vitruuii De architectura libri decem*, Leipzig.

Kronenberg, L. (2009) *Allegories of Farming from Greece and Rome: Philosophical Satire in Xenophon, Varro, and Virgil*, Cambridge.

Kube, J. (1969) *Τέχνη und ἀρετή: Sophistisches und platonisches Tugendwissen*, Berlin.

Kuchma, V. V. (1984) "Some Disputed Questions Concerning the *Strategemata* of Frontinus," *Vestnik drevnej istorii (Journal of Ancient History)* 171: 45–55.

Kühnert, F. (1961) *Allgemeinbildung und Fachbildung in der Antike*, Berlin.

Kullmann, W., J. Althoff, and M. Asper, eds. (1998) *Gattungen wissenschaftlicher Literatur in der Antike*, Tübingen.

LaBarge, S. (2006) "Aristotle on *Empeiria*," *AncPhil* 26: 23–44.

Lachmann, K., F. Blume, and A. Rudorff (1848–52) *Die Schriften der römischen Feldmesser*, 2 vols., Berlin.

Lagopoulos, A. P. (2009) "The Semiotics of the Vitruvian City," *Semiotica* 175: 193–251.

Lallot, J. (2003) *La grammaire de Denys le Thrace*, 2nd ed., Paris.

Lammert, F. (1939) Review of G. Bendz, *Die Echtheitsfrage des vierten Buches der frontinischen Strategemata (1938)*, *Philologische Wochenschrift* 59: 234–9.

Langlands, R. (2008) "'Reading for the Moral' in Valerius Maximus: The Case of *severitas*," *CCJ* 54: 160–87.

Langslow, D. R. (2000) *Medical Latin in the Roman Empire*, Oxford.

Langslow, D. R. and B. Maire, eds. (2010) *Body, Disease and Treatment in a Changing World: Latin Texts and Contexts in Ancient and Medieval Medicine*, Lausanne.

Lane Fox, R. (2018) "Aineias the Author: Who, Where and When?" in M. Pretzler and N. Barley, eds., *Brill's Companion to Aineias Tacticus*, Leiden: 33–48.

Lao, E. (2011) "Luxury and the Creation of a Good Consumer," in Gibson and Morello 2011: 35–56.

Laterza, G. (2018) "*Pectora fenestrata et aperta* (Vitr. *DA* III pr.): Analisi di un'immagine epistemica," *MD* 81: 191–200.

Laughton, E. (1978) "Humour in Varro," in J. Collart, ed., *Varron, grammaire antique et stylistique latine*, Paris: 105–111.

Le Bohec, Y. (1998) "Que voulait Onesandros?" in Y. Burnand, Y. Le Bohec, and J.-P. Martin, eds., *Claude de Lyon, empereur romain*, Paris: 169–79.

Le Gall, J. (1975) "Les romains et l'orientation solaire," *MEFRA* 87: 287–320.

Le projet de Vitruve: Objet, destinataires et reception du De architectura (1994), Rome.

Leeman, A. D. et al. (1981–2008) *M. Tullius Cicero: De oratore libri III. Kommentar*, Heidelberg.

Lefas, P. (2000) "On the Fundamental Terms of Vitruvius' Architectural Theory," *BICS* 44: 179–97.

Lehoux, D. (2007) *Astronomy, Weather, and Calendars in the Ancient World*: Parapegmata *and Related Texts in Classical and Near Eastern Societies*, Cambridge.

(2012) *What did the Romans Know?: An Inquiry into Science and Worldmaking*, Chicago.

Lenoir, M. (1979) *Pseudo-Hygin: Des fortifications du camp*, Paris.

Letzner, W., ed. (2013) *Die Wasserversorgung im antiken Rom: Sextus Iulius Frontinus*, 2nd ed., Munich

Leumann, M. (1977) *Lateinische Laut- und Formenlehre*, Munich.

Levithan, J. (2013) *Roman Siege Warfare*, Ann Arbor, MI.

Lewis, M. J. T. (2001) *Surveying Instruments of Greece and Rome*, Cambridge.

(2012) "Greek and Roman Surveying and Surveying Instruments," in R. J. A. Talbert, ed., *Ancient Perspectives: Maps and Their Place in Mesopotamia, Egypt, Greece & Rome*, Chicago: 129–62.

Li Causi, P., R. Marino, M. Formisano, and E. Romano (2015) *Marco Tullio Cicerone: De oratore*, Alessandria.

Lindermann, J.-O. (2013) "Varro und Isidor in den *gromatici veteres*," *Philologus* 157: 119–40.

(2014) "*Locus, ager, spatium*: Wortuntersuchungen zum Raumbegriff der *gromatici veteres*," in Knobloch and Möller 2014: 199–213.

(2016) "Echt oder unecht? Frontin als gromatischer Autor aus der Sicht der Philologie." *eTopoi Journal for Ancient Studies* 5: 18–45.

(2018) "Philologische Erläuterungen," in Lindermann et al. 2018: 10–63.

(2022) *Hygini liber gromaticus de limitibus constituendis: Historisch-kritische Edition und Erläuterungen*, Darmstadt.

Lindermann, J.-O., E. Knobloch, and C. Möller (2018) *Hyginus: Das Feldmesserbuch. Ein Meisterwerk der spätantiken Buchkunst*, Darmstadt.

Linderski, J. (1986) "The Augural Law," *ANRW* II.16.3: 2146–312.

(1999) "Effete Rome: Sallust, *Cat.* 53.5," *Mnemosyne* 52: 257–65.

Liou, B., M. Zuinghedau, and M.-T. Cam (2003) *Vitruve: De l'architecture. Livre VII*, 2nd ed., Paris.

Lloyd, G. E. R. (1966) *Polarity and Analogy: Two Types of Argumentation in Early Greek Thought*, Cambridge.

(1979) *Magic, Reason, and Experience: Studies in the Origin and Development of Greek Science*, Cambridge.

(1983) *Science, Folklore and Ideology: Studies in the Life Sciences in Ancient Greece*, Cambridge.

(1991a) "The Definition, Status, and Methods of the Medical Τέχνη in the Fifth and Fourth Centuries," in A. C. Bowen, ed., *Science and Philosophy in Classical Greece*, New York: 249–60.

(1991b) *Methods and Problems in Greek Science*, Cambridge.

(2011) "*Techne* and *Dao*: Observations on the Concepts of Art, Techniques and Methods in Ancient Greece and China," in A. Roselli and R. Velardi, eds., *L'insegnamento delle* technai *nelle culture antiche*, Pisa: 13–22.

Lloyd, G. E. R and N. Sivin (2002) *The Way and the Word: Science and Medicine in Early China and Greece*, New Haven, CT.

Long, A. A. (1992) "Stoic Readings of Homer," in R. Lamberton and J. J. Keaney, eds., *Homer's Ancient Readers: The Hermeneutics of Greek Epic's Earliest Exegetes*, Princeton, NJ: 41–66.

Long, A. A. and D. Sedley (1987) *The Hellenistic Philosophers*, 2 vols., Cambridge.

Long, P. O. (1999) "Openness and Empiricism: Values and Meaning in Early Architectural Writings and in Seventeenth-Century Experimental Philosophy," in P. Galison and E. Thompson, eds., *The Architecture of Science*, Cambridge, MA: 79–103.

Lonie, I. M. (1981) *The Hippocratic Treatises* On Generation, On the Nature of the Child, Diseases IV, Berlin.

Lucarini, C. M. (2010) "Ad Onasandri *Strategicum*." *MH* 67: 222–7.

(2014) "Ad Columellam." *Mnemosyne* 67: 648–59.

Lundström, V. (1928) "Småplock ur Columellas språk," *Eranos* 26: 116–18.

Lyngby, H. (1931) *Textkritiska studier till Celsus'* Medicina, Gothenburg.

Maas, E. (1925) "*Eunuchos* und Verwandtes," *RhM* new series 74: 432–76.

Maestre Maestre, J. M., L. Charlo Brea, and A. Serrano Cueto, eds. (1997) *Estudios sobre Columela*, Cádiz.

Mahaffy, J. P. (1889) "The Work of Mago on Agriculture," *Hermathena* 7: 29–35.

Maganzani, L. (1997) *Gli agrimensori nel processo privato romano*, Rome.

Magdelain, A. (1969) "L'*auguraculum* de l'*arx* a Rome et dans d'autres villes," *REL* 47: 253–69.

Malloch, S. V. J. (2015) "Frontinus and Domitian: The Politics of the *Strategemata*," *Chiron* 45: 77–100.

Manetti, D. and A. Roselli (1994) "Il ruolo della tradizione nei libri chirurgici di Celso," in Sabbah and Mudry 1994: 103–21.

Mankin, D. (2011) *Cicero: De oratore. Book III*, Cambridge.

Mann, J. E. (2012) *Hippocrates: On the Art of Medicine*, Leiden.

Manolaraki, E. (2015) "*Hebraei liquores*: The Balsam of Judaea in Pliny's *Natural History*," *AJPh* 136: 633–67.

(2018) "Senses and the Sacred in Pliny's *Natural History*," *ICC* 43: 207–33.

Mansfeld, J. (1983) "*Techne*: A New Fragment of Chrysippus," *GRBS* 24: 57–65.

Marquardt, J. (1886) *Das Privatleben der Römer*, 2nd ed. (rev. A. Mau), 2 vols., Leipzig.

Marrou, H.-I. (1937) *Saint Augustin et la fin de la culture antique*, Paris.

(1948) *Histoire de l'éducation dans l'antiquité*, 7th ed., Paris.

(1968) "Les arts libéraux dans l'Antiquité classique," in *Arts libéraux et philosophie au Moyen Âge*, Paris: 5–27.

Marsden, E. W. (1969) *Greek and Roman Artillery: Historical Development*, Oxford.

(1971) *Greek and Roman Artillery: Technical Treatises*, Oxford.

Martin, R. (1971) *Recherches sur les agronomes latins et leurs conceptions économiques et sociales*, Paris.

(1985) "État présent des études sur Columelle," *ANRW* II.32.3: 1959–79.

(1988) "Agriculture et religion: Le témoignage des agronomes latins," in Porte and Néraudau 1988: 294–305.

(1995) "*Ars an quid aliud?* La conception varronienne de l'agriculture." *REL* 73: 80–91.

Marx, F. (1915) *A. Cornelii Celsi quae supersunt*, Leipzig.

Masterson, M. (2004) "Status, Pay, and Pleasure in the *De architectura* of Vitruvius," *AJPh* 125: 387–416.

Mastrorosa, I. (2000) "Similitudini, metafore e lessico militari nella trattatistica retorica latina: Cicerone e Quintiliano," in Sconocchia and Toneatto 2000: 277–310.

Mastrorosa, I., C. Santini, and A. Zumbo, eds. (2002) *Letteratura scientifica e tecnica di Grecia e Roma*, Rome.

May, J. M. and J. Wisse (2001) *Cicero: On the Ideal Orator*, New York.

Mayer, R. (2005) "Creating a Literature of Information in Rome," in M. Horster and C. Reitz, eds., *Wissensvermittlung in dichterischer Gestalt*, Stuttgart: 227–41.

Mazzarino, A. (1952) *Introduzione al* De agri cultura *di Catone*, Rome.

Mazzini, I. and F. Fusco, eds. (1985) *I testi di medicina latini antichi: Problemi filologici e storici*, Rome.

(1992a) "Caratteri della lingua del *De medicina* di A. Cornelio Celso," *RCCM* 34: 17–46.

(1992b) "Ippocrate in Celso," in J. A. López Férez, ed., *Tratados hipocráticos: Estudios acerca de su contenido, forma e influencia*, Madrid: 571–83.

(1994) "Le *auctoritates* nei testi medici dell'antichità, in particolare in Celso," in Vázquez Buján 1994: 119–32.

(1999) *A. Cornelio Celso: La chirurgia* (*Libri VII e VIII del* De medicina), Macerata.

McEwen, I. K. (2003) *Vitruvius: Writing the Body of Architecture*, Cambridge, MA.

McGill, S. (2012) *Plagiarism in Latin Literature*, Cambridge.

Meiggs, R. (1982) *Trees and Timber in the Ancient Mediterranean World*, Oxford.

Meinecke, B. (1941) "Aulus Cornelius Celsus—Plagiarist or *artifex medicinae*?" *Bulletin of the History of Medicine* 10: 288–98.

Meißner, B. (1999) *Die technologische Fachliteratur der Antike: Struktur, Überlieferung und Wirkung technischen Wissens in der Antike (ca. 400 v. Chr. – ca. 500 n. Chr.)*, Berlin.

(2003) "Mündliche Vermittlung und schriftliche Unterweisung in der antiken Berufsausbildung," in Horster and Reitz 2003: 153–75.

Melfi, R. D. (1972) *A Commentary on the Various References to the Jews in Latin Literature from Cicero to Seneca*, MA thesis, Ottawa.

Mette, H. J. (1936) Sphairopoiia*: Untersuchungen zur Kosmologie des Krates von Pergamon*, Munich.

Meyers, G. E. (2005) "Vitruvius and the Origins of Roman Spatial Rhetoric," *MAAR* 50: 67–86.

Mielke, L. (2018) "Zur Vereinbarkeit von *ratio* und *reuerentia* in Columellas Umgang mit Vergil," in Finkmann et al. 2018: 515–45.

(2019) *Spaliere für Silvinus: Charakterschulung in Columellas Werk über die Landwirtschaft*, unpublished PhD dissertation, Rostock.

(2023) *Spaliere für Silvinus: Charakterschulung in Columellas Werk über die Landwirtschaft*, Göttingen.

Miles, G. B. (1980) *Virgil's* Georgics*: A New Interpretation*, Berkeley, CA.

Miller, D. (2014) "Aristotle on How to Fell a Tree and Other Matters Involving Experience," in M.-K. Lee, ed., *Strategies of Argument: Essays in Ancient Ethics, Epistemology, and Logic*, Oxford: 124–45.

Miller, K. J. (2018) "From Critical Days to Critical Hours: Galenic Refinements of Hippocratic Models," *TAPA* 148: 111–38.

Milner, N. P. (1991) *Vegetius and the* Anonymus De rebus bellicis, unpublished PhD dissertation, Oxford.

(1996) *Vegetius: Epitome of Military Science*, 2nd ed., Liverpool.

Milnor, K. (2005) *Gender, Domesticity, and the Age of Augustus: Inventing Private Life*, Oxford.

Moatti, C. (1993) *Archives et partage de la terre dans le monde romain (IIe siècle avant–Ier siècle après J.-C.)*, Rome.

(2015) *The Birth of Critical Thinking in Republican Rome* (transl. J. Lloyd), Cambridge.

Möller, C. (1993) "Die *mercennarii* in der römischen Arbeitswelt," *ZRG* 110: 296–330.

(2018) "Juristische Erläuterungen," in Lindermann et al. 2018, 85–107.

Mollweide, C. (1813) "Erläuterung einer in den *Scriptoribus rei agrariae pag. 176 u. 177, edit. Goesii*, gegebenen Vorschrift, aus drey beobachteten ungleichen Schattenlängen die Mittagslinie zu finden," *Monatliche Correspondenz zur Beförderung der Erd- und Himmels-kunde* 28: 396–425.

Momigliano, A. (1944) Review of S. A. Cook, F. E. Adcock, and M. P. Charlesworth, *The Cambridge Ancient History. Vol. X: The Augustan Empire* (1934), *JRS* 34: 109–16.

Mommsen, T. (1892) "Zum römischen Bodenrecht," *Hermes* 27: 79–117.

(1905–13) *Gesammelte Schriften*, 8 vols., Berlin.

Morello, R. (2011) "Pliny and the Encyclopaedic Addressee," in Gibson and Morello 2011: 147–65.

Morgan, M. H. (1906*a*) "On the Language of Vitruvius," *Proceedings of the American Academy of Arts and Sciences* 41: 467–502.

(1906b) "Notes on Vitruvius," *HSCP* 17: 1–14.

(1909) "The Preface of Vitruvius," *Proceedings of the American Academy of Arts and Sciences* 44: 149–75.

Mortet, M. V. (1896) *Un nouveau texte des traités d'arpentage et de géométrie d'Epaphroditus et de Vitruvius Rufus publié d'après le ms. Latin 13084 de la Bibliothèque royale de Munich*, Paris.

Morton, A. G. (1986) "Pliny on Plants: His Place in the History of Botany," in French and Greenaway 1986: 86–97.

Mudry, P. (1977) "La place d'Hippocrate dans la préface du *De medicina* de Celse," in R. Joly, ed., *Corpus Hippocraticum*, Mons: 345–52.

(1980) "*Medicus amicus*: Un trait romain dans la médecine antique," *Gesnerus* 37: 17–20.

(1982) *La préface du* De medicina *de Celse*, Bern.

(1985) "Médecins et spécialistes: Le problème de l'unité de la médecine à Rome au 1er siècle ap. J. C.," *Gesnerus* 42: 329–36.

(1990a) "Le scepticisme des médecins empiriques dans le traité *De la médecine* de Celse: Modèles et modalités," in A.-J. Voelke, ed., *Le scepticisme antique: Perspectives historiques et systématiques*, Geneva: 85–96.

(1990b) "Médecine et vulgarisation: Remarques sur le problème de la vulgarisation médicale dans l'Antiquité," in *Médecine et communication: De l'acte médical à la vulgarisation scientifique*, Geneva: 15–31.

(1990c) "Réflexions sur la médecine romaine," *Gesnerus* 47: 133–48.

(1993a) "Le *De medicina* de Celse: Rapport bibliographique." *ANRW* II.37.1: 787–99.

(1993b) "L'orientation doctrinale du *De medicina* de Celse," *ANRW* II.37.1: 800–18.

(1994) "Maladies graves et maladies mortelles: Présence et évolution d'une notion hippocratique chez les auteurs médicaux latins et en particulier Celse," in Vázquez Buján 1994:133–43.

(1997) "Éthique et médecine à Rome: La préface de Scribonius Largus ou l'affirmation d'une singularité," in H. Flashar and J. Jouanna, eds., *Médecine et morale dans l'Antiquité*, Geneva: 297–336.

(1999) "Celso e la medicina nel mondo greco-romano," in G. Reggi, ed., *La cultura materiale antica: Aspetti, problemi e spunti per la scuola d'oggi*, Bellinzona: 163–77.

(2001) "Le jeu de la nature et du hasard: La construction du savoir médical dans le traité de Celse," in Courrént and Thomas 2001: 57–69.

(2003) "Le regard souverain ou la médecine de l'évidence," in I. Boehm and P. Luccioni, eds., *Les cinq sens dans la médecine de l'époque impériale*, Paris: 31–8.

(2004a) "Le chou de Pythagore: Présence des modèles grecs dans le *De agricultura* de Caton," *EL* 1–2: 25–45.

(2004b) "*Mirabilia* et *magica*: Essai de définition dans l'*Histoire naturelle* de Pline l'Ancien," in Bianchi et al. 2004: 239–52.

(2010) "*Ratio* et *coniectura* dans les textes médicaux latins," in Langslow and Maire 2010: 337–47.

Müller, K. (1882) *Geographici graeci minores*, 2 vols., Paris.

Müller, W. (1989) *Architekten in der Welt der Antike*, Zurich.

Murphy, T. (2004) *Pliny the Elder's* Natural History*: The Empire in the Encyclopedia*, Oxford.

Mynors, R. A. B. (1990) *Virgil: Georgics*, Oxford.

Naas, V. (2000) "L'*Histoire naturelle* de Pline l'Ancien est-elle une œuvre scientifique?" in L. Callebat and O. Desbordes, eds., *Science antique, science médiévale (autour d'Avranches 235)*, Hildesheim: 255–71.

(2001) "*Est in his quidem, tametsi mirabilis, aliqua ratio* (*NH*, IX, 178): Modes de construction du savoir et imaginaires de Pline l'Ancien," in Courrént and Thomas 2001: 15–34.

(2002) *Le projet encyclopédique de Pline l'Ancien*, Rome.

(2004) "*Opera mirabilia in terris* et *Romae operum miracula*: Dans l'*Histoire naturelle* de Pline l'Ancien," in Bianchi et al. 2004: 253–64.

(2011a) "Philhellénisme et identité romaine chez Pline l'Ancien (*nat.* 7,81–130)," in A. Bonadeo, A. Canobbi, and F. Gasti, eds., *Filellenismo e identità romana in età flavia*, Como: 35–58.

(2011b) "Imperialism, *mirabilia* and Knowledge: Some Paradoxes in the *Naturalis historia*," in Gibson and Morello 2011: 57–70.

(2015) "Certitude et méthode dans l'*Histoire naturelle* de Pline l'Ancien," in S. Gambino Longo, ed., *La certitude de l'Antiquité à la Renaissance*, Paris: 129–56.

(2019) "Aux origines de l'*Histoire naturelle*: Sources et structure de l'inventaire plinien," in A. Cohen-Skalli, ed., *Historiens et érudits à leur écritoire: Les œuvres monumentales à Rome entre République et Principat*, Bordeaux: 225–42.

Nadjo, L. (2005) "Un nom de l'arpenteur en latin: *Agrimēnsor*," *Journal of Latin Linguistics* 9: 383–92.

Nelsestuen, G. A. (2011) "Polishing Scrofa's Agronomical *eloquentia*: Representation and Revision in Varro's *De re rustica*," *Phoenix* 65: 315–51.

(2015) *Varro the Agronomist: Political Philosophy, Satire, and Agriculture in the Late Republic*. Columbus, OH.

(2016) "Storing Produce and Staging Dinner Parties: Fruit-Galleries and Genre in Varro's *De re rustica* 1," *ICS* 41: 21–40.

(2017) "Varro, Dicaearchus, and the History of the Roman *Res rusticae*," *BICS* 60: 21–33.

Nap, J. M. (1927) "Ad Catonis librum de re militari," *Mnemosyne* new series 55: 79–87.

Netz, R. (2003) "The Goal of Archimedes' *Sand-Reckoner*," *Apeiron* 36: 251–90.

(2017) "The Authority of Mathematical Expertise and the Question of Ancient Writing *more geometrico*," in König and Woolf 2017: 374–408.

Neugebauer, O. (1938–9) *Über eine Methode zur Distanzbestimmung Alexandria – Rom bei Heron*, Copenhagen.

(1975) *A History of Ancient Mathematical Astronomy*, Berlin.

Nichols, M. F. (2017) *Author and Audience in Vitruvius'* De architectura, Cambridge.

Nicolet, C. (1991) *Space, Geography, and Politics in the Early Roman Empire* (transl. Hélène Leclerc), Ann Arbor, MI.

ed. (1996) *Les littératures techniques dans l'antiquité romaine: Statut, public et destination, et tradition*, Geneva.

Noè, E. (1977a) "I proemi del *De re rustica* di Varrone," *Athenaeum* 55: 289–302.

(1977b)"L'agronomo Cneo Tremellio Scrofa," *NAC* 6: 119–33.

(2000) "*Columella artifex agricola*," *RAL* 9: 399–441.

(2002) *Il progetto di Columella: Profilo sociale, economico, culturale*, Como.

Norden, E. (1939) *Aus altrömischen Priesterbüchern*, Lund.

Novara, A. (1983) "Les raisons d'écrire de Vitruve ou la revanche de l'architecte," *BAGB* 1: 284–308.

(1994) "Faire œuvre utile: La mesure de l'ambition chez Vitruve," in *Le projet de Vitruve* (q. v.): 47–61.

(1999) "Démocrite dans le *De architectura* de Vitruve (Sur un nom souvent cité)," *Helmantica* 50: 587–610.

(2005) Auctor in bibliotheca*: Essai sur les textes préfaciels de Vitruve et une philosophie latine du livre*, Louvain.

Nutton, V. (1986) "The Perils of Patriotism: Pliny and Roman Medicine," in French and Greenaway 1986: 30–58.

(1993) "Roman Medicine: Tradition, Confrontation, Assimilation," *ANRW* II.37.1: 49–78.

(2013) *Ancient Medicine*, 2nd ed., New York.

Nyström, G. (1926) *Variatio sermonis hos Columella*, Gothenburg.

Oakley, S. P. (1997–2005) *A Commentary on Livy, Books VI–X*, 4 vols., Oxford.

O'Hara, J. J. (2017) *True Names: Vergil and the Alexandrian Tradition of Etymological Wordplay*, new and expanded ed., Ann Arbor, MI.
Oksanish, J. (2011) *Building the Principate: A Literary Study of Vitruvius'* De architectura, unpublished PhD dissertation, Yale University.
(2016) "Vitruvius and the Programmatics of Prose," *Arethusa* 49: 263–80.
(2019) *Vitruvian Man: Rome under Construction*, Oxford.
Oliveira, F. (1992) *Les idées politiques et morales de Pline l'Ancien*, Coimbra.
Ollero Granados, D. (1973) "Dos nuevos capítulos de A. Cornelio Celso (*De medicina* IV, 27, 1 D)," *Emerita* 41: 99–108.
Osorio, P. (2020) "Vergil's Physics of *Bugonia* in *Georgics* 4," *CPh* 115: 27–46.
Otto, A. (1890) *Die Sprichwörter und sprichwörtlichen Redensarten der Römer*, Leipzig.
Otto, W. (1903) "Über die lateinischen Wörter auf *-īca, -īcus, -īcius, -īx* und Verwandtes," *IF* 15: 9–53.
Paniagua Aguilar, D. (2006) *El panorama literario técnico-científico en Roma (siglos I–II D.C.):* et docere et delectare, Salamanca.
(2007) "La arenga militar desde la perspectiva de la tradición polemológica grecolatina," *Talia Dixit* 2: 1–25.
(2010–11) "Frontino, agrimensura ed esegesi tardoantica del testo tecnico nel commento dello Pseudo-Agennio Urbico," *Incontri di filologia classica* 10: 29–79.
Paparazzo, E. (2008) "Pliny the Elder on Metals: Philosophical and Scientific Issues," *CPh* 103: 40–54.
(2011) "Philosophy and Science in the Elder Pliny's *Naturalis historia*," in Gibson and Morello 2011: 89–111.
Parmeggiani, G. (2015) "The Anaximean Authorship of the *Rhetoric to Alexander* in Light of Anaximenes' Indirect Tradition," *RSA* 45: 9–29.
Pascucci, G. (1982) "La lettera prefatoria di Plinio alla *Naturalis historia*," in Alfonsi and Ronconi 1982: 171–97.
Pazzini, A., M. T. Malato, R. Trifogli, and V. Tavone Passalacqua (1953) *Fonti Celsiane*, Rome.
Peachin, M. (2004) *Frontinus and the* curae *of the* curator aquarum, Stuttgart.
Pease, A. S. (1920–3) *M. Tulli Ciceronis De diuinatione*, 2 vols., Urbana, IL.
(1955) *M. Tulli Ciceronis De natura deorum libri III*. Cambridge, MA.
Pellati, F. (1921) "Vitruvio e la fortuna del suo trattato nel mondo antico," *RFIC* 49: 305–35.
(1927) "*Quod significatur et quod significat*: Saggio d'interpretazione di un passo di Vitruvio," *Historia* 1: 53–9.
(1951) "La dottrina degli elementi nella fisica di Vitruvio," *Rinascimento* 2: 241–59.
Perl, G. (1980) "Cn. Tremelius Scrofa in Gallia Transalpina: Zu Varro, *RR* 1,7,8," *AJAH* 5: 97–109.
Peters, W. (1972) *Untersuchungen zu Onasander*, published PhD dissertation, Bonn.
Peterson, J. (2006) "Map Conventions in Some Diagrams of the *agrimensores*," in A. Gonzalès and J.-Y. Guillaumin, eds., *Autour des* Libri coloniarum*: Colonisation et colonies dans le monde romain*, Besançon, 151–61.

Petrocelli, C. (2008) *Onasandro: Il generale*, Bari.

Peyras, J. (2003) "Colonies et écrits d'arpentage du Haut-Empire," in Garrido-Hory and Gonzalès 2003: II 103–55.

Piccoli, F. (2004) *Il vino nel Nord Italia in epoca Romana: Storia della coltivazione della vite, della produzione e del commercio del vino in Cisalpina*, Verona.

Pigeaud, J. (1972) "Un médecin humaniste: Celse. Notes sur le *prooemium* du *De medicina*," *LEC* 40: 302–10.

Pikulska, A. (2004) "Les arpenteurs romains et leur formation intellectuelle," *Revue International des Droits de l'Antiquité* 3rd ser., 51: 205–16.

Piras, G. (2017) "*Dicam dumtaxat quod est historicon*: Varro and/on the Past," *BICS* 60.2: 8–20.

Pisi, G. (1983) "Il medico amico in Seneca," *Pubblicazioni, Istituto di lingua e letteratura latina, Università degli Studi di Parma*, 7–47.

Pleket, H. (1967) "Technology and Society in the Graeco-Roman World," *Acta Historiae Neerlandica* 2: 1–27.

Plöger, H. (1975) *Studien zum literarischen Feldherrnporträt römischer Autoren des 1. Jahrhunderts v. Chr.*, unpublished PhD dissertation, Kiel.

Plommer, H. (1971) "The Circle of the Winds in Vitruvius i. 6," *CR* 21: 159–62.

(1973) *Vitruvius and Later Roman Building Manuals*, Cambridge.

Polanyi, M. (1966) *The Tacit Dimension*, Garden City, NY.

Polito, R. (1999) "On the Life of Asclepiades of Bithynia," *JHS* 119: 48–66.

Pollard, E. A. (2009) "Pliny's *Natural History* and the Flavian *Templum Pacis*: Botanical Imperialism in First-Century C.E. Rome," *Journal of World History* 20: 309–38.

Pomeroy, S. (1994) *Xenophon: Oeconomicus*, Oxford.

Porte, D. and J.-P. Néraudau, eds. (1988) *Hommages à Henri Le Bonniec*: Res sacrae, Brussels.

Possanza, M. (1990) "The Text of Lucretius 2.1174," *CQ* 40: 459–64.

Pouilloux, J., P. Roesch, and J. Marcillet-Jaubert (1987) *Salamine de Chypre: XIII*. Paris.

Powell, J. G. F. (1988) *Cicero: Cato Maior de senectute*, Cambridge.

Prioux, E. (2013) "Columelle et le *genus floridum*: Images programmatiques et parodie dans le livre X du *De re rustica*," *Dictynna* 10: 1–30.

Probst, O. (1905) Celsus und Plinius: In ihrem Verhältnis zum achten Buch der Encyclopädie Varros, unpublished PhD dissertation, Munich.

Purcell, N. (1983) "The *Apparitores*: A Study in Social Mobility," *PBSR* 51: 125–73.

(1990) Review of C. Nicolet, *L'inventaire du monde: Géographie et politique aux origines de l'empire romain* (1988), *JRS* 80: 178–82.

(2017) "'Such is Rome . . .': Strabo on the 'Imperial Metropolis,'" in D. Dueck, ed., *The Routledge Companion to Strabo*, New York: 22–34.

Rabouin, D. and B. Vitrac (2010) "Sur le passage mathématique de l'*Épinomis* (990c–992a)," *PhilosAnt* 10: 5–39.

Radermacher, L. (1951) Artium scriptores*: Reste der voraristotelischen Rhetorik*, Vienna.

Ratti, S. (1996) "Le substrat augustéen dans la *Constitutio limitum* d'Hygin le Gromatique et la datation du traité," *DHA* 22: 220–38.

(1998) "À propos de quelques difficultés gromatiques: Sur la datation d'Hygin le Gromatique, d'Hygin et sur les mots *decuria* et *pittacium* (Hygin 73 Th.)," *DHA* 24: 125–38.

Rawson, E. (1978) "The Introduction of Logical Organisation in Roman Prose Literature," *PBSR* 46: 12–34.

(1985) *Intellectual Life in the Late Roman Republic*, London.

Reay, B. (1998) *Cultivating Romans: Republican Agricultural Writing and the Invention of the* agricola, unpublished PhD dissertation, Stanford University.

(2005) "Agriculture, Writing, and Cato's Aristocratic Self-Fashioning," *CA* 24: 331–61.

Reeve, M. D. (1983) "*Agrimensores*," in L. D. Reynolds, ed., *Texts and Transmissions: A Survey of the Latin Classics*, Oxford: 1–6.

(forthcoming) "*Agrimensores* (Supplement)," in J. Stover, ed., *The Oxford Guide to the Transmission of the Latin Classics*, Oxford.

Rehm, Albert (1916) *Griechische Windrosen*, Munich.

Reinhardt, K. (1921) *Poseidonios*, Munich.

Reinhardt, T. and M. Winterbottom. (2006) *Quintilian: Institutio oratoria. Book* 2, Oxford.

Reischl, B. (1976) *Reflexe griechischer Kulturentstehungslehren bei augusteischen Dichtern*, unpublished PhD dissertation, Munich.

Reitz, C. (2017) "*Auctoritas* in the Garden: Columella's Poetic Strategy in *De re rustica* 10," in Formisano and van der Eijk 2017: 217–30.

Reitzenstein, R. (1884) *De scriptorum rei rusticae qui intercedunt inter Catonem et Columellam libris deperditis*, Berlin.

Reitz-Joose, B. (2016) "The City and the Text in Vitruvius' *De architectura*," *Arethusa* 49: 183–97.

Renehan, R. (1995) "Polus, Plato, and Aristotle," *CQ* 45: 68–72.

Rex, H. (2001) *Die lateinische Agrarliteratur von den Anfängen bis zur frühen Neuzeit*, unpublished PhD dissertation, Wuppertal.

Reynolds, J. (1986) "The Elder Pliny and His Time," in French and Greenaway 1986: 1–10.

Richter, W. (1978) *Gegenständliches Denken, archaisches Ordnen: Untersuchungen zur Anlage von Cato* De agri cultura, Heidelberg.

(1981–3) *Columella: Zwölf Bücher über Landwirtschaft*, 3 vols., Munich.

Riggsby, A. M. (2007) "Guides to the Wor(l)d," in König and Whitmarsh 2007: 88–107.

(2019) *Mosaics of Knowledge: Representing Information in the Roman World*, Oxford.

Rihll, T. E. (1999) *Greek Science*, Oxford.

Ritschl, F. (1877) "De M. Terentii Varronis Disciplinarum libris commentarius," in *Opuscula philologica*, III 352–402, Leipzig.

Roby, C. A. (2013) "*Natura machinata*: Artifacts and Nature as Reciprocal Models in Vitruvius," *Apeiron* 46: 419–45.

(2014) "Experiencing Geometry in Roman Surveyors' Texts," *Nuncius* 29: 9–52.

(2015) "Latin Didactic, Scientific, and Technical Literature," *Oxford Handbooks Online*, Oxford.

(2016) *Technical Ekphrasis in Greek and Roman Science and Literature: The Written Machine between Alexandria and Rome*, Cambridge.

Rodgers, R. H. (2004) *Frontinus: De aquaeductu urbis Romae*, Cambridge.

(2010) *L. Iuni Moderati Columellae Res rustica. Incerti auctori liber de arboribus*, Clarendon.

Rodríguez Neila, J. F. (1997) "Gades en tiempos de Columela," in Maestre Maestre et al. 1997: 35–82.

Roller, D. W. (2010) *Eratosthenes'* Geography, Princeton, NJ.

(2018) *A Historical and Topographical Guide to the* Geography *of Strabo*, Cambridge.

Romano, E. (1985) "Il proemio di Celso fra sapere tecnico e cultura umanistica," in Mazzini and Fusco 1985: 131–40.

(1987) *La capanna e il tempio: Vitruvio o dell'architectura*, Palermo.

(1994) "Dal *De officiis* a Vitruvio, da Vitruvio a Orazio: Il dibattito sul lusso edilizio," in *Le projet de Vitruve* (q. v.): 63–73.

(1997) "Fra astratto e concreto: La lingua di Vitruvio," in Gros et al. 1997: lxxix–xcv.

(2002) "Architettura," in Mastrorosa et al. 2002: 65–85.

(2003a) "Scienza e potere a Roma tra la tarda repubblica e il primo impero," in R. Uglione, ed., *Atti del convegno nazionale di studi intellettuali e potere nel mondo antico*, Alessandria: 167–87.

(2003b) "Il ruolo di Cicerone nella formazione di una cultura tecnica," in E. Narducci, ed., *Aspetti della fortuna di Cicerone nella cultura latina*, Florence: 92–111.

(2013) "*Litteras scire*: Sulle fonti del canone delle *artes* in Vitruvio," *Athenaeum* 101.1: 201–19.

(2016) "Between Republic and Principate: Vitruvius and the Culture of Transition," *Arethusa* 49: 335–51.

Roochnik, D. (1996) *Of Art and Wisdom: Plato's Understanding of Techne*, University Park, PA.

Roscher, W. H. (1906) *Die Hebdomadenlehren der griechischen Philosophen und Ärzte: Ein Beitrag zur Geschichte der griechischen Philosophie und Medizin*, Leipzig.

Rose, V. (1899) *Vitruuii De architectura libri decem*, Leipzig.

Rose, V. and H. Müller-Strübing (1867) *Vitruuii De architectura libri decem*, Leipzig.

Ross, D. O. (1987) *Virgil's Elements: Physics and Poetry in the* Georgics. Princeton, NJ.

Roth Congès, A. (1996) "Modalités pratiques d'implantation des cadastres romains: Quelques aspects (*Quintarios claudere. Perpendere. Cultellare. Varare*: La construction des cadastres sur une diagonale et ses traces dans le *corpus agrimensorum*)," *MEFRA* 108: 299–422.

(1998) "La *Varation dans les territoires* de M. Iunius Nypsius: Une application des propriétés des triangles semblables à la construction cadastrale," in M. Clavel-Lévêque and A. Vignot, eds., *Cité et territoire 2*, Paris: 165–84.

(2000) "*Artis copia*: Questions d'arpentage dans *La dioptre* d'Héron d'Alexandrie et dans les textes gromatiques romains," in G. Argoud and J.-Y. Guillaumin, eds., *Autour de* La dioptre *d'Héron d'Alexandrie*, Saint-Étienne: 107–47.

Rowland, I. D. (2005) "From Vitruvian Scholarship to Vitruvian Practice," *MAAR* 50: 15–40.

Rowland, I. D. and T. N. Howe (1999) *Vitruvius: Ten Books on Architecture*, Cambridge.

Ruffel, P. and J. Soubiran. (1962) "Vitruve ou Mamurra?" *Pallas* 11: 123–79.

Saastamoinen, A. (2003) "The Literary Character of Frontinus' *De aquaeductu*," in C. Bruun and A. Saastamoinen, eds., *Technology, Ideology, Water: From Frontinus to the Renaissance and Beyond*, Rome: 15–39.

Sabbah, G. and P. Mudry, eds. (1994) *La médecine de Celse: Aspects historiques, scientifiques et littéraires*, Saint-Étienne.

Sacks, K. S. (1990) *Diodorus Siculus and the First Century*, Princeton, NJ.

Sáez Fernández, P. (2013) "Cn. Tremelius Scrofa: Algunas precisiones sobre su obra agronómica," in Cid López and Garcia Fernández 2013: II 323–33.

Saliou, C. (2009) *Vitruve: De l'architecture. Livre V*, Paris.

Sallmann, K. (1971) *Die Geographie des älteren Plinius in ihrem Verhältnis zu Varro: Versuch einer Quellenanalyse*, Berlin.

(1984) "Bildungsvorgaben des Fachschriftstellers: Bemerkungen zur Pädagogik Vitruvs," in H. Knell and B. Wesenberg, eds., *Vitruv-Kolloquium des Deutschen Archäologen-Verbandes e. V.*, Darmstadt: 11–26.

(1986) "La responsabilité de l'homme face à la nature," *Helmantica* 37: 251–66.

Salmasius, C. [Claude Saumaise] (1629) *Plinianae exercitationes in Caii Iulii Solini Polyhistora item Caii Iuli Solini Polyhistor ex ueteribus libris emendatus*, Paris.

Sander, E. (1929) "Frontin als Quelle für Vegetius." *Philologische Wochenschrift* 40: 1230–1.

(1932) "Die Hauptquellen der Bücher I–III der *epitoma rei militaris* des Vegetius," *Philologus* 87: 369–75.

Santini, C. (1992) "Il prologo del *De aquae ductu V. R.* di Sesto Giulio Frontino," in Santini and Scivoletto 1990–2: II 917–24.

(2000) "Le traduzioni dei libri di Magone," in Sconocchia and Toneatto 2000: 367–80.

Santini, C. and M. Scivoletto, eds. (1990–2) *Prefazioni, prologhi, proemi di opere tecnico-scientifiche latine*, 2 vols., Rome.

Santon, J. T. (1998) *A Commentary on Book III of Columella's* De re rustica, unpublished PhD dissertation, Royal Holloway.

Šašel, J. (1981) "Castellum Larignum (Vitr. 2,9,15)," *Historia* 30: 254–6.

Saunders, T. J. (2001) "Dicaearchus' Historical Anthropology," in W. W. Fortenbaugh and E. Schütrumpf, eds., *Dicaearchus of Messana: Text, Translation, and Discussion*, New Brunswick, NJ: 237–54.

Scarborough, J. (1969) *Roman Medicine*, Ithaca, NY.
(1976) "Celsus on Human Vivisection at Ptolemaic Alexandria," *CM* 11: 25–38.
(1993) "Roman Medicine to Galen," *ANRW* II.37.1: 3–48.
Schanz, M. (1881) "Ueber die Schriften des Cornelius Celsus," *RhM* new series 36: 362–79.
Scheid, J. (2007) *Res gestae divi Augusti: Hauts faits du Divin Auguste*, Paris.
Scheidel, W. (1990) "Free-born and Manumitted Bailiffs in the Graeco-Roman World," *CQ* 40: 591–3.
Schellenberg, H. M. (2007) "Einige Bemerkungen zum *Strategikos* des Onasandros," in L. De Blois and E. Lo Cascio, eds., *Impact of the Roman Army (200 B.C.–A.D. 476): Economic, Social, Political, Religious, and Cultural Aspects*, Leiden: 181–91.
Scheller, E., and W. Frieboes (1906) *Aulus Cornelius Celsus über die Arzneiwissenschaft in acht Büchern*, 2nd ed., Braunschweig.
Schenk, D. (1930) *Flavius Vegetius Renatus: Die Quellen der* Epitoma rei militaris, Leipzig.
Schettino, M. T. (1998) *Introduzione a Polieno*, Pisa.
Schiefsky, M. J. (2005) *Hippocrates: On Ancient Medicine*, Leiden.
Schievenin, R. (2012) "Medici e musici (Vitruv. 1.15)," *Lexis* 30: 432–40.
Schindel, U. (1992) "Nachklassischer Unterricht im Spiegel der gromatischen Schriften," in Behrends and Capogrossi Colognesi 1992: 375–97.
Schiøler, T. (1994) "The Pompeii-*groma* in New Light," *ARID* 22: 45–60.
Schneider, H. (1989) *Das griechische Technikverständnis: Von den Epen Homers bis zu den Anfängen der technologischen Fachliteratur*, Darmstadt.
Schneider, J. G. (1794–7) *Scriptorum rei rusticae veterum latinorum*, 4 vols., Leipzig.
(1807–8) *Marci Vitruvii Pollionis De architectura libri decem*, 2 vols., Leipzig.
Schrijvers, P. H. (1989) "Vitruve I 1,1: Explication de texte," in Geertman and de Jong 1989: 49–54.
Schröder, B.-J. (1999) *Titel und Text: Zur Entwicklung lateinischer Gedichtüberschriften*, Berlin.
Schröter, R. (1960) *Studien zur varronischen Etymologie*, Wiesbaden.
Schuler, S. (1999) *Vitruv im Mittelalter: Die Rezeption von* De architectura *von der Antike bis in die frühe Neuzeit*, Cologne.
Schulte, H. K. (1935) Orator*: Untersuchungen über das ciceronianische Bildungsideal*, Frankfurt.
Schulz, F. (1961) *Geschichte der römischen Rechtswissenschaft*, Weimar.
Schulze, C. (1999) *Aulus Cornelius Celsus: Arzt oder Laie? Autor, Konzept und Adressaten der* De medicina libri octo, Trier.
(2001) *Celsus*, Hildesheim.
(2002) "Zum abrupten Schluss der *De medicina libri octo* des Aulus Cornelius Celsus," *RhM* 145: 223–36.
(2005) "Von wem stammt der Brief an Pullius Natalis?" *Hermes* 133: 486–95.
(2011) "Zum Autor der Schrift *Über die Kriegsführung gegen die Parther*," *Philologus* 155: 386–94.

Schumacher, L. (2001) *Sklaverei in der Antike: Alltag und Schicksal der Unfreien*, Munich.
Sciarrino, E. (2011) *Cato the Censor and the Beginnings of Latin Prose: From Poetic Translation to Elite Transcription*, Columbus, OH.
Scivoletto, N. (1992) "Le prefazioni nei *Rei rusticae libri* di Columella," in Santini and Scivoletto 1990–2: II 769–817.
Sconocchia, S. (1985) "Le fonti e la fortuna di Scribonio Largo," in Mazzini and Fusco 1985: 151–213.
(1993) "L'opera di Scribonio Largo e la letteratura medica latina del I sec. d. C." *ANRW* II.37.1: 842–922.
(1997) "Sui grecismi di Celso: I calchi," in Sconocchia and Toneatto 1997: 217–25.
Sconocchia, S. and L. Toneatto, eds. (1993) *Lingue tecniche del greco e del latino*, Trieste.
eds. (1997) *Lingue tecniche del greco e del latino II*, Bologna.
eds. (2000) *Lingue tecniche del greco e del latino III*, Bologna.
Scranton, R. L. (1974) "Vitruvius' Arts of Architecture," *Hesperia* 43: 494–9.
Sedley, D. (2003) *Plato's Cratylus*, Cambridge.
Serbat, G. (1986*a*) "Pline l'Ancien: État présent des études sur sa vie, son œuvre et son influence," *ANRW* II.32.4: 2069–2200.
(1986b) "Il y a Grecs et Grecs! Quel sens donner au prétendu antihellénisme de Pline?" *Helmantica* 38: 273–82.
(1995) *Celse: De la médecine. Tome I*, Paris.
Shanzer, D. R. (2005) "Augustine's Disciplines: *Silent diutius Musae Varronis?*" in K. Pollmann and M. Vessey, eds., *Augustine and the Disciplines: From Cassiciacum to Confessions*, Oxford: 69–112.
Sherk, R. K. (1974) "Roman Geographical Exploration and Military Maps," *ANRW* II.1: 534–62.
Sherwin-White, A. N. (1966) *The Letters of Pliny: A Historical and Social Commentary*, Oxford.
Short, W. M. (2008) "Thinking Places, Placing Thoughts: Spatial Metaphors of Mental Activity in Roman Culture," *I Quaderni del Ramo d'Oro On-Line* 1: 106–29.
Sidoli, N. (2011) "Heron of Alexandria's Date," *Centaurus* 53: 55–61.
Sinclair, P. (2003) "Rhetoric of Writing and Reading in the Preface to Pliny's *Naturalis historia*," in A. J. Boyle and W. J. Dominik, eds., *Flavian Rome: Culture, Image, Text*, Leiden: 277–99.
Skydsgaard, J. E. (1968) *Varro the Scholar: Studies in the First Book of Varro's* De re rustica, Copenhagen.
Smith, C. J. (1998) "Onasander on How to Be a General," in M. Austin, J. Harries, and C. J. Smith, eds., Modus operandi*: Essays in Honour of Geoffrey Rickman*, London: 151–66.
Smith, R. S. (2017) "Myth and Mythography in Pliny's Geography, *Naturalis historia* 3–6," *Polymnia* 3: 83–116.
Smith, T. G. (2003) *Vitruvius: On Architecture*, New York.

Smith, W. D. (1979) *The Hippocratic Tradition*, Ithaca, NY.

Solomon, J. (1980) *Cleonides: Ἐισαγωγὴ ἁρμονική*, unpublished PhD dissertation, University of North Carolina at Chapel Hill.

Soubiran, J. (1969) *Vitruve: De l'architecture. Livre IX*, Paris.

Spanier, E. (2010) *The Good Farmer in Ancient Rome: War, Agriculture and the Elite from the Republic to Early Empire*, unpublished PhD dissertation, University of Washington.

Sparshott, F. E. (1978) "Zeno on Art: Anatomy of a Definition," in J. M. Rist, ed., *The Stoics*, Berkeley, CA: 273–90.

Spencer, W. G. (1926) "Celsus' *De medicina*—A Learned and Experienced Practitioner upon What the Art of Medicine Could Then Accomplish," *Proceedings of the Royal Society of Medicine (Section of the History of Medicine)* 19: 129–39.

(1935–8) *Celsus: On Medicine*, 3 vols., Cambridge, MA.

Speranza, F. (1971–2) "Saserna e Sasernae," *Helikon* 11–12: 466–8.

(1974) *Scriptorum Romanorum de re rustica reliquiae, I: Ab antiquissimis temporibus ad aetatem Varronianam. Accedunt Magonis de agri cultura fragmenta*, Messina.

Spurr, M. S. (1986a) "Agriculture and the *Georgics*," *G&R* 33: 164–87.

(1986b) *Arable Cultivation in Roman Italy: c. 200 B.C.–c. A.D. 100*, London.

Stadter, P. A. (1978) "The *Ars tactica* of Arrian: Tradition and Originality," *CPh* 73: 117–28.

Steinmetz, H. (1907) *De ventorum descriptionibus apud Graecos Romanosque*, unpublished PhD dissertation, Göttingen.

Stevenson, W. (1995), "The Rise of Eunuchs in Greco-Roman Antiquity," *Journal of the History of Sexuality* 5: 495–511.

Stok, F. (1985a) "Un topos platonico nel *De medicina* di Celso," in Mazzini and Fusco 1985: 113–29.

(1985b) "Celso in Seneca?" *Orpheus* 6: 417–21.

(1993) "La medicina nell'enciclopedia latina e nei sistemi di classificazione delle *artes* dell'età romana," *ANRW* II.37.1: 393–444.

(1994), "Celso e gli *Empirici*," in Sabbah and Mudry 1994: 63–75.

(1997) "*Natura corporis*: Costituzioni e temperamenti in Celso e nella cultura dell'età imperiale," in Sconocchia and Toneatto 1997: 151–70.

(2016) "Schlaf und Schlafstörung in der Medizin des Celsus," *RQA* 111.1–2: 8–23.

Stoll, O. (2001) "Kontakt und Wandel: Wege der Vermittlung und Ausbreitung landwirtschaftlicher Technologien in der Antike," in P. Herz and G. Waldherr, eds., *Landwirtschaft im Imperium Romanum*, St. Katharinen: 285–318.

Striker, G. (2006) "Aristotle's Ethics as Political Science," in B. Reis, ed., *The Virtuous Life in Greek Ethics*, Cambridge: 127–41.

Strömberg, R. (1944) *Griechische Wortstudien: Untersuchungen zur Benennung von Tieren, Pflanzen, Körperteilen, und Krankheiten*, Gothenburg.

Svandrlik Civilia, R. (1993) "Su alcune questioni lessicali in Celso: Neologismi e loro sviluppo attraverso autori successivi," in Sconocchia and Toneatto 1993: 198–201.
Svenson-Evers, H. (1996) *Die griechischen Architekten archaischer und klassischer Zeit*, Frankfurt.
Suerbaum, W. (2004) *Cato Censorius in der Forschung des 20. Jahrhunderts: Eine kommentierte chronologische Bibliographie für 1900–1999 nebst systematischen Hinweisen und einer Darstellung des Schriftstellers M. Porcius Cato (234–149 v. Chr.)*, Hildesheim.
Syme, R. (1957) "The Origin of the Veranii," *CQ* 7.3/4: 123–5.
(1969) "Pliny the Procurator," *HSPh* 73: 201–36.
(1987) "Carrière et amis consulaires de Pline," *Helmantica* 38: 223–31.
(1991) "Consular Friends of the Elder Pliny," in A. R. Birley, ed., *Roman Papers VII*, Oxford: 496–511.
Talbert, R. J. A. (2017) *Roman Portable Sundials: The Empire in Your Hand.* Oxford.
Tarán, L. (1975) *Academica: Plato, Philip of Opus, and the Pseudo-Platonic* Epinomis, Philadelphia, PA.
Tarrant, R. J. (2012) *Virgil: Aeneid. Book XII*, Cambridge.
Tatum, W. J. (1995) "The Poverty of the Claudii Pulchri: Varro, *De re rustica* 3.16.1–2," *CQ* 42: 190–200.
Taub, L. C. (2017) *Science-Writing in Greco-Roman Antiquity*, Cambridge.
Taub, L. C. and A. Doody, eds. (2009) *Authorial Voices in Greco-Roman Technical Writing*, Trier.
Temkin, O. (1935) "Celsus' *On Medicine* and the Ancient Medical Sects," *Bulletin of the Institute of the History of Medicine* 3: 249–64.
Temkin, O. and C. L. Temkin, eds. (1967) *Ancient Medicine: Selected Papers of Ludwig Edelstein*, Baltimore.
Teuffel, W. S. and L. Schwabe (1891) *Teuffel's History of Roman Literature* (transl. G. C. W. Warr), London.
Thibodeau, P. (2011) *Playing the Farmer: Representations of Rural Life in Vergil's* Georgics, Berkeley, CA.
(2018) "Ancient Agronomy as a Literature of Best Practices," in Keyser and Scarborough 2018: 463–79.
Thiel, M. A. O. (1895) *Quae ratio intercedat inter Vitruvium et Athenaeum mechanicum*, Leipzig.
Thomas, E. (2007) *Monumentality and the Roman Empire: Architecture in the Antonine Age*, Oxford.
Thomas, R. (2000) *Herodotus in Context: Ethnography, Science and the Art of Persuasion*, Cambridge.
Thomas, R. F. (1982) *Lands and Peoples in Roman Poetry: The Ethnographical Tradition*, Cambridge.
(1988) *Virgil: Georgics*, 2 vols., Cambridge.
Thulin, C. (1905) *Die etruskische Disciplin*, Darmstadt.

(1913) *Opuscula agrimensorum veterum*, Corpus agrimensorum Romanorum 1.1, Leipzig.

Thurmond, D. L. (2017) *From Vines to Wines in Classical Rome: A Handbook of Viticulture and Oenology in the Roman West*, Leiden.

Timpanaro, S. (1986) *Per la storia della filologia virgiliana antica*, Rome.

Toneatto, L. (1994–5) Codices artis mensoriae*: I manoscritti degli antichi opuscoli latini d'agrimensura (V–XIX sec.)*, 3 vols., Spoleto.

(2002) "Agrimensura," in Mastrorosa et al. 2002: 1–28.

Toninato, P. (1993) "Metafore della medicina in Celso," in Sconocchia and Toneatto 1993: 202–7.

Touwaide, A. (1994) "La toxicologie dans le *De medicina*: Un système *asclépiado-méthodique*?" in Sabbah and Mudry 1994: 211–49.

Traglia, A. (1985) "Le *Res rusticae* di Varrone come opera letteraria," *Cultura e scuola* 94: 89–97.

Traina, G. (1988) "Da Dinocrate a Vitruvio: Tecnica, architettura, e società nel mondo ellenistico-romano," *CCC* 9: 303–49.

Tsouna, V. (2021a) "The Stoics on *Technê* and the *Technai*," in Johansen 2021: 166–90.

(2021b) "The Epicureans on *Technê* and the *Technai*," in Johansen 2021: 191–225.

Turra, G. (2017) "Gli *arborum bracchia*: Origini, varietà e riprese di un *topos*," *BStudLat* 47: 218–27.

Vallance, J. T. (1990) *The Lost Theory of Asclepiades of Bithynia*, Oxford.

(1993) "The Medical System of Asclepiades of Bithynia," *ANRW* II.37.1: 693–727.

van der Eijk, P. (1999) "The Methodism of Caelius Aurelianus: Some Epistemological Issues," in P. Mudry, ed., *Le traité des* Maladies aiguës *et des* Maladies chroniques *de Caelius Aurelianus: Nouvelles approches*, Nantes: 47–83.

Van Schaik, K. D. (2018) "*Coniecturalem artem esse medicinam*: Ambiguity and Uncertainty in Disease Diagnosis and Treatment in Celsus' *De medicina*," in M. Fontaine, C. J. McNamara, and W. M. Short, eds., Quasi labor intus*: Ambiguity in Latin Literature*, Gowanus, NY: 165–88.

Vázquez Buján, M. E., ed. (1994) *Tradición e innovación de la medicina latina de la antigüedad y de la Alta Edad Media*, Santiago de Compostela.

Vegetti, M. (1981) "Lo spettacolo della natura: Circo, teatro e potere in Plinio," *Aut aut* new series 184–5: 111–25.

Vela Tejada, J. (2004) "Warfare, History and Literature in the Archaic and Classical Periods: The Development of Greek Military Treatises," *Historia* 53: 129–46.

Vogüé, M. de (1866) "Inscriptions grecques inédites de l'île de Cypre," *RA* new series 13: 437–43.

von Cranach, P. (1996) *Die* opuscula agrimensorum veterum *und die Entstehung der kaiserzeitlichen Limitationstheorie*, Basel.

von Hesberg, H. (1988) "Die Veränderung des Erscheinungsbildes der Stadt Rom unter Augustus," in Hofter 1988: 93–115.

von Staden, H. (1975) "Experiment and Experience in Hellenistic Medicine," *BICS* 22: 178–99.

(1982) "Hairesis and Heresy: The Case of the *haireseis iatrikai*," in B. F. Meyer and E. P. Sanders, eds., *Jewish and Christian Self-Definition: Vol. 3, Self-Definition in the Graeco-Roman World*, London: 76–100.

(1989) *Herophilus: The Art of Medicine in Early Alexandria*, Cambridge.

(1990) "Incurability and Hopelessness: The Hippocratic Corpus," in P. Potter, G. Maloney, and J. Desautels, eds., *La maladie et les maladies dans la Collection hippocratique*, Quebec: 75–112.

(1991) "*Apud nos foediora uerba*: Celsus' Reluctant Construction of the Female Body," in G. Sabbah, ed., *Le latin médical: La constitution d'un langage scientifique. Réalités et langage de la médecine dans le monde romain*, Saint-Étienne: 271–96.

(1994a) "*Media quodammodo diuersas inter sententias*: Celsus, the 'Rationalists,' and Erasistratus," in Sabbah and Mudry 1994: 77–101.

(1994b) "Author and Authority: Celsus and the Construction of a Scientific Self," in Vázquez Buján 1994: 103–17.

(1996) "Liminal Perils: Early Roman Receptions of Greek Medicine," in F. J. Ragep and S. P. Ragep, eds., *Tradition, Transmission, Transformation*, Leiden: 369–418.

(1998). "The Rule and the Exception: Celsus on a Scientific Conundrum," in C. Deroux, ed., *Maladie et maladies dans les textes latins antiques et médiévaux*, Brussels: 105–28.

(1999) "Celsus as an Historian?" in P. J. van der Eijk, ed., *Ancient Histories of Medicine: Essays in Medical Doxography and Historiography in Classical Antiquity*, Leiden: 251–94.

(2000) "The Dangers of Literature and the Need for Literacy: A. Cornelius Celsus on Reading and Writing," in A. Pigeaud and J. Pigeaud, eds., *Les textes médicaux latins comme littérature*, Nantes: 355–68.

(2007) "Aulus Cornelius Celsus on *ars* and *natura*," in A. Ferraces Rodríguez, ed., *Tradición griega y textos médicos latinos en el período presalernitano*, A Coruña: 347–71.

(2010) "How Greek was the Latin Body? The Parts and the Whole in Celsus' *Medicina*," in Langslow and Maire 2010: 3–23.

Wachsmuth, C. (1860) "Über die Unächtheit des vierten Buchs der Frontinschen *Strategemata*," *RhM* new series 15: 574–83.

Walbank, F. W. (1957–79) *A Historical Commentary on Polybius*, 3 vols., Oxford.

Wallace-Hadrill, A. (1990) "Pliny the Elder and Man's Unnatural History," *G&R* 37: 80–96.

(2008) *Rome's Cultural Revolution*, Cambridge.

Walzer, R. and M. Frede (1985) *Galen: Three Treatises on the Nature of Science*, Indianapolis, IN.

Wardle, D. (2014) *Suetonius: Life of Augustus*, Oxford.

Watzinger, C. (1909) "Vitruvstudien," *RhM* new series 64: 202–23.

Webb, P. A. (2017) *The Tower of the Winds in Athens: Greeks, Romans, Christians, and Muslims. Two Millennia of Continual Use*, Philadelphia, PA.

Weinold, H. (1959) *Die dichterischen Quellen des L. Iunius Moderatus Columella in seinem Werke* De re rustica, unpublished PhD dissertation, Munich.

Wellmann, M. (1913) *A. Cornelius Celsus: Eine Quellenuntersuchung*, Berlin.

(1917) "Die Aufidiushypothese des neusten Celsus-Herausgebers," *Mitteilungen zur Geschichte der Medizin und der Naturwissenschaften* 16: 269–90.

(1925) "A. Cornelius Celsus," *Archiv für Geschichte der Medizin* 16: 209–13.

West, D. A. (1964) "Two Notes on Lucretius," *CQ* 14: 94–102.

Whately, C. (2020) "Military Manuals from Aeneas Tacticus to Maurice: Origins, Scholarship, Genre, Audience, and History," in J. T. Chlup and C. Whately, eds., *Greek and Roman Military Manuals: Genre and History*, New York: 17–38.

Wheeler, E. L. (1977) *Flavius Arrianus: A Political and Military Biography*, unpublished PhD dissertation, Duke University.

(1988) *Stratagem and the Vocabulary of Military Trickery*, Leiden.

(2004) "The Legion as Phalanx in the Late Empire (I)," In Y. Le Bohec and C. Wolff, eds., *L'armée romaine de Dioclétien à Valentinien Ier*, Paris: 309–58.

(2010) "Polyaenus: *Scriptor militaris*," in K. Brodersen, ed., *Polyainos: Neue Studien*, Berlin: 7–54.

(2016) Review of C. Matthew, *The Tactics of Aelian* (2012), *JRA* 29: 575–83.

White, D. J. (2013) *Columella* Res rustica *10: A Study and Commentary*, unpublished PhD dissertation, University of Florida.

White, K. D. (1970) *Roman Farming*. Ithaca, NY.

(1973) "Roman Agricultural Writers I: Varro and His Predecessors." *ANRW* I.4: 439–97.

(1977) *Country Life in Classical Times*, Ithaca, NY.

Whitehead, D. (2001) *Aineias the Tactician: How to Survive under Siege*, 2nd ed., Bristol.

(2008) "Fact and Fantasy in Greek Military Writers," *AAAntHung* 48: 139–55.

(2010) *Apollodorus Mechanicus: Siege Matters (Πολιορκητικά)*, Stuttgart.

(2016) *Philo Mechanicus: On Sieges*, Stuttgart.

Whitehead, D. and P. H. Blyth (2004) *Athenaeus Mechanicus: On Machines (Περὶ μηχανημάτων)*, Stuttgart.

Wilkinson, L. P. (1969) *The Georgics of Virgil: A Critical Survey*, new ed. with forward and bibliography by N. Rudd, Norman, OK.

Wilamowitz-Moellendorff, U. von (1911) "Die griechische Literatur des Altertums," in U. von Wilamowitz-Moellendorff et al., eds., *Die griechische und lateinische Literatur und Sprache*, 3rd ed., Leipzig: 1–318.

Wilson Jones, M. (2000) *Principles of Roman Architecture*, New Haven, CT.

(2014) "Greek and Roman Architectural Theory," in C. Marconi, ed., *The Oxford Handbook of Greek and Roman Art and Architecture*, Oxford: 41–69.

Wiseman, T. P. (1971) *New Men in the Roman Senate: 139 B.C.–A.D. 14*, Oxford

(2009) *Remembering the Roman People: Essays on Late-Republican Politics and Literature*, Oxford.

Wölfflin, E. (1875) "Frontins Kriegslisten." *Hermes* 9: 72–92.

Woodman, A. J. and C. S. Kraus (2014) *Tacitus: Agricola*, Cambridge.
Woodman, A. J. (2017) *The Annals of Tacitus: Books 5 and 6*, Cambridge.
Woolf, R. (2004) "Why Is Rhetoric Not a Skill?" *HPhQ* 21: 119–30.
Worthington, I. (2007) *A Companion to Greek Rhetoric*. Malden, MA.
Zainaldin, J. L. (2019) "*Epistula* 86 Once Again: Agriculture and Philosophy in Seneca's *Moral Letters*," *CPh* 114: 218–37.
(2020a) *Gargilius Martialis: The Agricultural Fragments*, Cambridge.
(2020b) *Roman* Technē*: The Growth and Structure of the Artes in the Early Roman Empire*, unpublished PhD dissertation, Harvard University.
(2022) "Columella *aduersus astrologos*: Weather Signs and Stars in *Res rustica* 11," *MD* 89: 157–68.
(2023) "The Agricultural Preface between Rome and China: The Virtues of Farming in Columella and Jia Sixie," *Hermes* 151: 71–104.
Zanker, P. (1987) *Augustus und die Macht der Bilder*, Munich.
Zollschan, L. (2011) "The Ritual Garb of the Fetial Priests," *MH* 68: 47–67.
Zurli, L. (1990) "Le *praefationes* nei *libri VIII de medicina* di A. Cornelio Celso," in Santini and Scivoletto 1990–2: I 295–337.

Index Locorum

For abbreviations and editions used, see the "Note on Texts and Abbreviations" at the front of the book. NB: only select citations provided in the footnotes are indexed here.

Index of Greek and Latin Words (Index Verborum)

Instances of Latin and Greek words are only indexed where they are defined or discussed. For convenience, all Latin words are printed in this index in assimilated spellings.

General Index (Index Nominum et Rerum)

Printed by Integrated Books International,
United States of America